Quantum Theory: A Two-Time Success Story

Daniele C. Struppa · Jeffrey M. Tollaksen
Editors

Quantum Theory: A Two-Time Success Story

Yakir Aharonov Festschrift

Springer

Editors
Daniele C. Struppa
Schmid College of Science and Technology
Chapman University
Orange, USA

Jeffrey M. Tollaksen
Schmid College of Science and Technology
Chapman University
Orange, USA

ISBN 978-88-470-5216-1 ISBN 978-88-470-5217-8 (eBook)
DOI 10.1007/978-88-470-5217-8
Springer Milan Heidelberg New York Dordrecht London

Library of Congress Control Number: 2013939094

Cover illustration: Magnetic Field of a Solenoid (Aharonov-Bohm effect), artwork by Paul Nylander (http://bugman123.com/)

Printed on acid-free paper

Springer is part of Springer Science+Business Media (www.springer.com)

Preface

This volume, "Quantum Theory: A Two-Time Success Story," is part of a number of new initiatives, started in 2012, to honor and advance the vision of Yakir Aharonov during the year of his 80th birthday. Starting with a conference held at Chapman University from August 16–18, 2012, these initiatives also included the launch of a new Institute for Quantum Studies,[1] the dedication of the Aharonov alcove in Chapman's library, and the introduction of a new journal called *Quantum Studies: Mathematics and Foundations*, with Yakir as the chief-editor and administered through the Institute.[2]

We all know that quantum mechanics is the most successful scientific theory in history and resulted in technological advances that drive our economy, such as the entire computer revolution, electronics, and the nuclear power industry. In addition, it impacts many other disciplines such as genetics, medicine and mathematics.

However, the foundations of the theory are so non-intuitive that there is no consensus as to the meaning of the theory. This may be impeding the development of new forms of quantum theory required for its extension to new frontiers such as cosmology, gravity, and high energy particle physics. It is to the advancement of those foundational questions to which this volume is dedicated. As Nobel Laureate David Gross emphasized in his chapter:

> "...the deep conceptual mysteries of quantum mechanics are still with us. Over most of [Yakir's] eighty years (for which this Festschrift is a celebration), Yakir successfully struggled to better understand these extraordinary aspects and to use them to construct new and surprising results."

Indeed, even after a career spanning 6 decades, Yakir's production of major scientific discoveries has continued to increase and deepen to this very day. Perhaps Sir Michael Berry said it best at the beginning of his contribution to this Festschrift:

> "for Yakir Aharonov on his 80th birthday: still quick, still deep, still subtle"

---

[1]The Institute's website is quantum.chapman.edu.

[2]The journals' website is www.birkhauser-science.com/QSMF.

Sir Michael Berry was paraphrasing his own article which he offered 20 years ago for Yakir's 60th birthday, an article which all but launched a new field of mathematics.

Yakir's contributions to physics are too many to mention. The contributors to this book tended to focus on a few of them which deserve special attention. Originally discovered and/or inspired by Yakir, they comprise a series of studies which set out a new interpretation of quantum theory. Elements of this interpretation include topological phases (the Aharonov-Bohm effect and its generalizations), "weak" measurements and "weak" values, time-symmetric boundary conditions, nonlocal measurements and relativistic causality, "modular" variables, and new axioms for quantum theory, to name a few.

We briefly focus on 2 of these items:

1. **Topological phases:** In 1959, Aharonov and Bohm discovered the AB effect[3] which revolutionized our understanding of the role of potentials in physics and appears in most modern texts on quantum mechanics. The impact of the AB effect has been huge and has continued to grow very rapidly. For the first time, they showed that a particle moving in a field-free region could be affected by a field in a disjoint region. Such an effect is alien to classical physics; indeed, it is a defining property of the quantum world. Numerous experiments have verified the effect, and recent novel techniques allow precise measurements of the shifts in electron interference patterns that demonstrate the phase (the AB phase) picked up by a charged particle moving around a solenoid. The AB phase is ubiquitous in modern physics—including cosmology, particle physics, non-abelian gauge theories, condensed matter chemical and molecular physics, and laser dynamics. Generalizations of the AB phase to non-abelian gauge theories, such as the Wilson and t'Hooft loops, are important tools for studying the issues of confinement and spontaneous symmetry breaking. The topological quantum phase explains charge quantization, the quantum Hall effect, the Josephson junction and many effects in the new field of mesoscopic physics where tiny electronic circuits exhibit quantum behavior. The AB phase plays a crucial role in electron microscope holography.
2. **Time-symmetry:** Aharonov, Bergmann and Lebowitz[4] suggested a two vector time symmetric formulation of quantum mechanics. Aharonov, Albert, and Vaidman[5] used ABL in conjunction with weak intermediate measurements performed on pre- and post-selected ensembles to introduce the notion of the "weak values." These values can be surprisingly large and have proven to be very useful tools for analyzing various physical phenomena, for constructing efficient devices for high precision measurements, etc. Phenomena which were thought to be unmeasureable have now been seen using this new approach. As a new paradigm for

---

[3] Y. Aharonov, D. Bohm, "Significance of Electromagnetic Potentials in the Quantum Theory," Physical Review, 115, 485 (1959).

[4] Y. Aharonov, P.G. Bergmann, and J.L. Lebowitz, Phys. Rev. 134, B1410 (1964).

[5] Y. Aharonov, D. Albert, L. Vaidman, Phys. Rev. Lett. 60 (1988).

the design of sensors, it is being broadly applied to precision Doppler frequency measurements, gravitational detectors, etc.[6] This also led to a new branch of mathematics, originally referred to as "Super-Fourier" by Sir Michael Berry, which manifests in super-oscillations and Quantum walks.[7]

These core themes of Yakir's life-work also impact perhaps the most controversial—yet the most important—aspect of his research which concerns the issue of time in quantum mechanics. As Nobel Laureate Sir Anthony Leggett said at the dedication of the Institute and library alcove at Chapman University:

> "...throughout all [previous scientific] revolutions in history, I think there's one assumption that has not really been seriously challenged and that is precisely that the past can affect the present and the present can affect the future and not vice versa. Personally, I believe if there is a really really major revolution in physics in the next 50 or 100 years, then it will involve the overthrow of that principle. If that happens, I think Yakir Aharonov will be seen to have played a major role in the preparatory work leading to that revolution."

A quick readers-note: whenever possible, we, the editors, have added footnotes to the first page of each chapter in order to refer the reader to relevant on-line videos of talks given by the respective chapter's author at the Aharonov-80 conference in 2012 at Chapman University. Finally, we have organized this book into six sections to allow the reader to more easily find particular subjects which were inspired by Yakir:

Part I: Quantum Mechanics and Reality
Part II: Building Blocks of Nature
Part III: Time and Cosmology
Part IV: Universe as a Wavefunction
Part V: Nonlocality
Part VI: Weak Values
Part VII: Mathematics of Weak Measurements
Part VIII: Weak Measurement Experiments
Part IX: Yakir Aharonov: Thinking Quantum

Acknowledgements

The editors are grateful to Springer for agreeing to publish this volume and to Chapman University for sponsoring the conference and this Festschrift. Also,we are much indebted to the Program Committee who played key roles in organizing the conference held at Chapman University in honor of Yakir Aharonov's 80th birthday from August 16–18, 2012. In particular, we would like to thank Doug Dechow, Erika Curiel, Ashley Melton, Lev Vaidman, Schmuel Nussinov, Kyle Lee, Shannon Halverson, Sheri Ledbetter, Susanna Branch, Essraa Nawar, and Casey Coleman for a variety of assistance in the conference and/or the publication of this book. Finally,

---

[6] O. Hosten, P. Kwiat, Science, 319, 787 (2008).

[7] Y. Aharonov, L. Davidovich, N. Zagury, Phys. Rev. A, 48 (1993) 1687.

we wish to thank Yakir Aharonov for contributing two chapters to this Festschrift and for generally inspiring this entire project along with the novel research reported here and the contributors themselves. To this day, he continues to deepen and enrich his profound impact on the physics world, as David Albert so eloquently put it:

> "...I couldn't think of anybody ...in their twenties—who were even remotely as brave, or as open, or as creative, or as experimental, or as overcome with wonder, or as bursting with life, or as constantly and resolutely expecting the impossible, or (in brief) as young, as Yakir. But even I could not have imagined at the time that thirty years later he would turn out to be younger still."

Yakir, we dedicate this book to you with love and friendship and look forward to another 80 years.

Orange, California
April 26, 2013

Daniele C. Struppa
Jeffrey M. Tollaksen

CHAPMAN UNIVERSITY | INSTITUTE FOR QUANTUM STUDIES

Contents

List of Chapters in Alphabetical Order (by First Author)

Part I
Quantum Mechanics and Reality

Chapter 1
A Century of Quantum Mechanics

David Gross

Abstract Approximately one hundred years after its formulation, quantum mechanics is the most successful of all the frameworks discovered to describe physical reality. In this paper I review its successes; namely that it works, that it makes sense and that it is hard to modify. Finally, I discuss where, at the frontiers of knowledge, our present quantum mechanical framework might breakdown. [*Editors note*: For a video of the talk given by Prof. Gross at the Aharonov-80 conference in 2012 at Chapman University, see quantum.chapman.edu/talk-1.]

One hundred years ago, twenty-four physicists met at the Hotel Metropole in Brussels; they were invited by Ernst Solvay to participate in a new kind of scientific congress. One of the first international scientific meetings, the Solvay conferences, were characterized by a highly restricted invitation list and an unusual mixture of short talks and long discussions. Solvay played a unique and important role in the development of twentieth century physics—most notably in the quantum revolution whose birth overlapped the initiation of these meetings. In the interim 100 years, quantum mechanics made great leaps and bounds (in addition to quantum field theory which grew out of quantum mechanics and relativity and on which I will mainly focus my paper). Yet the deep conceptual mysteries of quantum mechanics are still with us. Over most of his eighty years (for which this Festschrift is a celebration), Yakir successfully struggled to better understand these extraordinary aspects and to use them to construct new and surprising results.

Quantum mechanics emerged in the period between 1900, when Planck first quantized the energy of radiating oscillators, and 1925–1926 with Heisenberg, Schrodinger, Born and Dirac's formulation of the principles of quantum mechanics,

---

D.G. is Permanent Member and holder of the Frederick W. Gluck Chair in Theoretical Physics at the Kavli Institute for Theoretical Physics, Professor of Physics.
D.G. is 2004 Nobel Prize Winner in Physics.

This talk was first presented by Prof. Gross at the 25th Solvay Conference on Physics, "The Theory of the Quantum World," held in Brussels, October 19–22, 2011.

D. Gross (✉)
Department of Physics, University of California, Santa Barbara, USA
e-mail: gross@kitp.ucsb.edu

D.C. Struppa, J.M. Tollaksen (eds.), *Quantum Theory: A Two-Time Success Story*,
DOI 10.1007/978-88-470-5217-8_1, © Springer-Verlag Italia 2014

and thus is approximately one century old. The development of quantum mechanics and its application to atomic theory and the structure of matter dominated the first five Solvay conferences, culminating in the most famous 1927 Solvay meeting, where the meaning of quantum reality was heatedly debated between the pioneers and the revolutionaries of quantum mechanics.

The first Solvay conference, one hundred years ago to the month, addressed the central problem of physics at that time: Was the quantum structure of nature truly unavoidable? Lorentz's opening address at the first Solvay conference reverberates with the anguish that this master of classical physics felt at the first glimpses of the quantum world:

> Modern research has encountered more and more serious difficulties when at-tempting to represent the movement of smaller particles of matter and the connection between these particles and phenomena that occur in the ether. At the moment, we are far from being completely satisfied that, with the kinetic theory of gases gradually extended to fluids and electron systems, physicists could give an answer in ten or twenty years. Instead, we now feel that we reached an impasse; the old theories have been shown to be powerless to pierce the darkness surrounding us on all sides.

We Face no Such Crisis Today Quantum mechanics is the most successful of all the frameworks that we have discovered to describe physical reality. It works, it makes sense, and it is hard to modify. The order of this list of successes is in the order of importance that most physicists demand of a physical theory: It works, it makes sense, and it is hard to modify. I shall start with the second point.

Quantum mechanics does make sense, although the transition, a hundred years ago, from classical to quantum reality was not easy. It took time to learn how to get out of phase space and to live in Hilbert space. Some of the boldest pioneers of quantum theory (notably Einstein) resisted the replacement of classical determinism with a theory that often can only make probabilistic predictions. Even harder to get used to was the idea that in quantum mechanics one can describe a system in many different and incompatible ways, and that there is no unique exhaustive description. The freedom one has to choose among different, incompatible, frameworks does not influence reality—one gets the same answers for the same questions, no matter which framework one uses. That is why one can simply "shut up and calculate." Most of us do that most of the time. Different, incompatible aspects cannot both enter a single description. If one errs by mixing incompatible descriptions or histories, we produce paradoxes.

By now, especially with the consistent (or decoherent) histories approach, initiated by R. Griffiths, and further developed by Gell-Mann, Hartle, Omnes, Zurek and others, we have a completely coherent and consistent formulation of quantum mechanics that corresponds to what we actually do in predicting and describing experiments and observations in the real world. For most of us there are no problems.

Nonetheless, there are dissenting views. Experimentalists continue to test the predictions of quantum theory, and some theorists continue to question the foundations. Most interesting to me is the growing understanding as to how the classical framework emerges from quantum mechanics, especially interesting as our experimental friends continue to astonish us with their ability to control and manipulate quantum systems while preserving their quantum coherence. How can we explain measurements without invoking the absurd collapse of the wave function? How does classical physics emerge from quantum reality?

Quantum mechanics is more powerful and richer than classical mechanics, for, after all, classical physics is just a limiting, special case of quantum physics. In re-cent years we have also become aware of the increased computational power of quantum mechanical states. Entanglement, the strange new feature of quantum states, can be efficiently used to amplify computation, and has motivated an intensive effort to develop a quantum computer. This goal might take many decades to realize, but meanwhile the effort has provided enormous stimulation to atomic and condensed matter physics.

The dream of a quantum computer is only conceivable because of the enormous advances made in recent years towards greater control and understanding of matter, down to the scale of individual atoms: mesoscopics, atomic traps, quantum optics, and spintronics. A new field is developing that might be called quantum engineering, with enormous potential for both technological innovation and for use as a marvelous tool for the experimental exploration, and the simulation, of fascinating states of quantum matter. These tools enable not only the study of the static phases of complicated many body systems, but also of their dynamics and non-equilibrium behavior.

Quantum Mechanics Works It works not just for simple systems such as single atoms and molecules, but also for collections of 10^{23} atoms, sometimes strongly interacting, over an enormous range of energies. It explains not just the anomalies in the classical description of blackbody radiation and the specific heat of solids at low temperatures (that stimulated early developments), but also the detailed properties of ordinary matter, such as conductors, insulators, semiconductors as well as more exotic materials.

The quantum theory of matter (many-body theory) and the quantum theory of fields share many common features; indeed they are essentially the same thing. Thus, critical developments in condensed matter physics and in elementary particle physics towards the end of the twentieth century often occurred in parallel. In these developments, symmetry principles played a fundamental role. But if the secret of nature is symmetry, much of the texture of the world is due to mechanisms of symmetry breaking. Magnetism and chiral symmetry breaking are two important examples of the spontaneous breaking of a global symmetry.

One of the most important quantum phenomenon—that of superconductivity—was discovered by Onnes 100 years ago and discussed at the first Solvay conference. Parenthetically, Rutherford's discovery of the nucleus of atoms, made also in 1911, was not discussed, although Rutherford attended! It took almost half a century (until 1957) for Bardeen, Cooper and Schrieffer to come up with a full understanding of this first example of the spontaneous breaking of a local symmetry, which later played a fundamental role in the understanding of the weak nuclear force—the so-called Higgs mechanism—with the final confirmation coming from the LHC. Even today, unconventional superconductors are still a great mystery at the frontiers of the understanding of quantum states of matter. It now appears that there are new forms of matter—labeled not by symmetry but by topology. The important question, "What are the possible quantum phases of matter?" remains wide open.

Quantum Mechanics Works It works at distances that are a billion times smaller than the size of the atom, well within the nucleus and its constituent quarks. It works for energies that are a trillion times larger than atomic energies. From the beginning it was clear that quantum mechanics fit together seamlessly with special relativity and with Maxwell's theory of the electromagnetic field, despite a few technical difficulties that took some time to resolve. The resulting edifice, the quantum theory of fields, re-solved the perplexing duality of particles and waves, and, in what I regard as one of the most amazing successes of theoretical physics, predicted anti-matter, the first examples of which were soon discovered.

Quantum field theory has been tested with extraordinary precision. Much of the incredible precision that physics is able occasionally to achieve rests on quantum features of nature, such as the identity of indistinguishable particles and the existence of discrete sharp states. I cannot refrain from noting one of the most amazing of these precision tests, that of the measurement of the anomalous magnetic moment of the electron:

$$a_e = (g_e - 2)/2 = .00115965218085 + / - .00000000000076,$$

a test of Quantum Electrodynamics (QED) to almost one part in 10^{12}, sensitive to all the components of the standard model, but especially QED (the comparison involves 5 loop quantum effects). Quantum field theory works and has been tested over an incredible range of physical phenomena—from the edge of the galaxy (10^{27} cm) to the nano-nano centimeter scale, over forty-five orders of magnitude. In fact, we know of no reason why the framework of quantum field theory could not continue to be adequate until we reach the Planck scale (10^{-33} cm), where quantum effects of gravity become important.

Quantum Mechanics Works It provides the explanation, not only of the structure of atoms and molecules, but also of the structure of the nucleus, and the nature of the strong and weak nuclear forces. In a reductionist sense, the standard model of elementary particles (with 3 families of quarks and leptons, charged under 3 gauge groups that generate three forces) is an amazing theory, powerful enough to encompass almost all of the known forces that act on the known particles of nature (with the exception of dark matter and the right-handed partner of the neutrino). The standard model is so extraordinarily successful that we currently strain, so far unsuccessfully, to find deviations.

Finally, Quantum Mechanics is Hard to Modify Our present fundamental framework, quantum field theory, appears under no threat from observation or experiment, and seems to be completely adequate for the understanding of macroscopic and microscopic physics, from the edge of the universe to the nano-nano meter scale. It is very difficult to construct consistent alternatives to this framework that agree with observation. But no framework, no theory, is likely to survive untouched forever. Where might our present quantum mechanical framework breakdown and how?

Hints from observation and from experiment point to physics beyond the standard model. The existence of dark matter, the non-vanishing neutrino masses and the many unanswered questions regarding quark and lepton masses and their mixing require non-standard-model physics; but the necessary modifications do not necessarily force us to abandon the framework of quantum field theory. More hints come from trying to extend our standard theory to new regimes of energy and distance and from challenging our concepts with thought experiments.

The extrapolation of the standard model to high energy, or equivalently short distance, suggests that the atomic and nuclear forces are unified at very high energy. Such unification does not necessarily suggest a breakdown of the framework of quantum field theory; we can construct grand unified gauge theories. However, the fact that the implied unification scale is so close to the Planck scale, where the quantum nature of gravity becomes essential, is an important hint that the grand synthesis must include quantum gravity. Traditional quantum field theory appears to be at a loss to consistently describe gravity, due to the uncontrollable quantum fluctuations of the metric at the Planck scale. In the search for a unified theory of standard model forces, we have been led to string theory, which also automatically includes gravity and yields a consistent extension and quantization of classical Einstein gravity.

String theory was originally thought to break with traditional quantum field theory in important ways, but recently we have realized that string theory and quantum field theory are not mutually exclusive. Quantum field theory, in the old fashioned sense, is not sufficient to contain gravity. But it is part of a bigger framework that includes extended objects, strings, membranes, and higher dimensional "branes". The formulation in terms of strings is often best understood—thus "string theory". String theory always describes dynamical space-time-gravity. On the other hand, some string theory quantum states can be usefully described in terms of quantum field theory. This insight has been inspired by the remarkable duality between supersymmetric gauge theory in four dimensions (or more generally conformal field theory) and string theory in an AdS background. Even the theoretical framework we use for the standard model, consisting of quantum gauge theory with fundamental fermions and a few scalars has (many of us believe) a dual description) in terms of a string theory with highly curved extra dimensions. A close cousin of Quantum Chromodynamics, endowed with extra (super) symmetry, is undoubtedly identical to string theory in AdS space. So string theory and quantum field theory are part of a larger quantum mechanical framework, whose structure and extent are still being explored.

Finally, there are indications that once again we might be forced to modify our most fundamental of physical concepts, that of space and time. Many of us are more and more convinced that space is an emergent, not fundamental, concept. We have many examples of interesting quantum mechanical states, for which we can think of some (or all) of the spatial dimensions as emergent. Together with emergent space, we have the emergent dynamics of space and thus emergent gravity. But it is hard to imagine how time could be emergent? How would we formulate quantum mechanics without time as a primary concept? Were time to be emergent, our understanding of quantum mechanics would have to change.

To describe nature and to make predictions, we need more than just the framework of quantum mechanics, or of quantum field theory, or of quantum string theory. We need a particular dynamical principle, a Hamiltonian that determines the time development, and we also need an initial state. So what picks the dynamics? Quantum field theory offers little guide, except symmetry. String theory, in which all parameters are dynamical, appeared at first to offer the hope of providing a unique answer. But this hope appears to be a mirage. String "theory" does not provide such a principle; rather it consists of a set of tricks to find consistent quantum states, often constructed in a perturbative semiclassical expansion. And there are many such quantum states, an infinite number in fact, perhaps 10500 that resemble our universe. Some believe that this is the complete story, and that all of these universes might exist somewhere in a multiverse, and that to make predictions we must resort to arguing that our patch of the multiverse is particularly suited for our existence.

Since a theory of quantum gravity is a dynamical theory of space-time, we must finally come to grips with quantum cosmology. Here it makes no sense to separate the observer and the observed, and we are faced with many puzzling conceptual issues. What picks the initial condition? The final condition? In addition, we are challenged by astrophysics. In the last hundred years, we have learned much about the universe, including a detailed description of most of its history. The outstanding mysteries that remain—the dynamics of inflation, the mystery of the big bang and the accelerated expansion—represent serious challenges to our theoretical framework.

So what is the whole picture? We are faced today not with a crisis but with confusion at the frontiers of knowledge. Fundamental physics today is in a state more analogous to the one that prevailed in 1891, rather than in 1911. In 1891, with all the successes of classical physics—mechanics, electrodynamics, kinetic theory and statistical mechanics—physics appeared in fine shape. Who could have dreamed of the conceptual revolutions that lay in store?

We are unlikely to come to a resolution during this meeting. The most we can hope for is that our discussions will clarify the issues and most importantly stimulate the advances that are necessary. In any case it should be lots of fun.

Chapter 2
Realism and the Physical World

A.J. Leggett

Abstract I consider the extent to which the applicability of the concept of classical realism is constrained, irrespective of the validity or not of the quantum formalism, by existing experiments both in the EPR-Bell setup, including recent experiments testing "nonlocal realistic" theories, and in the area of "macroscopic quantum coherence". Unless we are willing to sacrifice one or more other intuitively plausible notions such as the conventional "arrow of time", it appears impossible, in either context, to maintain the classical notion of realism. [*Editors note*: For a video of the talk given by Prof. Leggett at the Aharonov-80 conference in 2012 at Chapman University, see quantum.chapman.edu/talk-16.]

Ever since the earliest days of quantum mechanics (QM), it has been appreciated that there are major difficulties in reconciling the account it gives of the behavior of the microscopic world of atoms and electrons with the classic notion of "realism," which crudely speaking is the postulate that physical objects have definite properties, and occupy definite states, independently of whether or not they are observed. This was, of course, explicitly realized by Nils Bohr, whose solution was to postulate that microscopic objects (such as electrons, photons, and atoms) are simply not the kind of thing that can possess properties independently of the macroscopic apparatus used to observe them; thus, for example, in the standard Young's slits experimental setup, an electron which is inspected to see which of the two slits in the intermediate screen it passed through is simply not the same object as an electron which passed through uninspected, and thus it is not at all surprising that the

A.J.L. is 2003 Nobel Prize Winner in Physics.

The material in this article is reproduced with permission from Leggett, A. J. "Realism and the physical world," Reports on Progress in Physics, Volume: 71, February 2008, IOP Publishing LTD.

A.J. Leggett
Department of Physics, National University of Singapore, Singapore, Singapore

A.J. Leggett (✉)
University of Illinois at Urbana-Champaign, Urbana, IL, USA
e-mail: aleggett@illinois.edu

D.C. Struppa, J.M. Tollaksen (eds.), *Quantum Theory: A Two-Time Success Story*,
DOI 10.1007/978-88-470-5217-8_2, © Springer-Verlag Italia 2014

diffraction pattern (or lack of it) which is produced on the final screen differs in the two cases. If one accepts, as Bohr seems to, a division of the world into the "microscopic" regime of atoms and electrons and the "macroscopic" level of measuring apparatus, which is postulated to behave classically, then this picture seems internally consistent, and it certainly seems to have satisfied a generation of physicists and even of philosophers.

However, as pointed out by Schrödinger in 1935 in his famous "Cat" paper [1], there is no good reason to accept this division of the world into a microscopic regime where QM reigns and a macroscopic one governed by classical physics; QM is a very "totalitarian" theory, and if it applies to individual atoms and electrons, then it should prima facie equally apply to the macroscopic objects made up of them, including any devices which we have set up as measuring apparatus. Thus, we are forced to conclude, with Schrödinger, that under appropriate circumstances the description of a macroscopic object (measuring apparatus, cat . . .) is inconsistent with the hypothesis that it is in a definite macroscopic state (in the case of the cat, is either "dead" or "alive").

It is worth stating the "cat" ("measurement," or better "realization") paradox a little more formally. At the microscopic level, the formalism of QM assigns probability amplitudes to various possible behaviors of the system (e.g., in a Young's slits experiment, to passage through each of the two slits in the intermediate screen). In view of the experimentally observed phenomenon of interference at the final screen (the total diffraction pattern is not the sum of the two patterns which would result from passage through one slit, the other being blocked), we can prima facie draw, in a situation in which the assigned amplitude at each slit is nonzero, the (negative) conclusion that it is incorrect to say that each electron of the ensemble in question passed either through the upper or through the lower slit. The same conclusion can be drawn for other microscopic-level examples when we have simultaneous nonzero amplitudes for different behaviors and see the effects of interference between them, e.g. oscillations in the $K_o - \bar{K}_o$ system. Now, when we extrapolate the QM formalism to the macroscopic level of cats and detectors, it is unarguable that in suitable circumstances (such as those postulated by Schrödinger in his original paper) it assigns simultaneous nonzero amplitudes to two or more macroscopically distinct states. Of course, it is generally agreed that, as a result of the decoherence predicted, under any normally realistic conditions, by the QM formalism itself, it is no longer possible to see the effects of interference between the two or more "branches" of the superposition. However, the quantum formalism is exactly the same at the microscopic and the macroscopic level; if, therefore, a given interpretation is excluded in the former case, it cannot become permitted in the latter! The above formulation shows clearly that the phenomenon of decoherence, while it may no doubt be an essential ingredient in any future resolution of the realization paradox, cannot by itself constitute such a resolution.

In the rest of this essay, I shall not be concerned with the numerous attempts to resolve the Cat paradox within the assumption of the universal validity of QM which have appeared in the literature over the last 70 years. Rather, I shall use the paradox as a motivation for the question: Irrespective of the validity or not of QM, what can we say *from experiment* about the validity, or not, of the concept of realism as applied to the physical world? And what can we expect to be able to say in the foreseeable future?

Two cautions before we start: First, we will never, by any finite sense of experiments, be able to establish unequivocally the truth either of realism or of any specific alternative to it (e.g. QM). This is a matter of simple logic: the fact that theory T predicts experimental consequence E, and that experiment finds E, *does not* establish that T is correct. (The contrary assertion is known to logicians as the fallacy of "affirmation of the consequent": that so many papers in experimental physics appear to commit this fallacy probably signals that there is an extra unspoken assumption in the authors' minds, namely something like "E is so surprising that it is highly unlikely that it would be predicted by any theory other than T." Depending on context, such an implicit argument may no doubt be somewhat persuasive, but it is hardly logically conclusive.) On the other hand, it may be possible to establish unequivocally, by experiment, the *falsity* of a set of propositions (the fact that theory T predicts experimental consequence E, and that "not-E" is found, does logically establish the falsity of T).

Secondly, I deliberately said in the last sentence "a *set* of propositions." I know of no set of experiments, and can imagine none, which could establish that the hypothesis of realism *taken in isolation* is false. Indeed, the Bohm-de Broglie "pilot wave" (or "hidden variable") interpretation of the QM formalism reproduces all the standard experimental predictions of that formalism, yet claims to maintain the concept of realism; thus, even if we should find that the experiments continue forever to verify the predictions of QM, realism *per se* would not be refuted. (Whether the additional assumptions which have to be made in the Bohm-de Broglie interpretation effectively devalue the concept of "realism" to the point where it is as it were no longer recognizable, is a matter of opinion.) Thus, we shall never attempt to test realism alone, but always in conjunction with one or more other prima facie plausible assumptions about the world.

The best known set of experiments which examine the question of realism are those which stem from the theoretical work of Einstein, Podolsky and Rosen [2] and of Bell [3], and are usually known as "EPR-Bell" experiments. They refer to measurements made on a pair of systems which have interacted in the past but are now separated to a distance such that, according to the postulates of special relativity, the outcome of a measurement as one system cannot be causally influenced by the choice of what to measure on the other. A schematic diagram of an idealized EPR-Bell experiment is shown in Fig. 2.1. The essential points are: (1) The atomic source emits photons in pairs: while in real life most of the pairs are not emitted back-to-back, we concentrate on the subensemble which is. (2) At each "station" (S1 or S2) a randomly activated device makes a choice to switch the corresponding photon into one of two different measuring devices, which will measure different properties (e.g. the device $M_{\mathbf{a}}$ consists of a polarizer $P_{\mathbf{a}}$ with transmission axis set in direction **a** (and the non-transmitted photons reflected), plus detectors $D_{\mathbf{a}}^{(+)}$, $D_{\mathbf{a}}^{(-)}$ to

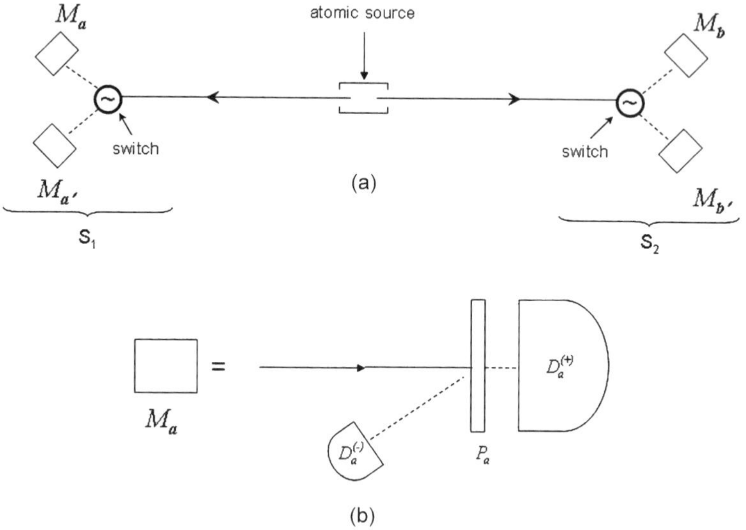

Fig. 2.1 (**a**) Schematic diagram of an EPR-Bell experiment. (**b**) Details of the "measurement device" M_a. The devices $M_{a'}$, M_b, $M_{b'}$ are similar mutatis mutandis

register the transmitted and the reflected photons respectively (see Fig. 2.1); the device $M_{\mathbf{a}'}$ is similar except that the transmission axis of the polarizer is set in direction $\mathbf{a}'$.) (3) The spatial dimensions are such that the "event" of detection at station S_2 is spacelike separated from the "event" of switching (choice of measurement) at S_1 (and vice versa). (4) The detectors are 100 % efficient. No existing experiment satisfies (at least to my knowledge) all of the above conditions simultaneously, but it is useful to consider this idealized version as a basis for discussion.

Let us first set up a notation to describe the data obtained in this experiment. Suppose that a given photon 1 is switched into the measuring device $M_{\mathbf{a}}$, and the "transmission" counter $D_{\mathbf{a}}^{(+)}$ clicks: then we *define* the variable A to take the value $+1$ for that photon. If the "reflection" counter $D_{\mathbf{a}}^{(-)}$ clicks, we define A to take the value -1. We take it as an experimental fact (of course easily verified) that for each photon switched into $M_{\mathbf{a}}$ either $D_{\mathbf{a}}^{(+)}$ or $D_{\mathbf{a}}^{(-)}$ clicks (but not both); hence for each photon switched into $M_{\mathbf{a}}$, either $A = +1$ or $A = -1$. Similarly, for photons 1 switched into device $M_{\mathbf{a}'}$, we define the variable A' to take the value $+1(-1)$ accordingly as the counter $D_{\mathbf{a}'}^{(+)}$ $D_{\mathbf{a}'}^{(-)}$ clicks. A similar definition is made for the variables B and B' measured on photon 2 at station S_2. Note carefully that if, for example, photon 1 is switched into device $M_{\mathbf{a}'}$, then while the variable A' is defined for this photon the variable A is not (and vice versa). In this way, one can of course obtain the mean value $\langle A \rangle$ of A, for the subensemble of photons 1 which is switched into $M_{\mathbf{a}}$ (let's denote averages taken on this subensemble by $\langle A \rangle_{\mathbf{a}}$); similarly we can measure $\langle A' \rangle_{\mathbf{a}'}$, $\langle B \rangle_{\mathbf{b}'}$, etc. More interestingly, we can measure the *correlation*

$$\langle AB \rangle_{\mathbf{ab}} \equiv \frac{N_{++} + N_{--} - N_{+-} - N_{-+}}{N_{++} + N_{--} + N_{+-} + N_{-+}} \tag{2.1}$$

where N_{++} is the number of pairs for which both A and B were measured to be $+1$, etc. Note that this correlation (average) is defined *on a particular subensemble*, namely that set of pairs for which photon 1 was switched into $M_{\mathbf{a}}$ and photon 2 into $M_{\mathbf{b}}$. Although this essay is not concerned with QM as such, we note as an aside that theory gives unambiguous predictions for the quantity $\langle AB\rangle_{\mathbf{ab}}$ for any given type of atomic source: for instance, if the pair of atomic transitions involved in the emission of the photon pair is of the "0^+" type (no change of atomic angular momentum or parity) then a simple argument shows that the prediction is

$$\langle AB\rangle_{\mathbf{ab}} = \cos(2\theta_{\mathbf{ab}}) \tag{2.2}$$

where $\theta_{\mathbf{ab}}$ is the angle between the polarizer settings **a** and **b** at S_1 and S_2 respectively.

The class of theories about the physical world with whose predictions the experimental data are most commonly compared is that of "objective local theories" (OLT's) [4]. This class is defined by the conjunction of three postulates, which may be labeled crudely as

(1) locality
(2) induction
(3) realism.

Locality (often denoted "Einstein locality" in the literature) is the postulate that an event (in the sense of special relativity) cannot be causally affected by any past events which lie outside its past light cone (i.e. which cannot transmit information to the spacetime point of the event in question by any signal whose velocity is less than or equal to that of light). This is, of course, a fundamental postulate of the theory of special relativity as the latter is usually formulated.

Induction is, crudely speaking, the postulate that "causality propagates only forward in time," that is, that an event cannot be affected by any future events, whether or not the latter lie inside its light cone. Since within the framework of special relativity this postulate can be derived as a consequence of locality and "transitivity of causality," it is often not listed separately; however, since we wish to examine the implications of the experimental data with a minimum of a priori assumptions, it is convenient to list it explicitly. In the present context, its significance is that the statistical properties of the complete ensemble of emitted pairs are determined only by conditions at the source. In particular, the photons of the "**ab**" subensemble do not know in advance that they will be switched into $M_{\mathbf{a}}$ and $M_{\mathbf{b}}$ respectively, so the statistical properties of this subensemble cannot be affected by this fact and so must be identical to those of the ensemble of all pairs.

Let's move then to the trickiest, and in the present context most interesting, ingredient in an OLT, namely "realism" (or "objectivity"). A crude definition of this concept would be that "each photon of a given pair possesses (a complete set of) properties in its own right." In other words, each photon 1 (say) carries with it information which is sufficient to decide, either deterministically or statistically, how it will behave *both* if switched into $M_{\mathbf{a}}$ *and* if switched into $M_{\mathbf{a}'}$. It is natural to think of the information in question as embodied in a set of "hidden" variables, thus OLT's are often loosely known as "hidden-variable" theories, but this is not essential to the argument to be developed below. If one phrases the concept of realism in this way, so that the properties of interest are possessed by microscopic objects (photons, as in other variants of the experiment atoms, etc.), it is natural to add to the word "realism" the adjective "microscopic." Thus, it is *microscopic realism*, in conjunction with locality and induction, which is tested by the Bell-EPR experiments.

There is however an interesting alternative formulation of postulates (3), which is usually called the hypothesis of "macroscopic counterfactual definiteness" (MCFD). To explain it, let us imagine that a given photon 1 was actually switched into $M_{\mathbf{a}'}$, thereby realizing a value of A'. Given the "random" nature of the switching process, it could equally well have been switched into $M_{\mathbf{a}}$, and would then have realized a value of A. The postulate of macroscopic counterfactual definiteness then states that the value of A which "would have" been realized is definite. That is, it can be treated as a "fact" about the physical world. It is fairly clear that microscopic realism implies macroscopic counterfactual definiteness (if the photon carried with it sufficient "instructions" to determine the value of A, then we can say that this value "would have" occurred), but the converse is not true: it is entirely conceivable that properties are indeterminate at the microscopic level but become determinate when as it were amplified, as in a photodetector, to some level of "macroscopicness"; indeed, many people hold (in the opinion of the present author incorrectly) that the formalism of QM itself somehow achieves this result. In any case, it is clear that the hypothesis of MCFD is somewhat weaker than that of microscopic realism, so that any experiment which excludes the former must automatically exclude the latter. It is also worth noticing that the postulate of MCFD is insensitive to whether the way in which the instructions carried by the photon decide the value of A is deterministic or only statistical.

It is well known that any theory satisfying the conjunction of postulates (1)–(3) (with either form of (3)) must predict values of the *experimentally measured* correlations $\langle AB\rangle_{\mathbf{ab}}$, which satisfy the celebrated "CHSH" inequality, derived by Clauser et al. [5] by extending the result of Bell:

$$\langle AB\rangle_{\mathbf{ab}} + \langle AB'\rangle_{\mathbf{ab'}} + \langle A'B\rangle_{\mathbf{a'b}} - \langle A'B'\rangle_{\mathbf{a'b'}} \leq 2 \tag{2.3}$$

A simple proof (one of many) goes as follows:

1. By postulate (3) (in either form) the values of A and A' simultaneously exist for each photon 1, and similarly values of B and B' exist for each photon 2.
2. By postulate (1), the value of A cannot be affected by whether it was B or B' which was chosen for the measurement in S_2 and vice versa.
3. Therefore, the quantities AB, AB', $A'B$, $A'B'$ exist for each photon pair, with A, A', B, B' each taking *the same* value ± 1 in each of the combinations in which it occurs.
4. By trivial algebra it then follows that for each pair $AB + AB' + A'B - A'B' \leq 2$.
5. If $\langle AB\rangle_{all}$ indicates the average over the complete ensemble of pairs emitted by the photon source, then (4) leads directly to the result

$$\langle AB\rangle_{all} + \langle AB'\rangle_{all} + \langle A'B\rangle_{all} - \langle A'B'\rangle_{all} \leq 2$$

6. Finally, by postulate (2), $\langle AB\rangle_{ab}$ is identical to $\langle AB\rangle_{all}$, etc., and thus we obtain the CHSH inequality, Eq. (2.3).

Thus, if an ideal EPR-Bell experiment were to produce results which violate the inequality (2.3) (in particular, were it to reproduce the QM prediction (2), which violates (3) for (e.g.) settings such that $\mathbf{a}\cdot\mathbf{b}=\mathbf{a}'\cdot\mathbf{b}=\mathbf{a}\cdot\mathbf{b}'=\pi/8,\ \mathbf{a}'\cdot\mathbf{b}'=3\pi/8$), then we should know for sure that one of the postulates (1)–(3) cannot hold in the real world. Unfortunately, there is at present no single experiment which satisfies all of the defining criteria for an "ideal" experiment; there are always various "loopholes" (imperfect detector efficiencies, lack of adequate spatial separation, questionable "randomness" in the switching process, etc.). However, while no existing experiment has blocked *all* existing loopholes simultaneously, with one exception each of them individually has been blocked in at least one experiment, so that it would seem to require a very peculiar conspiracy of nature to allow an OLT to be maintained. The exception is what is sometimes called the "collapse locality" loophole: if one postpones the "event" of realization of a particular outcome at each of the stations to a sufficiently late stage (perhaps long after the detector has, in our usual way of thinking, produced a macroscopic output) then the "events" at the two stations would not have fulfilled the condition of spacelike separation in any existing experiment, and indeed to fulfill it would require conditions which border on science fiction: cf. for example Ref. [6]. For the purposes of the present discussion, let me from now on assume that such an "ideal" (loophole-free) Bell-EPR experiment will someday be done, will turn out consistently with the QM prediction (2) and will thereby definitively invalidate the whole class of OLT's. What are the implications?

Of the three defining postulates (1)–(3) of the class of OLT's, the most impervious to challenge would seem at first sight to be (2), in the sense that once we give up our "common-sense" notions concerning the "arrow of time" it seems very difficult to continue to do physics at all in the mold to which we have been accustomed. Actually, in the present writer's opinion it is entirely possible, indeed probable, that the next major revolution in physics will force us to do just that, but unfortunately it is in the nature of scientific revolutions that their content is difficult or impossible to forecast, so speculation along these lines seems pointless at the present time. We are then left with postulates (1) and (3), and it is an amusing sociological observation that while popular writers on the subject of the EPR-Bell experiments almost without exception choose to give up (1) (locality), professional physicists usually opt to sacrifice realism. Is there any *experimental* way of deciding the question?

A partial answer comes from a very recent experiment and the associated theory. Consider the standard "ideal" setup of Fig. 2.1, and relax postulate (1) by allowing arbitrary nonlocal effects in the detection process. On the other hand, we will in some sense strengthen postulate (3), in the following sense: we require that the ensemble of pairs of photons emitted by the source is a disjoint union of subensembles in each of which each photon of the pair has a definite polarization, a statement which is operationally defined by the requirement that the photon obeys Malus's law, i.e. for any given photon there exists a (complex) unit polarization vector $\boldsymbol{e}$ such that when it is presented with a polarizer set with (complex) transmission "axis" $\boldsymbol{c}$, the probability of transmission is $|\boldsymbol{e}*\cdot\boldsymbol{c}|^2$. (In such a theory the transmission/rejection of any individual photon is determined inter alia by nonlocal effects, but the statistics for the ensemble of photons, obtained by averaging over these effects and others, is required to satisfy Malus's law.) Using the elementary inequalities, applicable for variables A, B taking values ± 1

$$-1 + |\langle A \rangle + \langle B \rangle| \leq \langle AB \rangle \leq 1 - |\langle A \rangle - \langle B \rangle|$$

it is straightforward to show [7] that such a theory, called a "crypto-nonlocal hidden-variable (CNLHV) theory" must predict inequalities for the experimental correlations which are violated by the QM predictions. These inequalities are however of a different nature from the Bell-CHSH inequalities, and a test of them requires a set of measurements with elliptically polarized analyzers which had not been previously done. Recently, the experimental group in Vienna carried out just such an experiment [8], finding agreement with the predictions of QM and violation of the CNLHV predictions by several standard deviations. Actually, to obtain this violation it was necessary for the authors of Ref. [8] to assume that the experimental correlations, while a function of the *relative* angle $\mathbf{a} \cdot \mathbf{b}$ of the polarizer settings, are independent of the "center-of-mass" variable, i.e. invariant under simultaneous rotation of **a** and **b** through the same angle, something which is certainly plausible but is not directly tested. However, in very recent work both these authors [9] and the Singapore group [10] have generalized the CNLHV inequalities and conducted further experiments so as to eliminate this loophole; once again the data are consistent with the QM predictions and violate those of the CNLHV class of theories, in the Vienna case by more than 80 standard deviations. While those experiments are subject to some of the same "loopholes" as the standard EPR-Bell ones, the outcome makes it extremely plausible that CNLHV theories cannot describe the physical world. What is the significance of this result? Can we regard it as establishing (subject, of course, to the above reservations concerning induction) that it is indeed realism rather than locality which has to be sacrificed? It is certainly suggestive in this respect; on the other hand, a critic might argue that by formulating our "realistic" postulate in a way which requires the realism to refer to the properties of individual photons of the pair rather than to the pair as a whole, we have in effect smuggled the concept of locality back in again. Perhaps the lesson is that while the concept of "local realism" is clear-cut, to try to analyze it in terms of its two prima facie components may not in the end be a particularly meaningful exercise. Certainly, in QM itself the two concepts, or rather their absence, in some sense appear blended, in that once one has the (non-realistic) concept of a quantum superposition then the idea of applying it to the coupled state of two spatially remote objects is an entirely natural development.

I now turn to a different class of experiments, which attempts to examine the question of realism at a level much closer to our own everyday consciousness than that of electrons and atoms. Since I wrote a review [11] of this area of research in 2002 and not much has changed qualitatively since then, I will be rather brief. The experiments of this type done to date look, explicitly or implicitly, for evidence that the predictions of the QM formalism continue to work when the states involved are quantum superpositions of two or more states which are in some intuitively reasonable sense "macroscopically" distinct; in most cases, the evidence for this conclusion comes from the time-dependent behavior, in particular from the "Ramsey-fringe" effects which are the characteristic signature of a quantum superposition. The systems in question range from fullerenes and other complex molecules to magnetic biomolecules, quantum-optical systems and superconducting devices such as SQUIDs; for details see Ref. [11]. It should be emphasized that from the point of view of the "logic" of realism this class of experiments is less advanced than the EPR-Bell ones, in the following sense: In the EPR-Bell case the existing experiments not only establish that the QM predictions are correct, but also (modulo the "loopholes", see above) refute those of the class of OLT's (and, now, of CNLHV theories). In the case of the present class of experiments (sometimes called "macroscopic quantum coherence" (MQC) experiments), what has been established so far is that *if the raw data are interpreted according to the QM formalism*, then the states generated must be coherent quantum superpositions of macroscopically distinct states rather than classical mixtures of such states (the latter description would be entirely consistent with a picture in which each system of the ensemble has reached a definite state). One might then go on to argue as follows: "The notion of QM superpositions at any given level is inconsistent with the assumption of realism at that level: we see evidence for QM superposition at the level of SQUIDs (etc.): therefore, realism is false at the level of SQUIDs." As pointed out above, this argument, while perhaps plausible, is logically unsound; in order to exclude realism at the level of SQUIDs, we need to do an experiment which goes beyond the existing ones (just as in the EPR-Bell case, it was necessary to do more than simply verify that the QM predictions were correct for a few randomly chosen polarizer settings). Such an experiment was proposed by Garg and the present author in Ref. [12], and I believe that a number of experimental groups are currently working towards it.

The general structure of the experiment is conceptually very similar to that of the EPR-Bell experiments. One deals with a single system, e.g., a SQUID ring in an appropriate external magnetic flux, which possesses a dichotomic variable $Q(t)$ (in practice the value of the circulating current in the ring), i.e. one which can take only one of two discrete values which we label by convention $+1$ and -1 respectively. We verify in a preliminary experiment that measurement of $Q(t)$ does indeed always produce the value ± 1. We then note that for this simple "2-state" system, if it is ideally isolated from its environment (a term which in this context includes any internal dissipative mechanism), QM predicts, independently of the density matrix of the system (or more accurately of the "time ensemble" formed by a sequence of experiments starting from the same initial conditions) 2-time correlations of the form

$$\langle Q(t)\, Q(t')\rangle = \cos\Delta(t-t') \tag{2.4}$$

where Δ is a tunneling frequency between the two states $Q = \pm 1$. The formal similarity of Eq. (2.4) to Eq. (2.2) above suggests that it might be possible to devise a CHSH-type inequality for this system, with the different times of measurement playing the role of the different polarizer settings, and we shall now see that this is so.

The class of theories whose predictions we propose to compare with those of QM and with experiment go under the name of "macrorealism." Similarly to the case of OLT's, this class of theories is defined by the conjunction of three postulates:

(1) macroscopic realism per se
(2) noninvasive measurability
(3) induction.

The third postulate needs no special comment; it plays essentially the same role as in the EPR-Bell case, assuring us that the outcome of a measurement on the system cannot be affected by what will or will not be measured on it later. The assumption of macroscopic realism simply says that: a macroscopic object which has available to it two or more macroscopically distinct states is at "almost all" times in one of these states (the "almost all" is necessary because one has to allow, in such a theory, for transits between the states; however, this complication can be taken into account in the analysis, see Ref. [12]). In the present context, the postulate the postulate of macroscopic realism is equivalent to the statement that $Q(t)$ possesses either the value $+1$ or the value -1 at (almost) all times t, irrespective of whether or not it is actually measured (we saw above that when measured, it certainly realizes one or other of these two values).

In the present case, in contrast to the EPR-Bell experiments, all measurements are made on a single system, so there is no question of invoking locality. Instead, one uses in a rather similar role the postulate of "noninvasive measurability." This postulate (which is emphatically *not* a QM one!) states that it is possible, at least in principle, to perform a measurement of $Q(t)$ in such a way that neither the state of the system at time t nor its subsequent behavior is influenced by the measurement. Given the generally very invasive nature of most measurement procedures on macroscopic solid-state systems such as SQUIIDs, this postulate might at first sight seem rather unlikely to be fulfilled in a real-life situation. However, we can make it a lot more plausible by invoking an "ideal negative result" procedure: We arrange our measuring apparatus so that if $Q(t)$ is (say) $+1$ it is triggered, while if $Q(t) = -1$ nothing happens. We then do a series of runs in which Q is measured at some time t_o; these runs on which $Q(t_o)$ is measured to be $+1$ we throw away, the rest we keep. We then "invert" the measuring setup, so that a value of $Q(t)$ equal to -1 triggers it while $Q(t) = +1$ does not; this time we keep the runs on which $Q(t_o)$ is measured to be $+1$ and throw the rest away. In this way we can construct the 2-time correlations which we are giving to require in the argument below. (Note that it is only the first measurement of any pair which needs to be noninvasive.) Conceptually, the postulate of noninvasive measurability seems a very natural corollary to that of macroscopic realism: would there be any sense in affirming that the system "really has" the value $Q = +1$ at some time t, if a measuring device which is activated only when $Q = -1$ could nevertheless affect its behavior? Of course, in real life it may well be difficult to guarantee that "nothing happens" in the literal sense, and one may have to be content with replacing it by the statement that whatever happens when $Q(t)$ has the "negative" value is reliably calculable. (On a possible procedure for setting up an ideal-negative-result measurement, and further discussion, see Ref. [13].)

Given the three postulates above, the argument for a CHSH inequality of the form

$$\langle Q(t_1)\, Q(t_2)\rangle_{t_1t_2} + \langle Q(t_2)\, Q(t_3)\rangle_{t_2t_3} + \langle Q(t_3)\, Q(t_4)\rangle_{t_3t_4} - \langle Q(t_1)\, Q(t_4)\rangle_{t_1t_4} \leq 2 \tag{2.5}$$

(where the subscript $\langle\,\rangle_{t_it_j}$ indicates that the averages are taken, as they must be, on the subensemble on which measurements are made at times t_i and t_j (only)) goes through analogously to the EPR-Bell case: postulates (1) and (2) justify the statement that provided the first measurement of any pair is conducted in an ideally non-invasive way, $Q(t_i)$ has for each run a definite value which is unaffected, according to postulate (3), by what will or will not be measured subsequently. Thus, for each run the quantities $Q(t_i)\, Q(t_j)$ exist, with $Q(t_i)$ having the same value for any t_j. The trivial algebra of step (4) of the proof of the original CHSH theorem goes through, so Eq. (2.5) is satisfied provided the averages are replaced by $\langle Q(t_i)\, Q(t_j)\rangle_{all}$, where "all" indicates that they are taken over all runs. Finally, postulates (2) and (3) together justify the replacement of $\langle Q(t_i)\, Q(t_j)\rangle_{all}$ by the experimentally measured correlation $\langle Q(t_i)\, Q(t_j)\rangle_{t_it_j}$, giving the inequality (2.5).

I will not discuss here the rather delicate question of exactly how "macroscopically distinct" are the two circulating-current states realized in a typical SQUID ring; this is at least partly a matter of subjective definition, and in the present context it is sufficient to note that by any reasonable definition these two states differ in the behavior of, at a minimum, several million electrons. Rather, I would like to close by asking the question: If the proposed experiment is done, and comes out according to QM (hence in violation of the inequality (2.5)), what will we have learned over and above what is already known from the EPR-Bell experiments?

To answer this question, it is essential to appreciate that the hypothesis of macroscopic counterfactual definiteness (MCFD), which is refuted by the EPR-Bell experiments, while it is as noted weaker than that of microscopic realism, is stronger than the postulate of macroscopic realism. MCFD is a statement about a macroscopic event which "would have" happened (or not) in a situation which did not in fact arise; by contrast, macroscopic realism is a statement about what actually happens in a situation which is realized. Of course, the hypothesis of macroscopic realism does presumably imply that a second macroscopic object, set up explicitly as a measuring device, "would have," had it been activated at time t_i, have given the output corresponding to $Q(t_i)$, and thus, one could, if desired, replace postulate (1) by an MCFD formulation similar to the one given in the microscopic case. However, it is clear that MCFD is a situation where the measured object is macroscopic is a much weaker assertion than MCFD in the microscopic (e.g. EPR-Bell) case, and thus its denial is a much stronger statement. Hence the experiment proposed in Ref. [12], should it be done and come out according to the predictions of QM, will be a considerably stronger denial of realism than follows from the existing EPR-Bell work.

Acknowledgements This work was supported in part by the National Science Foundation through grant no. NSF-EIA-03-50842.

References

1. E. Schrödinger, Naturwissenschaften **23**, 807 (1935)
2. A. Einstein, B. Podolsky, N. Rosen, Phys. Rev. **47**, 777 (1935)
3. J.S. Bell, Physics **1**, 195 (1965)
4. J.F. Clauser, A. Shimony, Rep. Prog. Phys. **41**, 1881 (1978)
5. J.F. Clauser, M.A. Horne, A. Shimony, R.A. Holt, Phys. Rev. Lett. **23**, 880 (1969)
6. A.J. Leggett, J. Phys. A **40**, 3141 (2007)
7. A.J. Leggett, Found. Phys. **33**, 1469 (2003), erratum, in press (2007) (the erratum inter alia improves the RHS of the final inequality (4.9) to $2 - \frac{4}{\pi}|sin\chi|$)
8. S. Gröblacher, T. Paterek, R. Kaltenbaek, C. Brukner, M. Zukowski, M. Aspelmeyer, A. Zeilinger, Nature **446**, 871 (2007). Corrigendum Ibid., **449**, 252 (2007)
9. T. Paterek, A. Fedrizzi, S. Gröblacher, T. Jennewein, M. Zukowski, M. Aspelmeyer, A. Zeilinger, Phys. Rev. Lett. **99**, 210406 (2007)
10. C. Branciard, A. Ling, N. Gisin, C. Kurtsiefer, A. Lamas-Linares, V. Scarani, Phys. Rev. Lett. **99**, 210407 (2007)
11. A.J. Leggett, J. Phys. Condens. Matter **14**, R415 (2002)
12. A.J. Leggett, A. Garg, Phys. Rev. Lett. **54**, 857 (1985)
13. A.J. Leggett, Found. Phys. **18**, 939 (1988)

Chapter 3
Each Instant of Time a New Universe

Yakir Aharonov, Sandu Popescu, and Jeff Tollaksen

Abstract We present an alternative view of quantum evolution in which each moment of time is viewed as a new "universe" and time evolution is given by correlations between them. [*Editors note*: for a video of the talks given by Prof. Aharonov at the Aharonov-80 conference in 2012 at Chapman University, see quantum.chapman.edu/talk-3 and quantum.chapman.edu/talk-30 and by Prof. Popescu, see quantum.chapman.edu/talk-31.]

3.1 Introduction

Since the dawn of civilization, mankind tried to understand the meaning of the inexorable flow of time. One of the philosophical ideas, whose origins can be traced back to Eraclitus from Ephesus, states that the universe gets re-created again and again, at every instant of time. "You never bathe twice in the same river" said Eraclitus. Every instant new water, every instant a new universe. This idea, of course, is different from the usual one in which we view the universe as unique, and the objects which inhabit it as just changing their state in time.

Going from philosophy to physics, the way physics is presently formulated is based on the idea of unique universe and evolving objects. But is it possible to reformulate physics to incorporate the idea of a new universe for each instant? As far as classical physics is concerned, the reformulation is rather trivial. We find however

Y. Aharonov · J. Tollaksen (✉)
Institute for Quantum Studies and Schmid College of Science and Technology, Chapman University, Orange 92866, CA, USA
e-mail: tollakse@chapman.edu

Y. Aharonov
School of Physics and Astronomy, Tel Aviv University, Tel Aviv, Israel

S. Popescu
H.H. Wills Physics Laboratory, University of Bristol, Tyndall Avenue, Bristol BS8 1TL, UK

S. Popescu
Institute for Quantum Studies, Chapman University, Orange 92866, CA, USA

D.C. Struppa, J.M. Tollaksen (eds.), *Quantum Theory: A Two-Time Success Story*,
DOI 10.1007/978-88-470-5217-8_3,

Fig. 3.1 An artistic illustration of our central idea of this chapter: an alternative view of quantum evolution in which each moment of time is viewed as a new "universe" and time evolution is given by correlations between them

that quantum mechanically things are more complicated. The standard formalism of quantum mechanics appears not to allow such a reformulation. It turns out nevertheless that the reformulation is possible if we use the two-states formalism [1, 2].

3.2 Toy Models: The Difficulty

In preparation for discussing the time evolution from this alternative point of view, let us start by asking a much simpler question. Consider first classical mechanics.

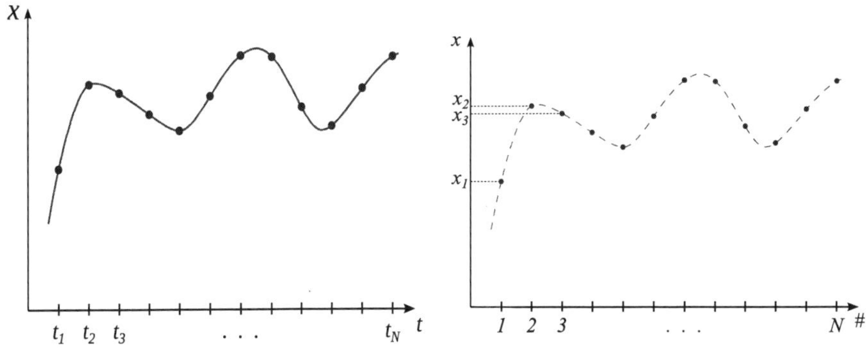

Fig. 3.2 (**a**) A single classical particle at N moments in time, (**b**) N classical particles at a given time τ. The position of particle i is the same as the position of the original particle at time t_i (see part (**a**))

Suppose a particle evolves such that its trajectory is $x(t)$ (Fig. 3.2.a). Consider now a set of N time moments, t_1, t_2,..., t_N. Is it possible to prepare instead of this single particle a set of N particles such that if we perform at some given time τ measurements on these N particles we'll get the same information as we would have obtained by making measurements at $t_1, t_2, \ldots, t_N$ on the original single particle considered before?

The solution is quite simple: One has to prepare the N particles (Fig. 3.2.b) such that

$$\begin{aligned} x_1(\tau) &= x(t_1) \\ x_2(\tau) &= x(t_2) \\ &\vdots \\ x_N(\tau) &= x(t_N). \end{aligned} \tag{3.1}$$

When N increases and the time intervals $t_{i+1} - t_i$ decrease, we could say that the N particles lay down, at a single moment of time (at τ) the entire history of the original particle. One particle at N times is thus equivalent to N particles at one time.

Can we do the same for a quantum mechanical particle?

Consider the simple case of a spin 1/2 particle, prepared in some initial state $|\psi\rangle$ and having the Hamiltonian $H = 0$. In this case the time evolution of the particle is trivial,

$$|\psi(t)\rangle = |\psi\rangle. \tag{3.2}$$

Could we now prepare N spin 1/2 particles such that if we perform measurements on them at some time τ we get the same information as we could get by measuring the state of the original particle at N different time moments, $t_1, t_2, \ldots, t_N$? Since the state of the original particle at all these moments is $|\psi\rangle$, one would suppose that this task can be accomplished (Fig. 3.3) by preparing the N particles each in the same state $|\psi\rangle$, that is

$$|\Psi\rangle_1 |\Psi\rangle_2 \cdots |\Psi\rangle_N. \tag{3.3}$$

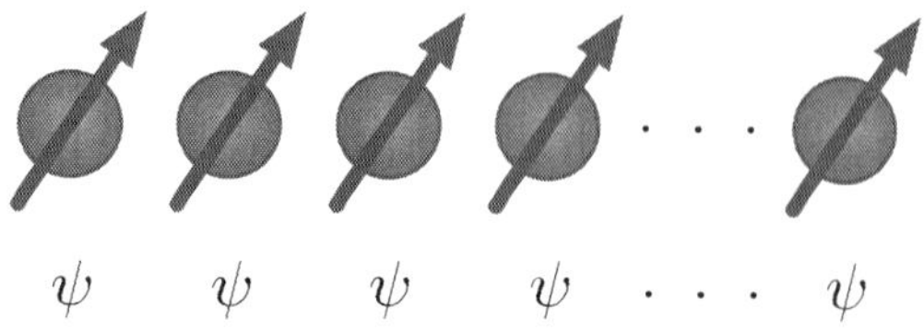

Fig. 3.3 N spin 1/2 particle at time τ, each prepared in the state ψ. They do not constitute however an appropriate mapping of N time moments of a single spin 1/2 particle

But this mapping is not appropriate for many reasons. First of all, the state (3.3) contains *too much* information. Indeed, suppose somebody prepared the original particle in state ψ and gave it to us without telling us what the state is. Then, as we have a single particle (i.e. a single copy of an unknown state ψ), there is no way of learning what its state is. However, in (3.3) we have N particles, all in the same state—by making different measurements on the different copies and looking at the statistics of the results we can learn the state (better and better as N becomes larger).

Second, the time evolution (3.2) contains subtle correlations, which usually are not noticed, and which do not appear in the state (3.3). Suppose, for example, that the state $|\psi\rangle = |\sigma_z = 1\rangle$, i.e. the spin is polarized "up" along the z axis. It is generally considered that since the particle is at every moment in a definite state of the z-spin component, the z-spin component is the only thing we know with certainty about the particle—no other spin component commutes with σ_z hence it is not well-defined. However, there are *multi-time* variables whose values are known with certainty, given the evolution (3.2). For example, although the x spin component is not well defined when the spin is in the $|\sigma_z = 1\rangle$ state, we know that it is constant in time, since the Hamiltonian is zero. Thus, in particular, the two-time observable

$$\sigma_x(t_2) - \sigma_x(t_1) = 0. \tag{3.4}$$

As described in [2, 3], this observable can be measured in the following way. Following von Neuman's measuring formalism, consider a measuring device whose pointer position is denoted by q and its canonical conjugate momentum p and let the interaction between the spin and the measuring device be described by the interaction Hamiltonian

$$H_{int} = -\delta(t - t_1)p\sigma_x + \delta(t - t_2)p\sigma_x. \tag{3.5}$$

We also assume that the Hamiltonian of the measuring device at all other times is zero (i.e. the pointer doesn't move by itself). Due to the strong coupling between the spin and the measuring device during the measurement (the delta function coupling) we can neglect during the measurement any other interaction affecting the spin. From the Heisenberg equations of motion we obtain

$$\frac{dq}{dt} = i[q, H_{int}] = \big(\delta(t - t_2) - \delta(t - t_1)\big)\sigma_x(t). \tag{3.6}$$

The equation can be integrated easily, by noting that σ_x doesn't change in the two (infinitesimally) short interaction times ($t_1 - \epsilon$ to $t_1 + \epsilon$ and $t_2 - \epsilon$ to $t_2 + \epsilon$) when the measuring interaction takes place since it commutes with the interaction Hamiltonian. Hence, integrating (3.6) we obtain

$$q(t_2+\epsilon) - q(t_1-\epsilon) = \sigma_x(t_2) - \sigma_x(t_1), \tag{3.7}$$

in other words, the difference between the final and initial positions of the pointer indicates the value of the two-time observable $\sigma_x(t_2) - \sigma_x(t_1)$. Crucially, this is a measurement which tells *only* the value of $\sigma_x(t_2) - \sigma_x(t_1)$ but not the value of $\sigma_x(t_1)$ or $\sigma_x(t_2)$ separately. Now, when the Hamiltonian acting on the spin is zero, $\sigma_x(t_1) = \sigma_x(t_2)$ and therefore the measuring device will indicate

$$q(t_2+\epsilon) - q(t_1-\epsilon) = \sigma_x(t_2) - \sigma_x(t_1) = 0. \tag{3.8}$$

It is also important to note that, given that the spin Hamiltonian is zero, after the measurement is finished, i.e. after t_2, the spin is brought back in the initial state, regardless of what this state is. Indeed, suppose that the initial state of the spin is $|\psi\rangle$. By decomposing $|\psi\rangle$ in the x-basis, in the Schrodinger representation the evolution of the spin and measuring device is given by

$$\begin{aligned} |\psi\rangle|q=0\rangle &= (\alpha|\uparrow\rangle + \beta|\downarrow\rangle)|q=0\rangle \\ &\to \alpha|\uparrow\rangle|q=-1\rangle + \beta|\downarrow\rangle|q=1\rangle \\ &\to \alpha|\uparrow\rangle|q=0\rangle + \beta|\downarrow\rangle|q=0\rangle = |\Psi\rangle|q=0\rangle \end{aligned} \tag{3.9}$$

where the two right pointing arrows describe the evolution of the spin and measuring device at t_1 and t_2 respectively. (Note that the first interaction shifts the pointer with the value $-\sigma_x$ while the second shifts the pointer with $+\sigma_x$.)

Coming back to the N spins in the state (3.3), there is no correlation whatsoever in between the x components of, say, particles 1 and 2 which were intended to describe the original particle at times t_1 and t_2. More over, it is not only the x spin component for which the original particle presents such multi-time correlations, but all spin components. That is, (3.4) generalizes to

$$\sigma_{\hat{n}}(t_1) = \sigma_{\hat{n}}(t_2) = \cdots = \sigma_{\hat{n}}(t_N) \tag{3.10}$$

for any direction $\hat{n}$. (Following the above procedure, we would then obtain $\sigma_{\hat{n}}(t_i) - \sigma_{\hat{n}}(t_j) = 0$ for any i and j.) Obviously, the $\hat{n}$ spin components of the N spins in state (3.3) (except for the z components) are not correlated in this way.

We reached thus the conclusion that the N particles in the state (3.3) *do not* describe faithfully the behavior of the original particle at $t_1, \ldots, t_N$. The question is now whether there exists any state of N particles that could be such a faithful representation? It is easy to see that the answer is "no". Indeed, there is no state of N spins such that

$$\sigma_{\hat{n}}^1 = \sigma_{\hat{n}}^2 = \cdots = \sigma_{\hat{n}}^N \tag{3.11}$$

for every direction $\hat{n}$. At best, one may find a two particle state—the singlet state-for which the spins are *anti-correlated* instead of correlated i.e.

$$\sigma^1_{\hat{n}} = -\sigma^2_{\hat{n}} \tag{3.12}$$

for every $\hat{n}$. And even anti-correlation cannot be extended to more than two particles. It appears thus that we reached a dead end.

It is tempting to think that the reason we arrived at this dead end is that we didn't take into account that quantum mechanically measurements modify the state of the measured system. We could say that the state of the original particle is constant in time, (3.2), and thus it is modeled by the N spins in state (3.3) *only* if no measurements are performed. If measurements are performed, the time evolution of the original particle is no longer given by (3.2), so we shouldn't expect to model it by (3.3). But this is actually *not* the true reason for our difficulties. Indeed, even if we don't actually measure $\sigma_x(t_2) - \sigma_x(t_1)$ but merely *compute* it for the evolution (3.2), we see that it has not the same value as if we compute it for the state (3.3). So problems arise already at this stage—the state (3.3) simply doesn't contain the correlations which the free evolution of the original particle prescribes.

3.3 Toy Models: The Solution

We now arrived at a crucial point. Although a state of N spin 1/2 particles with complete correlations among all their spin components as required by (3.11) doesn't exist in the usual sense, there exist pre- and post-selected states [2] with this property. As we show now, N spin 1/2 particles in a suitably prepared pre- and post-selected state can, at one time, τ, mimic N moments of time in the evolution of a single spin 1/2 particle.

The procedure we will use has three steps. At time $\tau - \epsilon$ we prepare an initial state. At time τ we perform the measurements that are supposed to simulate the evolution of the original single spin. At time $\tau + \epsilon$ we perform an additional measurement; only if this measurement is successful, we deem our simulation procedure to have succeeded. Specifically, consider $2N - 1$ spin 1/2 particles. N of them will be used as "spins", and we denote them by $S_0, S_1, \ldots, S_N$. They will simulate N time moments of the evolution of our original spin. For example, a measurement at time τ on the spin S_k should simulate the measurement on the original spin at time $t = t_k$. Furthermore a two time measurement on the original spin, say at t_k and t_l will be simulated by a measurement on spins S_k and S_l and so on. The other $N - 1$ spins are ancillas, which we denote by $A_1, A_2, \ldots, A_N$. They are used for helping in our procedure, however, no measurements will be performed on them at τ. We arrange the "spins" and ancillas as illustrated in Fig. 3.4.

At $\tau - \epsilon$ we prepare the particles in the initial state

$$|\psi\rangle_{S_0}|\Phi\rangle_{A_1,S_1}\cdots|\Phi\rangle_{A_N,S_N} \tag{3.13}$$

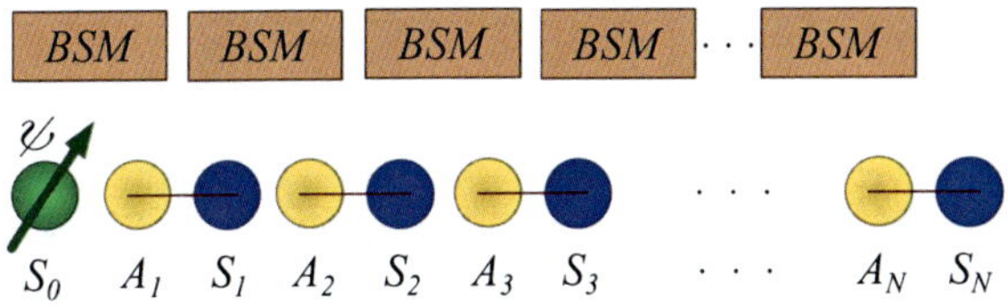

Fig. 3.4 $N+1$ "spins" and N ancillas. At time $\tau-\epsilon$ spin S_0 is prepared in state ψ while ancilla A_k is maximally entangled with spin S_k (the maximal entanglement is represented by the *continuous line* connecting the ancilla with the spin.) At time $\tau+\epsilon$ a "Bell state" measurement (BSM) is performed on spin S_{k-1} and ancilla A_k. The experiment is deemed successful if and only if each of the BSMs yields the outcome corresponding to the maximally entangled state $|\Phi\rangle_{S_{k-1},A_k}=\frac{1}{\sqrt{2}}\sum_{i=0}^{1}|i\rangle_{S_{k-1}}|i\rangle_{A_k}$. In case of a successful experiment, if measurements were performed on spins $S_0,\ldots S_{N-1}$ at time τ, their results simulate measurements performed at N moments of time on a single spin prepared in state ψ and evolving with a Hamiltonian equal to zero

where $|\Phi\rangle$ is a maximally entangled state of the ancilla A_k and the associated spin, S_k, namely $|\Phi\rangle_{A_k,S_k}=\frac{1}{\sqrt{2}}\sum_{i=0}^{1}|i\rangle_{A_k}|i\rangle_{S_k}$ and where $|i\rangle$, $i=0,1$ represent some arbitrary base vectors. At a later time, $\tau+\epsilon$ we subject the pairs of spins composed by spin S_{k-1} and ancilla A_k to a measurement of an operator that has the maximally entangled state $|\Phi\rangle_{S_{k-1},A_k}=\frac{1}{\sqrt{2}}\sum_{i=0}^{1}|i\rangle_{S_{k-1}}|i\rangle_{A_k}$ as one of its nondegenerated eigenstates (for example the "Bell" operator). Now, as it is easy to directly verify, in the case when all these measurements performed at $\tau+\epsilon$ yield the outcome corresponding to this state, then measurements performed at τ on the "spins" reproduce the same statistics as measurements on the original single spin.

The above procedure may seem rather convoluted. However, it has a very simple interpretation in the language of pre- and post-selected states [2, 3]. The procedure simply prepared a particular pre- and post-selected state of the N spins $S_1,\ldots,S_N$. The defining characteristic of this state is that the post-selected state of one particle is completely correlated to the pre-selected state of the next particle as illustrated in Fig. 3.5. Technically this is possible because post-selected states propagate backwards in time and behave as complex conjugates of pre-selected states. This accounts for the correlations that cannot be created when we consider a pre-selected only state (i.e. a state prepared at $\tau-\epsilon$).

The idea of pre- and post-selected states and the above way of preparing them was discussed in detail in [2]. Using the notation $\Phi_{k+1,k}^{\tau_-,\tau_+}$ for the *maximally entangled* two-time state

$$\Phi_{k+1,k}^{\tau_-,\tau_+}=\sum_i |i\rangle_{S_{k+1}}^{\tau_-}\ {}_{S_k}^{\tau_+}\langle i| \tag{3.14}$$

where $\tau_\pm=\tau\pm\epsilon$, from the results in [2] it follows that in our case the pre-and post-selected state of the spins is

$$\Phi_{N,N-l}^{\tau_-,\tau_+}\cdots\Phi_{2,1}^{\tau_-,\tau_+}\Phi_{1,0}^{\tau_-,\tau_+}|\Psi\rangle_{S_0}^{\tau_-}. \tag{3.15}$$

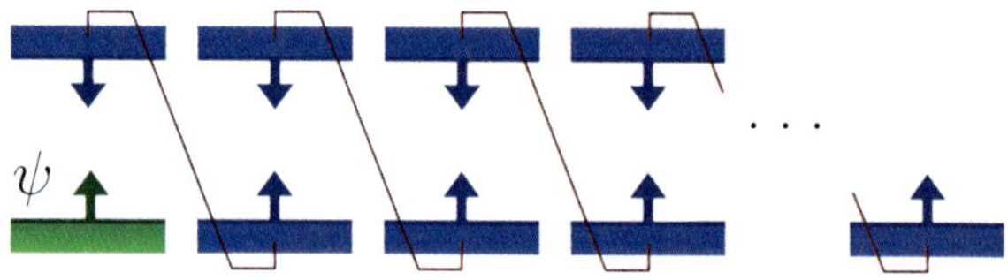

Fig. 3.5 Correlation between the spins $S_0 \dots S_N$. The post-selected state of one particle is correlated to the pre-selected state of the next particle. *Upward arrows* denote "pre-selected" state, which point towards the future while downward arrows denote "post-selected" states that point towards the past

Note that this pre-and post-selected state refers solely to the spins; the ancillas were there only to help prepare this state.

The pre- and post-selected state (3.15) explicitly shows the two main characteristics of the time evolution of the original spin. On one hand, at the initial time t_0 the original spin was prepared in the state $|\Psi\rangle$. To this corresponds the fact that in (3.15) the pre-selected state of S_0 is the same as that of the original spin. On the other hand, we also know (3.10) that the spin components along any direction, although undefined, are constant in time, that is, they are fully correlated. These correlations are realised in (3.15) via the complete correlations between the post-selected state of one spin and the pre-selected state of the next (3.14). (Indeed, it is easy to verify that the maximally entangled state (3.14) is invariant under a simultaneous change of basis for S_{k+1} and S_k.)

3.4 Every Moment of Time a New Universe

Up to this point we only dealt with a toy model. We now come to the main question, namely how to interpret the time evolution of a quantum particle from the point of view of the philosophical idea of "each moment of time a new universe". As far as classical physics is concerned, we could formalize this idea by associating a separate configuration-space to each moment of time. A moving particle would then correspond to one particle in each space, having their positions appropriately correlated. In effect, this would mean associating a different configuration space to each of the N particles in Fig. 3.2.b and "stacking" them one on top of the other along the time axis. Naively, one would expect that quantum mechanically this would correspond to associating to each moment of time a separate Hilbert space. The total Hilbert space would be therefore $\mathcal{H} = \mathcal{H}_0 \otimes \mathcal{H}_1 \otimes \cdots \otimes \mathcal{H}_N$. The problem however, as we saw before, is that no state in such a Hilbert space can account for the desired correlations. The solution follows from the above described mapping between the time evolution of the spin 1/2 particle and N spin 1/2 particles: we have to associate *two* Hilbert spaces to each moment of time and"stack" them on top of each other along the time axis (see Figs. 3.5 and 3.6.a).

We consider time as being discrete, made out of finite time intervals, stacked one on top of the other like little bricks. To each moment of time we associate two

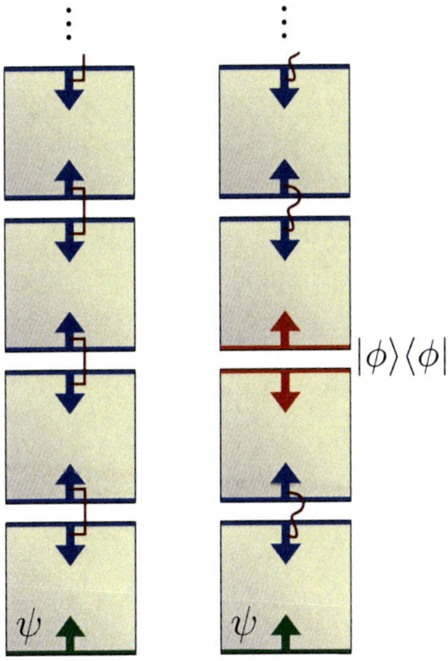

Fig. 3.6 (**a**) Each moment of time is a little "brick". The Hilbert space at the future boundary of one time moment is maximally entangled—with complete correlation—with the Hilbert space at the past boundary of the next time moment (*straight line*). The state ψ is associated only to the moment t_0 where it was prepared. (**b**) A more general time evolution. A measurement with a collapse on state ϕ disentangles the two subsequent moment of time. Non-trivial unitary time evolution at all other time is represented by maximal entanglement but between appropriately rotated bases (*squiggled line*)

Hilbert spaces: a Hilbert space of ket vectors at the time boundary towards the past and a Hilbert space of bra vectors at the time boundary towards the future. The total Hilbert space is therefore of the form $\mathcal{H}_N \otimes \cdots \otimes \mathcal{H}_1^\dagger \otimes \mathcal{H}_1 \otimes \mathcal{H}_0^\dagger \otimes \mathcal{H}_0$. The state of a quantum system during each instant of time is thus determined by two wavefunctions. One of them is fixed at the past time boundary of the time interval and is "evolving" towards the future, the other one is fixed at the future boundary of the time interval and it is "evolving" towards the past. We shall call these two states "pre-selected" and "post-selected".

Consider now the example we presented at the beginning of this paper: At time t_0 the spin was prepared in state $|\Psi\rangle$ and then the evolution is trivial, i.e. the Hamiltonian is zero. In our new formalism the preparation at time t_0 corresponds to having at time t_0 the forward evolving state $|\Psi\rangle_{t_0} \in H_0$. At all other moments of time we know that all spin components are completely correlated. We describe this by taking the wavefunctions at the common boundary of two subsequent moments (i.e. the post-selected state of the earlier moment and the pre-selected state of the later moment) to be *maximally entangled* and *completely correlated*: $\sum_i |i\rangle_{t_1\,t_0}\langle i|$ and so on at all times. Here the vectors $|i\rangle_{t_1}$ and ${}_{t_0}\langle i|$ form arbitrary orthonormal bases in $\mathcal{H}_1$ and $\mathcal{H}_0^\dagger$ respectively and represent the same physical state (such as $|\sigma_z = 1\rangle$ and

$\langle\sigma_z = 1|$). Since all time moments are correlated this way, the time evolution now looks like a chain of connected time intervals. Hence the total state is

$$\sum_k |k\rangle_{t_N t_{N-1}} \langle k| \cdots \sum_i |j\rangle_{t_2 t_1} \langle j| \sum_i |i\rangle_{t_1 t_0} \langle i| \, |\Psi\rangle_{t_0}. \tag{3.16}$$

The example above can be easily generalized to arbitrary quantum evolutions which consist of unitary evolutions and measurement induced collapses (Fig. 3.6.b). When the Hamiltonian is non-zero subsequent moments of time continue to be maximally entangled, but the correlation is now "skewed". That is, the correlation between two moments of time is now of the form

$$\sum_i |u_i\rangle_{t_2 t_1} \langle i| \tag{3.17}$$

where

$$|u_i\rangle_{t_2} = U_{2,1} |i\rangle_{t_2}. \tag{3.18}$$

Here $U_{2,1}$ represents a unitary transformation acting on $\mathcal{H}_2$, the pre-selected Hilbert space at time t_2, numerically equal to $U(t_2, t_1)$, the unitary that describes the evolution of the particle from t_1 to t_2 in the standard quantum description.

Note that, similarly to $\sum_i |i\rangle_{t_1 t_0} \langle i|$ the state (3.17) is also a maximally entangled state between the two moments of time, and also leads to full correlations. The only difference is that the correlations are now not between an arbitrary vector ${}_{t_1}\langle\xi|$ and the corresponding vector $|\xi\rangle_{t_2}$ but between ${}_{t_1}\langle\xi|$ and $U_{2,1}|\xi\rangle_{t_2}$.

On the other hand when a measurement occurs it completely disturbs all the observables that do not commute with the measured observable—their values before the measurement are no longer correlated with their values after the measurement. At the same time a collapse means the introduction of a new boundary condition for times following the collapse as well as a post-selection for times previous to the collapse. Suppose an instantaneous ideal von Neuman measurement took place at time t_k and let $|\phi\rangle$ be the eigenstate corresponding to the observed eigenvalue. In our alternative formalism this collapse is implemented simply by $|\phi\rangle_{t_{k+1} t_k} \langle\phi|$.

We have now reached our desired alternative description of time evolution. To summarize:

- Each moment of time is indeed one "Universe" but it has associated to it not one but two Hilbert spaces, one corresponding to the "past" boundary of this time moment and one to its "future" boundary
- Time evolution is represented by correlations between subsequent moments of time, more precisely between the "future" boundary of the earlier time moment and the "past" boundary of the later time moment.
- Unitary time evolution is implemented by maximal entanglement between subsequent moments of time.

- A measurement induced collapse destroys the entanglement and effectively decouples the entire time evolution up to that moment by what happens later; technically, the state before the moment of collapse is in a direct product with the state after it.
- A partial collapse, such as one due to an incomplete measurement will result in entanglement but less than maximal.

3.5 Probabilities

In the above section we described the time evolution of a quantum system in the "each moment of time a new universe" paradigm. To complete our description there is one more item to address: how to calculate probabilities for the different events.

The probability for a particular history to happen can be computed from the "history" state in a straightforward way, with one subtlety. As it stands now, the history is open ended, both towards the remote past and the remote future; without putting boundary conditions at those two end, nothing can be said about the overall probability of the history. The standard experimental questions however remove the need for these "cosmological" implications by completely separating a piece of time from its past and future by making complete measurements. The most well-known case is preparing at initial time t_0 the system in some state $|\Psi\rangle$ and then performing a measurement at the final time t_f and asking what is the probability to find the system in state $|\phi\rangle$. Looking at our description above, we see that indeed these two measurements cut out a piece of time, i.e. the state starting at t_1 and ending at t_2 is completely disentangled from the rest. For example, if the Hamiltonian is zero, this piece is:

$$_{t_f}\langle\phi|\sum_k |k\rangle_{t_N\,t_{N-1}}\langle k|\cdots\sum_i |j\rangle_{t_2 t_1}\langle j|\sum_i |i\rangle_{t_1\,t_0}\langle i|\;|\Psi\rangle_{t_0}. \tag{3.19}$$

To find the probability to obtain $|\phi\rangle$ when the system was prepared in $|\Psi\rangle$ all we do is the following: for each moment of time we contract the vectors belonging to the past and future boundary conditions, i.e. we take the scalar product between the bra and ket vectors corresponding to the same time t. The result is the scalar product $\langle\phi|\Psi\rangle$; the absolute value of this, $|\langle\phi|\Psi\rangle|^2$ is the probability we are looking for. It is also immediate to show that in case the Hamiltonian is non-zero, our recipe leads to $|\langle\phi|U(t_f,t_0)|\Psi\rangle|^2$, the expected result.

3.6 Discussion

We would like to emphasize what is arguably the most important property of this alternative view of time evolution, namely that there is a complete harmony in between the actions to which the system is subjected and their representation: If a measurement and its associated collapse occur at time t it is *there and only there* that this information is present—the state Φ prepared by the collapse appears only at this time. When there are more measurements, each measurement and the state associated to its outcome appear at the time of measurement and only there. At other times, when no measurement is performed all we know is that the time correlations are preserved, and this is what our formalism shows.

Our description is in stark contrast with the usual one in which once we prepare a system in a state $|\psi\rangle$ and we leave it undisturbed, then at every subsequent moment of time the state continues to be $|\psi\rangle$. As far as we are concerned however, the state $|\psi\rangle$ doesn't characterize *directly* any other moment of time except when it was prepared; it does influence the physics at these other moments, but it does so only indirectly, via a chain of time correlations. What does directly characterize a time when no measurement is performed is that it is an unbroken link in a chain of correlations, nothing more than this; what propagates along the chain is a completely independent issue.

It is very interesting to ponder more carefully the difference between a measurement and a unitary evolution from our point of view. What we see is a certain *complementarity* between kinematics and dynamics. When a measurement is performed we know the state at each of the two subsequent moments of time when the measurement took place:

$$|\phi\rangle_{t_{k+1}\,t_k}\langle\phi|. \tag{3.20}$$

On the other hand, when a unitary evolution takes place, the state at each of the two subsequent moments of time is completely uncertain, the state at one moment being entangled with the state at the next moment

$$\sum_i |u_i\rangle_{t_{k+1}\,t_k}\langle i|. \tag{3.21}$$

Furthermore, we note that every measurement is effectively an uncertain time evolution. This is a fact that, as far as we know, is very rarely mentioned in discussions about quantum measurements. Yet, it is quite obvious. Indeed, as is well known all the observables that do not commute with the measured one are randomized up to some extent, hence their Heisenberg equations of motion must show an uncertain evolution. In its turn, this is due to the fact that during the measurement the Hamiltonian of the system is uncertain. Indeed, in the standard von Neumann measurement formalism (as used above in (3.5)) in order to measure an observable A and register its value in the indication q of a pointer we use an interaction Hamiltonian of the form

$$H_{int} = \delta(t)Ap \tag{3.22}$$

where p is the canonical momentum conjugate to the position q of the pointer. Since the initial state of the pointer is well defined, say $|q = 0\rangle$, the momentum of the pointer, p has a large uncertainty $\Delta p = \infty$. In its turn, since p enters the interaction Hamiltonian, it means that as far as the system is concerned, its Hamiltonian is uncertain during the measurement.

The above observations, although not very commonly known, are nevertheless rather straightforward. What our new formalism shows however, is something more subtle: although a measurement is equivalent to an uncertain evolution, the collapse on a particular eigenstate of the measured observable is equivalent to a well-defined *superposition of different time evolutions*[1]. Indeed, take for example a measurement of the σ_x, the x component of the spin of a spin 1/2 particle performed at t_k. Suppose we found $\sigma_x = 1$. According to our formalism the quantum state at the time of measurement is

$$|\uparrow\rangle_{t_{k+1}\,t_k}\langle\uparrow|. \tag{3.23}$$

This can be viewed as the superposition of two unitary time evolutions

$$\begin{aligned} |\uparrow\rangle_{t_{k+1}\,t_k}\langle\uparrow| = &\frac{1}{2}\left(|\uparrow\rangle_{t_{k+1}\,t_k}\langle\uparrow| + |\downarrow\rangle_{t_{k+1}\,t_k}\langle\downarrow|\right) \\ &+ \frac{1}{2}\left(|\uparrow\rangle_{t_{k+1}\,t_k}\langle\uparrow| - |\downarrow\rangle_{t_{k+1}\,t_k}\langle\downarrow|\right). \end{aligned} \tag{3.24}$$

Hence our picture suggests a new kind of complementarity between having information about the state of a system versus having information about the dynamics: if one does not know the state, then our picture describes the dynamics as complete correlation. If one obtains information about the state, then the multi-time correlations between conjugate operators are made uncertain, i.e. one loses information about the dynamics in that interval of time. And as for a proper conjugacy relationship, there is a continuous graduation between the extremes. To see this complementarity, consider a partial measurement of σ_x in which the measuring device gives the correct answer (i.e. $\sigma_x = \pm 1$) with probability $|\alpha|^2$ and the wrong answer with probability $|\beta|^2$ and does this in a way which minimizes the disturbance to the state. This is obtained when the measuring device interacts with the spin via the unitary evolution

$$\begin{aligned} |\uparrow\rangle|0\rangle_M &\rightarrow |\uparrow\rangle(\alpha|1\rangle_M + \beta|-1\rangle_M) \\ |\downarrow\rangle|0\rangle_M &\rightarrow |\downarrow\rangle(\alpha|-1\rangle_M + \beta|1\rangle_M) \end{aligned} \tag{3.25}$$

where $|0\rangle_M$, $|1\rangle_M$ and $|-1\rangle_M$ are different states of the measuring device. Obtaining the value $+1$ corresponds in our picture to partially destroying the complete correlation between the moments when the measurement occurred and leading to only non-maximal correlations[2]:

$$\alpha|\uparrow\rangle_{t_2 t_1}\langle\uparrow| + \beta|\downarrow\rangle_{t_2 t_1}\langle\downarrow|, . \tag{3.26}$$

[1] In a completely different context, superpositions of time evolutions were discussed in [4] and experimentally verified in [5].

[2] In the standard language this outcome corresponds to the Krauss operator $\alpha|\uparrow\rangle\langle\uparrow| + \beta|\downarrow\rangle\langle\downarrow|$.

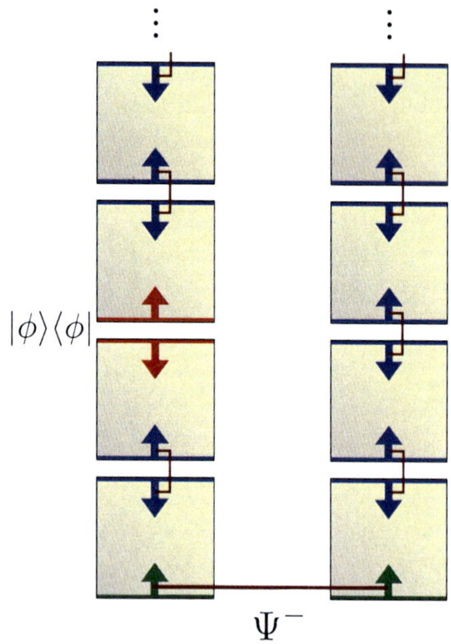

Fig. 3.7 Two entangled spin 1/2 particles. Entanglement characterizes solely time t_0 where entanglement is produced. All other times are characterized by trivial time evolution, i.e. maximal entanglement between subsequent moments of time; there is however no entanglement between the particles associated to these times. Alice's measurement disentangles the time moments of her particle but have no effect on Bob's particle

We see that in the case $\alpha = \beta$, we have complete correlation, thus modeling the dynamics. As $\alpha \to 1$ and $\beta \to 0$ we obtain more and more knowledge about the state, while the entanglement, i.e. the dynamics, becomes more and more uncertain.

3.7 Measurements on EPR States

It is very interesting to analyze using our point of view the time evolution of two spin 1/2 particles in a singlet state. The evolution is illustrated in Fig. 3.7. At t_0 the two particles A and B are prepared in the singlet state

$$|S\rangle_{AB,t_0} = \frac{1}{\sqrt{2}}|\uparrow\rangle_{A,t_0}|\downarrow\rangle_{B,t_0} - \frac{1}{\sqrt{2}}|\downarrow\rangle_{A,t_0}|\uparrow\rangle_{B,t_0}. \tag{3.27}$$

Then each particle evolves separately. That is, the time moments describing particle A are maximally entangled with each other and the time moments describing the evolution of B are maximally entangled with each other. There is however no entanglement between particles A and B at any other subsequent moment. This situation continues until we disturb the particles.

Suppose now that at time T Alice performs a measurement on particle A. For example, suppose she measures σ_x^A and finds the value $+1$. According to our point of view, for particle A the entanglement between the moments of time T and $T+\epsilon$ is broken and in its place we have a direct product state $|\uparrow\rangle_{A,T+\epsilon}\ {}_{A,T}\langle\uparrow|$. On the other hand, nothing happens to particle B—its time moments continue to be maximally entangled.

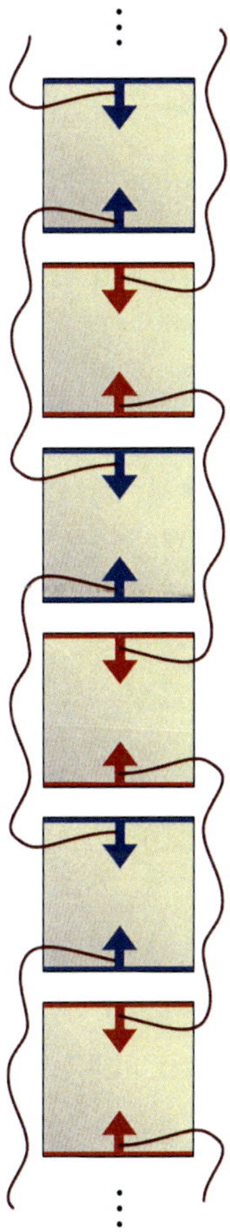

Fig. 3.8 Two independent "lives" lived in parallel by the same particle

This conclusion seems to contradict the standard quantum mechanical description. Indeed, in the usual description the state is $|S\rangle_{AB}$ from time t_0 until time T and at time T the state collapses into the direct product state $|\uparrow\rangle_A|\downarrow\rangle_B$. In particular, the collapse is symmetric with respect to who produces it: The time evolution is the same whether Alice measured σ_x^A and found $+1$ or Bob measured σ_x^B and found -1. In our description however, if Alice performs the measurement, the time evolution of her particle is affected and nothing happens to Bob's, while the opposite would be true if Bob were to perform the measurement. One can check directly however that all *observational* consequences, i.e. the probabilities for all measurements, are the same in both descriptions. Our point of view however has two main advantages.

First of all, it is relativistically covariant *at the level of states*. Of course, both views are relativistically covariant at the level of observed results. The standard way however is *non-covariant* as far as the state description is concerned. Indeed, the collapse occurs both at Alice and at Bob at time T, i.e. simultaneously in the reference frame in which we chose to work. Had we chosen a different reference frame, the moment at which the collapse occurs for Bob's particle could have been different. On the other hand, in our description, nothing happens to Bob's particle when Alice performs a measurement, so no covariance problems arise.

We would like to emphasize however that the relativistic covariance at the level of wave-functions does not necessarily require to consider each moment of time a new universe; it is already present in a simpler version of time evolution, with a "single universe" but with two wave-functions, one propagating forward and the other backward in time [6, 7].

A second interesting feature of our description is that it makes it clear that the evolution of Alice's particle is *different* from Bob's particle, while in the standard description they looked the same. Indeed, in the standard description they appear symmetric—they are in a singlet until time T and then they collapse together on a direct-product state. In our description is clear that for particle A the time moments before and after T are not maximally entangled while for particle B they are. This difference could be checked if in addition to the measurement at time T Alice also measures a two-time variable, say $\sigma_z^A(t_1) - \sigma_z^A(t_2)$ for $t_0 < t_1 < T < t_2$. Since the spin components along the z direction before and after T are not correlated, Alice could obtain $+2$, 0 or -2. On the other hand, if Bob were to measure $\sigma_z^B(t_1) - \sigma_z^B(t_2)$ he would obtain with certainty the value 0.

3.8 Extensions

Many more interesting situations are possible. An amusing one is illustrated in Fig. 3.8. Here every moment of time is fully correlated with the second next. In effect this particle has a "double life"—the even time moments describe a particle whose time evolution is $|\psi(t)\rangle = |\psi_1\rangle$ and the odd moments describe a particle whose time evolution is $|\psi(t)\rangle = |\psi_2\rangle$. As long as we do not take any action to connect them, such as a two-time measurement involving an odd and an even moment, the two lives of this particle do not interact with each other. It is interesting to speculate if such things exist in nature, and what their meaning would be.

Acknowledgements S.P. acknowledges support from the European Research Council ERC Advanced Grant NLST, EU grant Q-Essence and the Templeton Foundation. We thank Ildi Popescu for her artistic assistance with Fig. 3.1.

References

1. Y. Aharonov, P.G. Bergmann, J. Lebowitz, Phys. Rev **134**, B1410 (1964)
2. Y. Aharonov, S. Popescu, J. Tollaksen, L. Vaidman, Phys. Rev. A **79**, 052110 (2009)
3. Y. Aharonov, D.Z. Albert, Phys. Rev. D **29**, 223 (1982)
4. Y. Aharonov, J. Anandan, S. Popescu, L. Vaidman, Phys. Rev. Lett **64**, 2965 (1990)
5. D. Suter, M. Ernst, R.R. Ernst, Molecular Physics **78**(1), 95–102 (1998)
6. Y. Aharonov, D. Albert, Phys. Rev. D **29**, 228 (1982)
7. Y. Aharonov, D. Albert, S. D'Amato, Phys. Rev. D **32**, 1976 (1985)

Part II
Building Blocks of Nature

Chapter 4
The Brout-Englert-Higgs Mechanism and Its Scalar Boson

Francois Englert

Abstract The Brout-Englert-Higgs (BEH) mechanism extending spontaneous symmetry breaking to gauge fields had a considerable impact on both theoretical and experimental elementary particle physics. It is corroborated by the discovery of the Z and the W, and by the precision electroweak tests. The recent detection by the LHC at CERN of a particle identifiable to its massive BEH scalar boson not only constitutes a direct verification of the mechanism, but detailed analysis of its different decay processes may yield indications on the world hitherto hidden beyond the Standard Model. These topics are discussed with emphasis on conceptual issues. [*Editors note*: for a video of the talk given by Prof. Englert at the Aharonov-80 conference in 2012 at Chapman University, see quantum.chapman.edu/talk-2.]

To my friend Yakir.

4.1 Spontaneous Breaking of a Global Symmetry

4.1.1 Chiral Symmetry Breaking

Spontaneous symmetry breaking was introduced in relativistic quantum field theory by Nambu in analogy with the BCS theory of superconductivity [1]. The problem studied by Nambu [2] and Nambu and Jona-Lasinio [3, 4] is the spontaneous breaking of the chiral $U(1)$ symmetry of massless fermions resulting from the arbitrary relative (chiral) phase between their decoupled right and left constituent neutrinos.

F. Englert (✉)
Service de Physique Theorique, Universite Libre de Bruxelles, Campus Plaine, C.P. 225, Bruxelles, Belgium
e-mail: fenglert@ulb.ac.be

F. Englert
The International Solvay Institutes, Campus Plaine C.P. 231, Boulevard du Triomphe, 1050 Bruxelles, Belgium

D.C. Struppa, J.M. Tollaksen (eds.), *Quantum Theory: A Two-Time Success Story*,
DOI 10.1007/978-88-470-5217-8_4, © Springer-Verlag Italia 2014

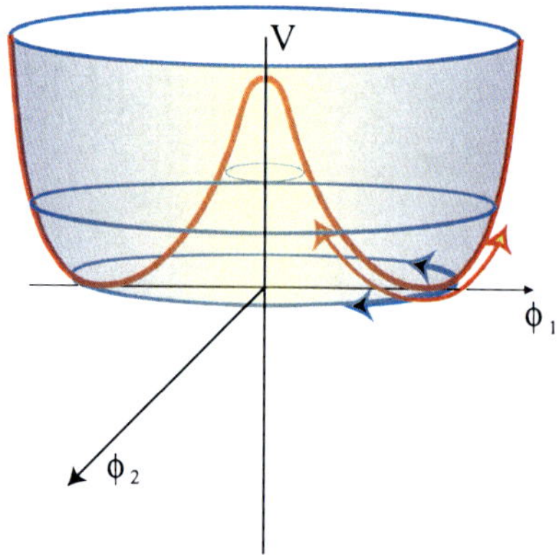

Fig. 4.1 Spontaneous symmetry breaking in the Goldstone model

They then generalize to include isospin. Fermion mass cannot be generated from a chiral invariant interaction in a perturbation expansion but may arise through a (non-perturbative) self-consistent fermion condensate: this breaks the chiral symmetry spontaneously. Nambu [2] showed that such spontaneous symmetry breaking (SSB) is accompanied by a massless pseudoscalar. This is interpreted as the chiral limit of the (tiny on the hadron scale) pion mass. Such interpretation of the pion constituted a breakthrough in our understanding of strong interaction physics. In the model of Refs. [3, 4], it is shown that SSB also generates a massive scalar boson.

4.1.2 The Simple Goldstone $U(1)$ Model

The significance of the massless boson(s) and of the massive scalar boson(s) occurring in SSB is well illustrated in a simple model devised by Goldstone [5]. A complex scalar field ϕ experiences a potential $V(\phi^*\phi)$. The Lagrangian density,

$$\mathcal{L} = \partial^\mu \phi^* \partial_\mu \phi - V(\phi^*\phi) \quad \text{with } V(\phi^*\phi) = -\mu^2\phi^*\phi + \lambda(\phi^*\phi)^2, \ \lambda > 0, \tag{4.1}$$

is invariant under the $U(1)$ group $\phi \to e^{i\alpha}\phi$. The $U(1)$ symmetry is spontaneously broken by the expectation value of the ϕ-field acquired, at the classical level, at the minimum of the potential $V(\phi^*\phi)$ depicted in Fig. 4.1.

Writing $\phi = \langle\phi\rangle + \varphi$ and $\phi = (\phi_1 + i\phi_2)/\sqrt{2}$, the $U(1)$ symmetry breaking is revealed by selecting the expectation value $\langle\phi\rangle$ to lie in some direction, say ϕ_1, of the (ϕ_1, ϕ_2) plane. The quadratic terms in φ_1 and φ_2 yield the mass squared of their respective fields, namely, using the condition $\langle\phi\rangle^2 = \mu^2/2\lambda$ at the minimum,

$$m^2_{\varphi_1} = 2\mu^2 \qquad m^2_{\varphi_2} = 0. \tag{4.2}$$

Thus φ_2 describes a massless boson, φ_1 a massive one, and the "order parameter" $\langle\phi_1\rangle$ may be viewed as a condensate of φ_1 bosons. Their significance is brought to light in Figs. 4.2 and 4.3 depicting respectively classical φ_1 and φ_2 waves on the background $\langle\phi_1\rangle$.

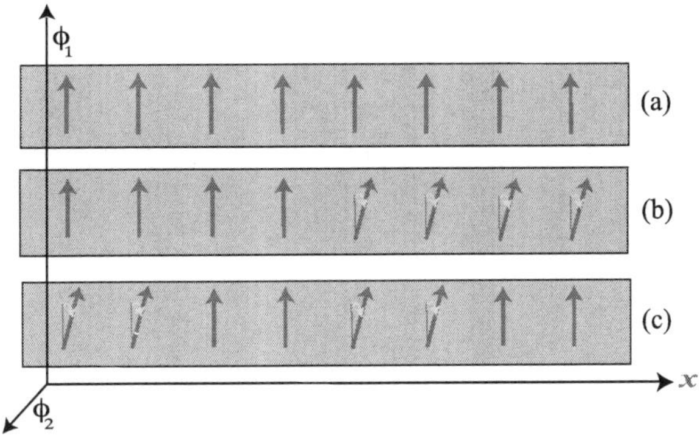

Fig. 4.2 Massless Nambu-Goldstone mode φ_2

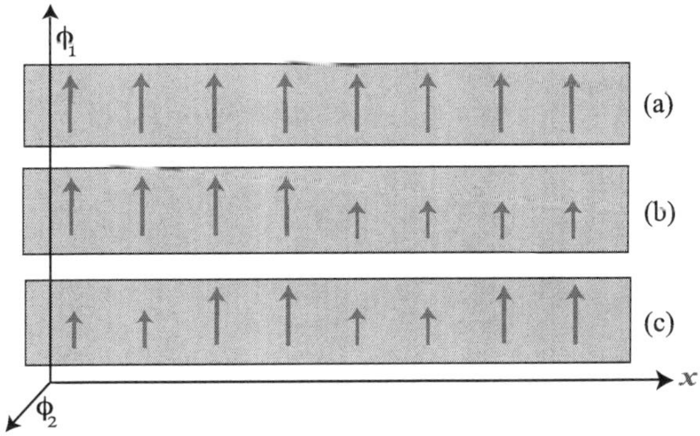

Fig. 4.3 Massive scalar mode φ_1

Figure 4.2a represents schematically a lowest energy state (a "vacuum") of the system: a constant non-zero value of the field $\phi_1 = \langle\phi_1\rangle$ pervades space-time. Figure 4.2b depicts the excitation resulting from the rotation of half the fields in the (ϕ_1, ϕ_2) plane. This costs only an energy localized near the surface separating the rotated fields from the chosen vacuum. SSB implies indeed that rotating all the fields would cost no energy at all: one would merely trade the initial chosen vacuum for an equivalent one with the same energy. This is the characteristic *vacuum degeneracy* of SSB. Figure 4.2c mimics a wave of φ_2. Comparing Fig. 4.2c with Fig. 4.2b, we see that as the wavelength of the wave increases indefinitely, its energy tends to zero. The wave generates in that limit a motion along the valley of Fig. 4.1. Quantum excitations carried by the wave reach thus zero energy at zero momentum and the mass m_{φ_2} is zero, in agreement with Eq. (4.2). Figure 4.2 can easily be generalized to more complex spontaneous symmetry breaking of *continuous* symmetries. Massless bosons are thus a general feature of such SSB already revealed by Nambu's discovery of the massless pion resulting from spontaneous chiral symmetry breaking. They will be labeled *massless Nambu-Goldstone (NG) bosons*. Formal proofs corroborating the above simple analysis can be found in the literature [6].

Figure 4.3 depicts similarly a classical wave corresponding to a stretching of the vacuum fields. These excitations in the ϕ_1 direction describe fluctuations of the order parameter $\langle\phi_1\rangle$. They are volume effects and their energy does not vanish when the wavelength becomes increasingly large. They correspond in Fig. 4.1 to a climbing of the potential. The quantum excitations φ_1 are thus now massive, in agreement with Eq. (4.2). These considerations can be again extended to more general SSB (even to discrete ones) to account for order parameter fluctuations. Lorentz invariance imposes that such massive excitations are necessarily scalar particles. They were also already present in Refs. [3, 4] and will be denoted in general as *massive scalar bosons*.

The above considerations are restricted to spontaneous symmetry breaking of *global* continuous symmetries. Global means that the symmetry acts everywhere in space-time: for instance in the $U(1)$ Goldstone model the parameter α in $\phi \to e^{i\alpha}\phi$ is independent of the space-time point x. We now discuss the extension from global to local symmetries.

4.2 The Symmetry Breaking Mechanism for Gauge Fields

4.2.1 From Global to Local Symmetry

The global $U(1)$ symmetry in Eq. (4.1) is extended to a local one $\phi(x) \to e^{i\alpha(x)}\phi(x)$ by introducing a vector "gauge field" $A_\mu(x)$ transforming under such local "gauge transformations" as $A_\mu(x) \to A_\mu(x) + (1/e)\partial_\mu\alpha(x)$. The Lagrangian density becomes

$$\mathcal{L} = D^\mu\phi^* D_\mu\phi - V(\phi^*\phi) - \frac{1}{4}F_{\mu\nu}F^{\mu\nu}, \tag{4.3}$$

where in Eq. (4.1) one replaces ∂_μ by the "covariant derivative" $D_\mu\phi = \partial_\mu\phi - ieA_\mu\phi$ and introduces the gauge invariant field strength $F_{\mu\nu} = \partial_\mu A_\nu - \partial_\nu A_\mu$ to account for the kinetic energy of the gauge field.

Local invariance under a semi-simple Lie group $\mathcal{G}$ is realized by extending the Lagrangian Eq. (4.3) to incorporate "non-abelian" Yang-Mills gauge vector fields A^a_μ. These transform under infinitesimal transformations of the group as $\delta A^a_\mu(x) = \epsilon^c(x) f_{acb} A^b_\mu(x) + (1/e)\partial_\mu\epsilon^a(x)$ where f_{acb} are structure constants. One gets

$$\begin{aligned} \mathcal{L}_{\mathcal{G}} &= (D^\mu\phi)^{*A}(D_\mu\phi)^A - V - \frac{1}{4}F^a_{\mu\nu}F^{a\,\mu\nu}, \\ (D_\mu\phi)^A &= \partial_\mu\phi^A - eA^a_\mu T^{a\,AB}\phi^B \\ F^a_{\mu\nu} &= \partial_\mu A^a_\nu - \partial_\nu A^a_\mu - ef^{abc}A^b_\mu A^c_\nu. \end{aligned} \tag{4.4}$$

Here, $(D_\mu\phi)^A$ are covariant derivatives, $F^a_{\mu\nu}$ are field strengths and ϕ^A belongs to the representation of $\mathcal{G}$ generated by $T^{a\,AB}$. The potential V is invariant under $\mathcal{G}$.

The local abelian or non-abelian gauge invariance of Yang-Mills theory hinges *apparently* upon the massless character of the gauge fields A_μ, hence on the long-range character of the forces they transmit, as the addition of a mass term for A_μ in the Lagrangian Eq. (4.3) or (4.4) destroys gauge invariance. But short-range forces such as the weak interaction forces seem to be as fundamental as the electromagnetic ones. To reach a basic description of such forces one is tempted to link this fact to gauge fields masses arising from spontaneous broken symmetry. However the problem of SSB is very different for global and for local symmetries.

4.2.2 The BEH Mechanism

This section is based on the field-theoretic approach of Ref. [7]. In view of slips often made about the content and the dates of the 1964 papers quoted in this section, references to these papers are detailed.[1]

4.2.2.1 Breaking by Scalars

Let us first examine the abelian case $U(1)$ as realized by the complex scalar field ϕ exemplified in Eq. (4.3). The interaction between the complex scalar field ϕ and the gauge field A_μ is

$$-ie\left(\partial_\mu\phi^*\phi - \phi^*\partial_\mu\phi\right)A^\mu + e^2 A_\mu A^\mu \phi^*\phi. \tag{4.5}$$

As in the Goldstone model of Sect. 4.1.2, the SSB Yang-Mills phase is realized by a non vanishing expectation value for $\phi = (\phi_1 + i\phi_2)/\sqrt{2}$, which we choose to be in the ϕ_1-direction. Thus

$$\phi = \langle\phi\rangle + \varphi, \tag{4.6}$$

with $\phi_1 = \langle\phi_1\rangle + \varphi_1$ and $\phi_2 = \varphi_2$, where as previously φ_2 and φ_1 are respectively the NG massless boson and the massive scalar boson.

In the covariant gauges, the free propagator of the field A_μ is

$$D^0_{\mu\nu} = \frac{g_{\mu\nu} - q_\mu q_\nu/q^2}{q^2} + \eta\frac{q_\mu q_\nu/q^2}{q^2}, \tag{4.7}$$

[1] Detailed references of all the 1964 papers quoted in the text:

	Article	Reception date	Publication date
1	F. Englert and R. Brout [7] *Phys. Rev. Lett.* **13(9)** (1964) 321	26/06/1964	31/08/1964
2	P.W. Higgs [9] *Phys. Lett.* **12** (1964) 132	27/07/1964	15/09/1964
3	P.W. Higgs [13] *Phys. Rev. Lett.* **13(16)** (1964) 508	31/08/1964	19/10/1964
4	G.S. Guralnik, C.R. Hagen and T.W.B. Kibble [10] *Phys. Rev. Lett.* **13(20)** (1964) 585	12/10/1964	16/11/1964

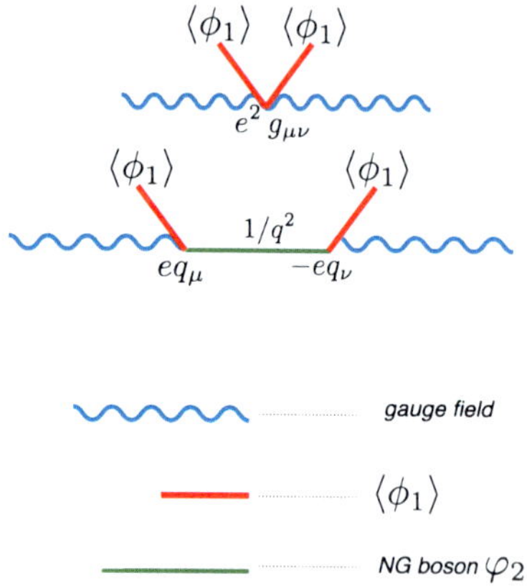

Fig. 4.4 Tadpole graphs of SSB. Abelian gauge theory

where η is a gauge parameter. In what follows, we shall choose the Landau gauge defined by $\eta = 0$.

The polarization tensor $\Pi_{\mu\nu}$ of the gauge field in lowest order perturbation theory around the self-consistent vacuum is given by the tadpole graphs of Fig. 4.4,

We see that, as a consequence of the contribution from the NG boson, the polarization tensor is transverse

$$\Pi_{\mu\nu} = (g_{\mu\nu} q^2 - q_\mu q_\nu)\Pi(q^2), \tag{4.8}$$

and yields a singular polarization scalar $\Pi(q^2)$ at $q^2 = 0$,

$$\Pi(q^2) = \frac{e^2 \langle\phi_1\rangle^2}{q^2}. \tag{4.9}$$

From Eqs. (4.7), (4.8) and (4.9), the dressed gauge field propagator becomes

$$D_{\mu\nu} = \frac{g_{\mu\nu} - q_\mu q_\nu / q^2}{q^2 - M_V^2}, \tag{4.10}$$

which shows that the A_μ-field gets a mass M_V,

$$M_V^2 = e^2 \langle\phi_1\rangle^2. \tag{4.11}$$

The transversality of the polarization tensor Eq. (4.8) results from the contribution of the NG boson and agrees with a Ward identity which guarantees that gauge invariance is preserved. Thus the mass of the gauge field A_μ acquired through the absorption of the NG boson is gauge invariant.

The generalization of these results to the non-abelian case described by the action Eq. (4.4) is straightforward. Writing the generators in terms of the real components of the fields, one gets the mass matrix

$$(M_V^2)^{ab} = -e^2 \langle\phi^B\rangle T^{a\,BC} T^{b\,CA} \langle\phi^A\rangle, \tag{4.12}$$

and the dressed gauge boson propagators have the same form as Eq. (4.10) in terms of the diagonalized mass matrix. As in the abelian case, the would-be NG bosons are absorbed by the gauge fields and generate gauge invariant masses in $\mathcal{G}/H$. Long-range forces only survive in the subgroup $\mathcal{H}$ of $\mathcal{G}$ which leaves invariant the non-vanishing expectation values $\langle\phi^A\rangle$.

The introduction of gauge fields and hence local symmetries resulted in the absorption of the NG boson in the gauge field propagator and in the generation of gauge field mass. These results are encoded in Eqs. (4.8), (4.11) and (4.12). Such consequences of local symmetry seems at odd with the appearance of massless NG bosons in global symmetries and calls for an elucidation of the concepts involved in extending the symmetry from global to local. This will now be done by unraveling the significance of the results of Sect. 4.2.2.1 for the NG boson and for the scalar boson. To avoid notational complications, I shall mostly consider the $U(1)$ extension from the global Goldstone model to its local counterpart, although the discussion in the Sections below apply in general to the non-abelian case as well .

4.2.2.2 The Fate of the Massless NG Boson

The diagrams of Fig. 4.4 show that the NG boson is absorbed in the gauge field propagator. This yields the required longitudinal polarization of the massive gauge field encoded in the numerator of Eq. (4.10) on the mass shell $q^2 = M_V^2$. The massless NG boson actually disappears entirely from the physical spectrum. This is an immediate consequence of gauge invariance. Consider indeed Fig. 4.2. As explained in Sect. 4.1.2, the massless NG mode originates in global SSB from the vacuum degeneracy: the energy of the excitations depicted in Fig. 4.2b and 4.2c tend to zero in the limit of infinite wavelength because they generate in that limit a vacuum equivalent to the original one under a symmetry operation. But local symmetry means that the configurations of Fig. 4.2b and 4.2c carry no energy at all! They are thus simply redundant description of the *same* gauge invariant vacuum, a redundancy not unexpected when fields are described by potentials A_μ. *Therefore there is no vacuum degeneracy, no spontaneous symmetry breaking and thus no massless NG boson!*[2]

An apparent symmetry breaking, akin to the Goldstone model $U(1)$ SSB, appears when one *chooses* a fixed orientation of the average scalar field $\langle\phi\rangle$, e.g. $\langle\phi\rangle = \langle\phi_1\rangle$. But this description is only a *convenient* gauge choice. It allows for the conventional assignment of group quantum numbers (such as isospin) to particles in perturbation theory. I shall therefore qualify also as SSB the mechanism generating mass for gauge fields but one should keep in mind that the symmetry is not intrinsically broken, a fact that renders the disappearance of massless NG bosons obvious. Their degrees of freedom are recovered in the longitudinal polarization of the massive gauge fields.[3]

---

[2] A more detailed description of the distinctive features of global and local SSB can be found in Ref. [8]. Formal proofs for the absence of massless NG bosons were given by Higgs [9], and then by Guralnik, Hagen and Kibble [10]. These proofs do not make use explicitly of the unicity of the gauge invariant vacuum.

[3] A non relativistic precursor of this effect was found by Anderson [11] in condensed matter physics. Namely in superconductivity the massless mode of the broken $U(1)$ symmetry disap-

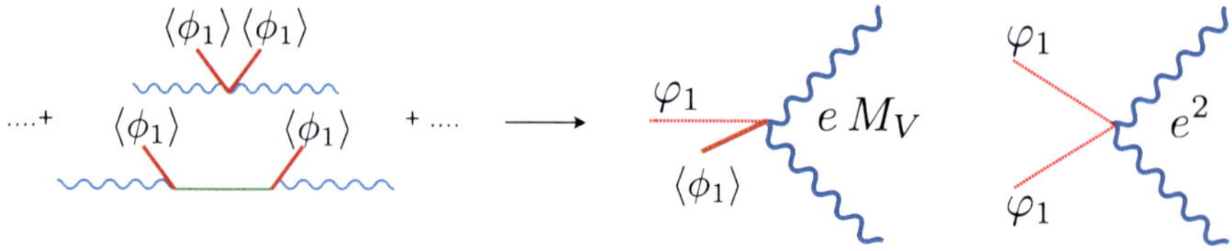

Fig. 4.5 Coupling of the BEH scalar boson φ_1 to massive gauge bosons

4.2.2.3 The Fate of the Massive Scalar Boson

A glance at Fig. 4.3 shows that the stretching of (classical) scalar fields are independent of local rotations of the ϕ-field in the (ϕ_1, ϕ_2) plane. This translates the fact that the modulus of the ϕ-field is gauge invariant. Hence the scalar bosons survives the gauging and their classical analysis is identical to the one given for the Goldstone model in Sect. 4.1.2.

The coupling of the BEH scalar boson to the massive gauge bosons follows from the graphs in Fig. 4.4. Using Eq. (4.6) one gets the two tree-level vertices of Fig. 4.5 where the heavy wiggly lines on the right hand side represent (tree-level) dressed massive gauge propagators. The vertex couplings follow from Eq. (4.11).

4.2.2.4 Dynamical Symmetry Breaking

The symmetry breaking giving mass to gauge vector bosons may also arise from a fermion condensate. If a spontaneously broken global symmetry is extended to a local one by introducing gauge fields, the massless NG bosons disappear as previously from the physical spectrum and their absorption by gauge fields renders these massive.

4.2.2.5 The Renormalization Issue

The interest in the symmetry breaking mechanism stems from the fact that it provides, as does quantum electrodynamics, a taming of quantum fluctuations. This allows the computation of the quantum effects necessary to cope with precision experiments. In other words, the theory is "renormalizable", in contradistinction to the theory of genuine non-abelian massive vector fields.

The massive vector propagator Eq. (4.10), which is also valid in the non-abelian case by diagonalizing the mass matrix Eq. (4.12), differs from a conventional free massive vector propagator. The numerator of the former is transverse for all momenta while the numerator of the latter, $g_{\mu\nu} - q_\mu q_\nu / M_V^2$, is only transverse on the

---

pears by being absorbed by electron density oscillations, namely by the longitudinal "massive" plasma mode.

mass shell $q^2 = M_V^2$. The soft behavior at large q^2 of the propagator Eq. (4.10) and the gauge invariance condition Eq. (4.8) are reminiscent of quantum electrodynamics. This suggested that the SSB mechanism renders charge vector meson theories renormalizable [12].

However there is a catch. The pole at $q^2 = 0$ in Eq. (4.10) has a negative residue and therefore is potentially violating unitarity. A glimpse into the solution of the problem appears from comparing our approach to the one of Higgs [13]. Higgs obtained most of our results from the classical equations of motion. In addition, he showed how to eliminate all contributions of the massless NG boson in that limit by the following field transformation

$$A_\mu - \frac{1}{e\langle\phi_1\rangle}\partial_\mu\phi_2 = B_\mu, \tag{4.13}$$

where B_μ satisfies the conventional classical equations of motion of a massive vector field. In terms of propagators Eq. (4.13) becomes the *identity*

$$\frac{g_{\mu\nu} - q_\mu q_\nu/q^2}{q^2 - M_V^2} - \frac{1}{M_V^2}\frac{q_\mu q_\nu}{q^2} = \frac{g_{\mu\nu} - q_\mu q_\nu/M_V^2}{q^2 - M_V^2}. \tag{4.14}$$

The term in the right hand side of Eq. (4.14) is indeed the conventional massive vector propagator of Higgs B_μ-field which displays no unwanted pole at $q^2 = 0$. It constitutes a "unitary gauge" propagator. It does not share the soft high q^2 behavior of the "renormalizable gauge" propagator Eq. (4.10). Gauge invariance should allow the use of either propagator, and the theory is thus expected to be both renormalizable and unitary. How can this happen?

The answer lies in the second term of Eq. (4.14). Let us couple Eq. (4.10) to an external (non-conserved) current associated to the SSB gauge symmetry. The second term in Eq. (4.14) describes the coupling of the Goldstone boson to its divergence. Note that the pole contribution of the Goldstone is cancelled on-shell by the unphysical q^2 pole of the propagator Eq. (4.10), leaving only off-shell contributions in agreement with the fact that the massless Goldstone boson has to disappear from the physical spectrum. Thus the identity Eq. (4.14) indicates that the off-shell contributions of the Goldstone are needed to restore unitarity in the renormalizable gauge. As an example, one easily verifies at the tree level that, taking into account the Goldstone contribution, the identity Eq. (4.14) ensures the equivalence of the renormalizable and the unitary gauges in the electroweak theory discussed below.

Although these arguments suggest that the mechanism can be consistent, it is a highly non trivial affair to show that the fully interacting theory is renormalizable and unitary. This was proven by 't Hooft and Veltman [14, 15], who thereby established the quantum consistency of the SSB mechanism.[4]

---

[4]See also Refs. [16–18].

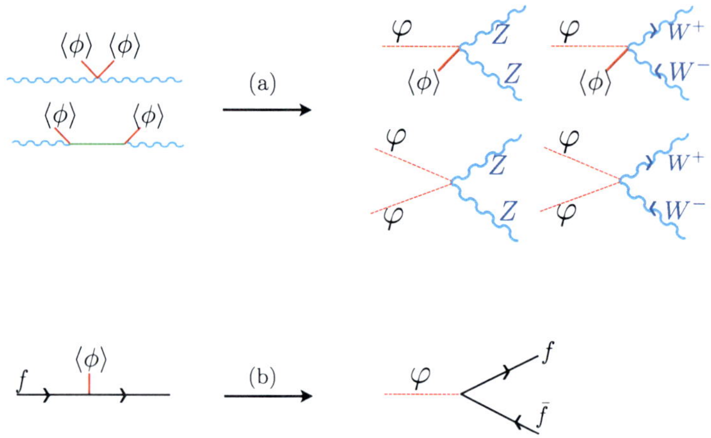

Fig. 4.6 Coupling of the BEH scalar boson φ to massive gauge bosons and to elementary fermions

4.3 The Electroweak Theory and the Standard Model

The Fermi theory of weak interactions, formulated in terms of a four Fermi point-like current-current interaction, was well-defined in lowest order perturbation theory and successfully confronted many experimental data. However, it is clearly inconsistent in higher orders because of uncontrollable divergent quantum fluctuations. In order words, in contradistinction to quantum electrodynamics, the Fermi theory is not renormalizable. This difficulty could not be solved by smoothing the point-like interaction by a massive, and therefore short-range, charged vector particle exchange (the W^+ and W^- bosons): theories with massive charged vector bosons are not renormalizable either. The solution to the problem came from the electroweak theory for weak and electromagnetic interactions [19–21], based on the SSB mechanism applied to the chiral group $SU(2) \times U(1)$ with $SU(2)$ acting on left-handed fermions only. Together with the colour $SU(3)$ theory of strong interactions, this is the "Standard Model" of elementary particles.

The generators and coupling constants are $gA^a_\mu T^a$ and $g'B_\mu Y/2$. The scalar field ϕ is a doublet of $SU(2)$ and its $U(1)$ charge is $Y = 1$. Breaking follows from a Goldstone type potential. It is characterized by $\langle\phi\rangle = 1/\sqrt{2}\ \{0, v\}$ and $Q = T^3 + Y/2$ generates the unbroken subgroup. Q is identified with the electromagnetic charge operator. The only residual massless gauge boson is identified with the photon and the electric charge e is usually expressed in terms of the mixing angle θ as $g = e/\sin\theta, g' = e/\cos\theta$. The expectation value $\langle\phi\rangle$ generates the masses of all known elementary fermions through Yukawa couplings.

Using Eqs. (4.11) and (4.12) one gets the mass matrix

$$|\mu^2| = \frac{v^2}{4}\begin{vmatrix} g^2 & 0 & 0 & 0 \\ 0 & g^2 & 0 & 0 \\ 0 & 0 & g'^2 & -gg' \\ 0 & 0 & -gg' & g^2 \end{vmatrix} \tag{4.15}$$

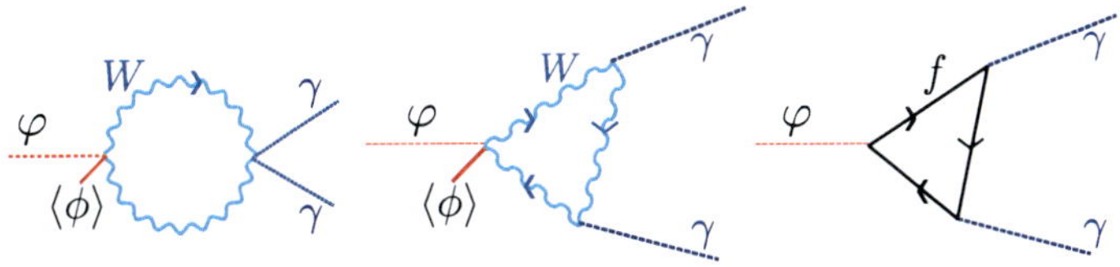

Fig. 4.7 Coupling of the BEH scalar boson φ to photons

whose diagonalization yields the eigenvalues

$$M_{W^+}^2 = \frac{v^2}{4} g^2 \qquad M_{W^-}^2 = \frac{v^2}{4} g^2 \qquad M_Z^2 = \frac{v^2}{4} (g'^2 + g^2) \qquad M_A^2 = 0. \quad (4.16)$$

The discovery of the Z and W bosons in 1983 and the precision experiments testing the quantum consistency of the theory corroborate the validity of the BEH mechanism. The couplings of the BEH scalar to the massive W and Z bosons follow from Fig. 4.5 and are depicted in Fig. 4.6a. Its coupling to elementary fermions similarly follows from the Yukawa couplings and are shown in Fig. 4.6b. The coupling to the massless photons occur at the loop level as indicated in Fig. 4.7.

4.4 The CERN Discovery

The recent discovery at the Large Hadron Collider (LHC) at CERN of a particle with mass 125 GeV consistent with the Standard Model scalar boson is a direct proof of the correctness of the BEH mechanism. In addition it will (and probably already does) shed light on a fundamental question with deep implications on the structure of the constituents of our universe: is this boson an elementary particle (at least to testable scales) or is it a phenomenological description of a composite object?

The SSB mechanism described in Sect. 4.2 could be realized by an elementary scalar condensate (Sect. 4.2.2.1) or dynamically (Sect. 4.2.2.4) in which case the scalar boson would, at best, be a bound state.

At first sight, the absence of previously known elementary scalar particle and the fact that a neutral scalar condensate could be, as is often the case in condensed matter physics, only a phenomenological description of a more complex dynamics, might have suggested a dynamical realization of the mechanism. This was comforted by the fact that simple dynamical models, such as Technicolor, can be constructed. Technicolor generate gauge vector boson masses, but its extension to produce elementary fermion masses is more problematic: giving mass to the fermions dynamically, which is natural in this perspective, might require additional groups which have then to confront many experimental constraints. As a rule, full dynamical symmetry breaking is very laborious and the corresponding phenomenological scalar(s) may have higher masses.[5] As there seems to be no indication in the CERN

[5]For a review on dynamical electroweak symmetry breaking see Ref. [22].

experiment of such different origins for fermion and gauge boson masses, dynamical origin for the scalar boson appears highly unlikely. In addition the relatively low mass ($\simeq$ 125 GeV) of this boson in a region where there appear to be no new fundamental constituents needed to generate it dynamically reinforces the conviction that *the 125 GeV boson should be viewed as an elementary particle, at least in the same sense as all other known "elementary" objects.*

This means that the complexity at high energy expected in dynamical models should not be present. On the other hand the simplicity of the Standard Model scalar is marred by the introduction of an elementary scalar not directly related to the fermionic content of the theory and submitted to the ad hoc Goldstone-type potential. These drawbacks could to some extend be avoided if some hidden supersymmetry, broken at larger energy scales, would be present. It would ensure, independent of the usual rather weak argument of "naturalness", that elementary scalars do appear and are accompanied by fermions at different masses. Although there is at present no indication of these supersymmetric partners, we cannot yet exclude their existence even at energies available at the LHC. Only future experiments will tell.

It is of course important to detect any possible extension of the Standard Model. A first glimpse on new physics could already appear in the detailed analysis of the different decay processes of the BEH boson into known particles. Quantum loop effects, such as those already responsible in lowest order for its decay into two photons, may indeed be sensitive to the possible existence of particles beyond the Standard Model ones.

Acknowledgements This work was supported in part by IISN-Belgium and by the Belgian Federal Science Policy Office through the Interuniversity Attraction Pole "Fundamental Interactions" I am very grateful to Jean-Marie Frère for illuminating discussions.

References

1. Y. Nambu, Quasi-particles and gauge invariance in the theory of superconductivity. Phys. Rev. **117**, 648 (1960)
2. Y. Nambu, Axial vector current conservation in weak interactions. Phys. Rev. Lett. **4**, 380 (1960)
3. Y. Nambu, G. Jona-Lasinio, Dynamical model of elementary particles based on an analogy with superconductivity I. Phys. Rev. **122**, 345 (1961)
4. Y. Nambu, G. Jona-Lasinio, Dynamical model of elementary particles based on an analogy with superconductivity II. Phys. Rev. **124**, 246 (1961)
5. J. Goldstone, Field theories with "superconductor" solutions. Nuovo Cimento **19**, 154 (1961)
6. J. Goldstone, A. Salam, S. Weinberg, Broken symmetries. Phys. Rev. **127**, 965 (1962)
7. F. Englert, R. Brout, Broken symmetry and the mass of gauge vector mesons. Phys. Rev. Lett. **13**, 321 (1964)
8. F. Englert, Broken symmetry and Yang-Mills theory, in *50 Years of Yang-Mills Theory*, ed. by G. 't Hooft (World Scientific, Singapore, 2005), pp. 65–95. hep-th/0406162
9. P.W. Higgs, Broken symmetries, massless particles and gauge fields. Phys. Lett. **12**, 132 (1964)

10. G.S. Guralnik, C.R. Hagen, T.W.B. Kibble, Global conservation laws and massless particles. Phys. Rev. Lett. **13**, 585 (1964)
11. P.W. Anderson, Random-phase approximation in the theory of superconductivity. Phys. Rev. **112**, 1900 (1958)
12. F. Englert, Fundamental problems in elementary particle physics, in *Proceedings of the 1967 Solvay Conference* (Wiley-Interscience, New York, 1967), p. 18
13. P.W. Higgs, Broken symmetries and the masses of gauge bosons. Phys. Rev. Lett. **13**, 508 (1964)
14. G. 't Hooft, Renormalizable lagrangians for massive Yang-Mills fields. Nucl. Phys. B **35**, 167 (1971)
15. G. 't Hooft, M. Veltman, Regularization and renormalization of gauge fields. Nucl. Phys. B **44**, 189 (1972)
16. B.W. Lee, J. Zinn-Justin, Spontaneously broken gauge symmetries. Phys. Rev. D **5**, 3121 (1972)
17. B.W. Lee, J. Zinn-Justin, Spontaneously broken gauge symmetries. Phys. Rev. D **5**, 3137 (1972)
18. B.W. Lee, J. Zinn-Justin, Spontaneously broken gauge symmetries. Phys. Rev. D **5**, 3155 (1972)
19. S.L. Glashow, Partial-symmetries of weak interactions. Nucl. Phys. **22**, 579 (1961)
20. S. Weinberg, A model of leptons. Phys. Rev. Lett. **19**, 1264 (1967)
21. A. Salam, Elementary particle physics, in *Proceedings of the 8th Nobel Symposium*, ed. by N. Svartholm (Almqvist and Wiksell, Stockhlom, 1968), p. 367
22. C.T. Hill, E.H. Simmons, Strong dynamics and electroweak symmetry breaking. Phys. Rep. **381**, 235 (2003). hep-ph/0203079

Chapter 5
NET = T.O.E.?

Shmuel Nussinov

Abstract We suggest a picture where all of physical reality is restricted to one connected net consisting of a huge number of points, links, and gauge fluxes on the latter. [*Editors note*: for a video of the talk given by Prof. Nussinov at the Aharonov-80 conference in 2012 at Chapman University, see quantum.chapman.edu/talk-24.]

5.1 Introduction

Local quantum (gauge) field theories: QED, its $SU(2) \times U(1)$ electroweak generalization and QCD have been remarkably successful and may unify at high energies into one simple gauge theory with a common gauge couplings. Yet the theory does not explain the number, spectrum and mixing of the quarks and leptons. Also quantum gravity is non-renormalizable, bringing in a fundamental length scale $\ell_p = G_{Newton}^{1/2} \approx 10^{-33}$ cm or energy $m_p \approx \ell_p^{-1} \approx 10^{19}$ GeV. Space-time and the underlying degrees of freedom could drastically change below ℓ_p. Super-strings may provide a fundamental Planck scale theory where (Super) space-time coordinates are (fermionic) bosonic fields on the string. In the following we sketch a very different approach which we term the "net".

5.2 Motivations for NET

Much of modern (and ancient) science is an ongoing clash between the atomistic and continuum points of view. Molecular theory and quantum mechanics were victories of the first, but field theories are connected with the second approach. Indeed Local fields $\phi_\alpha(\mathbf{x}, t)$ with time evolution determined by local $\phi^n(\mathbf{x}, t)$ interactions and

S. Nussinov (✉)
Institute for Quantum Studies, Chapman University, Orange, CA, USA
e-mail: nussinov@gmail.com

S. Nussinov
Sackler School Faculty of Sciences, Tel Aviv University, Ramat Aviv, Tel Aviv 69978, Israel

D.C. Struppa, J.M. Tollaksen (eds.), *Quantum Theory: A Two-Time Success Story*,
DOI 10.1007/978-88-470-5217-8_5, © Springer-Verlag Italia 2014

"almost local" Laplacians, guarantee causal wave propagation and special relativity. However all continuum theories appear to have an inherent difficulty connected with the infinite number of degrees of freedom DOF. Thus consider a local field theory where an ultraviolet cutoff Λ_c has been introduced. For renormalizable theories this does not change physics at energies much lower than the cut-off. Still when embeded in an expanding universe the introduction of a cutoff is problematic.

If all momenta $K \leq \Lambda_c$ are discarded and all others are kept, then the total number of field degrees of freedom (free field modes) in a given volume $V = L^3$ is:

$$N_{DOF}(\text{in } V) = (\kappa_{max} L)^3 = \Lambda_c^3 V \tag{5.1}$$

The volume of a co-moving large region expands between the present time t_{Hubble} and later time t by:

$$V(t_H) \to V(t) = [a(t)/a(t_H)]^3 V(t_H). \tag{5.2}$$

This expansion preserves the number of particles of each species (protons, photons, neutrinos, etc.) and red shifts (by a_{t_H}/a_t) their momenta.

However, mass parameters of the microscopic theory, such as m_e, m_μ, $m_{q_i}^{(0)}$, Λ_{QCD}, m_W and $m_{Planck} \approx (G_{Newton})^{-1/2}$ (which is the natural candidate for the ultimate cutoff), do not scale with $a(t)^{-1}$. Indeed severe experimental lower bounds on

$$\tau_G = (\dot{G}_N/G_N)^{-1}; \qquad \tau_e = ((\dot{m})_e/m_e)^{-1}; \qquad \tau_{\alpha(em)} = [\dot{\alpha}_{em}/\alpha_{em}]^{-1} \tag{5.3}$$

stating that all these times vastly exceed t_H, have been established. Consequently, the cutoff Λ_c should also not change with $a(t)$ and hence the total number of degrees of freedom in the local fields in a fixed, co-moving volume increases according to Eqs. (5.1), (5.2) like $a(t)^3$. This, however, may lead to an inconsistency. Suppose we consider a hypothetical co-moving large volume which is isolated from the rest of the universe at some epoch when $a(t) \gg ct$ = horizon size, and that we want to describe it completely in a quantum mechanical framework. This description could be extremely complicated, involving a functional which depends on the fields Φ_a at all points in a very fine and large lattice:

$$\mathbf{x}(\mathbf{n}) = \Lambda_c^{-1}(n_x\hat{e}_x + n_y\hat{e}_y + n_z\hat{e}_z) - L\Lambda_c \leq n_x, n_y, n_z \leq L\Lambda_c \tag{5.4}$$

inside the volume of interest:

$$-L \leq x, y, z \leq L. \tag{5.5}$$

However, no such unitary, quantum mechanical description is possible in principle: As the universe expands, so does the dimensionality of the Hilbert space with more and more degrees of freedom being continuously generated.

The Holographic principle stating that the maximal number of degrees of freedom in the volume L^3 is proportional to the area $N_{DF} \approx (L/\ell_p)^2$ rather than the volume ameliorates, but does not resolve, this difficulty. Another alternative which we will not pursue is that the underlying theory is truly conformal allowing only for angular variables but no mass or length parameters. While a most elegant and useful idealization for many field theories at high energies it does not seem to be realized in nature.

Lacking a quantum theory with multiplying degrees of freedom we are left with two alternative "extreme continuum" and "extreme atomistic" approaches.

The first approach, adopted in string theories, disallows any discretization or introduction of cutoffs which is redundant if the complete theory is finite. Indeed any such discretization destroys reparametrization invariance—a vast "gauge group" whose strict preservation is crucial-just like that of usual gauge invariance in ordinary renormalizable gauge theories.

Our present suggestion is based on the other, extremely "discrete" approach where all physical reality consists of a finite and fixed number of degrees of freedom: the vertices and links of a "net". These degrees of freedom evolve according to simple "local" rules. Hopefully the resulting evolution (of our finite number of degrees of freedom) will give rise to a 3-D space, to (general) covariance and gravity, to the observed fields and to a correct cosmology.

As noted above, a strong motivation for superseding local field theories are the difficulties associated with sub-Planckian (gravitational) physics. We resolve this by postulating that the Planck length ℓ_p, time t_p, and mass, m_p, are the minimal length and time possible, and the maximal energy that can be associated with one elementary degree of freedom [1]. Thus $\Delta x \leq \ell_p$, $\Delta t \leq t_p$ and elementary excitations with $K \geq m_p$ do not exist.

We demand that the theory be quantum mechanical at all levels. Indeed, the desire to have a Hilbert space of fixed dimensionality motivated our model to start with.

While gravity and the 3-D space itself are derived concepts in our net some form of gauge interactions is taken to be fundamental. Indeed, the basic forms in gauge theories: $\int A_\mu dx^\mu$, $\int\int F_{\mu\nu} dx^\mu dx^\nu$ do not depend on the existence of metric and can be defined for discrete lattices or even general networks of Links.

Cosmology strongly motivates searches for new alternative schemes. Astroparticle physics, combining the Standard Big Bang (SBB) and standard particle models, successfully predicts Helium and other light element abundances. Field theories with spontaneous symmetry breakings can generate phase transitions in the early universe and induce inflation which, in turn, explained the horizon, Flatness, and Entropy puzzles. However the tiny positive cosmological constant, the nature of dark matter and the lack of good models for the baryon asymmetry and for inflation remain outstanding puzzles. Conventionally, we first develop a microscopical theory which is as complete as possible, and then apply it to macrophysics and cosmology. This constrains the microphysics parameters as manifested e.g. in bounds on neutrino masses or neutrino species required in order to avoid overclosure or too rapid expansion at the time of nucleosynthesis. It is however conceivable that a satisfactory theory of both microphysics and cosmology can only emerge jointly from a common underlying structure which incorporates a novel form of Planck scale physics.

5.3 The Net

The primitive degrees of freedom of our "net" are taken to be a set of $N_v = N$ vertices, N_L links, and some "gauge connections" along these links. The resulting connected network (or graph) is our toy model universe. The various connected nets define an (orthogonal) set of states in Hilbert space. and the wave function of the universe is a linear superposition of those:

$$|\Psi_{universe}\rangle = \sum_{nets} A^{\psi}_{net} |net\rangle \tag{5.6}$$

As for the gauge degrees of freedom we allow only the fundamental $k \times k$ representations of an SU(k) gauge group to reside on each link—maintaining a discrete finite number of degrees of freedom.

We first specify the kinematics of our net: the allowed vertices and links and the definition of "length" along the net. We then proceed to the dynamics—the rules of evolution of the gauge degrees of freedom which are the conventional rules and the evolution of the net itself.

A priori both the kinematics and dynamics of the net could be very complex: any vertex could be joined to any number of other vertices and the evolution of the system could involve arbitrary link switchings.

Many very complex systems evolve towards a "fixed point" of a second-order phase transition that the renormalizable four-dimensional field theories are in its "long-wavelength universality class."

We will not adopt this point of view and assume that well-defined kinematics and dynamics exist. Since these embody the fundamental degrees of freedom and the basic laws of physics, respectively, we cannot attempt to specify them. Rather, we will describe general constraints on and guesses of rules which make the net evolve in the desired manner.

5.3.1 *Kinematics*

First we specify the valency i.e., the number of links incident on each vertex. Assuming an underlying $SU(K)$ gauge group we allow some K fold vertices on which we can place K "incoming" inwardly-directed "electric" fluxes, or K "outgoing" fluxes. The fluxes are analogous to those in QCD which emanate from quarks and terminate on antiquarks. In particular, baryons in $SU(K)$ contain a baryonic junction point where the K fluxons emanating from the K quarks meet, coupling them to a gauge singlet via an ϵ symbol.

$$B = \epsilon_{\alpha_i \ldots \alpha_K} \Psi^{\alpha_1} \cdots \Psi^{\alpha_K} \tag{5.7}$$

We will introduce fermions later but for now assume that we have no "free links" i.e. external lines. In addition to the K-fold "baryonic" vertices (Fig. 5.1). It is natural to have mostly four-fold vertices. Indeed junctions of three links where each

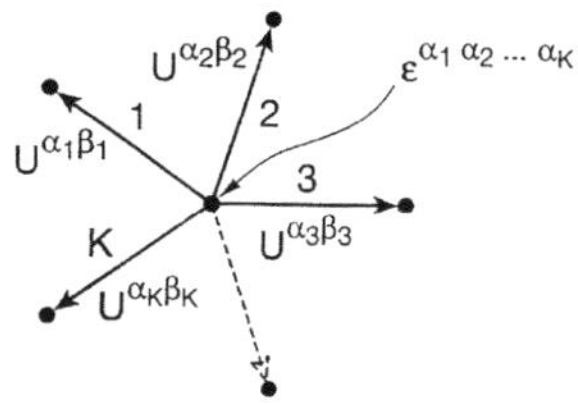

Fig. 5.1 K-fold "baryonic vertex"

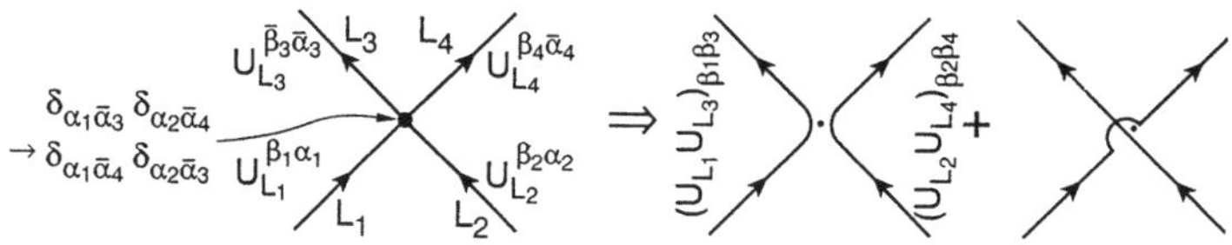

Fig. 5.2 Four-fold "mesonic vertex"

carries the fundamental K representation of $SU(K)$ are forbidden by the K color conservation whereas four fold vertices allow for two incoming $\alpha_1\alpha_2$ and two outgoing $\bar{\alpha}_3\bar{\alpha}_4$ directed links so that the relevant "color" coupling is:

$$\delta_{\alpha_1\alpha_3}\delta_{\alpha_2\alpha_4} + \delta_{\alpha_1\alpha_4}\delta_{\alpha_2\alpha_3} \tag{5.8}$$

We next define the distance d_{ab} between two vertices V_a and V_b as the minimal number of links along a path connecting the two vertices a definition which satisfies the triangular inequality:

$$d_{ab} + d_{bc} \geq d_{ac} \tag{5.9}$$

To conform with the assumption of the preceding chapter the physical distance between neighboring vertices is taken to be the minimal-Planck length.

$$\text{Planck length} = \ell_p = 10^{-33}\ \text{cm} \tag{5.10}$$

Points in ordinary physical space are usually separated by a vast number of Planck lengths and hence there should be a correspondingly large number of links on the shortest path connecting them. Certain "pathologies" arise when, with the exception of one or few very short paths joining two vertices on the net, all other joining paths have a very large number of links. A more stable definitions of ℓ_{ab} accounts not only for the shortest path but also for the number of relatively shot paths.

Next we address time and evolution. Intrinsic local times associated with the "physical" rates of evolution in almost "flat regions" to be defined later of the net may allow four-dimensional Lorentz invariance in those regions. On cosmological scales and particularly near "creation" ($t = 0$), space and time are radically different and an overall "net" time may be useful.

Let us use discrete "updating" of the net, as in Monte Carlo calculations with a fixed, short time interval, Δt, between consecutive updates which we take to be the Planck time,

$$\Delta t = t_p = \ell_p / c \approx 10^{-44}\ \text{sec.} \tag{5.11}$$

Fig. 5.3 An elementary operation of switching two equal orientation links which originate from two n.n. vertices and terminate on two n.n. vertices

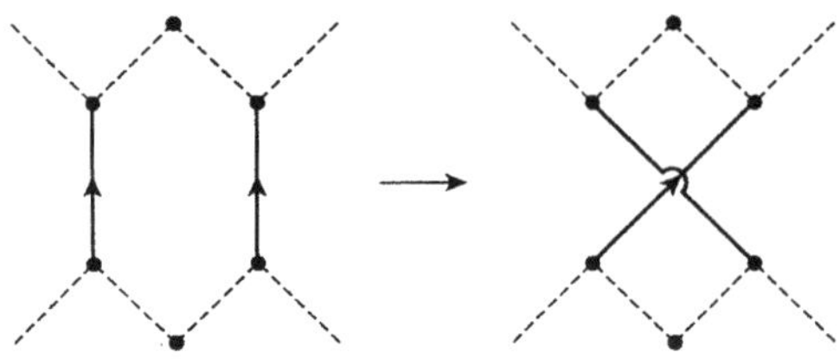

5.3.2 Dynamics

The "update rules" or the transfer matrix T (or the Hamiltonian H) of our quantum mechanical net:

$$T \equiv \mathbf{e}^{i\Delta t H/\hbar} \approx 1 - it_p H/\hbar \tag{5.12}$$

constitute the dynamics. Consider first the link-switching dynamics of the net. In the $\{net\}$ basis, $\langle \{n'\}|T|\{n\}|\rangle$ is the probability amplitude that a net $\{n\}$ at time t will transform into a net $\{n'\}$ at time $t' = t + t_p$. The wave function of the net (=universe) starts at $t = 0$ in a state $\Psi_0\rangle$ and after a large number $\tau = t/t_p$ of evolutions becomes

$$|\Psi_t\rangle = T^\tau |\Psi_0\rangle = \mathbf{e}^{iHt/\hbar}|\Psi_0\rangle(17) \tag{5.13}$$

The dimension of the vector Ψ or matrices $T(H)$ is huge, equaling the number of all allowed nets which grows with N, the total number of vertices, at least as $N!$.

To describe the present (or future!) three-dimensional universe, the net must contain, as we show in the following, a very large number of vertices:

$$N = N_v \geq 10^{180} \text{ (or } 10^{250}) \tag{5.14}$$

so that $N!$ is truly huge.

The allowed $\langle \{net'\}|T|\{net\}\rangle = T_{nn'}$ are restricted by demanding "locality". Distances and nearest neighbors (n.n.) and next-to-nearest neighbors (n.n.n.), etc. are defined for vertices in a single net but we can use them to define neighboring nets. Thus consider the particular "elementary" links switch illustrated in Fig. 5.3 for the case when these links emanate from and terminate on n.n. four-valent vertices of Eq. (5.8).

We will allow in T only transitions of this type, namely only matrix elements connecting a given net to nets which are "nearest in net space," i.e., obtainable by one operation of the kind shown. (Clearly T will have also diagonal matrix elements connecting identical nets.)

An important advantage of such "local evolution" is that it avoids faster than light signals.

To demonstrate this, and for many other reasons, we introduce a special net $\{Net^{(vac)}\}$ characterizing the "vacuum" state of our theory. In general the physical states of the universe need not be diagonal in the $\{net\}$ basis and any such state, and $|\Psi^{(vac)}\rangle$ in particular is a superposition of nets of the general form of Eq. (5.6). The superposition is often dominated by a specific net and by neighboring nets obtained by local flippings of links. This is motivated by the vacuum of ordinary interacting

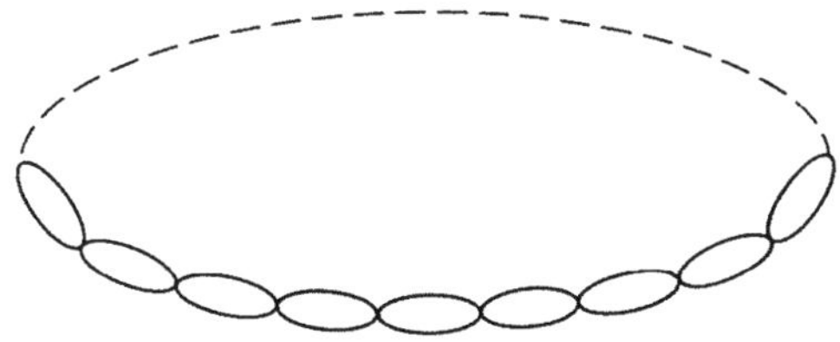

Fig. 5.4 The "bubble chain" as a 1-d $\{Net^{(vac)}\}$

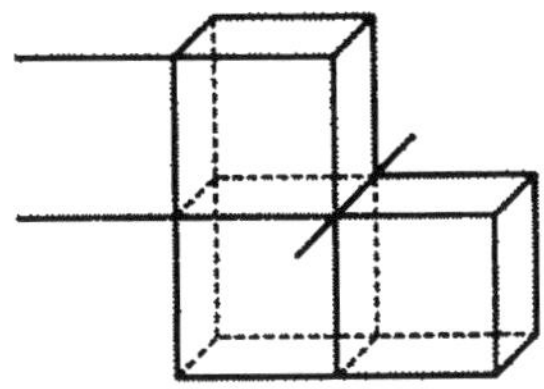

Fig. 5.5 Cubic 3-d lattice

quantum field theories which is a superposition of the "true" empty vacuum, and states with various local "vacuum fluctuations".

The vacuum has no special points or regions differing from others and is therefore translational and rotational invariant. Like all other states describing the evolving net (or physical universe), it cannot be eigenstate of the net hamiltonian as it would then evolve just by picking up overall phases.

We define $\{Net^{(vac)}\}$ by requiring maximal symmetry and following the above discussion all vertices in the closed net are taken to be 4-valent.

Such completely symmetric nets with one-, two- and three-dimensional features are the bubble chain shown in Fig. 5.4, the square lattice and the diamond lattice (which for ease of drawing in Fig. 5.5 was replaced by the simple cubic 3-d lattice) respectively. All lattices are closed in a toroidal fashion as shown for the 1-d bubble chain.

Generic nets may have arbitrary Hausdorf dimensions and could be fractal. Such nets are not translationally invariant and have lower symmetries than the integer dimensional nets discussed here. Such a completely symmetric net is postulated to be the dominant component of the vacuum-the state to which the universe ultimately evolves.

The size of the 3-d universe in this final vacuum state is:

$$L = N^{1/3}\,\ell_p \approx 10^{-33} N^{1/3}\ \text{cm} \tag{5.15}$$

At the present time, say, $t \approx 10^{10}$ years $\approx 3 \cdot 10^{17}$ sec, the size of the universe is $L \approx R_{Hubble} \propto ct \approx 10^{28}$ cm. If we would identify the present state of the universe with $\{Net^{(vac)}\}$, then from Eq. (5.15) $N^{1/3} \approx 60$ and $N \approx 10^{180}$. However, the present universe is very different from the ultimate vacuum. It is still expanding and even accelerating a bit. Also, there are inhomogenities and structures such as the baryons, which, by direct experiments, have a very long lifetime;

$$\tau_{proton} \geq 10^{32}\ \text{cm yrs}. \tag{5.16}$$

Conceivably, the "true" vacuum can be achieved only after all the baryons decay; namely, at time $t > \tau_{proton}$. (The preferred decay mode being from a bound hydro-

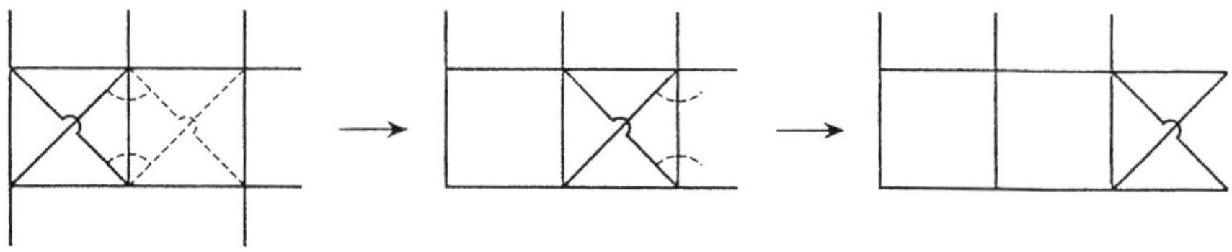

Fig. 5.6 A causal propagation—at a speed of one link length ℓ_p per one updating time t_p—of a lattice imperfection (in this case a quartet of anomalous "5-fold" vertices) via local link flips

gen atom: $p + e^- \to$ two neutrinos so as to get rid also of the electrons.) At such time, $R = ct \approx 10^{50}$ cm and Eq. (5.11) implies we find the large number of degrees of freedom quoted above in Eq. (5.10).

Let us consider now a local disturbance of $\{Net^{(vac)}\}$ where the latter is depicted as a two-dimension square lattice. This disturbance, say, some locally "disconnected" or "anomalous" set of vertices, can "travel" along the lattice of $\{Net^{(vac)}\}$ by having at each updating (i.e., at each time interval $\Delta t = t_p$) some n.n. pairs of vertices disconnect and reconnect at the location of the disturbance. See Fig. 5.6.

Thus the "perturbation" can travel one link length ℓ_p in one updating time interval $\delta t = t_p$, namely at a velocity which, be definition, never exceeds $c \equiv \ell_p / t_p$ ensuring causal propagation.

Had the link switching in the net been a classical "cellular automata type" process, the resulting shifting in both the vertical and horizontal direction leads to "diffusion" rather than proper wave propagation, and/or relativistic motion of particles (wave packets). To ensure the latter we need to multiply the various propagation paths by appropriate phases and then superpose the paths. This is consistent with our choice of a quantum mechanical evolution (5.13).

The updating of the net could be done by switching one pair of links at a time in the whole net. This is not consistent with the Q.M. evolution (Eq. (5.13)) where all transitions are tried at any updating, i.e., during each time interval and also is too slow. Just as in the case of ordinary computers, we can, in general, do "parallel computations". In the present context this means simultaneously flipping links in various parts of the net. Such simultaneous flipping can be done freely in disjoint regions of the regular vacuum net. Suppose, however, that we proceed with two sets of consecutive, local, link switches representing the propagation of two wave packets or "particles" which simultaneously approach a particular link, or pair of n.n.n. links. We then encounter a "conflict" or "frustration" as different link switches would best serve to propagate the first or second "disturbance". A computational analogy of this is the conflict between two parallel processors trying to simultaneously address the same memory unit. This conflict would slow down the evolution at this point. If physical time is fixed by the local "rate of evolution" this appear to slow down or red-shifted, due to interactions. More generally "matter", unlike empty space, is unstreched, crumpled regions of the net—that gradually mesh with the regular and flat remaining regions of the net that constitute the "vacuum". If the dislocation of Fig. 5.6 hits a disordered region it will meanders around and slow down in the disordered region. This is a-fortiori the case in quantum mechanics due to Anderson like localization. Fermat's principle that light follows the path of minimal time—a corollary of its stationary phase and maximal coherence—then implies the gravitational light bending and consequent time delays.

5.3.3 The Initial State

Specifying the initial conditions for a classical system or the initial quantum state is important, particularly when the system in question is the whole universe. Thus, in the standard big bang (SBB) the universe expands, despite gravity which tends to halt the expansion, due to "initial conditions". In the present case, the final state of the universe is dominated by the symmetric translational invariant $|net^{(vac)}\rangle$.

Quantum mechanics allows for a time symmetric formulation [2] where both initial and final states are specified. If, these two states are almost orthogonal to each other, then the ensemble of pre—and post-selected systems exhibits unusual behavior at intermediate times. It was shown that "weak measurements" of the usual hermitian observables can yield values lying outside the spectral range of the operator considered. These reflect the novel "potentials and forces of pre- and post-selection" required in order to guide the system along its unusual path. We may have to resort to this formulation in which the "destiny state" of the universe is also prescribed [3].

The conventional alternative is to specify only the initial state $|\Psi^{(0)}\rangle$ and [prove??/assume?/hope] that the dynamics encoded in $\langle n'|T|n\rangle$, evolves after a sufficiently long time, $|\Psi^{(0)}\rangle$ to $\|Psi^{vac}\rangle$.

While $|\Psi^{vac}\rangle$ was chosen by maximal symmetry considerations, the initial state (or rather the net configuration dominating it) will be highly hierarchical and with partial symmetry. An example of such a net is the "doubled tree" of Fig. 5.7. It is obtained by first constructing a tree: the "root" vertex V_0 connects to four first-generation vertices, $V_1^{(i)}$ $i = 1 \dots 4$; each of the $V_1^{(i)}$ vertices "grows" three second-generation vertices, $V_2^{(i)}$ $i = 1 \dots 12$, etc. This process repeats for g generation. There are $4 \cdot 3^g$ vertices of the g'^{th} generation. We connect them via $4 \cdot 4^g$ links to an image, inverted tree. This yields then one connected net with altogether $N > 4 \cdot 3^{g+1} - 2$ vertices and $2N$ links.

$\{net^{(0)}\}$ is remarkably compact. The maximal distances between the antipodal vertices, V_0, V_0', is only $d(V_0, V_0') \equiv L(\{net^{(0)}\}) = 2g + 1$. Thus, the complete size of the initial net (or universe) is

$$L(\{net^{(0)}\}) = L(t = 0) \approx 2(\ell g_3 N)\ell_p \approx 10^4 \ell_p \quad (5.17)$$

and all the $N \approx 10^{250}$ degrees of freedom are nestled within this tiny volume.

5.3.4 "Toy Scenario"

A basic feature of our toy net universe is its ongoing fantastic expansion from the microscopic, almost Planckian, initial size to present (and future) cosmological di-

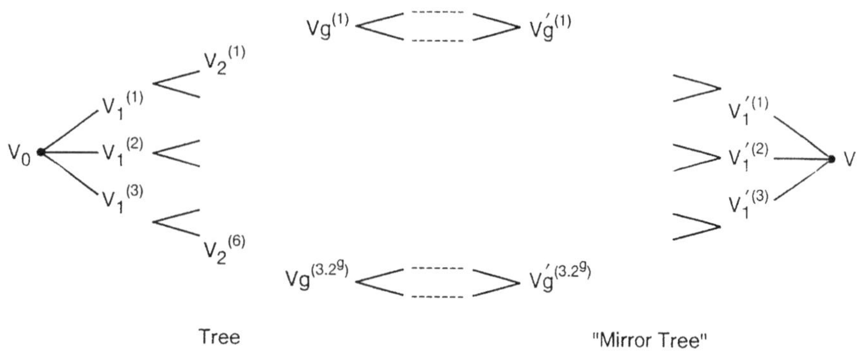

Fig. 5.7 $Net^{(0)}$. The folded-over tree diagram. For convenience of drawing we used 3-fold rather than 4-fold vertices

mensions achieving Inflation "for free". Indeed the horizon and flatness problems are resolved right away by our postulated initial state $\{net^{(0)}\}$ and (hopefully to be derived) final state $|net^{(vac)}\rangle$, respectively. Thus the extreme smallness, $\approx 10^3\,\ell_p$ of the initial universe (not to mention the existence of one overall wave function, $|\Psi_{t=0}\rangle$ of the net and its gauge degrees of freedom, i.e., of the universe), avoids the first problem. The final completely stretched and flat vacuum state—out of which not only "monopole-like" wrinkles have been ironed out but possibly all baryons as well—would resolve the second issue, if we can show that on a coarse grained scale the present universe is close to $\{net^{(vac)}\}$.

It is important however, that gravity be of a secondary, induced nature. Indeed, in ordinary space-time-gravity, the initial $\{net^{(vac)}\}$ encounters right away fatal difficulties.

We have a huge number of degrees of freedom, $N \approx 10^{250}$ packed within a tiny, almost Planckian, volume. Endowing each degree of freedom (link of the net) with a small energy (w) causes, if gravity exists at $t = 0$, an immediate collapse of $\{net^{(vac)}\}$ into a black hole.

To avoid then gravity from nipping, so to speak, our emergent universe in the bud, it should not be there at $t = 0$! This is quite consistent with the fact that 3-d space did not exist at $t = 0$ either. We will have to demonstrate that eventually (or even rather soon) the "partially stretched" net will possess gravity and Robertson-Walker-like cosmology.

Whatever effects gravity will have it should not overcome the "primary" drive of expansion of the net universe towards $\{net^{(vac)}\}$. Still gravity should generate all fairly stable and long lived gravitationally bound structures—a highly non trivial issue which we revisit towards the end of this section.

In some way we can view the initial highly hierarchical state $\{Net^{(0)}\}$ as the "false vacuum" in inflationary scenarios whereas the final flat symmetric net $|Net^{(vac)}\rangle$ is, by its very definition, the "true vacuum". Inflation is driven here not by "negative pressure", but rather by the net evolving towards a more symmetric state. We cannot use energy considerations, not only because energy-momentum is not yet defined. Indeed had $|\Psi^{(0)}\rangle$ or any later state, $\mathbf{e}^{i\frac{H}{\kappa}t}|\Psi^{(0)}\rangle$ been an eigenstate of the net Hamiltonian H, it evolves only via an overall phase factors. Still the final "vacuum" like net is far more stable than the initial doubled over tree state. As elaborated in the next section we need to allow cutting of fluxon carrying links in order to naturally incorporate fermions. Cutting the initial very hierarchical double tree net just at two symmetric locations near each of the two antipodal roots separates it into almost two equal parts and a whole giant "baby universe" will then separate and float away. This is definitely not the case for the final vacuum net—or even for intermediate sufficiently stretched nets where any such cutting is just a "vacuum" fluctuation corresponding to the creation of Fermion anti fermion pair.

Note that we have achieved eternal inflation without finely-tuned special "Inflation Theories". Inflation is driven by the peculiar connections in the initial net and hence can be viewed as Planck time/Quantum era inflation. Despite the overall tiny $10^3 l_P$ size of the initial "doubled tree" net—the line separating the two mirror trees cuts across a huge O(N(V)) number of links. Thus in another loose sense one may be tempted to identify the two trees with the two colliding branes in some models [4]. Such niceties aside, the on-going expansion in the net scenario can lead to a very serious difficulty. While on cosmological scales we observe (accelerating/Hubble +) expansion, objects or structures on all scale $\sim 10^2$–10^3 times smaller than R(Hubble) do not expand. At the higher end starting with clusters of galaxies this is due to gravity. Once the escape velocity $(GM/R)^{1/2}$ from a structure of size R mass M exceeds HR, the differential Hubble expansion across the structure, it becomes "Autonomous" and no longer participates in the cosmological expansion. In particular the well measured solar system and other multi-stellar configuration have to a very high accuracy fixed sizes. Lower size structures like rocks, grains, atoms and nuclei are held together by the electric $U(1)$ and the $SU(3)$ QCD gauge groups. The latter were introduced into the net via "small" Wilson loops—which in the limit of a regular 3 d net should become essentially the Wilson action/Kogut Susskind Hamiltonian. In the continuum limit of the theory i.e. on scales (far!) larger than l_{planck} we assume that the original $SU(K)$ theory breaks down leaving the direct product of the confined $SU(3)$ color and the (hard to get Xiral) $-SU(2)_L XU(1)$ E.W. gauge group and the resulting local field theories yield the above electric and nuclear/QCD forces.

However gravity is generated in the net framework by local changes of the metric i.e. the inter-vertex distances as defined above, by flipping near-by links. Since this involves two links each of which is a vector-spin 1 object, in a symmetric manner, we expect the resulting gravitons to carry spin 2 or 0. The latter scalar/dilaton part corresponds to local change of scale which is absent in the final ideal symmetric vacuum net. A proper continuum limit is expected to generate à-la Feynman the theory of GR by exchanging gravitons.

As in the standard approach gravity reflects the changing metric/topology of the net via link flips. In particular a set of flips forced by the pre and post selected, initial and final, nets is driving the early and present inflation. Furthermore matter is identified here as a region of space where fermions and gauge fields reside and with a substantial deficit of the standard, valence = 4, vertices and hence is far from being a part of ideal regular vacuum like net. This then suggests that the gravitational binding of two (or more) chunks of matter manifest via some special connections between vertices of the net in each of the two regions. These connections correspond to irregular "deviant" links which "jump" over the vast number of links in the relatively flat and regular region of "empty space" separating the two bodies in question. The number of such connections emanating from region A is roughly the number of special vertices in say region A which very crudely is correlated with the "total gravitational mass of A". since such connections of A to the rest of the net tends to fix this region and impede its motion which cannot be achieved by local flips of links we have here—in an extremely "Machian" fashion the equivalence of inertial and gravitational mass. When diluted by the many regular and long paths between vertices in A and those outside A the fewer deviant links emanating out of A smoothly modify the metric around it by a small amount. If matter around A is spherically symmetric or if A is surrounded by vacuum we expect—just like for the electrostatic field lines around a charge—that the special outgoing lines will be distributed in a spherically symmetric manner and hence the $1/R^2$ law.

5.3.5 Introducing the Fermions

Scalar fields are naturally discretized by placing them at the vertices of a, say, square lattice $\Phi(\mathbf{x}) \rightarrow \Phi(\mathbf{n})$. The kinetic (derivative) part of the Hamiltonian:

$$\sum(\Phi_{\mathbf{n}} - \Phi_{\mathbf{n}+\hat{i}})^2 + (\Phi_{\mathbf{n}} - \Phi_{\mathbf{n}+\hat{i}})^2 \approx \int (\nabla\Phi(x))^2 d^3x \tag{5.18}$$

allows, via $\Phi_n \Phi^+_{\mathbf{n}+\hat{i}}$, the scalar particle to hope from a vertex $\mathbf{n}$ to its nearest neighbors $\mathbf{n}+\hat{i}$. In lattice gauge theories the A_i fields or the corresponding connections

$$U_{n,n+\hat{i}} = \exp i(A_i \Delta x_i) \tag{5.19}$$

lie along the link connecting $\mathbf{n}$ and $\mathbf{n}+\hat{i}$. $\boldsymbol{\Phi}_n\, U_{\mathbf{n},\mathbf{n}+\hat{i}}\, \Phi_{n+i}$ is then the gauge invariant hopping term, reflecting the gauge invariant, continuum kinetic term $\int (D_\mu \Phi)^2$. In this formulation we naturally associate the field strength $F_{ij}(x) \equiv (D_i, D_j)$ with plaquettes.

$$U_{n,i,j} = U_{n,n+\hat{i}}\, U_{n+\hat{i},n+\hat{j}}\, U^+_{n+\hat{i}+\hat{j},n+\hat{i}}; , U^+_{n+\hat{j},n} \tag{5.20}$$

The plaquettes in turn allow the U's to hope from a link to a neighboring link. Higher order antisymmetric forms $F_{\mu\nu\lambda}(x)$, if appearing in the theory can be associated with three, etc., dimensional volume elements.

Spin 1/2 fermions have no such natural locations on static lattices. Our dynamic net which allows motion of links by rearrangements may offer such a location by identifying fermions with cut links. Specifically we will allow cutting links which carry the $SU(K)$ flux thereby mimicking the production of say quark-anti-quark

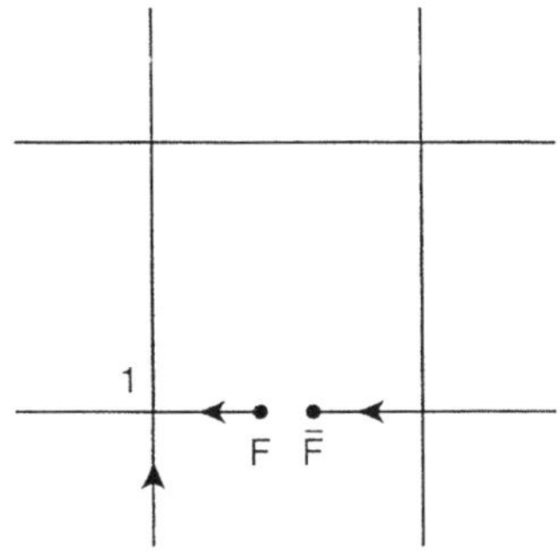

Fig. 5.8 A "half link" Fermion F and anti-Fermion $\bar{F}$ generated by cutting one link of our net. The *arrows* indicate $SU(K)$ $\kappa, \bar{k}$ (fundamental and anti-fundamental)

pair in a chromoelectric flux tube which then separates into two smaller parts. (See Fig. 5.8.)

Each member of the pair can then be viewed as "an external line" incident on our original vacuum diagram carrying into (or out of) it the gauge quantum numbers of the $SU(K)$ fundamental representation K.

The odd statistics of "half link" fermions is suggested by reversing the original argument for QCD color [5]. If both are at the same baryonic vertex, then it is manifest in the $\epsilon_{a_i \ldots a_k}$ couplings (Eq. (5.7)). In general the two half-links, or fermions, F and F', are at different vertices and we "parallel transport" F to some baryonic K-fold vertex, B, along a path P connecting F to B and F' to the same B along the path P'. The structure of the bosonic vertices (Eq. (5.8) and Fig. 5.1.B) ensures that this transporting of F yields an $U_{(P)}$ factor which is the path ordered product of the U matrices along the path P:

$$U_{(P)} = \Pi_1^n \, U_{(1)} U_{(2)} \ldots U_{(n)} \tag{5.21}$$

Similarly, F' picks up a

$$U_{(P')} = \Pi_1^{n'} \, U'_{(1)} U'_{(2)} \ldots U'_{(n')} \tag{5.22}$$

factor. After exchanging F, F', thereby obtaining the (-1) factor due to the ϵ symbol at B, we transport F backtracking along the path P' to the vertex where the half-link F' was originally incident collecting a $U^+(P')$ factor. Likewise, we transports F' backwards along P to the original location of F collecting $U^+(P)$. This completes the exchange of F and F' the U factors cancel leaving just the required overall minus sign:

$$U^+(P)U^+(P')(-1)U(P')U(P) = -1 \tag{5.23}$$

Until we demonstrate that (locally) our net possesses Lorentz and, in particular, rotation invariance, the spin of the half-link fermion cannot be ascertained. The following suggests that the spin should be 1/2.

We can join a half-link and its "conjugate" half-link on a neighboring vertex by "welding together" the free endpoints of the two so as to re-form the original full link-as indicated in Fig. 5.9.

If we allow the additional operation of disconnecting a half-link from its original vertex of incidence we have yet an alternative way of combining the two half-links.

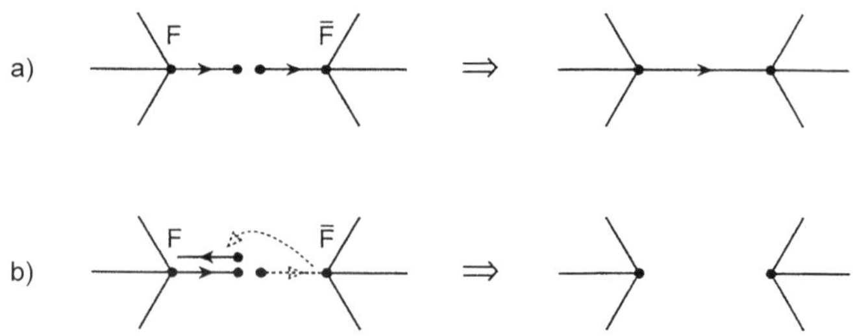

Fig. 5.9 The two ways of adding the spin of $F\bar{F}$ corresponding to $1/2 \otimes 1/2 \rightarrow 1+0$

We can then disconnect and flip backward one half-link so as to retrace backward the other half-link flux and annihilate it. In the end, we have transformed the original two n.n. three valent vertices which have no direct, one-link connection (Fig. 5.9(b)). To match $SU(K)$ quantum numbers we will have to assume that scalar fields in the fundamental representative of $SU(K)$ are now located at the amputated vertices. We may eventually try to associate these with "Higgs" scalar fields.

Recalling that links (vertices) are associated with spin $s = 1(0)$, the two operations of "parallel" and "backward" joining of half-links could have a simple interpretation as addition of angular momenta,

$$1/2 \times 1/2 = 1 \oplus 0 \tag{5.24}$$

if indeed the half-links carry spin-1/2.

5.4 Further Comments and Summary

To make the Net scenario viable we need to indicate how it can yield Lorentz invariance,relativistic dynamics, gravity, the observed pattern of gauge interactions and the three families of quarks and leptons and mimic the SBB cosmology over most of its history. The present rather crude picture cannot achieve or even meaningfully address most of these goals. In this last section we recall beside the glaring faults some nice features that the net picture may offer. Specifically we comment on:

- The unsatisfactory introduction of non-gravitational gauge + matter parts. and a possible way for obtaining Einstein Maxwell (or YM) interaction.
- The lack of satisfactory simple beginning—the need to multiply degrees of freedom (DOF) early on.
- The primary role of entropy/computationability, possible connection with the suggestion of T. Toffoli re the emergence of Lorentz Invariance and also of the principles of minimal action in the limit of a large number of degrees of freedom.
- a suggestive picture of black holes
- a possible understanding of the direction of time and of the Non-unitary collapse in the NET.
- The merging of space and matter and their common origin in the net.

I. While gravity and even space-time are supposed to emerge from the pattern of flipping near-by links in the net—most of particle/field dynamics was grafted onto the net by allowing Yang-Mills fluxes to reside on closed loops. To keep a finite number of degrees of freedom we allowed only the fundamental fluxes of say an $SU(K)$ gauge group. This implied that vertices of valence four are needed to allow non zero fluxes in all directions. with most vertices of this type we can have the nice feature that 3 dimensional approximately flat space naturally arises as the one with the highest dimension in which a regular -diamond like—lattice is allowed. Still the simple rules for the flux flow of minimal fluxes along the basic links illustrated in Fig. 5.2 above are actually of a $U(1)^K$ so that specific "colors" red, blue, etc can be continuously traced and conserved (or circulate in a closed loop). On a bigger coarser scale we can have color a and anti-color flowing along parallel links and thus generate fluxes in the adjoint of $U(K)$, or thanks to the baryonic vertices, of $SU(K)$. Still lacking the encoding of the specific group via the Clebsch-Gordan coefficients at the vertices as e.g in the Penrose spin nets, we can only hope that the correct group structure will emerge. All the above strongly suggest that we should look for a more natural construction of the gauge DOF replacing the above Wilson ex-machina introduction. It is still encouraging that the gauge DOF—the closed flux loops on the net—naturally interact with gravity namely the dynamical topology of the net as follows. The need to conserve $SU(K)$ limits our freedom in flipping links. Since the maximal stability under the local link flips was the criterion for selecting the symmetric lattice shape as the "lowest most stable, vacuum state" this contributes an extra energy. More generally, the presence of matter such a gauge loops and/or fermions i.e., flux carrying half links does prevent achieving in their local vicinity the ideal net—or equivalently "curves" space.

II. The basic principle underlying the net—that the total number of DOF is finite and fixed does imply that the starting net (aka the universe) already possess a huge number ($\sim 10^{250}$) of DOF's i.e vertices/links. This is very artificial though in the spirit of the "landscape" one can argue that it is needed to create our habitable universe. Still the very special tree structure of each half of the nascent universe suggests that at the time of "creation" we abandon unitary evolution and allow a tree like exponential growth not just of the physical size but also of the total number of DOF's.

III. The above asymmetric treatment of time and space makes Lorentz invariance in locally flat regions of the net hard to achieve. This and the lack of coherent dynamics are obvious and serious deficiencies. In this connection we note that in his work on "Action, or the fungibility of computations" Tommaso Toffoli [6] suggested a new principle of maximal number of "Computations" that the system can perform underlies the action principles of physics. This would reduce not just thermodynamics—but all of physics to basic "laws of large numbers". He then showed how this yields Lorentz invariance. We have noted that the principles underlying the evolution of the net from its initial hierarchical to the final regular form involve maximizing the number of possible near-by variants—or if the system is endowed with some temperature—the entropy or (the negative of the) free energy. Since furthermore, the actual step by step evolution seems to require non-trivial computations Toffoli's philosophy and its results may apply in the net approach.

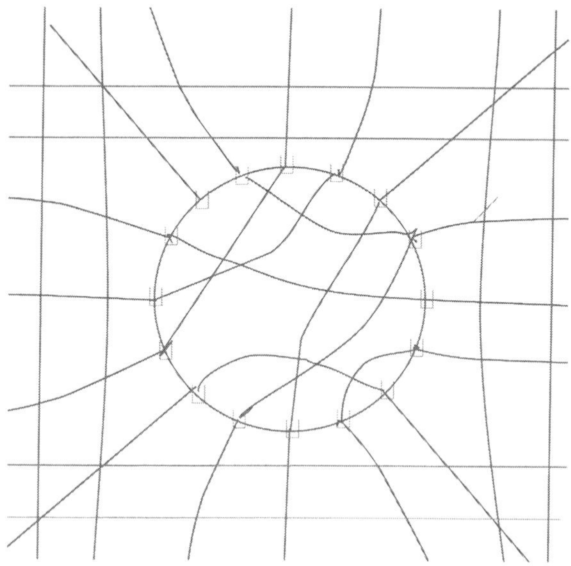

Fig. 5.10 Two-dimensional illustration of a black-hole

IV. Having gravitational binding manifest via "irregular" links connecting net vertices which are far apart in an almost regular back-ground suggests the following picture for the most tightly bound structure—the black holes. It obtains by having the inward pointing links emanating from half of all the $N = A/l_p^2$ vertices on the surface of a black hole of some radius R connect at random to the remaining half. Clearly this construct has:

(1) Entropy proportional to its area.
(2) no small closed loops and hence cannot accommodate gauge particles and matter in general—apart of coarse from conserved charge.
(3) while macroscopic (or even astrophysical $R \sim 10^8$–10^9 km for galactic BH's), $2\pi R$ perimeter and $4\pi R^2$ area of the B.H. obtain by walking around it, the B.H. has a one Planck thickness and no "inside"—realizing in a most extreme fashion the Fire-wall paradigm (see Fig. 5.10).
(4) After many local flips the deviant internal "Spaghetti" like links may wind up with both ends terminating at near-by points and eventually be emitted as $\sim N$ hawking photons.

It would be nice to have reliable estimates of the mass lifetime and external gravitational fields of B.H's. Still the above few points are very suggestive.

V. The evolution in the above scenario of the net/universe from the initial nascent Planckian state to the final regular vacuum—is clearly very directional—and provides an "arrow of time". Also constructing fermions by cutting links yields, if an appropriate set of links are cut, a non-unitary evolution by having one or more vertices leave the net. Since cutting happens only for flux carrying links i.e in regions where matter resides and is more likely for macroscopic systems this may underlie a possible mechanism for the collapse of the wave function.

VI. The distinction between space and matter fade away in the net picture. There is no qualitative difference but rather a matter of the degree to which we have "irregularities" in the underlying net and "physical" matter is replaced by the abstract connections between points realizing Wheelers' "it from bit" concept.

Prolog: The following proverb suggests the importance of recognizing the overall frame-work before making up theories. On the skin of a giant elephant, in the neighborhood of one of it's many scars lives a colony of intelligent ticks which can make only very tiny excursions within the immediate vicinity of their home scar. Still by ingeniously measuring the distances to other scars they found that the universe is a) curved and b) expanding. While these are sound conclusions, out-side observers have a much simpler explanation.

Acknowledgements The above ideas occurred to me ~ 15 years ago during a summer institute at the university of South Carolina and were, albeit most indirectly inspired by the work of Pawel Mazur and subsequent work by Casher and myself. Had it not been for the unique, free, atmosphere in these institutes around Yakir I doubt that I would have come across it. For this, for the time symmetric formulation of Q. Mechanics with a destiny state, for the opportunity to present this "idea for an idea" of the net on the occasion of his 80'th birthday and for innumerable hours of intellectual challenge and joy I am deeply indebted to Yakir. I am grateful to Sandu Popescu who pointed to me the work of T. Toffoli and to Lev Vaidman, the late Jeeva Anandan, Benni Reznik, Alonso Botero, Francoise Englert, Ted Jacobson Jeff Tollaksen and David Eichler (who was willing to listen and did not insist on describing his own radical approach) for their patience and to Aharon Casher for helping understand the increasing dimensionality of Hilbert space in an expanding universe.

References

1. A. Casher, S. Nussinov, Is the Planck momentum attainable? TAUP-2455-97, 39 pp. (1997). hep-th/9709127
2. Y. Aharonov, E. Gruss, Two time formulation of quantum mechanics (2005). quant-ph/0507260
3. Y. Aharonov, E. Gruss, Two-time interpretation of quantum mechanics (2005). quant-ph/0507269
4. J. Khoury, B.A. Ovrut, P. Steinhardt, N. Seiberg, Phys. Rev. D **65**, 086007 (2002)
5. O.W. Greenberg, Phys. Rev. Lett. **13**, 598 (1964)
6. T. Toffoli, in *The Action Principle and the Fungebility of Computations*. ed. by A. Hey. The Feynman Lectures on Computations (Addison-Wesley, Reading, 1998)

Part III
Time and Cosmology

Chapter 6
The Limits of Black Hole Complementarity

Leonard Susskind

Abstract Black hole complementarity, as originally formulated in the 1990's by Preskill, 't Hooft, and myself is now being challenged by the Almheiri-Marolf-Polchinski-Sully firewall argument. The AMPS argument relies on an implicit assumption—the "proximity" postulate—which says that the interior of a black hole must be constructed from degrees of freedom that are physically near the black hole. The proximity postulate manifestly contradicts the idea that interior information is redundant with information in Hawking radiation, which is very far from the black hole. AMPS argue that a violation of the proximity postulate would lead to a contradiction in a thought-experiment in which Alice distills the Hawking radiation and brings a bit back to the black hole. According to AMPS the only way to protect against the contradiction is for a firewall to form at the Page time. But the measurement that Alice must make, is of such a fine-grained nature that carrying it out before the black hole evaporates may be impossible. Harlow and Hayden have found evidence that the limits of quantum computation do in fact prevent Alice from carrying out her experiment in less than exponential time. If their conjecture is correct then black hole complementarity may be alive and well. My aim here is to give an overview of the firewall argument, and its basis in the proximity postulate; as well as the counterargument based on computational complexity, as conjectured by Harlow and Hayden.

6.1 Introductory Remarks

I have known Yakir for a very long time, and whenever I get confused about quantum mechanics, my first reaction is to ask how he would think about the problem. I am very confused right now. A puzzle has come up involving quantum mechanics that threatens to undermine everything we thought we know about black holes. I offer it in hopes that Yakir will figure out what is going on.

L. Susskind (✉)
Stanford Institute for Theoretical Physics and Department of Physics, Stanford University, Stanford, CA 94305-4060, USA
e-mail: susskind@stanford.edu

D.C. Struppa, J.M. Tollaksen (eds.), *Quantum Theory: A Two-Time Success Story*,
DOI 10.1007/978-88-470-5217-8_6, © Springer-Verlag Italia 2014

The claim of black hole complementarity (BHC) [1] is that information is not invariantly localized. Under certain conditions a bit can appear to be "here" to one observer, and far away to another. The ambiguous nature of localization was codified in BHC and the holographic principle. John Preskill, Gerard 't Hooft, and I championed this view in the early and mid 1990's. Since then it has become an accepted principle, particularly after gauge-gravity dualities were discovered.

BHC has been challenged by Almheiri, Marolf, Polchinski and Sully (AMPS) [2] who argued that the postulates of BHC are mutually inconsistent; in particular, they claim that the purity postulate, the assumption of semiclassical QFT outside the black hole, and the no-drama postulate lead to a contradiction.

If the firewall argument is correct then it may represent a step backward to a more traditional idea of information-localization, but at the cost of a breakdown in the concept of a smooth horizon [3].

6.1.1 Postulate 5

When asked if BHC means that information behind the horizon is meaningless I've generally answered no; information in the interior of a black hole is meaningful, but it is redundant with information in the exterior of the black hole. At early time, before there's been much evaporation, the redundancy is between the interior and the stretched horizon. Later, after a great deal of evaporation, the redundancy is between the interior and the Hawking radiation. The interior is meaningful, but it, and the Hawking radiation, should not be counted as independent.

One implicit assumption of the AMPS papers explicitly contradicts the above statement of BHC. The authors assume, as in the original BHC paper, that the degrees of freedom of the interior must be constructed from exterior degrees of freedom. But they also assume that those exterior degrees of freedom are physically near the black hole. In other words AMPS assume that the interior is constructed from the near-horizon degrees of freedom, and that the far-away Hawking radiation is not involved. Just to give the assumption a name, I will call it Postulate 5: the *proximity* postulate.

The proximity postulate directly contradicts the statement that "after a great deal of evaporation, the redundancy is between the interior and the Hawking radiation." AMPS were of course aware of this, and an important part of their argument is devoted to justifying the proximity postulate.

6.2 Aspects of Entanglement

In order to make the written version of this lecture self-contained I have included a section on various aspects of entanglement.

There are two situations in which large amounts of entanglement are known to occur. The first has to do with the properties of the ordered ground states of quantum

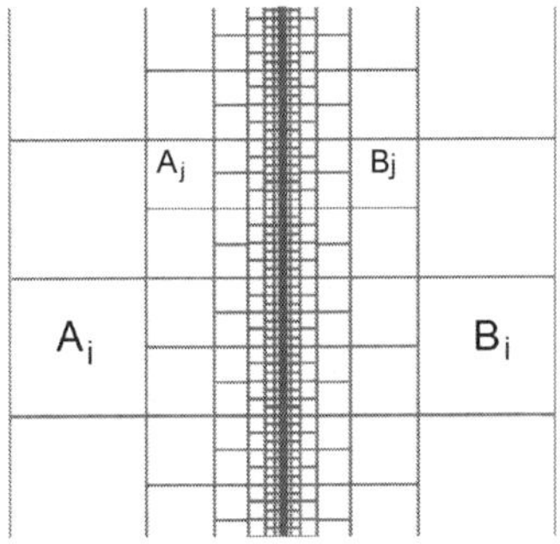

Fig. 6.1 Dividing space into two entangled half-planes. The entanglement of a conformal theory can be envisioned in terms of mirror-image Bell pairs formed from cells *on the right and left side*

field theories including condensed matter systems. The second is almost the complete opposite; it involves entanglement that occurs as a result of complete randomness. The first case is fairly familiar; nearby subsystems tend to be highly entangled as a result of energy considerations. This type of entanglement leads to the area law for entanglement entropy, the reason being elementary; the number of lattice points adjacent to a given region is proportional to the surface area of the region. If one thinks of entanglement as the sharing of Bell pairs, then the Bell pairs in this first type of entanglement are well localized and the components of a pair are not distantly separated.

6.2.1 *Ordered Ground States*

A typical example of an ordered ground state is the vacuum of a conformal field theory. If we divide space into a left and right half, the two halves will be entangled with an divergent entanglement entropy proportional to the area of the dividing plane.

$$S = \frac{A}{l^2} \tag{6.1}$$

where l is an ultraviolet cutoff. A rough picture of the entanglement can be provided by dividing the space on either side into cells in a scale-invariant way, as in Fig. 6.1. The purple line in the figure representing the boundary between the entangled regions has been drawn thickened to represent the cutoff length l.

In each cell a degree of freedom can be defined by averaging the field over the cell. The degree of freedom in a cell at a distance ρ from the dividing-surface are therefore field-modes with wavelength of order ρ. The entanglement across the surface can be approximated by saying that mirror image cells are entangled. The locality and scale-invariant character of the entanglement can be roughly modeled by thinking of the cells as qubits which are entangled in Bells pairs, A_i entangled with B_i, as in Fig. 6.1. Each entangled Bell pair contributes a single bit of entanglement entropy.

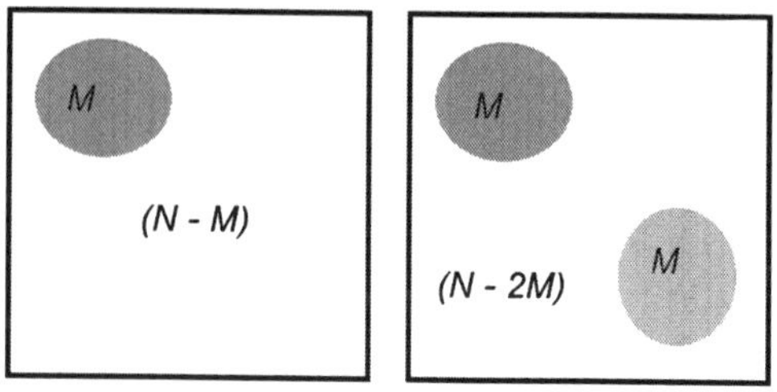

Fig. 6.2 *On the left side* an N-qubit system is divided into a small subsystem with M qubits and a big subsystem with $N-M$ qubits. *On the right side* the big subsystem is further divided into a second subsystem with N qubits and a remainder of $N-2M$ qubits

6.2.2 Scrambled Systems

The other less familiar situation is entirely different in character; it occurs when energy is not a consideration at all. It is the entanglement of a scrambled system [4–7]. The shared Bell pairs in this type of entanglement are extremely de-localized; they are diffused over the entire system. Since it plays a large role in what follows, I will spend some time explaining scrambling entanglement. A good example is based on a random system of a large number, N, of qubits. In a particular basis (sometimes called the computational basis) each qubit has two basis states labeled 0 or 1.

Begin with a highly non-typical state such as

$$|\Psi_0\rangle = |0000000\ldots 00\rangle. \tag{6.2}$$

To scramble the state, randomly pick a unitary operator U from some ensemble of $2^N \times 2^N$ unitary matrices. A simple ensemble is the maximally random Haar ensemble which indeed scrambles, but it is also overkill—a point we will come back to.

The scrambled state is defined by,

$$|\Psi\rangle = U|\Psi_0\rangle \tag{6.3}$$

With overwhelming probability $|\Psi\rangle$ has the scrambled property; namely, any small subsystem has essentially no information. A small subsystem means any subset of qubits fewer than half the total number. If $M < N/2$, then a system of M qubits is small. In the left side of Fig. 6.2 an N-qubit system is divided into an M-qubit subsystem, and an $(N-M)$-qubit subsystem.

The remarkable property of scrambled systems is that with overwhelming probability the amount of information in a small subsystem is negligible. The precise meaning of this statement is that for almost all matrices U the entanglement entropy a small subsystem M is very close to maximal,

$$S_M = M \log 2. \tag{6.4}$$

(From now on I will drop the factor log 2 and measure entropy in bits.) The equality sign in (6.4) is not quite exact but the error is less than a single bit, and generally much less than that. I will ignore this small discrepancy in what follows.

Another way to say the same thing is that the density matrix of the small subsystem M is extremely close to the maximally mixed density matrix,

$$\rho_M = 2^{-M} I \tag{6.5}$$

where I is the unit matrix in the state-space of the M qubit system. Again, the equality sign is correct up to negligible errors in the large N limit.

It follows that the scrambled state $|\Psi\rangle$ in (6.3) can be written in the form,

$$|\Psi\rangle = \sum_i |i\rangle_s |\phi_i\rangle_b \tag{6.6}$$

where the states $\sum_i |i\rangle_s$ represent a basis for the small M-qubit system, and the $|\phi_i\rangle_b$ represent states in big subsystem of $(N - M)$ qubits. Moreover, the fact that the density matrix of the small subsystem is maximally mixed implies that the $|\phi_i\rangle_b$ are orthonormal.

The $|\phi_i\rangle_b$ are not a complete basis for the big $(N - M)$-qubit system. They only span a subspace of dimension 2^M. We can think of the $|\phi_i\rangle_b$ as the basis states for a subsystem of M qubits that lives in the larger $(M - N)$ qubit subsystem. This subsystem is most certainly not a collection of the original defining qubits. However, it is unitarily equivalent to such a subsystem. To make this precise we can take any M-qubit subsystem from the $(N - M)$ system. This is shown in the right side of Fig. 6.2. Thus we have three subsystems. The first is the original small M-qubit subsystem. Next is a second small subsystem which belongs to the $(M - N)$ qubit system. Finally there are the left over $N - 2M$ qubits, also belonging to the big subsystem. The point is that any vector of the form (6.6) is close to a vector that can be expressed by a two step process. First define a state in which the two small subsystems are maximally entangled, and the third subsystem factors off.

$$|\Phi\rangle = \sum_i |i\rangle_s |i\rangle_{s'} |00000\ldots\rangle \tag{6.7}$$

where s' refers to the second small subsystem, and $|00000\ldots\rangle$ denotes the state of the remaining $(N - 2M)$ qubits. In such a state the small subsystem is manifestly maximally entangled with a subsystem of the big subsystem.

To obtain $|\Psi\rangle$ from $|\Phi\rangle$ we apply a unitary scrambling operator V on the big $(N - M)$ qubit subsystem.

$$|\Psi\rangle = V|\Phi\rangle. \tag{6.8}$$

The operator V is the product of a scrambling operator on the big subsystem and the unit operator in the small subsystem. What V does is to scramble the M qubits, that are entangled with the small subsystem, and hide them among the larger $(N - M)$ qubits of the big subsystem. One point to bear in mind is that the matrix V depends on the state $|\Psi\rangle$. In other words V is a function of U.

States of this type are not special. Almost all states of the original N-qubit system will have this behavior, however they are divided into a small and big subsystems.

I suggested earlier that the matrix U may be drawn from an ensemble which is less random than the uniform Haar measure. A Haar-random matrix will certainly produce scrambled states, but it is very difficult to achieve Haar-randomness; indeed it can only be done by an exponential number of operations (exponential in N). However, scrambling can typically be achieved by a polynomial number of steps; a weaker type of randomness is sufficient. The matrix U need only be drawn from the ensemble of unitary 2-designs, U2. U2-randomness is equivalent to Haar randomness for any quadratic function of the density matrix of the system, and is enough to scramble. The main point about U2 is that it is much easier to average over U2 than to average over the Haar ensemble. Averaging over the Haar-measure typically takes exponential time, while U2 averaging can be accomplished in polynomial time.

On the face of it, the two situations in which large amounts of entanglement occur; namely highly ordered ground states, and highly random scrambled states, seem to have little to do with each other. However, in black hole physics the two come together in a surprising way. The highly ordered ground state seen by a freely falling observer at the horizon is "dual" to a highly scrambled state seen by an observer who stays outside the black hole. Moreover, the entanglements of the ordered infalling state are dual to the entanglements of the random thermal state of the exterior description.

In some ways scrambled states resemble maximally mixed states (density matrices proportional to the unit matrix), but there are subtle fine-grained differences. For each scrambled state there are observables which distinguish it both from other scrambled states, and from mixed states. I will refer to them as fine-grained properties, by contrast with coarse-grained properties that do not encode such distinctions.

Consider a maximally impure state for the entire system described by the density matrix

$$\rho = 2^{-N} I \tag{6.9}$$

where I is the 2^N-dimensional unit matrix. The maximally impure state ρ shares the property with $|\psi\rangle$ that small subsystems have no information, but in the impure case, large subsystems also have no information. Moreover ρ does not give rise to massive entanglement. No two subsystems are entangled in ρ.

6.2.3 *Coarse-Grained and Fine-Grained*

In the theory of large chaotic systems there are quantities that are so ridiculously hard to keep track of that we are inclined to think of them as completely meaningless. An example would be the memory of the initial position $x(0)$ of a particular particle in a sealed box of gas, after an exponentially long time t. The principles of classical mechanics imply that $x(0)$ can be expressed in terms of the coordinates and momenta of all the particles at time t, but this information is of no practical value.

It is so fine-grained that the tiniest perturbation in either the initial conditions or the Hamiltonian of the box will radically alter the connection between the degrees of freedom at time 0 and time t. We call such quantities fine-grained. Coarse-grained quantities are the opposite; they are relatively insensitive to such tiny perturbations. For example the number of particles in a small sub-volume averaged over a one-second interval is coarse-grained. It will be important to realize that the quantities that the AMPS argument deals with are analogous to the very fine-grained details of a box of gas.

Properties of a scrambled qubit system fall into two classes, coarse-grained properties are sensitive to the difference between $|\Psi\rangle$ and ρ; and fine-grained properties which are not. All quantities that are made out of fewer than half the qubits are coarse grained. Fine grained observables are always made out of more than half the qubits. Any test of the entanglement between small and large subsystems is fine-grained.

For the simple qubit systems we've discussed, typical fine-grained quantities are trivial. Each qubit can be described by the usual Pauli operators $\sigma(\mathbf{n})$ where n label the qubit. Any operator in the entire system can be expressed as a sum of products these Pauli operators. Suppose we consider an operator made out the M qubits of a small subsystem. Since the density matrix of any small subsystem is proportional to the identity, the expectation value of any product of the M qubit operators is zero. This is true for both a scrambled pure state and the maximally mixed state.

On the other hand when we come to operators involving more than half the qubits, things change. Consider a particular qubit—say qubit 1—in the small subsystem of M qubits that we discussed earlier. In the scrambled state that qubit is entangled with a hidden qubit in the big subsystem. Suppose we call the Pauli operators for the hidden qubit $\tau(\mathbf{1})$. By an appropriate choice of conventions we can assume that $\sigma(\mathbf{1})$ and $\tau(\mathbf{1})$ are, to a high approximation, in a singlet state, Therefore

$$\langle\Psi|\sigma(\mathbf{1})\cdot\tau(\mathbf{1})|\Psi\rangle = -3 \tag{6.10}$$

But in the maximally mixed state

$$\langle\Psi|\sigma(\mathbf{1})\cdot\tau(\mathbf{1})|\Psi\rangle = 0. \tag{6.11}$$

By measuring $\sigma(\mathbf{1})\cdot\tau(\mathbf{1})$ and obtaining a number far from -3 one could distinguish[1] $|\Psi\rangle$ from ρ.

The operator $\tau(\mathbf{1})$ is not one of the original qubit operators. It is an extremely complicated combination of all $(N-M)$ qubits of the big subsystem, and therefore the operator in (6.10) is made out of more than half the original qubits. From the fact that the expectation value is different in the scrambled and mixed states we recognize it as representing a fine-grained property.

[1]One can distinguish the maximally mixed state from a particular scrambled state in this manner. But one could not distinguish whether the state was pure without specifying the particular state. In other words this is not a one-shot method to determine purity.

Fine-grained operators are extremely dependent on the scrambled state in the following sense. If we pick the particular combination of original qubit operators that define $\sigma(\mathbf{1}) \cdot \tau(\mathbf{1})$ for the state $|\Psi\rangle$ and evaluate it in another scrambled state $|\Psi'\rangle$ the result will be negligible (exponentially small) [8, 9]. So to test out if a state is pure by measuring a fine-grained operator, one has to know in advance exactly what scrambling dynamics has taken place.

Fine grained quantities for large systems generally don't play any role in practical many-body physics. The quantities that interest us usually don't depend on whether a large system is in a pure state or a thermal ensemble unless the system is in the ground state. Fine-grained observables are much too difficult to measure. Thus we have very little experience with fine-grained physics, but the questions that will occupy us in this lecture are of the most fine-grained kind.

6.2.4 Distillable Entanglement

6.2.4.1 Pure States

It's important to have a quantitative concept of how entangled two subsystems are. We imagine a system of qubits and consider two subsystems will be denoted by $\mathscr{B}$ and $\mathscr{H}$. To define the amount of entanglement between $\mathscr{B}$ and $\mathscr{H}$ I will introduce a concept from quantum information theory called *distillable entanglement* [10] represented by the symbol D. I will not attempt to be too precise in its definition; in essence it is the number of Bell pairs shared by two subsystems.

First of all, if we insist on exactly maximal entanglement, then the generic answer is zero, but if we relax the tolerance a bit, the generic answer is lots. An exact Bell pair is a pair of qubits, B and H_B, such that:

(1) The density matrix of the union of the two is pure. In other words the Von Neumann entropy of the union $B \cup H_B$ is exactly zero.
(2) The density matrices of the individual subsystems B and H are each maximally random, i.e., proportional to the identity matrix. We can also say that the Von Neumann entropy of each subsystem is maximal and equal to 1.

To relax these conditions we can introduce a small parameter ϵ and require that:

(1′) The density matrix of the union of the two is almost pure. In other words the Von Neumann entropy of the union $B \cup H_B$ is less than ϵ.
(2′) The density matrices of the individual subsystems are almost maximally random. The Von Neumann entropy of each subsystem is almost maximal and greater than $1 - \epsilon$.

This defines a regulated version of a Bell pair.

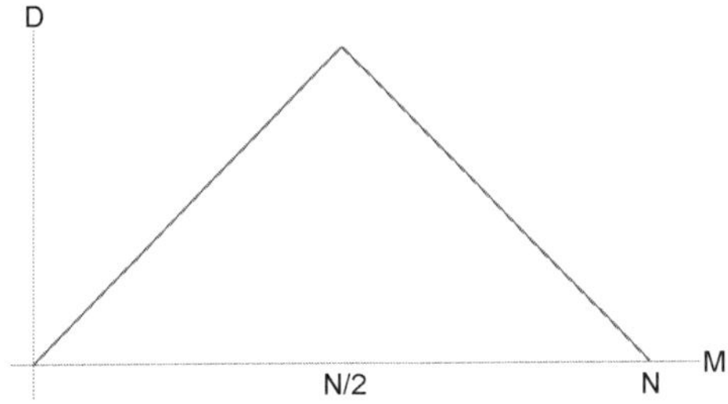

Fig. 6.3 The Page curve for distillable entanglement

6.2.4.2 The Page Transition

Consider a scrambled system of N qubits, and divide it into two subsystems with M and $(N-M)$ qubits. The following is true. Given any small ϵ, there is an $N(\epsilon)$, such that if the total number of qubits is greater than $N(\epsilon)$, then the number of regulated Bell pairs is equal to the smaller of M and $(N-M)$. A more general way to express this is that the number of approximate Bell pairs is equal to the entanglement entropy of the subsystems. For a large black hole the numbers are so big that the difference between regulated and exact Bell pairs is not important.

In Fig. 6.3 the distillable entanglement is plotted as a function of M. As M increases from zero, the number of Bell pairs increases linearly until it reaches a maximum at $M = N/2$. At that point the curve exhibits a sudden change in the slope, and by symmetry, it decreases to zero when $M = N$. This sharp transition or cusp-like behavior was first observed by Page [4, 5] and the curve is sometimes referred to as the Page curve.

6.2.4.3 Mixed States

Let's consider the distillable entanglement between $\mathscr{H}$ and $\mathscr{B}$, assuming the total system is not in a pure state. Counting the distillable entanglement is more difficult if the bipartite system under consideration is not pure, because the concept of entanglement entropy does not exist. Nevertheless the distillable entanglement D is defined, for our purposes as follows:

Consider a unitary operator U constructed as the tensor product of a $2^{N_H} \times 2^{N_H}$ matrix in the Hilbert space of the subsystem $\mathscr{H}$, and the identity matrix in $\mathscr{B}$. Apply it to the density matrix ρ_{BH} of the $\mathscr{H}\mathscr{B}$ system. The result is $\hat{\rho}_{BH}$.

$$\hat{\rho}_{BH} = U^{\dagger} \rho_{BH} U \tag{6.12}$$

Next, pair the $\mathscr{B}$-qubits with a subset of the $\mathscr{H}$-qubits and count the number of regulated Bell pairs. Finally maximize that number with respect to all $2^{N_H} \times 2^{N_H}$ unitary transformations. Basically what we are doing is unscrambling the $\mathscr{H}$ system and counting the Bell pairs. The process is called distilling and the resulting number of Bell pairs is the distillable entanglement.[2]

The distillable entanglement is very difficult to compute but we can bound it, and in some situations of interest the bound is almost saturated. The useful bound on D is given in terms of the mutual information of the $\mathscr{H}\mathscr{B}$ system. If S_B, S_H, and S_{BH} are the Von Neumann entropies of $\mathscr{B}$, $\mathscr{H}$, and $\mathscr{H}\mathscr{B}$ (the union of $\mathscr{H}$ and $\mathscr{B}$) we define

$$\mu = \frac{1}{2}(S_B + S_H - S_{BH}) \tag{6.13}$$

(Note that μ is defined to be half the usual mutual information.)

Suppose that the $\mathscr{H}\mathscr{B}$ subsystem is purified by a third subsystem $\mathscr{R}$. Then it follows that $S_{BH} = S_R$ and we can write

$$\mu = \frac{1}{2}(S_B + S_H - S_R) \tag{6.14}$$

It is known that distillable entanglement is bounded by μ [10],

$$D \leq \mu. \tag{6.15}$$

Another simple fact that helps in computing μ is that for a scrambled system the Von Neumann entropy of a small subsystem is always maximal [4, 5]. This means that for any subsystem of M qubits, its entropy is given by M as long as $M < N/2$.

Finally, there are two situations in which D equals μ or is very close to it. The first case is $\mu = 0$. Since both μ and D are never negative it follows that if μ vanishes, so does D.

The second less trivial case is when μ is maximal or close to it. This happens when $\mu \approx M$. In that case $D \approx M$.

These facts will be helpful in explaining the firewall argument.

6.3 Complementarity and the Firewall Argument

6.3.1 Degrees of Freedom

Let's begin with an account of the firewall argument as described in [3]. The argument, which assumes the proximity postulate, is based on a simplified model of a black hole that starts by dividing the black hole geometry into four regions as in Fig. 6.4. The first three called $\mathscr{R}$, $\mathscr{B}$, and $\mathscr{H}$ form the exterior of the black hole;

[2]There are several notions of distillable entanglement including various "one-shot" definitions. Generally they all allow transformations on both subsystems. Since the definition I am using restricts the search for Bell Pairs by allowing only transformations on $\mathscr{H}$, it gives a smaller answer than other definitions.

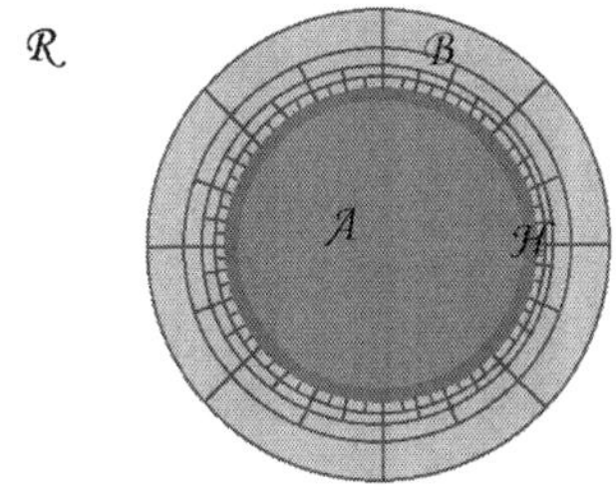

Fig. 6.4 The Black hole geometry is divided into four regions, $\mathscr{R}$, $\mathscr{B}$, $\mathscr{H}$, and $\mathscr{A}$. Region $\mathscr{B}$ is shown divided into thermal cells, each with a single bit of entropy

in other words the regions outside the horizon. The most distant region lies beyond the photon sphere at $r = \frac{3R_s}{2}$ (R_s is the Schwarzschild radius). This outer region is called $\mathscr{R}$ for radiation. Detectors in $\mathscr{R}$ unambiguously detect Hawking radiation. The degrees of freedom in $\mathscr{R}$ are very low energy and low angular momentum massless quanta.

Moving inward, the *zone* labeled $\mathscr{B}$ is the next region, which lies between the photon sphere and the stretched horizon. This region is well-approximated by Rindler Space. The degrees of freedom of $\mathscr{B}$ are quantum fields but with a short-distance cutoff at the string scale l_s, where string effects become important. The zone lies between the photon sphere and a proper distance l_s from the horizon.

The remaining portion of the black hole exterior is the stretched horizon labeled $\mathscr{H}$. The degrees of freedom of $\mathscr{H}$ are not field theoretic; the main thing I will assume is that they are fast scramblers [4–7]. Possibly they form a matrix system as in Matrix theory [11–14].

In the static Schwarzschild frame, unlike the infalling frame, the zone $\mathscr{B}$ and the stretched horizon $\mathscr{H}$ are in thermal equilibrium at a non-zero temperature. The dimensionless coordinate temperature is $\frac{1}{2\pi}$ and the local proper temperature, in the zone, varies according to

$$T = \frac{1}{2\pi\rho}.$$

Here ρ is the proper distance from the horizon. At the stretched horizon the proper temperature is $\frac{1}{2\pi l_s}$.

The fact that the $\mathscr{H}\mathscr{B}$ system is thermalized is the basis for the random qubit model. Pure states of a complex system which have a finite energy per degree of freedom are a lot like scrambled states. To make the connection explicit, one can subdivide a thermal system into cells, in such a way that each cell has about one bit of entropy. For low temperature systems the thermal cells are large, while for hot systems they are small. If the system is rescaled so that there is one thermal cell per unit volume, then the temperature will be rescaled to order unity, and each thermal cell can be thought of as a single qubit of a scrambled system.

As we will see, the similarity between one side of Fig. 6.1 and the cells of Fig. 6.4 represent the duality between the ground-state entanglements of a pure state and the scrambled entanglements of the thermal state describing the $\mathscr{H}\mathscr{B}$ system.

The total entropy of the black hole includes the entropy in $\mathscr{H}$ as well as that in $\mathscr{B}$. The entropy in field-modes of $\mathscr{B}$ can be computed. It would be divergent without the cutoff, but with a cutoff at $\rho = l_s$ it's of order

$$S_B \sim \frac{R_s^2}{l_s^2}. \tag{6.16}$$

We can also write it as [15],

$$S_B = g^2 S_{BH} \tag{6.17}$$

where g is of order the four-dimensional string coupling constant, and S_{BH} is the total black hole (Bekenstein Hawking) entropy. If we assume $g \ll 1$ then zone entropy is a small numerical fraction of S_{BH}.

If the black hole is formed by collapse, then while it is young the radiation degrees of freedom can be ignored. But as it evaporates entropy is transferred from the $\mathscr{H}\mathscr{B}$ system to $\mathscr{R}$. For simplicity we can assume that the total entropy is conserved although in practice it increases by a factor of about 3/2 [16].

In visualizing the field theory modes in the zone, it's helpful to transform from proper distance ρ to the tortoise coordinate u where

$$u = \log \frac{\rho}{l_s}.$$

The horizon is at $u = -\infty$ but the stretched horizon is at $u = 0$. Thus in the zone the physical range of u is from zero to the photon sphere at $u = \log \frac{2MG}{l_s}$. For simplicity consider a massless scalar field ϕ. If the field is weakly interacting then we can express it in terms individual angular momentum modes in which case the wave equation in the zone takes the form

$$\nabla^2 \phi + \frac{l_s^2}{R_s^2} L^2 e^{2u} \phi = 0 \tag{6.18}$$

where: R_S is the Schwarzschild radius of the black hole: l_s is the string length scale: L is the angular momentum of the mode. The expression $\frac{l_s^2}{R_s^2} L^2 e^{2u}$ is the usual centrifugal barrier which inside the photon sphere is attractive.

In tortoise coordinates the dimensionless coordinate temperature is $\frac{1}{2\pi}$ and the thermal wavelength Δu is order 1. The modes that are excited in the equilibrium state of the black hole can be classified by angular momentum, and by the radial tortoise coordinate u. The u-axis can be coarse-grained into thermal cells with a spread of order $\Delta u = 1$.

For a given angular momentum the range of u runs from $u = 0$ at the stretched horizon, to a value determined by the centrifugal barrier. Using the fact that the temperature is of order unity, the modes become frozen out of the thermal ensemble when $e^u > \frac{R_s}{L l_s}$. Thus each L mode lives in a "box" defined by

$$0 < u < \log \frac{R_s}{l_s L} \tag{6.19}$$

It follows that for each total angular momentum L there are $(2L+1)\log\frac{R_s}{l_s L}$ effective field modes in the zone, and each field mode carries about a single bit of entropy. Roughly speaking each of these modes is a qubit. The qubits can be thought of as each inhabiting a thermal cell as in Fig. 6.4.

The simplified model represents the black hole as a system of N qubits where N is the Bekenstein-Hawking entropy shortly after collapse. The qubits are assigned to the subsystems $\mathcal{H}$, $\mathcal{B}$, and $\mathcal{R}$ according to,

$$N = N_H + N_B + N_R. \tag{6.20}$$

At any given time the black hole entropy is

$$S_{BH} = N_H + N_B \tag{6.21}$$

and the entropy in the radiation is[3]

$$S_R = N_R. \tag{6.22}$$

The fraction of the black hole degrees of freedom carried by $\mathcal{B}$ is

$$\frac{N_B}{S_{BH}} = g^2. \tag{6.23}$$

We can assume that the black hole is formed in a pure state and that the internal dynamics quickly scrambles it, long before any appreciable amount of radiation has been emitted.

These qubit modes do not describe everything that can happen in the zone. For example, a particle can have an energy much higher than the thermal energy $\sim \frac{1}{\rho}$. Such modes describe infalling particles that fall in from infinity with energy larger than the Hawking temperature, but they contribute a negligible amount to the entropy of the black hole. When a particle with high energy falls in, in the static frame it disturbs the equilibrium for a short time. But in the scrambling time of order $R_s \log R_s$ the energy is distributed into the thermalized modes of the stretched horizon $\mathcal{H}$.

The model that I'll use for evaporation is very simple. One by one, qubits are transferred from the black hole subsystem ($\mathcal{H}\mathcal{B}$) to the radiation subsystem. The entire system remains in a pure state but the $\mathcal{H}\mathcal{B}$ subsystem loses its purity. Eventually all the qubits are transferred to $\mathcal{R}$ and the black hole disappears. The final state of the radiation is pure, but it is highly scrambled.

The fourth region of the black hole is the interior—the region behind the horizon—called $\mathcal{A}$. As long as we keep away from the singularity, the degrees of freedom in $\mathcal{A}$ are similar to those in $\mathcal{B}$; they are field theoretic with a cut-off at l_s. However, the main point of BHC is that they are not new independent degrees of freedom. The degrees of freedom in $\mathcal{A}$ are constructions built out of the exterior degrees of freedom in $\mathcal{H}$, $\mathcal{B}$, and $\mathcal{R}$.

[3]This formula is correct for the coarse-grained entropy of the radiation. For the fine-grained Von Neumann entropy, it is correct up to the Page time.

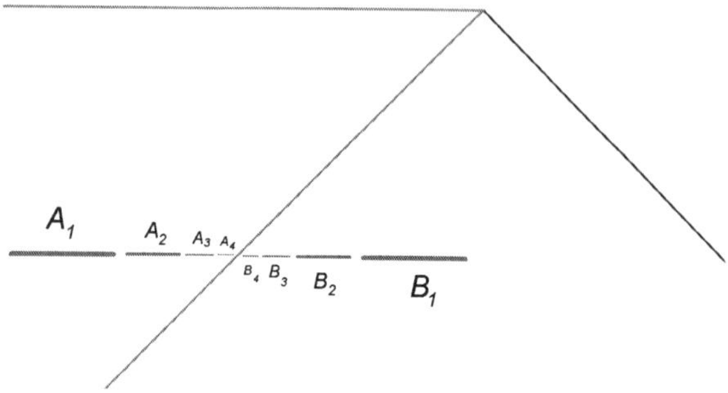

Fig. 6.5 The pairing of A and B modes can be carried out on a space-like surface in an infalling frame

6.3.2 *Entanglement of $\mathscr{A}$ and $\mathscr{B}$*

To make their argument AMPS take the perspective of a freely infalling observer passing from the zone to the interior. According to BHC such an observer sees "no drama" at the horizon, meaning that she sees the ordered Minkowski vacuum of a field theory like the one in Sect. 2.1. The field modes in $\mathscr{A}$ and $\mathscr{B}$ must be entangled in approximate Bell pairs—an example of the ground-state entanglement dictated by energy considerations. Breaking the entanglements would lead to a large energy density in the infalling frame.

As we have seen, a mode in $\mathscr{B}$ can be characterized by a spherical harmonic and a radial tortoise coordinate, uncertain to about $\Delta u = 1$. The same is true of the modes in $\mathscr{A}$. In fact the $\mathscr{B}$ modes and $\mathscr{A}$ modes come in matched pairs of opposite angular momentum and similar tortoise distance from the horizon. This pairing can is displayed in Fig. 6.5.

Let's denote a particular mode in $\mathscr{B}$ by the notation B_i and the corresponding partner in $\mathscr{A}$ by A_i. The entanglement that AMPS assumes is between A_i and B_i. For simplicity I will follow AMPS's idealized assumption that the A and B can be treated as qubits and that in the infalling frame A_i is maximally entangled with its partner B_i. In other words, in the infalling frame the $\mathscr{A}$ $\mathscr{B}$ system consists of a collection of g^2N maximally entangled Bell pairs. AMPS explain that any significant disturbance to the entanglement of these Bell pairs constitutes a violent perturbation of the infalling vacuum, and would certainly destroy an infalling observer. Therefore the no-drama postulate [2] of BHC requires that the A, B pairs remain in a state of maximal entanglement as the black hole evaporates.[4]

This discussion of A, B entanglement in the infalling frame must be translatable to the language of the exterior degrees of freedom, since by assumption, the interior degrees of freedom are constructed from the exterior degrees of freedom. The exterior description is thermal, and by appropriately defining thermal cells, it can be thought of as a scrambled system. That is why I said earlier that the ordered entanglement of a ground state is dual to the scrambled entanglement of a random

[4]Entanglement is a necessary requirement but not sufficient. There are many entangled states but only one of them has the correct form to describe a smooth space between A and B.

(thermal) system. Duality between infalling ground state and exterior thermal state is of course not new, but the point I want to emphasize is the duality of two kinds of entanglement: the entanglement of ordered ground states, and the entanglement implicit in scrambled states.

Since most of the exterior degrees of freedom are in $\mathscr{H}$, we can assume that the A_i are constructs made of the $\mathscr{H}$ qubits. Identifying A_i in $\mathscr{H}$ is a matter of finding a unique subsystem of $\mathscr{H}$ which is maximally entangled with B_i (the partner of A_i). In general there is no guarantee that such a subsystem of $\mathscr{H}$ exists. However, for the case of a relatively young black hole we can be sure that it does.

By relatively young I mean that the black hole has already scrambled, but the evaporation is negligible. In that case the $\mathscr{H}\mathscr{B}$ system is in a pure but scrambled state. In other words; to a high degree of approximation, every small subsystem is described by a random density matrix proportional to the identity, but the overall state is pure.

Given that the state of $\mathscr{H}\mathscr{B}$ system is pure, there is an important consequence of this fact; namely, that to an equally high degree of approximation, every small subsystem is maximally entangled with the rest of the system. Since $\mathscr{B}$ is a small subsystem of the $\mathscr{H}\mathscr{B}$ system, it follows that $\mathscr{B}$ is maximally entangled with $\mathscr{H}$. Furthermore each qubit of $\mathscr{B}$ is almost exactly maximally entangled with a unique subsystem of $\mathscr{H}$.

$\mathscr{H}$ may be given to us as some sort of recognizable quantum system, such as matrix quantum mechanics [11–14]. But the subsystem of $\mathscr{H}$ that B_i is entangled with is unlikely to be easily recognizable from the defining degrees of freedom of $\mathscr{H}$. Scrambled systems hide their entanglements in extremely difficult codes. Typically the code involves a unitary *descrambling transformation* acting on the $\mathscr{H}$ subsystem. The transformation descrambles the hidden qubits that are entangled with specific qubits of $\mathscr{B}$. Nevertheless, the purity and scrambled nature of the $\mathscr{H}\mathscr{B}$ state is enough to insure that each B_i is partnered with a subsystem, H_{B_i}, of the stretched horizon. (The notation means that H_{B_i} is a subsystem of $\mathscr{H}$, and that it is maximally entangled with B_i.) By the monogamy of entanglement H_{B_i} is unique.

The conclusion of this line of argument is obvious.

(1) In the infalling frame B_i and A_i are maximally entangled.
(2) In the exterior frame B_i and H_{B_i} are maximally entangled.
(3) Maximal entanglement is monogamous.

Therefore it follows that A_i and H_{B_i} must be the same thing. The formal equation

$$A = H_B \tag{6.24}$$

expresses this identification.

Perhaps another way to say this is that $\mathscr{H}$ is the hologram at the horizon, that represents the interior $\mathscr{A}$. It is evident that the relation between the interior and exterior of the black hole is extremely fine-grained from the point of view of the exterior degrees of freedom. I'll refer to it as the $\mathscr{H} \Longleftrightarrow \mathscr{A}$ mapping.

6.3.3 Non-linearity of $\mathscr{H} \Longleftrightarrow \mathscr{A}$ Mapping

An issue that is bound to come up, is the non-linearity of the $\mathscr{H} \Longleftrightarrow \mathscr{A}$ mapping.[5] By non-linear I mean that the relation between A_i and operators in $\mathscr{H}$ depends on the initial state $|\Psi_0\rangle$. That's because the particular form of H_{B_i} is state-dependent. Although this does not imply an observable non-linear violation of quantum mechanics in either the exterior or infalling frames, it does seem to violate the linear spirit of quantum mechanics.

Such non-linearity of the $\mathscr{H} \Longleftrightarrow \mathscr{A}$ mapping is not completely new. It occurs in the simple *pull-back–push-forward* strategy [17–20] for young black holes [3]. Suppose that the black hole is made by sending in a shell of a particular composition. The shell could be a coherent electromagnetic wave, a similar gravitational wave, or a mix of the two.

To carry out the pull-back–push-forward procedure, the operator A_i has to be pulled back to the remote past using the low energy equations of motion in the infalling frame. Since the equations of motion have to be pulled back through the shell, the operator one obtains in the remote past will depend on the nature of the shell. That dependence will also be present after the operator is pushed forward. Therefore the dictionary between interior operators and operators in the Hawking radiation depends on the initial state through the dependence on the state of the shell. This is a mild form of the same kind of non-linear dependence.

Non-linear dependence in the $\mathscr{H} \Longleftrightarrow \mathscr{A}$ mapping is probably inevitable. However, linearity can be restored by embedding the system in a larger system. For example, if the Hilbert space is big enough to include the operators which create the shell, then instead of saying the map depends on the state, we would say that the operator A_i maps to an exterior operator in the joint Hilbert space of $\mathscr{H}$ and the additional factor describing the shell.

6.3.4 Evaporation

Assuming the proximity postulate, a necessary condition for an uncorrupted black hole interior is that the distillable entanglement between $\mathscr{B}$ and $\mathscr{H}$ should be equal to the number of qubits in $\mathscr{A}$. If the number is less than that, then there is not enough of an entanglement resource to define all the interior modes. Even worse, if $D = 0$ it is impossible to define any vacuum modes in $\mathscr{A}$. That's the case in which AMPS argue that the geometry is terminated at the horizon by a firewall. The AMPS argument can be formulated as a calculation which shows that the $\mathscr{H}\mathscr{B}$ distillable entanglement goes to zero before the black hole has evaporated.

I will adopt the very simple model of evaporation as in Page's description [4, 5], and the one used by Hayden and Preskill [6]. Represent the collection of N qubits

---

[5] I thank Raphael Bousso and Douglas Stanford for discussions about this point.

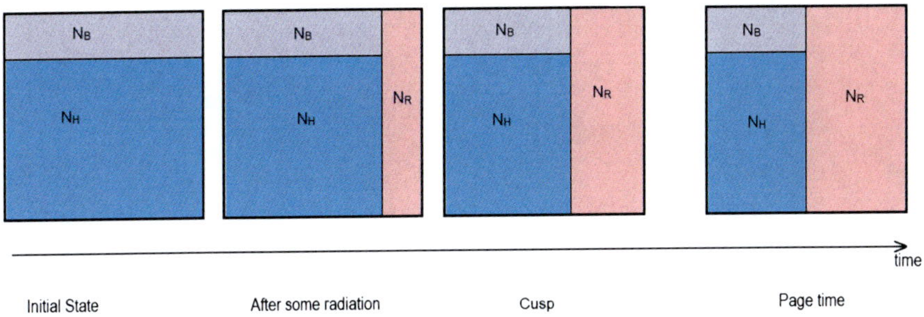

Fig. 6.6 Initially N qubits are distributed into boxes $\mathscr{H}$ (*blue*) and $\mathscr{B}$ (*purple*). As time evolves qubits get transferred to $\mathscr{R}$ (*pink*)

by a box, subdivided into three boxes $\mathscr{H}$, $\mathscr{B}$, and $\mathscr{R}$. At first the $\mathscr{R}$-box is empty and the N qubits occupy the boxes $\mathscr{B}$ and $\mathscr{H}$. The $\mathscr{B}$-box has g^2N qubits and the $\mathscr{H}$-box has $(1-g^2)N$. The total box is scrambled.

As the evaporation proceeds the $\mathscr{B}$ and $\mathscr{H}$ boxes shrink, while retaining the same relative size. The decreasing number of qubits is compensated by a growing number in the $\mathscr{R}$-box, the total number being kept fixed. Schematically this is shown in Fig. 6.6.

Obviously the state of the $\mathscr{H}\mathscr{B}$ system does not remain pure as qubits are transferred to $\mathscr{R}$. The question is what happens to the entanglements between $\mathscr{B}$ and $\mathscr{H}$? Does the existence of matched Bell pairs persist so that we can continue to identify the A_i with H_{B_i}?

6.3.5 *Loss of* $\mathscr{H}$, $\mathscr{B}$ *Entanglement*

In the infalling frame $\mathscr{A}$, $\mathscr{B}$ entanglement is essential for the no-drama postulate, i.e., for the existence of a smooth horizon. Translated to the exterior description, $\mathscr{H}$, $\mathscr{B}$ entanglement is essential. How much entanglement? In the infalling frame the number of Bell pairs is equal to $g^2 S_{BH}$. As we will see, that amount of entanglement persists for a long time as the black hole evaporates. But at some point the entanglement begins to diminish, and by the Page time it vanishes. This is the crux of the AMPS argument.

Armed with the facts of Sect. 2.3, it is easy to prove the following:

- As the black hole evaporates and N_R increases, the distillable entanglement between $\mathscr{B}$ and $\mathscr{H}$ remains maximal and equal to N_B, until a specific "cusp" time t_c. As a proportion of the black hole entropy the fractional distillable entanglement satisfies

$$\frac{D}{S_{BH}} = g^2 \quad (t < t_c) \tag{6.25}$$

To prove this we use (6.14) and observe that as long as $\mathcal{H}$ is greater than half the system, then

$$\begin{aligned} S_B &= N_B \\ S_R &= N_R \\ S_H &= S_B + S_R \end{aligned} \tag{6.26}$$

which gives

$$\mu_{BH} = N_B \tag{6.27}$$

Since this is the maximal value for μ_{BH} we can also write

$$D_{BH} = N_B$$

or

$$\frac{D}{S_{BH}} = \frac{N_B}{N_B + N_H} \tag{6.28}$$

which is equivalent to (6.25).

- This behavior continues until the point where N_H is half the total number of qubits. The cusp time t_c is defined by the condition that N_H is half the total number of qubits N. One may also write that at the cusp time $N_R = N_c \equiv N_H - N_B$.

  Note that the cusp time is earlier than the Page time t_p at which N_R becomes half the system.
- After the time t_c the fractional distillable entanglement decreases linearly with time. It vanishes at t_p and stays equal to zero until the black hole evaporates. To see this we note that between the cusp time and the Page time all three subsystems have less than half the total number of qubits. Therefore $S_B = N_B$, $S_H = N_H$, and $S_R = N_R$. It follows that

$$\mu_{BH} = \frac{1}{2}(N_B + N_H - N_R) \tag{6.29}$$

  which can be written in the form

$$\mu_{BH} = N_B - \frac{N_R - N_c}{2} \tag{6.30}$$

  In other words the mutual information begins to decrease relative to N_B once the cusp is passed. It is easy to see that it vanishes when $N_R = N_H + N_B$, i.e., at the Page time.

  From the fact that μ bounds D we see that the distillable entanglement between $\mathcal{H}$ and $\mathcal{B}$ also decreases to zero at the Page time.

The evolution of the $\mathcal{H}\mathcal{B}$ distillable entanglement is illustrated in Fig. 6.7.

The first fact indicates that D remains large enough so that the interior degrees of freedom can be defined for a long time. Indeed for small g the value of t_c is almost

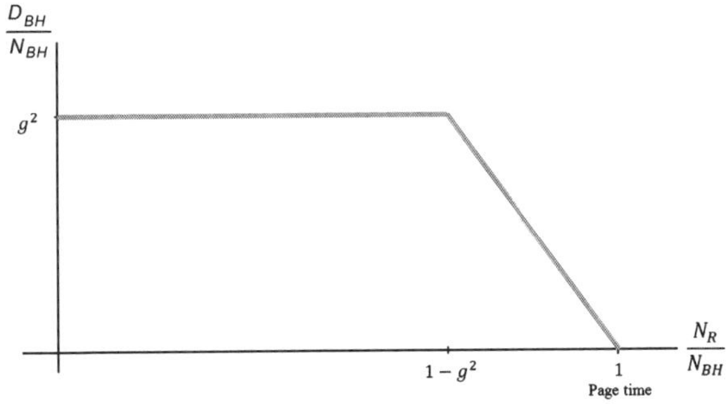

Fig. 6.7 The $\mathscr{H}\mathscr{B}$ fractional distillable entanglement stays constant until the cusp time. It then decreases until it vanishes at the Page time

the Page time. This is good news for a long-lived interior geometry, but we should be clear: distillable entanglement is a necessary condition for an uncorrupted region $\mathscr{A}$, but it may not be sufficient. It is quite possible for A_i and B_i to be maximally entangled but in the wrong Bell state.

But after the Page time there is no hope; the fine-grained quantities associated with $\mathscr{H}\mathscr{B}$ entanglement have disappeared altogether. As long as we insist that the interior be built from near-horizon degrees of freedom the evaporation will destroy the necessary entanglements and the space-time behind the horizon will be destroyed, at least if the proximity postulate is correct.

How this happens is a mystery, but one suggestion is that as the entanglement of $\mathscr{B}$ and $\mathscr{H}$ disappear, the singularity expands until, at the Page time, it intersects the horizon, so that an infalling observer will hit a brick wall instead of sailing past a locally undramatic point-of-no-return.

6.4 $\mathscr{A} = \mathscr{R}_{\mathscr{B}}$?

6.4.1 Redundancy of $\mathscr{A}$ and $\mathscr{R}$

One can argue that AMPS did not prove that the standard postulates of complementarity are inconsistent, but only that they are inconsistent with the proximity postulate. Turning it around, they proved that the first four postulates predict that the proximity postulate must be wrong, and that information in $\mathscr{A}$ must eventually become redundant with information in $\mathscr{R}$.

$\mathscr{R}$ does provide a resource for entangled Bell pairs. Indeed, after the Page time the degrees of freedom B_i continue to be entangled but with a subsystem of $\mathscr{R}$ instead of $\mathscr{H}$. It is not hard to prove that the distillable entanglement between $\mathscr{B}$ and the union $\mathscr{H}\mathscr{B}$ remains large enough to define partner modes for B_i. In Fig. 6.8 the fractional distillable entanglement of $\mathscr{R}\mathscr{B}$ is plotted alongside that of $\mathscr{H}\mathscr{B}$. The total is in fact conserved. Before the cusp time the Bell pairs are shared between $\mathscr{H}$ and $\mathscr{B}$. After the Page time they are shared between $\mathscr{R}$ and $\mathscr{B}$. Between the cusp and Page times they are partly shared with $\mathscr{H}$ and partly with $\mathscr{R}$, but the number of Bell pairs shared by $\mathscr{B}$ is constant.

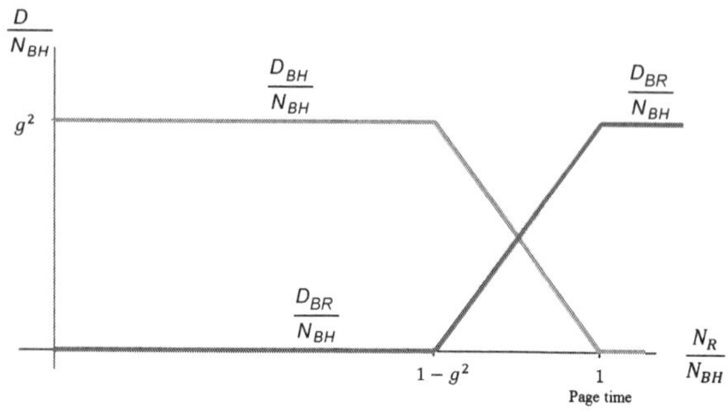

Fig. 6.8 As the $\mathscr{HB}$ distillable entanglement decreases, it is compensated by $\mathscr{RB}$ entanglement

After the Page time the degree of freedom that is maximally entangled with B_i lives in $\mathscr{R}$ and can be called R_{B_i}. The hypothesis that at late times $\mathscr{A}$ becomes redundant with the Hawking radiation would replace the $\mathscr{H} \iff \mathscr{A}$ mapping by an $\mathscr{R} \iff \mathscr{A}$ mapping,

$$A_i = R_{B_i}. \tag{6.31}$$

The relation between B_i and R_{B_i} is very fine-grained and depends in detail on the precise initial state and dynamics of the black hole.

An identification such as (6.31) would imply a radically greater localization-ambiguity than

$$A = H_B$$

(and would also eliminate the need for firewalls). Note however that such large scale delocalization of information is already present in any holographic theory.

Now let's turn to the reasons that AMPS rejected (6.31).

6.4.2 *Time Travel*

AMPS argued against (6.31) by invoking a thought experiment that leads to an apparent contradiction. The experiment involves an observer Alice, who is equipped with a very powerful quantum computer (QC). The input to the QC is the early half of the Hawking radiation. The output is a specific hidden qubit that Alice can hold and manipulate. It is assumed that Alice or her computer knows the exact initial state of the black hole, and the precise laws of evolution of all N degrees of freedom comprising the system.

Furthermore, according to the AMPS argument, she can collect the radiation and use her quantum computer to distill R_{B_i}, which by hypothesis is equal to A_i. If A_i refers to a field degree of freedom well after the Page time, then Alice knows information about A_i long before A_i has even happened. Alice then jumps into the black hole, carrying R_{B_i}, in time to meet the original A_i and its partner B_i.

The problem is that Alice can check whether her version of A_i (namely R_{B_i}) is entangled with B_i. If it is then, by the monogamy of entanglement, B_i cannot also

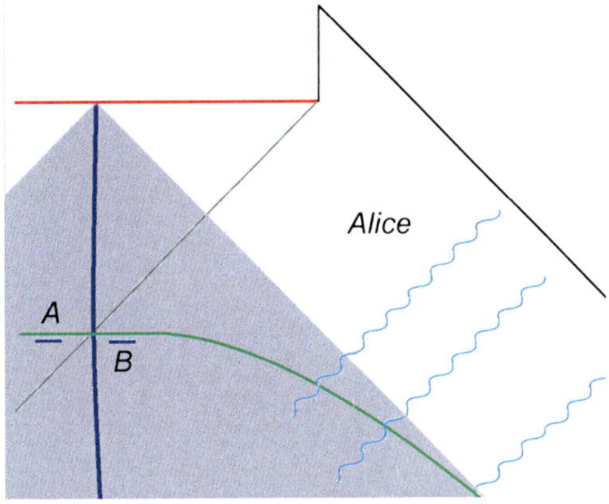

Fig. 6.9 Alice's causal patch is shown as the *light blue region*. The green spacelike surface asymptotes to the light-like boundary of Alice's causal patch. It encompasses A, B, and the early half of the Hawking radiation

be entangled with the original A_i, and thus a firewall must exist. In fact there is no need for her to check since she is already sure the quantum computer has accurately distilled R_{B_i}.

Another way to express the paradox is that Alice's experiment is analogous to past time-travel. The degree of freedom A_i has somehow appeared long in the past as R_{B_i}, and then traveled back to meet itself at A_i. In this form the firewall is playing the role of Hawking's *chronology protection agent*.

In [3] an possible alternative chronology protection agent was remarked on; "It may simply be physically impossible to distill R_B out of the Hawking radiation in time to bring it back to meet A." The same idea had also been expressed by Bousso and by Harlow.[6]

The decoding part of the experiment is clearly impractical. However, unless it violates a principle of physics such as locality, or the uncertainty principle, it must be allowed as a legitimate part of the AMPS argument. Therefore, to be consistent, the postulates of complementarity predict that the laws of physics absolutely forbid the distillation of R_{B_i} in less than the evaporation time.

6.4.3 Strong Complementarity

There is a formal version of the Alice's argument [21, 22]. As Bousso and Harlow discuss, we can follow Alice to the end of her world-line at point **p** on the singularity. The causal past of **p** defines Alice's causal patch shown in Fig. 6.9.

Her causal patch can be sliced by a family of space-like surfaces, one of which passes through the modes A, B, and asymptotes to the light-like boundary of the

[6]The possibility that Alice's experiment could not be carried out in the required length of time came up repeatedly in conversations between Raphael Bousso, Daniel Harlow, myself, and several others, soon after the AMPS paper appeared. The feeling at that time was that while difficult or impossible in practice, the principles of quantum mechanics do not forbid the experiment.

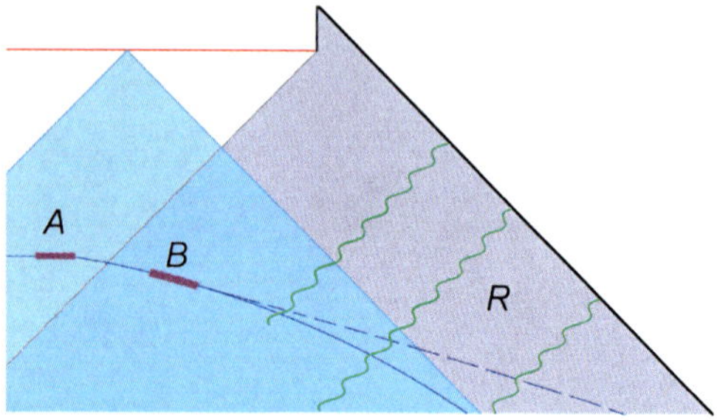

Fig. 6.10 This figure is the same as the previous figure with Bob's causal patch superimposed. The *dotted surface* is a space-like hypersurface that asymptotes to light-like infinity

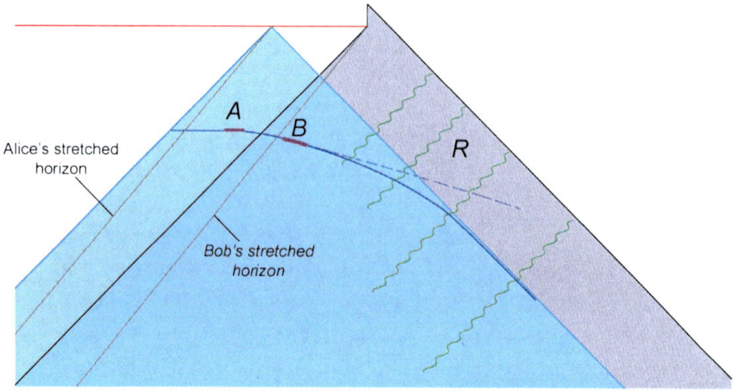

Fig. 6.11 Both Alice and Bob have stretched horizons but they are not at the same place

Alice's causal patch. Complementarity requires entanglement between B and A in Alice's infalling frame. Assuming that A and B are well after the Page time, the space-like slice passing through them also intersects more than half the outgoing Hawking radiation.

On the other hand the causal patch of Bob, who stays outside the black hole, contains B, as well as the outgoing radiation that was seen from Alice's patch (see Fig. 6.10), but it does not contain A. On Bob's space-like slice B must be entangled with the outgoing radiation, i.e., with R_B

The apparent contradiction is that photons (of the first half of the radiation) which pass through Alice's surface, also pass through Bob's. Thus it would seem that if B and R_B are entangled on Bob's space-like slice, the same must be true on Alice's, and this leads to the unallowed polygamous entanglement that AMPS base their argument on. However, things may be more subtle than this.

Consider Fig. 6.11 which is the same as (6.10) except that stretched horizons have been drawn in. Bob's quantum description is from the outside of the black hole and the radiation region, the zone, and his stretched horizon. These degrees of freedom are coupled and Bob's stretched horizon is dynamically involved in the production of Hawking radiation. On the other hand, in Alice's infalling frame Bob's stretched horizon is absent; it is replaced by a displaced stretched horizon that is shifted to the left. What is clear is that the two descriptions of the production of radiation cannot be exactly the same.

At the level of coarse grained properties of the radiation, the descriptions must match in the overlap region of the causal patches, out beyond the photon sphere where it is well understood that infalling and stationary observers see the same photons. The question is whether the two descriptions must match at the fine-grained level. In fact, Bousso and Harlow [21, 22] suggest that they do not. B is a coarse-grained object and therefore Alice and Bob should agree about it, but R_B is extremely fine-grained. Since in Alice's frame there cannot be multiple entanglement, R_B must not exist in her quantum description. In other words the radiation should not have the large-scale entanglements of a pure but scrambled state in her frame.

In Bob's frame purity requires those entanglements, but Bob's description does not include A.

But now it may be objected that we have an observable contradiction: Bob studies the photons that he sees and concludes that there is (B, R_B) entanglement. Alice studies the same photons and says there is not (B, R_B) entanglement.

Bousso and Harlow [21, 22] have argued against this conclusion, advocating a strong form of complementarity that can be described by saying each causal patch has its own quantum description. In Alice's quantum mechanics B is entangled with A and not with the outgoing radiation. In Bob's description B is entangled with R_B.

In fact we will see that the Harlow-Hayden conjecture protects against any real contradiction and allows the Bousso-Harlow strong complementarity to be consistent.

6.4.4 The Harlow-Hayden Conjecture

Hayden and Harlow (HH) have presented evidence suggesting that the Alice experiment does, in fact, violate fundamental physical principles [23]. They conjectured that the distillation of R_{B_i} in less than exponential time (exponential in N) is impossible with a QC that satisfies the principles of quantum mechanics and relativity. Satisfying quantum mechanics is, of course, part of being a quantum computer, but special and general relativity also impose limitations such as locality and holographic limitations on information.

If the HH conjecture is correct, then the operational limitation on Alice's experiment are shown in Fig. 6.12. If the extraction of R_{B_i} takes exponential time, then an observer can only distill R_{B_i} after she has reached the corner in the diagram where the black hole has already evaporated. In other words whatever mechanism implements the HH conjecture serves as a chronology protection agent.

The Harlow Hayden argument is technical [23] but let me discuss the issues. Here is the problem Alice faces: To begin with, the system of N qubits starts in a known pure state, schematically representing the initial state of the black hole. As an example the initial state could be $|\Psi_0\rangle = |00000000\ldots\rangle$. Subsequently the

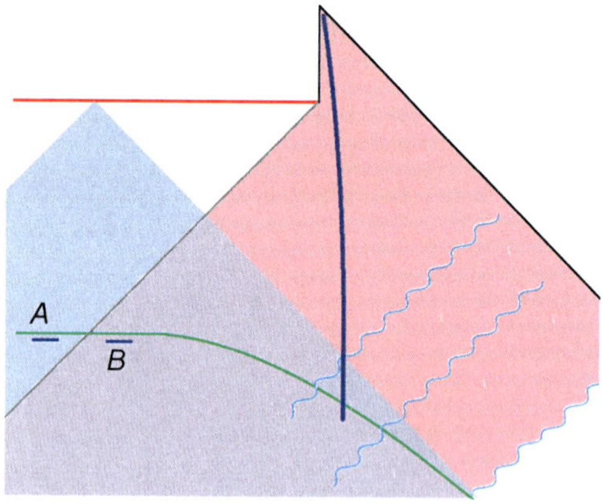

Fig. 6.12 If the HH conjecture is true, R_B cannot be distilled until the black hole has evaporated. The distillation can only be done by the time Alice arrives in the uppermost corner of the Penrose diagram

internal dynamics of the black hole, represented by a unitary scrambling operator U, scrambles the state,

$$|\Psi\rangle = U|\Psi_0\rangle \tag{6.32}$$

The matrix U is drawn from a statistical ensemble which is not strictly random. Scrambling does not require the state $|\Psi\rangle$ to be typical, but only that the density matrices of small subsystems be random (proportional to the identity to high accuracy). In the quantum-information jargon, it is sufficient that U be a unitary 2-design.

The unitary 2-design ensemble (U2) is very far from the Haar-random ensemble that would ordinarily define random matrices. To see how different they are, we can compare the number of 2-body interactions that must take place to implement U2, and the number needed to Haar-randomize. The answer is that the lower bound to obtain U2 randomness is $N \log N$ interactions (the bound is achievable), while Haar randomness requires an exponential number of interactions [6]. However, although U2 and Haar-randomness are very different, they lead to the same statistical properties for small subsystems, and for that reason U2 matrices scramble, and they can do so in the scrambling time of order $N \log N$.

Once the system is scrambled, evaporation separates $(N - M)$ qubits into $\mathscr{R}$. The case of interest is when $(N - M)$ is of order N and somewhat larger than $N/2$. The remaining black hole $(\mathscr{H} \cup \mathscr{B})$ is represented by M qubits.

Select one qubit from the remaining black hole and let it represent B_i. To be precise, the symbol B_i represents the set of operators that act on the qubit in the same sense that the symbol σ represents the three Pauli spin operators acting on a spin. The partner qubit R_{B_i} is hidden among the $(N - M)$ qubits in $\mathscr{R}$, and Alice's job is to isolate or distill it, and convert it to a qubit that she can hold and manipulate. To do that, she can make use of her QC to physically implement an unscrambling unitary V operator that acts on the subsystem of $(N - M)$ qubits. The purpose of V is to act on R_{B_i} and convert it to a particular qubit, for example the last qubit, q_0 which Alice is waiting to grab. Thus V is defined by

$$V^\dagger R_B V = q_0 \tag{6.33}$$

The operator V is of course not unique but this is not too important.

Alice does not have to search through all unitary operators V until she finds one that accomplishes the task. Knowing the initial state, and the scrambling matrix U, she can calculate V. In fact she could have done this long before the black hole formed. She has all of infinite past time to prepare for the experiment. Alice's limitation is not the difficulty of determining V, but of actually using her QC to implement V, in other word to physically apply it to the system of M qubits.

The original scrambling operator U is drawn from the U2 ensemble for N qubits, and only takes polynomial time to implement. If Alice had all N qubits to work with she could unscramble the system with the inverse matrix $U^\dagger$ which can be implemented in polynomial time. Because Alice has only $(N - M)$ qubits to work with, the matrix V is much more generic than a unitary 2-design, and more difficult to implement. This is analogous to the classical problem of time-reversing a complex chaotic system if even a small amount of information is lost before reversing it.

If V is sufficiently generic it will take an exponential number of gates (exponential in $(N - M)$) to implement V. Parallel computing does not help much. HH argue from experience with similar problems in quantum information theory, that the number of gates that have to operate in order to implement V is indeed exponentially in $(N - M)$. Therefore, they conjecture that it takes exponential time for Alice to distill R_{B_i}. (The exponential time is analogous to the recurrence time for a classical system.)

Given that the internal dynamics of a black hole can scramble in a short time, why then does it take so long to de-scramble? There is a close classical analog from the dynamics of complex chaotic systems. Consider a gas of N molecules in a sealed box. $(N - 1)$ of the molecules are identical of type a, and one is different of type b. Now start the system in some very non-generic state; for example all $(N - 1)$ a-type molecules at rest in one corner of the box, and the single b-type molecule at some other location x_b with a very large energy. Let the system evolve for a (polynomial) time until it appears to be in thermal equilibrium. The goal of the exercise is for Alice to recover x_b.

Alice is equipped with an incredibly powerful computer, which can if necessary integrate the equations backward, but her eyesight is somewhat blurry. The input to the computer is the final particles at some time after the system looks to be in equilibrium. The question is how long must it take the computer to recover the initial location of b?

The answer is—no longer than it took to come to equilibrium. If the computer is powerful enough it can simply reverse the process and bring the system back to its initial state. From that configuration, which is after all, special and easily recognizable, Alice just reads off x_b. By assumption that can be done in polynomial time.

One might ask why can't Alice look carefully at the system at the final time, and recognize what the initial state was? She can't because she is a bit myopic; the phase point is a bit blurry; and any small error will, if run backward, exponentially grow. But if the system is run all the ways back, the initial state is sufficiently distinct that Alice's vision is good enough, and she can recover x_b in polynomial time.

But now lets suppose that just before the system is run backward, we change the hamiltonian by dynamically decoupling M a-type particles. The computer must work with the remaining $(N - M)$ particles. Then running backward will not result in a distinctive configuration that allows Alice to read off the information we are looking for in polynomial time. Instead the system will stay in scrambled equilibrium for a recurrence time of order $\exp(N)$. Only after such a long time will the system fluctuate to a highly non-generic state, which Alice can read.

One might argue that the limitations of Alice's eyesight are irrelevant; she can just buy a better pair of eyeglasses. That's where quantum mechanics comes in. The uncertainty principle means that the fuzzyness is fundamental.

In the real case of decoding R_B, if Alice had all the degrees of freedom of $\mathscr{H}$, $\mathscr{B}$, and $\mathscr{R}$, she could run U backward in polynomial time (less than the evaporation time). But because she only gets to look at the qubits in $\mathscr{R}$ her job is much harder, and the guess is that it takes exponential time to distill R_B.

Considerations of this type led Harlow and Hayden to the following conjecture:

The minimum time that it takes to distill qubit R_{B_i}, is exponentially grows with N.

At the time of writing this talk, the Harlow-Hayden conjecture is still a conjecture. However, if true, it implies a fundamental physical constraint—call it computational complexity—that would prevent Alice from carrying out her experiment. For an ordinary Schwarzschild black hole the total time before it evaporates is of only of order N^3. Thus the truth of the HH conjecture would undermine the thought experiment designed to prove the existence of firewalls.

A possible way out of the computational complexity constraint was suggested in [3] and probably by several other people. Alice may try to slow down the evaporation process while her quantum computer is distilling R_{B_i}. One way to do that would be to surround the black hole by mirrors to keep it from radiating. But as Harlow and Hayden argue, this might not help Alice if the decoding time is exponential. An exponential time scale has multiple meanings for a complex closed system. For one thing the time scale for resolving tiny energy differences between neighboring states is of order $\Delta t = \frac{1}{\Delta E}$. For a system of entropy S this is equal to $e^S \sim e^N$. Of greater relevance, over such time scales Poincare recurrences will repeatedly occur, undoing and re-collapsing the black hole. It is unlikely that the identity of a mode A_i has any meaning over such long times.

Now let's return to Alice and Bob's Disagreement: Alice says no B, R_B entanglement in her frame. Bob says that there is B, R_B entanglement in his frame. That's of course the nature of complementarity, but can their disagreement lead to an observational conflict? The answer is no if the HH conjecture is correct; the difference between states with massive entanglement like $|\Psi\rangle$ and those with none like ρ can only be detected long after it is too late for Alice and Bob to communicate.

6.5 Conclusion

The AMPS paradox is currently forcing a rethinking of how, and where, information is stored in quantum gravity. The possible answers range from more or less conventional localization (proximity and firewalls) to the radical delocalization of $A = R_B$.

The argument in favor of the proximity postulate assumes the possibility of an "Alice experiment", which in some respects resembles time travel to the past. From this perspective, the firewall would function as a chronology protection agent, but at the cost of the destruction of the interior of the black hole. The Harlow-Hayden conjecture opens an entirely new perspective on chronology protection. It is based on the extreme fine-grained character of information that Alice needs to distill before "returning to the present."

Fine-grained information is something that has never been of much use in the past, given how hard it is to extract. But there is clearly a whole world of fine-grained data stored in the massive entanglements of scrambled pure states. That world is normally inaccessible to us, but if BHC is correct, it is accessible to an observer who passes through the horizon of a black hole. Thus complementarity implies a duality:

Ordinary coarse-grained information in an infalling frame is dual to the fine-grained information in the exterior description.

Or, as expressed earlier:

Ground-state entanglements in the infalling Minkowski vacuum, are dual to scrambled entanglements of the exterior thermal description.

The HH conjecture also represent a new principle, based on quantum-mechanical computational complexity:

Extracting fine-grained information cannot be done in less than exponential time, comparable to the time scale for Poincare recurrences. Therefore it cannot be communicated back to the black hole before it evaporates.

Whether the Harlow-Hayden conjecture and $A = R_B$ provide a way out of the AMPS puzzle, or lead to new contradictions is unknown as I write this.

References

1. L. Susskind, L. Thorlacius, J. Uglum, The stretched horizon and black hole complementarity. Phys. Rev. D **48**, 3743 (1993). hep-th/9306069
2. A. Almheiri, D. Marolf, J. Polchinski, J. Sully, Black holes: complementarity or firewalls? arXiv:1207.3123 [hep-th]
3. L. Susskind, The transfer of entanglement: the case for firewalls. arXiv:1210.2098 [hep-th]
4. D.N. Page, Average entropy of a subsystem. Phys. Rev. Lett. **71**, 1291 (1993). gr-qc/9305007
5. D.N. Page, Information in black hole radiation. Phys. Rev. Lett. **71**, 3743 (1993). hep-th/9306083
6. P. Hayden, J. Preskill, Black holes as mirrors: quantum information in random subsystems. J. High Energy Phys. **0709**, 120 (2007). arXiv:0708.4025 [hep-th]
7. Y. Sekino, L. Susskind, Fast scramblers. J. High Energy Phys. **0810**, 065 (2008). arXiv:0808.2096 [hep-th]

8. F. Dupuis, J. Florjanczyk, P. Hayden, D. Leung, Locking classical information. arXiv:1011.1612
9. N. Lashkari, D. Stanford, M. Hastings, T. Osborne, P. Hayden, Towards the fast scrambling conjecture. arXiv:1111.6580 [hep-th]
10. M.B. Plenio, S. Virmani, An introduction to entanglement measures. Quantum Inf. Comput. **7**, 1 (2007). quant-ph/0504163
11. T. Banks, W. Fischler, S.H. Shenker, L. Susskind, M theory as a matrix model: a conjecture. Phys. Rev. D **55**, 5112 (1997). hep-th/9610043
12. T. Banks, W. Fischler, I.R. Klebanov, L. Susskind, Schwarzschild black holes from matrix theory. Phys. Rev. Lett. **80**, 226 (1998). hep-th/9709091
13. I.R. Klebanov, L. Susskind, Schwarzschild black holes in various dimensions from matrix theory. Phys. Lett. B **416**, 62 (1998). hep-th/9709108
14. T. Banks, W. Fischler, I.R. Klebanov, L. Susskind, Schwarzschild black holes in matrix theory. 2. J. High Energy Phys. **9801**, 008 (1998). hep-th/9711005
15. L. Susskind, Some speculations about black hole entropy in string theory, in *The black hole*, ed. by C. Teitelboim, (1993), pp. 118–131. hep-th/9309145
16. D.N. Page, Comment on 'Entropy evaporated by a black hole'. Phys. Rev. Lett. **50**, 1013 (1983)
17. B. Freivogel, L. Susskind, A framework for the landscape. Phys. Rev. D **70**, 126007 (2004). hep-th/0408133
18. R. Bousso, L. Susskind, The multiverse interpretation of quantum mechanics. Phys. Rev. D **85**, 045007 (2012). arXiv:1105.3796 [hep-th]
19. G. Horowitz, A. Lawrence, E. Silverstein, Insightful D-branes. J. High Energy Phys. **0907**, 057 (2009). arXiv:0904.3922 [hep-th]
20. I. Heemskerk, D. Marolf, J. Polchinski, J. Sully, Bulk and transhorizon measurements in AdS/CFT. J. High Energy Phys. **1210**, 165 (2012). arXiv:1201.3664 [hep-th]
21. R. Bousso, Observer complementarity upholds the equivalence principle. arXiv:1207.5192 [hep-th]
22. D. Harlow, Complementarity, not firewalls. arXiv:1207.6243
23. D. Harlow, P. Hayden, Talks given at the SITP meeting on firewalls (2012). To be published

Chapter 7
Quantum Weak Measurements and Cosmology

P.C.W. Davies

Abstract The indeterminism of quantum mechanics generally permits the independent specification of both an initial and a final condition on the state. Quantum pre- and post-selection of states opens up a new, experimentally testable, sector of quantum mechanics, when combined with statistical averages of identical weak measurements. In this paper I apply the theory of weak quantum measurements combined with pre- and post-selection to cosmology. Here, pre-selection means specifying the wave function of the universe or, in a popular semi-classical approximation, the initial quantum state of a subset of quantum fields propagating in a classical background spacetime. The novel feature is post-selection: the additional specification of a condition on the quantum state in the far future. I discuss "natural" final conditions, and show how they may lead to potentially large and observable effects at the present cosmological epoch. I also discuss how pre- and post-selected quantum fields couple to gravity via the DeWitt-Schwinger effective action prescription, in contrast to the expectation value of the stress-energy-momentum tensor, resolving a vigorous debate from the 1970s. The paper thus provides a framework for computing large-scale cosmological effects arising from this new sector of quantum mechanics. A simple experimental test is proposed. [*Editors note*: for a video of the talk given by Prof. Davies at the Aharonov-80 conference in 2012 at Chapman University, see quantum.chapman.edu/talk-13.]

7.1 Quantum State Post-selection Combined with Weak Value Measurements

One of Yakir Aharonov's most significant contributions to quantum mechanics has been to uncover the hitherto neglected sector of weak measurements combined with post-selection. In this tribute to Aharonov's pioneering work, I wish to extend his ideas to cosmology, drawing inspiration from his paper with Gruss [1]. All cosmological observations and measurements are necessarily weak in the quantum sense.

P.C.W. Davies (✉)
Arizona State University, Tempe, AZ, USA
e-mail: paul.davies@asu.edu

D.C. Struppa, J.M. Tollaksen (eds.), *Quantum Theory: A Two-Time Success Story*,
DOI 10.1007/978-88-470-5217-8_7, © Springer-Verlag Italia 2014

For example, the red shift of a galaxy will be measured by observing the light of a very large number of photons emanating from a very large number of sources within the galaxy. Although the emission of a given photon by a given atom cannot be considered as a weak process as far as the atom is concerned, the use of a large ensemble of photons to measure a property of the entire galaxy does constitute a weak measurement. The quantum back-reaction of the photons on the relevant physical variable of the whole galaxy (in this case its momentum) is negligible.

Quantum weak measurements become more interesting when combined with pre- and post-selection. (For a review of this field, with references to earlier work, see [2].) In a typical laboratory quantum weak measurement experiment, a system is prepared in a well-defined initial state $|in\rangle$ at time $t = t_{in}$, for example, by making strong (projective) measurements of some observable A on an ensemble and selecting those members with the desired eigenstate. The system is then allowed to evolve, and at time t_{out} another strong measurement is made, of an observable B (B is not necessarily the same as A). If this procedure is repeated for a large ensemble $\{E\}$ of identically-prepared systems, i.e. systems all of which are in the same pre-selected eigenstate at $t = t_{in}$, and all of which are allowed to evolve with identical Hamiltonians H, then in general the final measured states will not be the same. Rather, they will be distributed over the set of eigenvalues of B with relative probabilities assigned by the Born rule. The experimenter then has the option of post-selecting, at time t_{out}, a sub-ensemble $\{e\}$ of the total ensemble for which all the members are in a particular eigenstate of B; call this state $|out\rangle$. The freedom to both pre- and post-select the state of a system is a key property of quantum mechanics, stemming from its indeterminism. (In a closed classical Hamiltonian system, the initial state plus the Hamiltonian suffices to completely determine the final state.) If, now, the experimenter carries out weak quantum measurements of some observable C on every member of the ensemble $\{E\}$ at one or more times t in the interval $t_{in} < t < t_{out}$, then the combined results of these weak measurements for the sub-ensemble $\{e\}$, satisfying both the pre- and post-selection criteria, when expressed as a statistical average, is the so-called *weak value*, given by

$$w(t) = \frac{\langle out|U^{\dagger}(t - t_{out})CU(t - t_{in})|in\rangle}{\langle out|U^{\dagger}(t - t_{out})U(t - t_{in})|in\rangle} \tag{7.1}$$

where U is the evolution operator e^{-iHt}. Weak values are not eigenvalues. Rather, they may be regarded as members of a decomposition of eigenvalues, weighted by the relative probabilities of finding the out states among $\{E\}$, and must be understood always in terms of statistical averages over large sub-ensembles: individual weak measurement results have no objective physical meaning. A good way to think about Eq. (7.1) is that the initial state is evolved forward in time from t_{in} to t, while the final state is evolved backward in time from t_{out} to t, and w is evaluated at time t as a composite of the forward- and backward-evolving wave functions. Note that w in general is not a real number, but both the real and imaginary parts have physical meaning [2]. Special interest attaches to the case that the denominator is small, which can occur if

$$Re(\langle out|U^{\dagger}(t - t_{out})U(t - t_{in})|in\rangle) \ll 1 \tag{7.2}$$

In that case, the weak values lie well outside the spectrum of eigenvalues, and can be interpreted in some cases as an amplification process.

7.2 The Wave Function at the End of the Universe: Post-selection in Cosmology

The concept of weak values has a clear operational meaning in a laboratory context, when the intervention of the experimenter is used to select and define appropriate in and out states. In this paper I wish to consider whether weak values with post-selection are physically meaningful in a cosmological context. In Sect. 7.1, I pointed out that cosmological observations necessarily involve weak measurements. Furthermore, observational values are derived by taking statistical averages over a large ensemble (e.g. one billion photons emanating from the same galaxy), thus producing a weak value in the quantum sense. Turning now to post-selection, in the cosmological case, the problem, expressed poetically, is who or what plays the role of "The Great Selector"? That is, how are the states $|in\rangle$ and $|out\rangle$ to be determined?

Questions of this sort are not new in quantum cosmology. There are many proposals for defining the initial quantum state of the universe, $|in\rangle$. Most appeal in one way or another to simplicity. A much-discussed example is the Hartle-Hawking so-called no-boundary wave function [3]. Other examples are to be found within the framework of the theory of quantum fields propagating in a classical background spacetime—a semi-classical approximation to a full theory of quantum cosmology. It is customary in this theory to take the initial quantum state to be a vacuum state of one form or another, e.g. the conformal vacuum (see, for example, [4]). The appeal to initial quantum state simplicity follows from the assumption that the universe started out physically simple and has evolved to states of greater complexity over time. Although it is far from obvious that this trend is correct [5], quantum cosmology has been practiced as a theoretical discipline for three decades without a serious challenge to the assumption of a simple initial quantum state. Often, such a state is described as "natural".

So much for pre-selection: what about post-selection—the wave function at the end of the universe, so to speak? Is there a natural $|out\rangle$ state too? Here we encounter a fundamental prejudice that runs through all of physics. Whereas it is "natural" to suppose that the universe started out in a "special" state, few physicists consider imposing a "natural" or "special" final condition on the universe. (Some exceptions are [6, 7]). The time asymmetry implicit in the discrimination between initial and final cosmological states has been the basis of much debate [8, 9, 11]. Attempts have been made [10] to construct explicitly time symmetric models of quantum cosmology, but these are considered little more than curiosities. However, in the context of weak values, there is no obligation when making a final state selection to impose time symmetry: the final state can be anything at all, so long as it is not orthogonal to the initial state evolved to the future spacetime boundary. One is therefore free to explore a number of "natural" final states in addition to the mirror image of the initial state. A proposal along these lines has been presented by Aharonov and Gruss [1].

The standard (asymmetric) treatment in quantum cosmology is to simply evolve the (simple, natural) $|in\rangle$ state into the (usually complicated) $|out\rangle$ state with the appropriate evolution operator:

$$|out\rangle = U(t_{out} - t_{in})|in\rangle \tag{7.3}$$

If (7.3) is substituted into Eq. (7.1), then the weak value reduces to the expectation value for observable C at time t . But suppose, within the context of the proposal to impose pre- and post-selection criteria independently, that

$$|out\rangle \neq U(t_{out} - t_{in})|in\rangle \tag{7.4}$$

then $w(t)$ will in general not coincide with $\langle C\rangle$ at time t. Indeed, it may differ from it dramatically in the case that condition (7.2) obtains.

By way of illustration, consider the two spacetime dimensional toy model of a scalar field with mass m propagating in an expanding universe with scale factor $a(t)$ and metric

$$ds^2 = dt^2 - a^2(t)dx^2 \tag{7.5}$$

Defining the conformal time η by

$$t = \int^t dt' = \int^\eta a(\eta')d\eta' \tag{7.6}$$

and the conformal scale factor C as

$$C(\eta) = a^2(\eta) \tag{7.7}$$

it follows that

$$ds^2 = a^2(\eta)(d\eta^2 - dx^2) \tag{7.8}$$

The specific choice

$$C(\eta) = A + B\tanh(\rho\eta); \quad A, B, \rho \text{ constants} \tag{7.9}$$

$$C(\eta) \to A \pm B, \quad \eta \to \pm\infty \tag{7.10}$$

describes an expanding universe that is asymptotically Minkowskian in both the far past and far future, but with different scale factors $A - B$ and $A + B$. The wave equation for the massive scalar field may be explicitly solved in terms of hypergeometric functions [12]. A complete set of modes is given by

$$\begin{aligned} u_k^{in}(\eta, x) &= (4\pi\omega_{in})^{-1}\exp\left(ikx - i\omega_+\eta - (i\omega_-/\rho)\ln[2\cosh(\rho\eta)]\right) \\ &\quad {}_2F_1\left(1 + (i\omega_-/\rho), i\omega_-/\rho; 1 - (i\omega_{in}/\rho); \tfrac{1}{2}(1 + \tanh\rho\eta)\right) \\ &\to (4\pi\omega_{in})^{-1/2}e^{ikx - i\omega_{in}\eta}, \quad \eta \to -\infty \end{aligned} \tag{7.11}$$

These modes coincide with standard exponential field modes in the asymptotic past. Therefore one may define a vacuum state using them. These modes will coincide with the standard quantum vacuum as $t \to -\infty$. Let us choose this vacuum state to be the "in" state of the field ($|in\rangle$) and denote it by $|0_{in}\rangle$. It corresponds to a quantum state in which there are no particles present in the "in" region (i.e. in the asymptotic past). Another complete set of field modes is given by

$$
\begin{aligned}
u_k^{out}(\eta, x) &= (4\pi\omega_{out})^{-1}\exp\left(ikx - i\omega_+\eta - (i\omega_-/\rho)\ln[2\cosh(\rho\eta)]\right) \\
&\quad {}_2F_1\left(1 + (i\omega_-/\rho), i\omega_-/\rho; 1 - (i\omega_{out}/\rho); \tfrac{1}{2}(1 + \tanh\rho\eta)\right) \\
&\quad \to (4\pi\omega_{out})^{-1/2} e^{ikx - i\omega_{out}\eta}, \quad \eta \to \infty
\end{aligned}
\tag{7.12}
$$

This second set of modes coincides with standard exponential field modes in the asymptotic future, $t \to \infty$, and may be used to define a second vacuum state $|0_{out}\rangle$, corresponding to a quantum state in which there are no particles present in the asymptotic future (the out region). These two sets of modes are not the same, so

$$
|0_{in}\rangle \neq |0_{out}\rangle. \tag{7.13}
$$

That is, the two vacuum states, $|0_{in}\rangle$ and $|0_{out}\rangle$, are inequivalent: each corresponds to a no-particle state in its respective region, as determined by the (null) response of an inertially-moving particle detector adiabatically switched on and off again in that region. If the quantum state of the universe is chosen to be the "in" vacuum $|0_{in}\rangle$, then although there will be no particles present in the "in" region, there will be a non-zero probability for particles to be present in the "out" region. Physically, (7.13) corresponds to particles being created by the expanding universe (see, for example, [4]). (Because we work in the Heisenberg representation and there are no interactions, the evolution is contained in the time-dependence of the field modes, so the operator U is trivial, and will be omitted in what follows.) The spectrum of created particles may readily be evaluated using the Bogoliubov transformation between the in and out modes

$$
u_k^{in}(\eta, x) = \alpha_k u_k^{out}(\eta, x) + \beta_k u_{-k}^{out*}(\eta, x) \tag{7.14}
$$

$$
\alpha_k = \left(\frac{\omega_{out}}{\omega_{in}}\right)^{1/2} \frac{\Gamma(1 - i\omega_{in}/\rho)\Gamma(-i\omega_{out}/\rho)}{\Gamma(-i\omega_+/\rho)\Gamma(1 - i\omega_+/\rho)} \tag{7.15}
$$

$$
\beta_k = \left(\frac{\omega_{out}}{\omega_{in}}\right)^{1/2} \frac{\Gamma(1 - i\omega_{in}/\rho)\Gamma(i\omega_{out}/\rho)}{\Gamma(i\omega_-/\rho)\Gamma(1 + i\omega_-/\rho)} \tag{7.16}
$$

$$
\alpha_{kk'} = \alpha_k\delta_{kk'}, \qquad \beta_{kk'} = \beta_k\delta_{-kk'} \tag{7.17}
$$

from which it follows that

$$
|\alpha_k|^2 = \frac{\sinh^2(\pi\omega_+/\rho)}{\sinh(\pi\omega_{in}/\rho)\sinh(\pi\omega_{out}/\rho)} \tag{7.18}
$$

$$
|\beta_k|^2 = \frac{\sinh^2(\pi\omega_-/\rho)}{\sinh(\pi\omega_{in}/\rho)\sinh(\pi\omega_{out}/\rho)} \tag{7.19}
$$

where the number of particles in mode k in the out region is given by Eq. (7.19).

In the foregoing treatment, the state $|0_{in}\rangle$ defines the pre-selection. In the Heisenberg representation, the state remains $|0_{in}\rangle$ throughout, including in the out region

$t \to \infty$. Thus, the choice of post-selected state is the same as the pre-selected state. However, we may readily generalize the treatment to the case where the post-selected state differs from the pre-selected state evolved forward to the "out" region. A natural choice of post-selected state is $|0_{out}\rangle$. We may then consider weak values of observables at times between in and out. For example, consider the particle number operator

$$N_i = \Sigma_i a_i^\dagger a_i, \tag{7.20}$$

where $a_i^\dagger$ and a_i are creation and annihilation operators respectively for particles in mode i. The weak value of this quantity is then given by

$$w_N = \langle 0_{out}|\Sigma_i a_i^\dagger(t) a_i(t)|0_{in}\rangle / \langle 0_{out}|0_{in}\rangle \tag{7.21}$$

To give w_N a well-defined meaning, the modes corresponding to the operators $a_i^\dagger$ and a_i need to be specified. One way to do this is to use the procedure of Hamiltonian diagonalization to define instantaneous modes and associated creation and annihilation operators at time t during the expansion [13]. The excitations of such modes are often referred to as quasiparticles. These modes will reduce to (7.11) and (7.12) in the in and out regions respectively. It therefore follows directly from Eq. (7.21) that w_N will reduce to zero in both the "in" region (because the state is $|0_{in}\rangle$ there) and the "out" region (because the state is $|0_{out}\rangle$ there). Observers measuring the weak values of instantaneous particle numbers present at time t will thus observe the values to rise from zero, peak around the time of maximum expansion rate, and fall towards zero again at late times. Explicit expressions for the weak values may be calculated from the mode functions and Bogoliubov transformations derived by Pavlov [13] using Hamiltonian diagonalization for a massive scalar field in an expanding universe.

The toy model serves to illustrate the key issue, within the theory of quantum fields propagating in a classical background spacetime, concerning weak measurements combined with post-selection in a cosmological context, namely, the choice of final state. In the example given, it seems very natural to choose the vacuum state in the "out" region, as well as the (different) vacuum state in the "in" region. In part this is due to the fact that the spacetime is asymptotically Minkowskian in both the far past and far future. In the context of the real universe, which is (presumably) not asymptotically Minkowskian, the choice is less clear. However, it is widely accepted that the very early universe was characterized by a period of inflation, when the spacetime was a very close approximation to de Sitter space, and for which a de-Sitter invariant vacuum state is a natural choice for the initial quantum state (a popular example is the so-called Bunch-Davies vacuum state; see [14]). In addition, it is now widely accepted that the universe is dominated by dark energy, and will approach a (different, slower) de Sitter like spacetime in the far future. Therefore, it is natural to choose the final state of the universe to be the corresponding de Sitter-invariant vacuum too. If the quantum field of interest is a conformally invariant free field, then the out vacuum is merely the in vacuum evolved into the out region, and quantum post-selection will be trivial, that is, weak values will coincide with expectation values. In general, however, the fields will not be a conformally invariant, and will involve particle-creating interactions, so the in and out vacuum states will not be equivalent. In that case quantum post-selection of a de Sitter vacuum will result in non-trivial large weak values. Depending on the specifics, there could be weak values at the present cosmological epoch (corresponding to a small denominator in Eq. (7.1)). These would show up as cosmological anomalies—what Aharonov has referred to as "quantum miracles".

To explore these quantum miracles further, I shall consider the case of anomalies in the gravitational field. To investigate this, we need to consider the back-reaction of the quantum fields on the gravitational field of the universe, in the case that we impose both pre- and post-selection on the quantum state.

7.3 Gravitational Back-reaction

In a semi-classical theory in which the background gravitational field is treated classically, and the quantum state is both pre- and post-selected, the appropriate source term to place on the right hand side of the gravitational field equations will be the weak value of the stress-energy-momentum tensor $T_{\mu\nu}$. In the case we choose vacuum states in the "in" and "out" regions, denoted in what follows by $|0_{in}\rangle$ and $|0_{out}\rangle$, the gravitational equation will be

$$G_{\mu\nu} + \text{higher order terms in curvature} = \langle 0_{out}|T_{\mu\nu}|0_{in}\rangle / \langle 0_{out}|0_{in}\rangle \tag{7.22}$$

where $G_{\mu\nu}$ is the Einstein tensor and the source term on the right hand side of Eq. (7.22) is seen to be the weak value of $T_{\mu\nu}$. Equation (7.22) was originally derived by DeWitt by adapting the Schwinger effective action theory of quantum electrodynamics to the gravitational case [15]. The effective action is defined as

$$W = i \ln(\langle 0_{out}|0_{in}\rangle) \tag{7.23}$$

from which it follows by variation of W with respect to the metric tensor $g^{\mu\nu}$ that

$$\frac{2}{(-g)^{1/2}} \frac{\delta W}{\delta g^{\mu\nu}} = \frac{\langle 0_{out}|T_{\mu\nu}|0_{in}\rangle}{\langle 0_{out}|0_{in}\rangle} \tag{7.24}$$

Applications of Eq. (7.22) were widespread in the literature in the 1970s (see, for example, [16–18]). In the case that there is pre-selection (of state $|0_{in}\rangle$) but no post-selection, the appropriate source term to use on the right hand side of the semi-classical gravitational field equations is the expectation value $\langle 0_{in}|T_{\mu\nu}|0_{in}\rangle$:

$$G_{\mu\nu} + \text{higher order terms in curvature} = \langle 0_{in}|T_{\mu\nu}|0_{in}\rangle \tag{7.25}$$

(see, for example, [4]). It is important to note that the source terms on the right hand sides of both Eqs. (7.22) and (7.25) are formally divergent and must be renormalized [4]. It is well known that the divergent terms of both expressions are identical. However, the finite residue will in general differ, and in the case that the denominator of the right hand side of Eq. (7.22) is very small, this difference could be extremely large. The upshot is that the cosmological gravitational field of a universe in which both the initial and final states are selected could be dramatically different from one in which only the initial state is selected, the difference amounting to a gravitational quantum miracle, and being perceived by an observer as a major departure in the cosmological dynamics from what might be expected on the basis of a natural initial state alone (such as the aforementioned Bunch-Davies vacuum of inflationary cosmology). On a historical note, there was much confused debate in the 1970s about which source term, (7.22) or (7.25), to use in the semi-classical gravitational field equations. Some leading researchers (for example, [16–19]) advocated what we would now term the weak value $\langle 0_{out}|T_{\mu\nu}|0_{in}\rangle / \langle 0_{out}|0_{in}\rangle$ derived from the effective action. However, the use of the expectation value $\langle 0_{in}|T_{\mu\nu}|0_{in}\rangle$ eventually prevailed. With the benefit of hindsight, we can now see that both camps were correct. When both pre- and post-selection are involved, the weak value is indeed the appropriate source term, but if there is only pre-selection then the expectation value should be used.

The (finite) difference between the weak value and the expectation value may be formally cast in terms of the Bogoliubov coefficients α and β between the states $|0_{in}\rangle$ and $|0_{out}\rangle$, as follows:

$$\langle 0_{out}|T_{\mu\nu}|0_{in}\rangle / \langle 0_{out}|0_{in}\rangle - \langle 0_{in}|T_{\mu\nu}|0_{in}\rangle = -i\,\Sigma_{i,j}\Lambda_{ij}T_{\mu\nu}(u^*_{in,i}, u^*_{in,j}) \quad (7.26)$$

where $T_{\mu\nu}(u^*_{in,i}, u^*_{in,j})$ is the bilinear expression for the stress-energy-momentum tensor, with the field amplitudes replaced by the mode expression for the "in" region (e.g. Eq. (7.11)), while

$$\Lambda_{ij} = -i\,\Sigma_k \beta_{kj}\alpha^{-1}_{ik} \quad (7.27)$$

[19]. There is an extensive literature on how to calculate both Bogoliubov transformations and renormalized expectation values $\langle 0_{in}|T_{\mu\nu}|0_{in}\rangle_R$ for a variety of quantum fields in a large number of cosmological models ([4], and references contained therein). Equation (7.26) can then in principle be used to calculate the corresponding weak values. For example, for the two dimensional expanding universe model discussed in Sect. 7.2, the renormalized expectation value of the stress-energy-momentum tensor was worked out by Bernard and Duncan [12]. In that example, the Bogoliubov coefficients are given by Eqs. (7.15) and (7.16). In practice, obtaining explicit expressions for renormalized stress tensors and Bogoliubov coefficients is very hard, and the manipulations involved in (7.26) and (7.27) would need to be performed numerically. However, an order of magnitude estimate may be given on dimensional grounds. For example, in the model of Sect. 7.2, significant particle production will take place only for large values of ρ, i.e. for rapidly expanding spacetimes, as one may infer from Eq. (7.19). On general grounds, one would also expect most quantum miracles to be of the same order of magnitude. In the real universe, particle production by the expanding universe is negligible at this epoch (typically one particle per Hubble volume per Hubble time), so gravitational anomalies resulting from imposing a vacuum post-selection condition on an otherwise free field are likely to be many orders of magnitude below observability. However, in the very early universe, when the rate of expansion was very high, quantum post-selection might lead to significant effects. For example, the graviton background

generated in the very early universe would propagate freely to $t \to \infty$, so imposing a graviton vacuum post-selection condition would produce a large change in the gravitational source term at early times. In the case of interacting fields, such as the electromagnetic field coupled to charged matter, the resulting particle production at the present epoch is very high, and although the situation is more complicated and harder to analyze, it seems likely that a vacuum condition imposed at $t \to \infty$ would produce significant effects even at the present epoch. In Sect. 7.5, I give one suggestion for how such an effect might manifest itself.

7.4 Quantum Miracle

As a simple illustration of a gravitational quantum miracle, consider a massless scalar field propagating in Minkowski space with pre-selected state

$$|\psi_{in}\rangle = \alpha|0\rangle + \beta|1_{\mathbf{k}}\rangle \tag{7.28}$$

for some momentum $\mathbf{k}$, and post-selected state

$$|\psi_{out}\rangle = \gamma|0\rangle + \delta|1_{\mathbf{k}}\rangle \tag{7.29}$$

where

$$|\alpha|^2 + |\beta|^2 = |\gamma|^2 + |\delta|^2 = 1. \tag{7.30}$$

The weak value of the total particle number operator N (see Eq. (7.20)) is then readily calculated:

$$w_N \equiv \langle\psi_{out}|N|\psi_{in}\rangle / \langle\psi_{out}|\psi_{in}\rangle = \beta\delta/(\alpha\gamma + \beta\delta) \tag{7.31}$$

It is easy to choose values of the coefficients $\alpha, \beta, \gamma, \delta$ to yield weird weak values for N (quantum miracles). For example, the choice $\alpha = \sqrt{3}/2, \beta = 1/2, \gamma = 2/\sqrt{7}, \delta = -\sqrt{3/7}$ gives

$$w_N = -1. \tag{7.32}$$

The concept of a negative particle number is clearly analogous to the three box problem discussed, for example in [2]. It is certainly very strange, but in the present example it attains a clear physical meaning when we calculate the weak values of $T_{\mu\nu}$ for the "in" and "out" states (7.28) and (7.29):

$$\langle\psi_{out}|T_{\mu\nu}|\psi_{in}\rangle / \langle\psi_{out}|\psi_{in}\rangle = \left(\alpha\gamma\langle 0|T_{\mu\nu}|0\rangle + \beta\delta\langle 1|T_{\mu\nu}|1\rangle\right) / (\alpha\gamma + \beta\delta). \tag{7.33}$$

The right hand side of Eq. (7.33) is divergent. Because the field is propagating in Minkowski space, the divergence is readily renormalized by subtracting $\langle 0|T_{\mu\nu}|0\rangle$, this being the standard vacuum expectation value for the field (which is conventionally set to zero). The finite part is then given, for the energy density and pressure components, by

$$\langle\psi_{out}|T_{00}|\psi_{in}\rangle_R / \langle\psi_{out}|\psi_{in}\rangle = \beta\delta\omega/(\alpha\gamma + \beta\delta) = -\omega \tag{7.34}$$

and

$$\langle\psi_{out}|T_{11}|\psi_{in}\rangle_R/\langle\psi_{out}|\psi_{in}\rangle = \beta\delta\mathbf{k}/(\alpha\gamma+\beta\delta) = -\mathbf{k} \tag{7.35}$$

where R denotes renormalized. The weak value for the energy density is negative, as is the weak value of the pressure component, which is a measure of the momentum of the state. A negative momentum implies that if a reflecting boundary (mirror) were introduced, then (the weak value of) the recoil of the boundary would also be negative, i.e. a weak measurement of the momentum of the mirror would indicate a shift toward the source of the incoming beam rather than away from it. If Eq. (7.33) renormalized is used as the source term in the gravitational field equations, Eq. (7.22), it will exert a negative gravitational effect, so that a weak measurement of the gravitational force would show a repulsion.

A dramatic example of what one might now call a gravitational quantum miracle has been known for nearly four decades, and was discovered by Boulware [16], who calculated the vacuum energy of a massless scalar field round a black hole using the Schwinger-DeWitt "in-out" effective action formalism, so that there is no radiation emanating from the black hole. That is, the "out" state is post-selected to be a quantum vacuum. (This contrasts with Hawking's treatment [20] in which the out state corresponds to a steady flux of thermal radiation—the so-called Hawking effect.) Boulware's treatment implies that the corresponding stress tensor, $\langle 0_{out}|T_{\mu\nu}|0_{in}\rangle/\langle 0_{out}|0_{in}\rangle$, diverges on the black hole event horizon, producing an infinite gravitational back-reaction [4]. Thus, imposing a vacuum post-selection condition on the universe could imply a major modification to the structure of black holes.

7.5 Experimental Cosmology

The possibility that the quantum state of the universe might be both pre- and post-selected represents a radical departure from standard theory, and one may legitimately wonder about the possibility of experimental testing. One way is to search for gravitational and non-gravitational quantum miracles in cosmological data. Another is to perform an experiment. In this section I will describe an adaptation of an experiment performed by Partridge as a test of the time-symmetric Wheeler-Feynman theory of electrodynamics [21].

Suppose it is the case, as proposed at the end of Sect. 7.2, that the end state of the universe is a de Sitter vacuum state. Let us further suppose that such a state applies to the electromagnetic field. Then, irrespective of the CMB, and of the myriad electromagnetic interactions that are currently taking place, and will subsequently take place, throughout the universe, no photon will survive into the asymptotic future, $t \to \infty$. If it were the case that cosmological material along the future light cone of the universe absorbed all emitted photons, then the future vacuum condition would have no discernible effect on electromagnetic phenomena at the current epoch. However, it is known to be the case that the density of matter in the universe is too low for all photons emitted at our current epoch to eventually be absorbed; that is, our future light cone is not a complete absorber of photons [22, 23]. If a laser beam is directed to an empty part of the night sky, away from the galactic plane, then there is not enough matter along the line of sight of the beam to absorb all the photons, a state of affairs enhanced by the apparent accelerating expansion of the universe that serves to further dilute the density of absorbing material along the future light cone. Consider, as postulated, that the final state of the universe is an electromagnetic vacuum. Propagate the modes of this field back in time from $+\infty$ to the present epoch. Some modes will encounter matter, for example, in the interstellar medium. But, given the non-opaqueness of our future light cone, a subset of modes will travel relatively undisturbed back in time to encounter the laser. (There will be no creation of photons by the expanding universe in this case on account of the fact that the electromagnetic field is conformally invariant and a Friedmann-Roberston-Walker universe is conformally flat.) Therefore, the laser will not create photons in those modes of the electromagnetic field because, by post-selection, they have photon occupation number 0. To an experimenter, it will appear that a laser directed to an empty part of the sky will suffer a lower power drain than a laser directed at an absorbing surface. The degree of reduction will depend on the opacity of the future light cone. In the ideal case of perfect transparency, the laser would emit no energy at all in the relevant direction. In practice, of course, the effect is likely to be small, owing to the spreading of the laser beam, and hard to discern, given the gross nature of assessing the power output of a laser by measuring its power input. Nevertheless, an experiment of this sort was performed by Partridge [21] using microwaves rather than a laser, with null results.

A more refined experiment to test the post-selected vacuum hypothesis would be to use pairs of entangled photons, which I collectively label A and B, from a continuously pumped source. Photons B are directed to a counter while their entangled partners A, traveling in the opposite direction, are permitted to leave the apparatus undisturbed. In the first part of the experiment, the escaped photons A are intercepted by an absorber (such as a black screen on the other side of the laboratory). Maintaining this configuration, a count rate is established for photons B which, by reason of the entanglement, also establishes the count rate for the emission of photons A. Next, the screen is retracted and photons A are permitted to escape into the sky, at a variety of orientations away from any obvious absorbing material (such as the dust of the Milky Way), and the count rate of photons B measured again. A test of the post-selection hypothesis is that the count rate of photons B in the latter configuration would be less than in the former, because the photon source cannot excite modes of the electromagnetic field that propagate undisturbed to the postulated vacuum state at $t \to \infty$. In a perfectly transparent universe with vacuum post-selection, no photons B would ever be detected in the latter configuration. Cosmology is an observational science, but this experiment, along with Partridge's original, constitutes a genuine exercise in experimental cosmology.

Acknowledgements I have greatly benefited from discussions with Alonso Botero, Jeff Tollaksen and Yakir Aharonov in preparing this paper. I would like to thank Katherine Lee and Saugata Chatterjee for their help reformatting and editing the paper.

References

1. Y. Aharonov, E. Gruss, Two-time interpretation of quantum mechanics (2005). arXiv: quant-ph/0507269
2. Y. Aharonov, D. Rohrlich, *Quantum Paradoxes* (Wiley-VCH, Weinheim, 2005)
3. J. Hartle, S. Hawking, Wave function of the universe. Phys. Rev. D **28**(12), 2960 (1983)
4. N.D. Birrell, P.C.W. Davies, *Curved Space. Quantum Fields* (Cambridge University Press, Cambridge, 1982)
5. P.C.W. Davies, C.H. Lineweaver, M. Ruse (eds.), *Complexity and the Arrow of Time* (Cambridge University Press, Cambridge, 2013)
6. F. Hoyle, J.V. Narlikar, Electrodynamics of direct interparticle action. I. The quantum mechanical response of the universe. Ann. Phys. **54**, 207–239 (1969)
7. F.J. Tipler, Cosmological limits on computation. Int. J. Theor. Phys. **25**(6), 617–661 (1986). doi:10.1007/BF00670475
8. S.W. Hawking, R. Penrose, *The Nature of Space and Time* (Princeton University Press, Princeton, 1996)
9. H. Price, *Time's Arrow and Archimedes' Point: New Directions for the Physics of Time* (Oxford University Press, New York, 1996)
10. M. Gell-Mann, J.B. Hartle, Complexity, entropy and the physics of information, in *Time Symmetry and Asymmetry in Quantum Mechanics and Quantum Cosmology, vol. VIII*, ed. by W.H. Zurek (Addison-Wesley, Reading, 1990), p. 425
11. D.N. Page, No time asymmetry from quantum mechanics. Phys. Rev. Lett. **70**, 4034–4037 (1993)
12. C. Bernard, A. Duncan, Regularization and renormalization of quantum field theory in curved space-time. Ann. Phys. **107**, 201–222 (1977)
13. Yu.V. Pavlov, Nonconformal scalar field in a homogeneous isotropic space and the method of Hamiltonian diagonalization (2000). gr-qc/0012082
14. T.S. Bunch, P.C.W. Davies, Quantum field theory in de Sitter space: renormalization by point-splitting. Proc. R. Soc. Lond. Ser. A **360**, 117 (1978)
15. B.S. DeWitt, *Dynamical Theory of Groups and Fields* (Gordon and Breach, New York, 1965)
16. D.G. Boulware, Quantum field theory in Schwarzschild and Rindler spaces. Phys. Rev. D **11**, 1404–1424 (1975)
17. J.B. Hartle, B.L. Hu, Quantum effects in the early universe. II. Effective action for scalar fields in homogeneous cosmologies with small anisotropy. Phys. Rev. D **20**, 1772–1782 (1979)
18. J.S. Dowker, R. Critchley, Effective Lagrangian and energy-momentum tensor in de Sitter space. Phys. Rev. D **13**, 3224–3232 (1976)
19. B.S. DeWitt, Phys. Rep. **19C**, 297 (1975)
20. S.W. Hawking, Particle creation by black holes. Commun. Math. Phys. **43**, 199–220 (1975)
21. R.B. Partridge, Absorber theory of radiation and the future of the universe. Nature **244**, 263–265 (1973)
22. P.C.W. Davies, Is the universe transparent or opaque? J. Phys. A, Gen. Phys. **5**, 1722–1737 (1972)
23. P.C.W. Davies, J. Twamley, Time-symmetric cosmology and the opacity of the future light cone. Class. Quantum Gravity **10**, 931 (1993). doi:10.1088/0264-9381/10/5/011

Chapter 8
The Quantum Mechanical Arrows of Time

James B. Hartle

Abstract The familiar textbook quantum mechanics of laboratory measurements incorporates a quantum mechanical arrow of time—the direction in time in which state vector reduction operates. This arrow is usually assumed to coincide with the direction of the thermodynamic arrow of the quasiclassical realm of everyday experience. But in the more general context of cosmology we seek an explanation of all observed arrows, and the relations between them, in terms of the conditions that specify our particular universe. This paper investigates quantum mechanical and thermodynamic arrows in a time-neutral formulation of quantum mechanics for a number of model cosmologies in fixed background spacetimes. We find that a general universe may not have well defined arrows of either kind. When arrows are emergent they need not point in the same direction over the whole of spacetime. Rather they may be local, pointing in different directions in different spacetime regions. Local arrows can therefore be consistent with global time symmetry. [*Editors note*: for a video of the talk given by Prof. Hartle at the Aharonov-80 conference in 2012 at Chapman University, see quantum.chapman.edu/talk-15.]

8.1 Introduction

In his 1932 book [1] von Neumann summarized the quantum mechanics of a subsystem of the universe that is sometimes measured but otherwise isolated. Two laws of evolution for the quantum state of the subsystem were postulated. The first is the Schrödinger equation that specifies how the state evolves in time when the subsystem is isolated:

$$i\hbar \frac{d|\hat{\psi}(t)\rangle}{dt} = H|\hat{\psi}(t)\rangle \qquad \text{(I)}. \tag{8.1}$$

J.B. Hartle
Santa Fe Institute, Santa Fe, NM 87501, USA

J.B. Hartle (✉)
Department of Physics, University of California, Santa Barbara, CA 93106-9530, USA
e-mail: hartle@physics.ucsb.edu

D.C. Struppa, J.M. Tollaksen (eds.), *Quantum Theory: A Two-Time Success Story*,
DOI 10.1007/978-88-470-5217-8_8,

The second law specifies how the state evolves when an 'ideal measurement' is carried out on the subsystem at time t_m. It is

$$|\hat{\psi}(t_m)\rangle \rightarrow \frac{s|\hat{\psi}(t_m)\rangle}{\| s|\hat{\psi}(t_m)\rangle \|} \quad \text{(II)} \tag{8.2}$$

where s is the projection onto the measurement outcome.[1]

The Schrödinger equation (I) is time reversible—it can be run both forward and backward in time. The second law of evolution (II) is not reversible. It operates only forward in time. That defines the quantum mechanical arrow of time.

It is commonly assumed that the quantum arrow of time coincides with the thermodynamic arrow defined by the direction in which total entropy is increasing. This identification of a fundamental quantum arrow with a classical one must have seemed natural in a theory which posited separate classical and quantum worlds with a kind of movable boundary between them. A second law describing an "irreversible act of amplification" from the quantum world to the classical one in a measurement was naturally connected with the second law of thermodynamics describing more general classical irreversible processes.

A thermodynamic arrow of time is not an inevitable feature of a classical world like ours governed by time-neutral dynamical laws. The fact that presently isolated subsystems are mostly evolving towards equilibrium in the same direction in time cannot be a consequence of time-neutral dynamical laws. Rather our thermodynamic arrow arises because the initial state of our universe is such that the progenitors of today's isolated subsystems were all far out of equilibrium a long time ago. As Boltzmann wrote over a century ago: "The second law of thermodynamics can be proved from the [time-reversible] mechanical theory, if one assumes that the present state of the universe... started to evolve from an improbable [i.e. special] state" [2]. Our thermodynamic arrow of time is an emergent feature of the particular initial condition of our universe.

Is the quantum mechanical arrow of time a fundamental property of quantum mechanics or can it also be seen as an emergent feature of our universe in a more general formulation of quantum mechanics free from arrows of time? Is a quantum mechanical arrow of time always codirectional with a thermodynamic arrow? Do arrows always point in one direction over the whole of spacetime or can they point in different directions in different spacetime regions? Such questions are the subject of this essay. To answer them, as Boltzmann noted, we are naturally led to cosmology.

It is almost certain that there will not be a thermodynamic arrow of time that points consistently in one direction over the whole of spacetime in the vast universes contemplated by contemporary inflationary cosmology. Rather the thermodynamic arrow may point in different directions in different regions of spacetime. A simple example that we will discuss in this paper is illustrated in Fig. 8.1.

The figure shows a curved two-dimensional slice of a four-dimensional homogeneous and isotropic cosmological spacetime embedded in a (Lorentz signatured)

---

[1] von Neumann called I and II, 2 and 1 respectively, but today the given convention is more used.

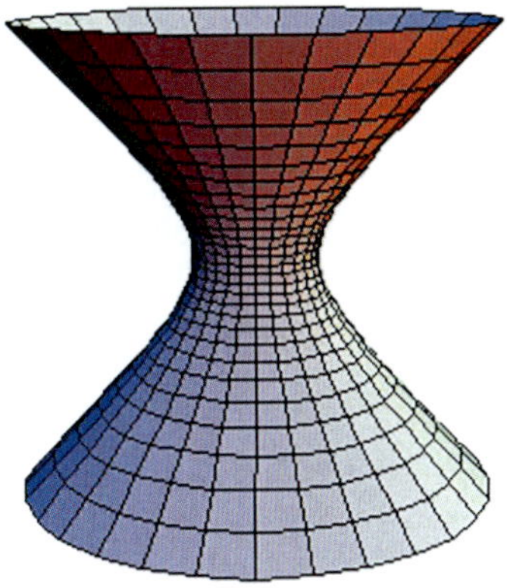

Fig. 8.1 A bouncing universe. The figure shows the geometry of a two-dimensional slice of a four-dimensional cosmological bouncing spacetime embedded in a Lorentz signatured three-dimensional flat space. Time is up. The universe is spatially closed. It has a large spatial volume at large negative times, collapses to smaller and smaller volumes until a minimum is reached (the bounce), and then expands to larger volumes at positive times. For simplicity we have assumed that the contraction and expansion are time symmetric. DeSitter space is a well known example

three-dimensional flat space. The universe begins at large radii at the bottom, contracts to smaller and smaller radii, bounces at a minimum radius and reexpands towards the top. DeSitter space is a classic example of such a bouncing spacetime.

Such bouncing geometries are among the classical spacetimes predicted by the no-boundary quantum state of the universe (NBWF) in simple models using a framework that includes quantum gravity [3–5]. In addition to predicting the behavior of homogeneous and isotropic classical backgrounds the NBWF also predicts the behavior of the quantum fluctuations in matter and geometry away from these symmetries. The key result for the present discussion is the following: The NBWF predicts that fluctuations are in a state of low excitation (i.e. small) at the bounce. On either side they grow away from the bounce, become classical, collapse gravitationally, and eventually create a large scale structure of galaxies, stars, and planets. The thermodynamic arrow of time therefore points in opposite directions on opposite sides of the bounce. The universe has two spacetime regions with opposing thermodynamic arrows. We can then ask: What are the laws of quantum mechanics for the quantum fluctuations? Do they have a quantum arrow of time, and which way does it point in the two regions?

We will answer these questions in the context of a family of generalizations of textbook quantum theory that are time-neutral—not preferring one time direction over another and without any built in quantum arrows of time. In a pioneering paper Aharonov, Bergmann, and Lebovitz [6] showed how to use initial and final conditions to construct such a time-neutral quantum mechanics of measured subsystems. We will answer these questions, not in this context, but in the more general time-neutral quantum mechanics of closed systems that are suitable for cosmology [7].

In a time-neutral formulation no arrows of any kind are built in. Arrows of time of a particular universe emerge from the conditions that specify it. As we have discussed, the thermodynamic arrow of time in our universe emerges from a special initial condition (and a final condition of indifference). We will show that quantum mechanical arrows of time can emerge in a similar way.

For the simple example illustrated in Fig. 8.1, we will find that in a suitable generalization of quantum mechanics there is both a quantum and thermodynamic arrow of time pointing away from the bounce on each side. The overall situation is time symmetric.

More generally we will conclude that quantum mechanical arrows of time are not an inevitable feature of quantum mechanics. Quantum mechanics can be formulated without them. The arrows of time that characterize the approximate quantum mechanics of measured subsystems obeying laws I and II in our universe arise in particular spacetime regions from the conditions that specify the universe and the region.

The paper is organized as follows: In Sect. 8.2 we review the work of Aharonov, Bergmann, and Lebovitz [6]. Section 8.3 introduces a time-neutral quantum mechanics of closed systems with initial and final conditions. All arrows of time arise from asymmetries between these two. Section 8.4 introduces the class of generalized quantum theories of which the one in Sect. 8.3 is but one example. Section 8.5 constructs a time-neutral generalized quantum theory for the quantum fluctuations in a bouncing universe illustrated in Fig. 8.1. Section 8.6 draws some brief conclusions.

8.2 Time-Neutrality in the Quantum Mechanics of Measured Subsystems

In a seminal paper Aharonov, Bergmann, and Lebovitz [6] showed how the quantum mechanics of measured subsystems could be formulated without an intrinsic arrow of time by allowing for final states as well as initial ones. We summarize the essence of their argument here in a notation that is analogous to that we will use for cosmology in subsequent sections.

Consider a subsystem of the universe whose states are vectors in a Hilbert space $\mathcal{H}_{\rm sub}$. Alternatives at a moment of time can be reduced to a set of yes/no questions. For instance asking for the position of a particle is equivalent to taking an exhaustive set of position intervals and asking whether the particle is in the first interval (yes or no), then the second interval (yes or no), etc. In the Heisenberg picture such yes/no alternatives at a time t are represented by an exhaustive set of exclusive projection operators $\{s_\alpha(t)\}$. For instance, the these might be projections onto an exhaustive set of ranges of position as discussed above. These operators satisfy

$$s_\alpha(t)s_{\alpha'}(t)=\delta_{\alpha\alpha'}s_\alpha(t), \qquad \sum_\alpha s_\alpha(t)=I. \tag{8.3}$$

showing that they are projections, that they are exclusive, and that they are exhaustive. The projection operators representing the same alternative at two different times t_1 and t_2 are connected by the Heisenberg equations of motion

$$s_\alpha(t_2) = e^{\frac{+ih(t_2-t_1)}{\hbar}} s_\alpha(t_1) e^{\frac{-ih(t_2-t_1)}{\hbar}}, \tag{8.4}$$

where h is the Hamiltonian of the subsystem in isolation.

Suppose a sequence of ideal measurements[2] is carried out on the subsystem by another subsystem at a sequence of times $t_1, t_2, \dots t_n$. The measurements are described by a sequence of *sets* of projections $\{s^k_{\alpha_k}(t_k)\}$, $k = 1, 2, \dots n$. The upper index allows for the contingency that the measurements might be of different quantities at different times—a measurement of position at time t_1, of momentum at time t_2, etc.

Suppose that the subsystem is in a (Heisenberg picture) state $|\psi_i\rangle$ in $\mathscr{H}$. Then the joint probability for a history of outcomes $\alpha_1, \alpha_2, \dots \alpha_n$ is [8, 9]

$$p(\alpha_n, \dots, \alpha_1) = \| s^n_{\alpha_n}(t_n) \dots s^1_{\alpha_1}(t_1)|\psi_i\rangle \|^2. \tag{8.5}$$

It is easy to work out that this compact formula for the joint probability for a sequence of ideal measurement outcomes follows from the two laws of evolution (8.1) and (8.2)—evolve, reduce, evolve, reduce, evolve.... The formula can be made more compact by defining $\alpha \equiv (\alpha_1, \alpha_2, \dots \alpha_n)$ and

$$c_\alpha \equiv s^n_{\alpha_n}(t_n) \dots s^1_{\alpha_1}(t_1). \tag{8.6}$$

Then

$$p(\alpha) = \|c_\alpha|\psi_i\rangle\|^2 = \| s^n_{\alpha_n}(t_n) \dots s^1_{\alpha_1}(t_1)|\psi_i\rangle \|^2. \tag{8.7}$$

A quantum mechanical arrow of time is manifest in (8.5) and (8.7). On one end of the chain of projections there is the state, and on the other end there is nothing.[3] Aharonov, Bergmann, and Lebovitz noticed that if one added a final state ψ_f corresponding to post-selection then the formula for the probabilities becomes

$$p(\alpha) = N|\langle\psi_f|c_\alpha|\psi_i\rangle|^2, \quad N \equiv |\langle\psi_f|\psi_i\rangle|^{-2}. \tag{8.8}$$

These formulae are symmetric in the initial and final states, in particular one can write (8.8) as

$$p(\alpha) = N|\langle\psi_f|c_\alpha|\psi_i\rangle|^2 = N|\langle\psi_i|c^\dagger_\alpha|\psi_f\rangle|^2. \tag{8.9}$$

That is, the probabilities are the same when the order of the projections is reversed and the notion of initial and final interchanged.

[2]An ideal measurement, sometimes called a projective measurement, is one that disturbs the subsystem as little as possible so that after the measurement its (Schrödinger picture) state is given by (8.2).

[3]The quantum mechanical arrow of time does not arise from the time-ordering of the projections. That could be reversed by a CPT transformation since field theory is invariant under CPT. But there would still be the state on one end and nothing on the other.

From this perspective, the quantum mechanical arrow of time arises from not specifying a final state. As Aharonov and Rohrlich say [10], "By imposing an initial and not a final condition we have already sent the arrow of time flying."

8.3 A Time-Neutral Formulation of the Quantum Mechanics of Closed Systems

8.3.1 A Model Quantum Universe

Cosmology provides not only the most general context for a discussion of arrows of time but also the most relevant one. That is because the observed arrows operate on cosmological scales and can be explained by cosmological conditions. For instance, as far as we know, the thermodynamic arrow of time extends over the whole of the visible universe and holds from the time of the big bang to the distant future. The evidence of the observations is that the universe was more ordered earlier than now and that disorder has been increasing ever since [11, 12]. That is the thermodynamic arrow of time. Similarly the electromagnetic arrow—the retardation of electromagnetic radiation—arises because the early universe has very little free electromagnetic radiation that today would be at readily accessible wavelengths [13]. The psychological arrow of time can be seen to follow from the other two [13].

To keep the discussion manageable, we consider a closed quantum system in the approximation that gross quantum fluctuations in the geometry of spacetime can be neglected.[4] closed system can then be thought of as a large (say $\gtrsim$ 20,000 Mpc), perhaps expanding, box of particles and fields in a fixed, flat, background spacetime (Fig. 8.2). Everything is contained within the box, in particular galaxies, planets, observers and observed, measured subsystems, and any apparatus that measures them. This is the most general physical context for prediction.

8.3.2 Time-Neutral Decoherent Histories Quantum Theory

The quantum mechanics of this model universe is formulated in a Hilbert space $\mathcal{H}$ that is vastly larger than the Hilbert space of any isolated subsystem it contains. However, the kinematics of the prediction of probabilities for histories bears many similarities to the quantum mechanics of measured subsystems as presented in the previous section.

The most general objective of a quantum mechanics of the universe is the prediction of the probabilities for sets of alternative coarse-grained time histories of its

[4]For the further generalizations that are needed for quantum spacetime see *e.g.* [14, 15]. For discussions of the arrows of time in contexts that include quantum spacetime see [5] and the earlier references therein especially [16].

Fig. 8.2 A simple model of a closed quantum system is a universe of quantum matter fields inside a large closed box (say, 20,000 Mpc on a side) with fixed flat spacetime inside. Everything is a physical system inside the box—galaxies, stars, planets, human beings, observers and observed, measured and measuring. The most general objectives for prediction are the probabilities of the individual members of decoherent sets of alternative coarse grained histories that describe what goes on in the box. That includes histories describing any measurements that take place there. There is no observation or other intervention from outside

contents. Alternatives at one moment of time are described by an exhaustive set of exclusive projection operators $\{P_\alpha(t)\}$ acting in $\mathscr{H}$. These satisfy [cf. (8.3)]

$$P_\alpha(t)P_{\alpha'}(t) = \delta_{\alpha\alpha'}P_\alpha(t), \quad \sum_\alpha P_\alpha(t) = I. \tag{8.10}$$

A set of alternative coarse-grained histories is specified by a sequence of such sets at a series of times $t_1, t_2, \ldots t_n$. An individual history corresponds to a particular sequence of events $\alpha \equiv (\alpha_1, \alpha_2, \ldots, \alpha_n)$ and is represented by the corresponding chain of projections:

$$C_\alpha \equiv P^n_{\alpha_n}(t_n)\ldots P^1_{\alpha_1}(t_1). \tag{8.11}$$

An immediate consequence of this and (8.10) is that

$$\sum_\alpha C_\alpha = I, \tag{8.12}$$

showing that the set of histories is exhaustive.

This description of histories is analogous to those in the quantum mechanics of measured subsystems [cf. (8.6)]. However, there are at least two crucial differences. First, there is no posited separate classical world. It's all quantum. Second, the alternatives represented by the P's are not restricted to measurement outcomes. They might, for example, refer to the orbit to the Moon when no one is looking at it, or to

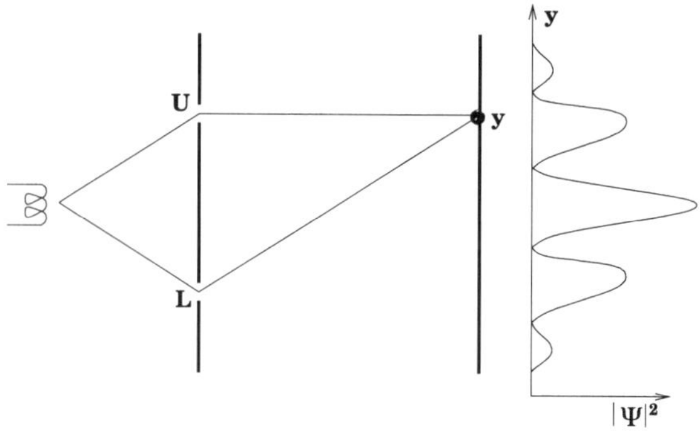

Fig. 8.3 The two-slit experiment. An electron gun at left emits an electron which is detected at a point y on a screen after passing through another screen with two slits. Because of quantum interference, it is not possible to assign probabilities to the alternative histories in which the electron arrives at y having gone through the upper or lower slit. The probability to arrive at y should be the sum of the probabilities of the two histories. But in quantum mechanics probabilities are squares of amplitudes and $|\psi_L(y) + \psi_U(y)|^2 \neq |\psi_L(y)|^2 + |\psi_U(y)|^2$. In a different physical situation where the electron interacts with apparatus that measures which slit it passed through, then quantum interference is destroyed and consistent probabilities can be assigned to these histories

the magnitude of density fluctuations in the early universe when there were neither observers nor apparatus to measure them. Laboratory measurements can of course be described in terms of correlations between two particular kinds of subsystems of the universe—one being measured the other doing the measuring. But laboratory measurements play no central role in formulating the theory, and are just a small part of what it can predict.[5]

A time-neutral decoherent histories quantum mechanics of our model universe with both initial and final conditions was formulated by Gell-Mann and the author in [7]. The formula for the probabilities for histories is

$$p(\alpha) = NTr(\rho_f C_\alpha \rho_i C_\alpha^\dagger), \quad N^{-1} \equiv Tr(\rho_f \rho_i). \tag{8.13}$$

Here, ρ_i and ρ_f are density matrices representing initial and final conditions. There is a clear analogy with (8.8). This expression is time-neutral because initial and final density matrices can be interchanged and the order of times in the C_α's reversed using the cyclic property of the trace.

However, (8.13) does not supply probabilities for all sets of alternative histories. The resulting probabilities might not be consistent with the usual sum rules of probability theory. Generally probabilities cannot be assigned to interfering alternatives in quantum theory. The two-slit experiment described in Fig. 8.3 is a simple example.

---

[5]Indeed, the quantum mechanics of measured subsystems in Sect. 8.2 is an approximation appropriate for measurement situations to the more general quantum mechanics of closed systems. See, e.g. [17] Sect. II.10.

In decoherent (or consistent) histories quantum theory probabilities are assigned only a set of alternative histories if the quantum interference between members of the set is negligible as a consequence of the initial and final conditions and the dynamics. The measure of quantum interference is provided by the decoherence functional:

$$D(\alpha, \alpha') = NTr(\rho_f C_\alpha \rho_i C_{\alpha'}{}^\dagger) \tag{8.14}$$

where $N^{-1} = Tr(\rho_f \rho_i)$. A set decoheres when the off diagonal elements of D are negligible. The diagonal elements then give probabilities (8.13) that are consistent with all the rules of probability theory.[6] Like (8.13), the decoherence functional (8.14) is time neutral.

In the quantum mechanics of closed systems decoherence replaces 'measured' as the criterion for when a set of histories can be consistently assigned probabilities. Measured histories decohere, but histories do not have to be of measurement outcomes in order to decohere. Decoherence is a more precise, more general, and more objective criterion than 'measured' and certainly more useful in cosmology.

8.3.3 *Emergent Arrows of Time*

As already mentioned, the expressions both for probabilities (8.13) and interference (8.14) are time-neutral. There is thus no distinction between 'initial' and 'final' that is not conventional. This formulation of quantum theory for our model universe is therefore free from any built in arrow of time.

If there is no arrow of time in the basic formulation of quantum theory, then the observed arrows of time observed in our particular universe can only arise from differences between the ρ_f and ρ_i that characterize it. We will then say that arrows of time *emerge* for our particular universe from ρ_f and ρ_i. We will discuss only quantum arrows and thermodynamic arrows, since, as already mentioned, other arrows are connected to these.

Our observations of the universe from laboratory to cosmological scales are consistent (so far) with one special condition that might be a pure state $\rho_i = |\Psi\rangle\langle\Psi|$ and a second condition of indifference[7] $\rho_f \propto I$. It is conventional to call the special condition 'initial', as we have done here, the second one 'final', and define the direction of increasing time from initial to final.

With these initial and final conditions the formula for the decoherence functional defining quantum mechanics in the box becomes

$$D(\alpha, \alpha') = Tr(C_\alpha \rho_i C^\dagger_{\alpha'}). \tag{8.15}$$

[6]For more complete expositions of decoherent histories quantum theory than the brief synopsis given here the reader can consult the classic expositions [18–20], a short tutorial in [21], or a review in [22].

[7]For some discussions of the observable information about the final condition, see, e.g. [7, 23–25].

In particular the probabilities for the histories in a decoherent set are:

$$p(\alpha_n, \ldots, \alpha_1) = \|C_\alpha|\Psi\rangle\|^2 = \|P^n_{\alpha_n}(t_n)\ldots P^1_{\alpha_1}(t_1)|\Psi\rangle\|^2. \tag{8.16}$$

This has a state on one end of the chain and nothing on the other. Thus, for $\rho_i = |\Psi\rangle\langle\Psi|$ and $\rho_f \propto I$ a quantum mechanical arrow of time emerges [cf. (8.7)]. It is not an arrow that is associated just with histories of measurement situations but more generally with any set of alternative histories of the universe.

With further assumptions on ρ_i we also recover the thermodynamic arrow. Suppose the usual entropy of chemistry and physics[8] is low for ρ_i. It will be maximal for $\rho_f \propto I$. It will therefore tend to increase from the time of the initial condition to that of the final one. That is the simplest characterization of the thermodynamic arrow.

However if both ρ_i and ρ_f are non-trivial then there is generally no clear definition of either a global thermodynamic or quantum mechanical arrow. For instance when ρ_i and ρ_f have comparable low entropies classical analyses [7, 27] suggests that the entropy could first rise and after a time decrease leading to a thermodynamic arrow that is local in time first pointing one way and then another.[9]

There is no clear meaning to a local quantum mechanical arrow but also no physical need for one. With non-trivial initial and final ρ's there is no notion of a single state at a moment of time from which either the future or the past could be predicted [7]. The theory is fully four-dimensional.[10]

8.4 Generalized Quantum Theory

The time-neutral formulation of quantum mechanics in the previous section is as notable for its simplicity as it is for its freedom from a built in quantum arrow of time. Formulating quantum theory has been reduced to just two specifications: (1) The sets of possible alternative coarse-grained histories $\{C_\alpha\}$, and (2) a decoherence functional (8.14) that measures the quantum interference between histories and specified their probabilities when the set decoheres.

---

[8]Entropy depends on coarse graining. The usual entropy is defined in terms of a coarse graining expressed in the variables that occur in the deterministic equations of classical physics like the Navier-Stokes equation. For more on its construction and its relation to the quasiclassical realms that are features of our universe see, e.g. [26].

[9]There will also generally not be a notion of state at a moment of time. However there might be a way of expressing the probabilities in terms of two state vectors similarly to [28].

[10]Advanced civilizations with large laboratories and enough money could in principle reverse the thermodynamic arrow of time over a region of spacetime in a universe like ours by pre- and post-selection of quantum states. If they selected initial states at one time indifferently, and states at a later time distributed according to a low entropy set of probabilities, they would have effectively have reversed both the quantum mechanical and thermodynamic arrows (e.g. [17], Fig. 8).

The decoherence functional of time-neutral quantum mechanics (8.14) is a generalization of that for usual quantum mechanics (8.15). But it is not the only generalization. The essential features of quantum mechanics are captured by any complex valued decoherence functional $D(\alpha, \alpha')$ that satisfies the following conditions [14, 29]:

i. Hermiticity: $D(\alpha, \alpha') = D^*(\alpha', \alpha)$,
ii. Normalization: $\sum_{\alpha\alpha'} D(\alpha, \alpha') = 1$,
iii. Positivity: $D(\alpha, \alpha) \geq 0$,

and, most importantly, consistency with the principle of superposition. This means the following: Partitioning a set of histories $\{C_\alpha\}$ into bigger sets $\{C_{\bar{\alpha}}\}$ is an operation of coarse graining. Every history C_α is in one and only one of the sets $C_{\bar{\alpha}}$ a fact that we indicate schematically by $\alpha \in \bar{\alpha}$. Then consistency with the principle of superposition means[11]:

iv. Principle of superposition:

$$D(\bar{\alpha}, \bar{\alpha}') = \sum_{\alpha \in \bar{\alpha}} \sum_{\alpha' \in \bar{\alpha}'} D(\alpha, \alpha').$$

Given a decoherence functional satisfying i-iv, the central formula of quantum mechanics which specifies both which sets of histories $\{C_\alpha\}$ decohere and their probabilities $p(\alpha)$ is:

$$D(\alpha, \alpha') \approx \delta_{\alpha\alpha'} p(\alpha). \tag{8.17}$$

Interference between histories vanishes when the decoherence functional is diagonal and the diagonal elements are the probabilities of the histories in a decoherent set. These probabilities satisfy all the usual rules of probability theory as a consequence of i–iv.

The decoherence functional of usual quantum mechanics (8.15) satisfies i–iv. All the other ways of satisfying these conditions give generalizations of usual quantum theory—generalized quantum theories. The decoherence functional (8.14) of the time-neutral formulation is one example. We will see another in the next section.

8.5 Bouncing Universe Models

The universe of quantum matter fields in a closed box that has been at the center of our discussion so far is a much oversimplified model for cosmology. It's chief deficiency is that it ignores gravity. A better, still manageable, kind of model describes quantum matter fields moving in a fixed, classical, cosmological background such

[11] The is just the usual superposition of amplitudes applied to a quantity D that is bilinear in amplitudes.

as the bouncing universe shown in Fig. 8.1. What kind of decoherence functional should we assume for such a model to study classical and quantum arrows of time? A model problem in quantum cosmology suggests the answer.

In [3, 4] both spacetime geometry and matter fields were treated quantum mechanically. Probabilities for different homogeneous and isotropic classical background spacetimes and the behavior of quantum matter fields in them were predicted from the no-boundary quantum state of the universe [30, 31] in a simple minisuperspace model. There were two key results that are relevant for the present discussion: (1) Some bouncing classical background spacetimes like that in Fig. 8.1 were predicted with non-zero probability.[12] (2) The predicted matter fields were small and simple at the bounce where the spatial volume is the least, and not at one or the other of the infinite volume ends of the spacetime [5]. That suggests that the quantum mechanics of matter fields in such spacetimes should not have initial and final density matrices but rather one density matrix ρ_0 at the bounce. We now produce a generalized quantum theory with this property.

Before starting on quantum mechanics it is worthwhile to consider the thermodynamic arrow of time in this model.[13] As discussed above, the matter field fluctuations are small near the bounce. They will therefore grow in the two time directions away from the bounce. Eventually fluctuations may grow large enough to collapse and dissipate giving rise to a large scale structure of galaxies, stars, planets, biota, IGUSes, civilizations, etc on both sides of the bounce.[14] The thermodynamic arrow of time is thus bidirectional in this model—pointing away from the bounce on both sides.

Generalized quantum mechanics for quantum fields in a bouncing universe can be constructed by specifying first the histories and then a decoherence functional obeying properties i-iv in the previous section. There will be many ways of doing this such as simply generalizing the time-neutral formulation of Sect. 8.3.2 with initial and final density matrices at the large ends of the expansion. But motivated by the quantum cosmology model described above, we are looking for a decoherence functional with a density matrix at the bounce.[15]

To specify the histories we arbitrarily label the two sides of the bounce as A and B as in Fig. 8.4. A given history will generally have a part on the A side and a part on the B side—generally different. On each side the parts of histories can be represented by chains of projections—$\{C^A_\alpha\}$ and $\{C^B_\beta\}$. We make the convention

[12] Backgrounds that are not time symmetric were also predicted, but for simplicity we are focusing on time symmetric ones.

[13] See [5] for a more detailed discussion within quantum cosmology and also [32, 33] for not unrelated ones outside of quantum cosmology.

[14] This large scale structure will generally not be the *same* on both sides of the bounce. Individual histories do not have to be time-symmetric. It is the ensemble of possible histories predicted by quantum mechanics that is time symmetric [5, 16].

[15] We should stress that we are not deriving this decoherence functional from the more general quantum cosmological model that includes quantum spacetime, but using that as a motivation to posit a particular kind of model.

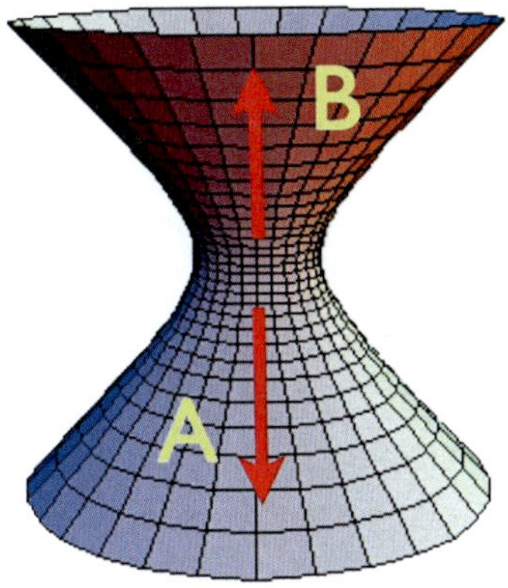

Fig. 8.4 A bouncing universe like that described in Fig. 8.1 divided into two sides A and B at the minimum volume three surface (the bounce). When the quantum mechanics of matter fields is described by a decoherence functional (8.19) each side has a coincident quantum and thermodynamic arrow. But the arrows point in opposite directions on opposite sides of the bounce. This is an example whose quantum mechanics is globally time neutral but with local arrows of time

that the projections in the chains are *time-ordered away from the bounce*. We have separately [cf. (8.12)]

$$\sum_\alpha C^A_\alpha = I, \qquad \sum_\beta C^B_\beta = I. \tag{8.18}$$

The following decoherence functional then suggests itself

$$D(\beta,\alpha;\beta',\alpha') = Tr\left(C^B_\beta \sqrt{\rho_0} C^{A\dagger}_\alpha C^A_{\alpha'} \sqrt{\rho_0} C^{B\dagger}_{\beta'}\right). \tag{8.19}$$

It is not difficult to verify that this satisfies requirements i–iv.

The generalized quantum theory defined by (8.19) is time neutral. The decoherence functional D is symmetric under interchanging A and B. It is perhaps the simplest generalized quantum theory with this property.[16]

Familiar results emerge if we consider histories just on one side, say B. The appropriate decoherence functional D^B results from coarse-graining (summing) over alternatives on the A side. Using (8.18) we find

$$D^B(\beta,\beta') = Tr(C_\beta \rho_0 C^\dagger_{\beta'}). \tag{8.20}$$

But this is just the expression (8.15). There will thus be a quantum mechanical arrow of time on side B and a coincident thermodynamic arrow. Similar results are obtained by following histories on side A and ignoring those on side B.

---

[16] It is not, however, the only one. For instance initial and final conditions represented by density matrices ρ_i and ρ_f at the large ends could have been incorporated in addition to ρ_0 in analogy with (8.14).

Thus the generalized quantum theory defined by the decoherence functional (8.19) exhibits *local* thermodynamic and quantum mechanical arrows that are codirectional on either side but point in opposite directions on opposite sides of the bounce. There are no global arrows—either quantum mechanical or thermodynamic—pointing consistently in one direction over the whole of the spacetime.

If we live in a bouncing universe questions naturally arise as to how much present events on our side are influenced by what occurred before the bounce and what we can infer about events on the far side from observations on our side. The answers to such questions are contained in the joint probabilities $p(\beta,\alpha)$ for correlations between histories on the far side of the bounce and histories on the far side.

We can anticipate that it will be difficult to find causal correlations between the two sides because the thermodynamic arrow points in opposite directions on opposite sides of the bounce [5]. The two sides are in each other's pasts as determined by the thermodynamic arrow. There is as much chance of events on the far side of the bounce influencing us, as we have of influencing events in our past by actions taken now.

In the simple case where the density matrix ρ_0 in (8.19) is pure, mutual influence is impossible. To see this write

$$\rho_0 = |\Psi\rangle\langle\Psi| = \sqrt{\rho_0}. \tag{8.21}$$

Then

$$D(\beta,\alpha;\beta',\alpha') = \langle\Psi|C^{B\dagger}_{\beta'}C^{B}_{\beta}|\Psi\rangle\langle\Psi|C^{A\dagger}_{\alpha'}C^{A}_{\alpha}|\Psi\rangle. \tag{8.22}$$

The immediate consequence is that the joint probabilities of a decoherent set of histories factor

$$p(\beta,\alpha) = p^{B}(\beta)p^{A}(\alpha), \tag{8.23}$$

and there is no correlation between events on one side and the events on the other. The far side might as well not exist.

8.6 Conclusions

Cosmology is the natural context for understanding the origin of the arrows of time our universe. Arrows operate over cosmological distances and can be explained by cosmological conditions.

No arrows of time need be built into a fundamental formulation of quantum mechanics of closed systems like the universe. Rather quantum mechanics can be formulated time-neutrally. The observed arrows of time are then emergent features of the asymmetries between conditions that specify our particular universe among the possibilities that the time-neutral theory allows. This general perspective allows a discussion of the different ways arrows to time can be exhibited by different universes specified by different conditions. In particular it allows a discussion of the connections between arrows that follow from those conditions.

This essay exhibited a number of cosmological models with different possibilities for the quantum mechanical and thermodynamic arrows in the framework of time-neutral generalized quantum theory in fixed background spacetimes. From these examples we can conclude that a given universe may not exhibit well defined arrows of either kind. Further, when arrows do emerge they need not consistently point in one direction over the whole of spacetime. Rather they may point in different directions in different regions of spacetime as the bouncing universe model cleanly illustrates. Local arrows can be consistent with global time-symmetry [5, 33].

In some examples there was a local thermodynamic arrow defined by the direction of local entropy increase but no obvious quantum mechanical arrow. In all examples, where both arrows were available locally they coincided in direction. (Of course, a few examples do not make a general result.)

From this perspective, other features of quantum mechanics such as states on spacelike surfaces, their unitary evolution and their reduction may also emerge only locally in a more general framework for quantum theory that is fully four-dimensional and time neutral [14].

Acknowledgements The author thanks Thomas Hertog for many discussions of the arrows of time in quantum cosmology and Murray Gell-Mann for discussions on the quantum mechanics of the universe over many decades. He thanks the Santa Fe Institute for supporting many productive visits there. The this work was supported in part by the National Science Foundation under grant PHY12-05500.

References

1. J. von Neumann, *Mathematische Grundlagen der Quantenmechanik* (Springer, Berlin, 1932). English transl. Mathematical Foundations of Quantum Mechanics by R.T. Beyer (Princeton University Press, Princeton, 1955)
2. L. Boltzmann, Zu Hrn. Zermelo's Abhandlung Über die mechanische Erklärung Irreversibler Vorgange. Ann. Phys. **60**, 392–398 (1897)
3. J.B. Hartle, S.W. Hawking, T. Hertog, Classical universes of the no-boundary quantum state. Phys. Rev. D **77**, 123537 (2008). arXiv:0803.1663
4. J.B. Hartle, S.W. Hawking, T. Hertog, The no-boundary measure of the universe. Phys. Rev. Lett. **100**, 202301 (2008). arXiv:0711.4630
5. J.B. Hartle, T. Hertog, Arrows of time in the bouncing universes of the no-boundary quantum state. Phys. Rev. D **85**, 103524 (2012). arXiv:1104.1733
6. Y. Aharonov, P. Bergmann, J. Lebovitz, Time asymmetry in the quantum process of measurement. Phys. Rev. **4B**, 1410–1416 (1964)
7. M. Gell-Mann, J.B. Hartle, Time symmetry and asymmetry in quantum mechanics and quantum cosmology, in *The Physical Origins of Time Asymmetry*, ed. by J. Halliwell, J. Pérez-Mercader, W. Zurek (Cambridge University Press, Cambridge, 1994). gr-qc/9309012
8. H.J. Groenewold, Information in quantum measurement. Proc. K. Ned. Akad. Wet., Ser. B **55**, 219 (1952)
9. E. Wigner, The problem of measurement. Am. J. Phys. **31**, 6 (1963)

10. Y. Aharonov, D. Rohrlich, *Quantum Paradoxes* (Wiley-VCH, Weinheim, 2005)
11. B. Basu, D. Lynden-Bell, A survey of entropy in the universe. Q. J. R. Astron. Soc. **31**, 359 (1990)
12. X. Zhao, Y. Li, Q. Zhu, D. Fox, Co-evolution of cosmic entropy and structures in the universe. arXiv:1211.1677
13. J.B. Hartle, The physics of now. Am. J. Phys. **73**, 101–109 (2005). gr-qc/0403001
14. J.B. Hartle, Spacetime quantum mechanics and the quantum mechanics of spacetime, in *Gravitation and Quantizations. Proceedings of the 1992 Les Houches Summer School*, ed. by B. Julia, J. Zinn-Justin. Les Houches Summer School Proceedings, vol. LVII (North Holland, Amsterdam, 1995). gr-qc/9304006
15. J.B. Hartle, Generalizing quantum mechanics for quantum spacetime, in *The Quantum Structure of Space and Time: Proceedings of the 23rd Solvay Conference on Physics*, ed. by D. Gross, M. Henneaux, A. Sevrin (World Scientific, Singapore, 2007). gr-qc/0602013
16. S.W. Hawking, R. Laflamme, G.W. Lyons, Origin of time asymmetry. Phys. Rev. D **47**, 5342–5356 (1993)
17. J.B. Hartle, The quantum mechanics of cosmology, in *Quantum Cosmology and Baby Universes: Proceedings of the 1989 Jerusalem Winter School for Theoretical Physics*, ed. by S. Coleman, J.B. Hartle, T. Piran, S. Weinberg (World Scientific, Singapore, 1991), pp. 65–157
18. R.B. Griffiths, *Consistent Quantum Theory* (Cambridge University Press, Cambridge, 2002)
19. R. Omnès, *Interpretation of Quantum Mechanics* (Princeton University Press, Princeton, 1994)
20. M. Gell-Mann, *The Quark and the Jaguar* (Freeman, New York, 1994)
21. J.B. Hartle, The quantum mechanics of closed systems, in *Directions in General Relativity, Volume 1: A Symposium and Collection of Essays in Honor of Professor Charles W. Misner's 60th Birthday*, ed. by B.-L. Hu, M.P. Ryan, C.V. Vishveshwara (Cambridge University Press, Cambridge, 1993). arXiv:gr-qc/9210006
22. P.C. Hohenberg, An introduction to consistent quantum theory. Rev. Mod. Phys. **2835**, (2844) (2010)
23. R. Laflamme, Time-symmetric quantum cosmology and our universe. Class. Quantum Gravity **10**, L79 (1993). arXiv:gr-qc/9301005
24. D. Craig, Observation of the final condition: extragalactic background radiation and the time symmetry of the Universe. Ann. Phys. **251**, (384) (1995). gr-qc/9704031
25. H. Price, *Time's Arrow and Archimedes' Point* (Oxford University Press, New York, 1996)
26. M. Gell-Mann, J.B. Hartle, Quasiclassical coarse graining and thermodynamic entropy. Phys. Rev. A **76**, 022104 (2007). arXiv:quant-ph/0609190
27. W.J. Cocke, Statistical time symmetry and two-time boundary conditions in physics and cosmology. Phys. Rev. **160**, 1165 (1967)
28. Y. Aharonov, L. Vaidman, The two-state vector formalism: an updated review, in *Lect. Notes Phys.*, vol. 734, (2008), pp. 399–447
29. C.J. Isham, Quantum logic and the histories approach to quantum theory. J. Math. Phys. **35**, 2157 (1994). arXiv:gr-qc/9308006
30. J.B. Hartle, S.W. Hawking, The wave function of the Universe. Phys. Rev. D **28**, 2960–2975 (1983)
31. S.W. Hawking, The quantum state of the Universe. Nucl. Phys. B **239**, 257–276 (1984)
32. F. Hoyle, J.V. Narlikar, Symmetric electrodynamics and the arrow of time in cosmology. Proc. R. Soc. Lond. Ser. A **277**, 1–23 (1964)
33. S. Carroll, J. Chen, Spontaneous inflation and the origin of the arrow of time. arXiv:hep-th/0410270

Part IV
Universe as a Wavefunction

Chapter 9
Collapse Miscellany

Philip Pearle

Abstract An introduction to the CSL (Continuous Spontaneous Localization) theory of dynamical wave function collapse is provided, including a derivation of CSL from two postulates, a new result. There follows a review of applications to a free particle, or to a 'small' rigid cluster of free particles, in a single wave-packet and in interfering packets: the latter result is new. [*Editors note*: for a video of the talk given by Prof. Pearle at the Aharonov-80 conference in 2012 at Chapman University, see quantum.chapman.edu/talk-11.]

9.1 Introduction

Standard quantum theory is readily applied to measurement situations. This requires additional (ad hoc—for this case only) information be supplied for each situation. One may hope the theory could be extended to describe reality independent of experiment and without the need for ad hoc information. Given a state vector for any physical system, no matter how large or complex, one may hope for a theory that specifies the state vectors corresponding to the possible realizable states of nature, and their probabilities of realization.

Why can't this be done with standard quantum theory? Given a state vector,

$$|\psi, t\rangle = \sum_n |a_n\rangle\langle a_n|\psi, t\rangle, \tag{9.1}$$

one would like to say that the $|a_n\rangle$ correspond to the possible realizable states of nature, and that $|\langle a_n|\psi, t\rangle|^2$ are their probabilities of realization.

There are two problems with this. The first is the *preferred basis problem*. Why $|a_n\rangle$ and not another orthonormal basis $|b_n\rangle$? No one has been able to specify the needed preferred basis. Of course, in experimental situations and special model situations this is possible, but it has not been possible in general.

P. Pearle (✉)
Department of Physics, Hamilton College, Clinton, NY 13323, USA
e-mail: ppearle@hamilton.edu

D.C. Struppa, J.M. Tollaksen (eds.), *Quantum Theory: A Two-Time Success Story*,
DOI 10.1007/978-88-470-5217-8_9, © Springer-Verlag Italia 2014

The second I like to call the *hopping problem*. If $|\langle a_n|\psi,t\rangle|^2$ is the probability of $|a_n\rangle$ being realized in nature at time t, and $|\langle a_m|\psi,t\rangle|^2$ is the probability of $|a_m\rangle$ $(m \neq n)$ being realized in nature at time $t+dt$, then

$$|\langle a_n|\psi,t\rangle|^2|\langle a_m|\psi,t+dt\rangle|^2$$

is the probability of occurrence of both these events which, of course, is not seen. The difficulty is that quantum theory just gives the probabilities of these events: it does not give the probability of transitions between these events.

The resolution chosen by the Founding Fathers was to restrict quantum theory to experimental situations, with empirically defined preferred bases, and to adopt the 'collapse postulate,' to ensure that a preferred basis state, once chosen by nature, would remain chosen. They thereby relinquished the hope to describe reality.

An alternative resolution, described here, does not give up on this hope. It alters quantum theory so that it may be realized. The Schrödinger equation is modified by adding a randomly fluctuating term to the Hamiltonian, to account for the probabilistic behavior of nature. There is a preferred basis built into this term. The state vector dynamically collapses toward one of these basis states, very slowly for micro-objects, very rapidly for macro-objects.

The preferred basis is essentially the mass density basis. This choice could not be made for standard quantum theory, since collapse to position eigenstates gives the particles infinite energy. It works here because the collapse never goes all the way. First, it would take infinite time for that to occur. Second, it doesn't happen because of the interaction of the usual Hamiltonian dynamics with the collapse dynamics (e.g., the example in Sect. 9.6.2).

9.2 Deriving CSL

For collapse dynamics, one wants the following behavior.

Given the state vector Eq. (9.1) at time $t=0$, the squared amplitudes $x_n(t) \equiv |\langle a_n|\psi,t\rangle|^2$ should fluctuate until eventually one amplitude becomes equal to 1, and the rest equal to 0: that is the collapse.

Moreover, with repeated evolutions, the nth amplitude should eventually reach 1 for a fraction $x_n(0)$ of the evolutions: that is the Born rule.

9.2.1 Gambler's Ruin Game

There is a rather precise analogy to this behavior, the 'gambler's ruin game' [1, 2]. Consider two gamblers, with \$100 between them. Gambler 1 starts with $\$X_1(0)$, gambler 2 with \$ $X_2(0) = \$100 - X_1(0)$. They toss a fair coin: 'fair' is crucial, making the game what mathematicians call a 'Martingale.' Heads, gambler 1 gives a dollar to gambler 2, tails, the reverse. The amount each possesses, $X_n(t_k)$, fluctuates. Eventually, the game ends, with one gambler in possession of all the money.

Define $P(X_1)$ as the conditional probability that gambler 1 eventually wins, given that he possesses $\$X_1$ at any time. Then,

$$P(X_1) = \frac{1}{2}P(X_1 - 1) + \frac{1}{2}P(X_1 + 1).$$

That is, when the coin is tossed, gambler 1 can either lose the toss but win from there, or win the toss and win from there. The solution of this difference equation is $P(X_1) = AX_1 + B$ (A and B are constants). With boundary conditions $P(0) = 0$ and $P(100) = 1$, the solution is $P(X_1) = X_1/100$.

Define $x_n(t_k) \equiv X_n(t_k)/100$. Thus, if gambler 1 starts out with a fraction $x_1(0)$ of the total amount of money, that is the probability he wins it all, i.e., the probability that $x_1(\infty) = 1$ (and, of course, likewise for gambler 2).

This is just the behavior we want for the squared amplitudes $x_n(t)$ characterizing a state vector which is the sum of two basis vectors $|a_1\rangle$, $|a_2\rangle$. Thus, the two basis vectors may be thought of as competing in a continuous version of the gambler's ruin game.

For, the gambler's ruin game is a zero-sum, discrete, fair, random walk with absorbing barriers (at $x_i = 0, 1$) in discrete time, for $x_n(t_k) \equiv X_n(t_k)/100$.

Collapse is a zero-sum, continuous, fair, random walk with absorbing barriers in continuous time, for $x_n(t) \equiv |\langle a_n|\psi, t\rangle|^2$.

The continuous limit of discrete random walk is Brownian motion. Thus, it is natural to regard the $x_n(t)$ as undergoing some kind of Brownian motion. We shall replace the probabilities associated with a single coin toss, and with a sequence of coin tosses respectively by

$$P(dB) \equiv \frac{1}{\sqrt{2\pi\lambda dt}} e^{-dB^2/2\lambda dt}, \tag{9.2a}$$

$$P\big(B(t)\big) \equiv \frac{1}{\sqrt{2\pi\lambda t}} e^{-B(t)^2/2\lambda t} \tag{9.2b}$$

where λ is a constant diffusion rate. Just as the gambler's ruin dollar count $X_n(t_k)$ depends upon a sequence of coin tosses, so we shall take the collapse $x_n(t)$ to depend upon $B(t)$.

9.2.2 Postulates

We shall derive the CSL collapse dynamics from two postulates:

1) gambler's ruin behavior.
2) a linear, real, Schrödinger equation.

2) could use some explanatory remarks.

In order to achieve the *linearity* part of 2), it shall be necessary to relinquish the condition that the state vector norm is 1. The derivation shows that a state vector evolution free from the unit-norm restriction is unique.

In standard quantum theory, the unit-norm condition is mandatory, since it ensures that probabilities add up to 1. Here there is no need for the unit-norm condition: probabilities are provided by $B(t)$'s probabilities. Physical information is carried by a state vector's direction in Hilbert space, not its norm. Thus, the invariance of standard quantum theory to multiplication of the state vector by a phase factor is generalized to invariance under multiplication by any complex number. To calculate expectation values, one can always normalize a state vector. The un-normalized state vector shall be denoted $|\phi, t\rangle$.

Regarding the *real* part of 2), we wish to consider a different dynamics from the usual Hamiltonian dynamics $d|\psi, t\rangle = -iHdt|\psi, t\rangle$. That utilizes a hermitian hamiltonian H, so we shall consider dynamics of the form $d|\psi, t\rangle = H_c dt|\psi, t\rangle$ where the 'collapse hamiltonian' H_c is hermitian (we note that the most general linear equation is of the form $d|\psi, t\rangle = -i(H + iH_c)dt|\psi, t\rangle$).

Now, for the purposes of this derivation, for simplicity and appropriate to the generalization of a classical game, we shall further restrict all quantities to be real numbers. Thus, we restrict H_c to be a real symmetric operator in the chosen basis $|a_n\rangle$. Once we find the unique form of H_c that allows postulate 1), it may readily be seen that complex state vector components and the most general hermitian H_c also allow postulate 1). Because the state vector components are real, we may write

$$x_n(t) = \frac{\langle a_n|\phi, t\rangle^2}{\langle \phi, t|\phi, t\rangle} = \frac{\langle a_n|\phi, t\rangle^2}{\sum_m \langle a_m|\phi, t\rangle^2}. \tag{9.3}$$

Also in order to achieve the linearity part of 2), the Schrödinger equation shall be linear in a Brownian motion $B'(t)$. However, the relation assumed between $B'(t)$ and $B(t)$ is allowed to be non-linear in the state vector:

$$dB'(t) = dB(t) + f(\mathbf{x})dt, \tag{9.4}$$

where f is an arbitrary real function of the $x_n(t)$. For, collapse violates the superposition principle, and this *somehow* requires a non-linearity. What 2) really means, then, is that we look to have a linear Schrödinger equation and isolate all the non-linearity in the probability [3].

The derivation, which is presented in the next few sections, is rather lengthy, and it involves stochastic differential equations—but the result does not. Some readers may wish to move immediately to Sect. 9.3 where the results obtained here are utilized.

Postulate 1) is implemented by an *Itô* equation for $x_n(t)$:

$$dx_n(t) = b_n(\mathbf{x})dB(t), \tag{9.5}$$

where b_n is an arbitrary real function of the $x_m(t)$'s.

We shall denote by an overline the ensemble average of a quantity, e.g., from Eq. (9.2a), $\overline{dB} = \int_{-\infty}^{\infty} d(dB)P(dB) = 0$.

It immediately follows from Eq. (9.5) that $\overline{dx_n(t)} = 0$. This says that it is a 'fair game' for each 'player' which, as we have said, is crucial for gambler's ruin behavior [4]. This is why Eq. (9.5) is chosen to be an Itô equation.

The other necessary condition for gambler's ruin behavior is the end-game condition, that $b_n(\mathbf{x}) = 0$ when one $x_n(t)$ is equal to 1 and the rest vanish. However, that does not need to be separately imposed since it automatically occurs, as shall be seen.

Postulate 2) is implemented by the *Stratonovich* Schrödinger equation for the un-normalized state vector amplitudes:

$$d|\phi, t\rangle = [RdB' + Sdt]|\phi, t\rangle$$

where R and S are arbitrary symmetric real operators, as discussed above. This is chosen to be a Stratonovich equation because the calculus manipulations (e.g., derivative of the product of functions) are the usual ones, which would not be the case for an Itô equation.

However, a rather tedious calculation (Appendix A) shows that R and S have to be diagonal in the $|a_n\rangle$ basis if postulate 2) is to imply postulate 1). Therefore, we shall write this equation as

$$d\langle a_n|\phi, t\rangle = [\alpha_n dB' + \beta_n dt]\langle a_n|\phi, t\rangle \tag{9.6}$$

where α_n, β_n are real constants.

9.2.3 Derivation

We proceed to find $dx_n(t)$ from Eq. (9.3). With use of Eqs. (9.4), (9.6), we obtain the Stratonovich equation

$$dx_n(t) = 2\big\{(\alpha_n - \alpha\cdot\mathbf{x})x_n(t)dB \\ + \big[(\alpha_n - \alpha\cdot\mathbf{x})f + (\beta_n - \beta\cdot\mathbf{x})\big]dt\big\}x_n(t). \tag{9.7}$$

where $\alpha\cdot\mathbf{x} \equiv \sum_m \alpha_m x_m(t)$

Now we may use the rule for converting a Stratonovich equation to an Itô equation, which in this case means adding

$$\frac{\lambda dt}{2}\sum_m 2(\alpha_m - \alpha\cdot\mathbf{x})x_m(t)\frac{\partial}{\partial x_m}2(\alpha_n - \alpha\cdot\mathbf{x})x_n(t)$$

to the right side of Eq. (9.7). The result is

$$dx_n(t) = 2\big\{(\alpha_n - \alpha\cdot\mathbf{x})dB$$

$$+\big[(\alpha_n - \alpha\cdot\mathbf{x})f + (\beta_n - \beta\cdot\mathbf{x})\big]dt$$
$$+\big[-\alpha^2\cdot\mathbf{x} + (\alpha\cdot\mathbf{x})^2 + (\alpha_n - \alpha\cdot\mathbf{x})^2\big]\lambda dt\big\}x_n(t), \tag{9.8}$$

where $\alpha^2\cdot\mathbf{x} \equiv \sum_m \alpha_m^2 x_m(t)$. In order that the Itô Eq. (9.8) (consequence of postulate 2) agree with the Itô Eq. (9.5) (consequence of postulate 1), the coefficient of dB and the coefficient of dt in both equations must be equal, so

$$b_n(\mathbf{x}) = 2(\alpha_n - \alpha\cdot\mathbf{x})x_n(t), \tag{9.9a}$$
$$(\alpha_n - \alpha\cdot\mathbf{x})f + (\beta_n - \beta\cdot\mathbf{x}) = \lambda\big[\alpha^2\cdot\mathbf{x} - (\alpha\cdot\mathbf{x})^2 - (\alpha_n - \alpha\cdot\mathbf{x})^2\big]. \tag{9.9b}$$

To find f and β_n, operate on Eq. (9.9b) with $\sum_n x_n\partial^2/\partial x_m^2$. Remembering that $\sum_n x_n = 1$, we obtain

$$2\alpha_m\frac{\partial}{\partial x_m}f = 4\lambda\alpha_m^2, \quad \text{or} \quad f = 2\lambda\alpha\cdot\mathbf{x} + c, \tag{9.10}$$

where c is an arbitrary constant. Putting Eq. (9.10) back into Eq. (9.9b) gives

$$(\beta_n - \beta\cdot\mathbf{x}) = -c(\alpha_n - \alpha\cdot\mathbf{x}) - \lambda\big(\alpha_n^2 - \alpha^2\cdot\mathbf{x}\big), \quad \text{or} \quad \beta_n = -c\alpha_n - \lambda\alpha_n^2, \tag{9.11}$$

the last step following from applying $\partial/\partial x_i$ to the first. Putting f from Eq. (9.10) and β_n from Eq. (9.11) into Eq. (9.7), we find that the latter is independent of c. Therefore, c has no physical effect. Its only effect is to add a constant drift to $B'(t)$ (see Eq. (9.4)), and since we are free to choose $B'(t)$ to be as simple as possible, we may take $c = 0$.

This concludes our derivation. We have found that postulates 1) and 2) lead to the *Schrödinger equation* (9.6) and the *Probability Rule* (9.4), in the forms

$$d\langle a_n|\phi,t\rangle = \big[\alpha_n dB' - \lambda\alpha_n^2 dt\big]\langle a_n|\phi,t\rangle, \tag{9.12a}$$
$$dB'(t) = dB(t) + 2\lambda\alpha\cdot\mathbf{x}dt, \tag{9.12b}$$

where the α_n are completely arbitrary constants and $\mathbf{x}$ is given by Eq. (9.3).

9.2.4 Schrödinger Equation

We may introduce an operator A defined by $A|a_n\rangle = \alpha_n|a_n\rangle$, in terms of which the Schrödinger equation (9.12a) may be written in basis-independent form,

$$d|\phi,t\rangle = \big[AdB' - \lambda A^2 dt\big]|\phi,t\rangle \quad \text{or} \quad |\phi,t\rangle = e^{AB'(t)-A^2\lambda t}|\phi,0\rangle. \tag{9.13}$$

Since the $|a_n\rangle$ are eigenstates of A with eigenvalues α_n, we could instead label them $|\alpha_n\rangle$. However, we shall rather replace α_n by a_n.

9.2.5 Probability Rule

Equation (9.12b) is the same form whether an Itô or Stratonovich equation. It may be written as

$$dB'(t) = dB(t) + 2\lambda dt\langle A\rangle(t) \quad \text{where } \langle A\rangle(t) \equiv \frac{\langle\phi, t|A|\phi, t\rangle}{\langle\phi, t|\phi, t\rangle}. \tag{9.14}$$

We shall use it to find the expression for the probability of $B'(t)$.

It follows from Eq. (9.14) that $\overline{dB'(t)} = 2\lambda dt\langle A\rangle(t)$, $\overline{(dB'(t))^2} = \lambda dt$, and all higher moments of $dB'(t)$ can be neglected as they are of higher order than dt. Therefore, $dB'(t)$ is a gaussian process, characterized by just these two moments. One may immediately check that the probability density, variously written as

$$\begin{aligned} P\big[dB'(t)\big] &= \frac{1}{\sqrt{2\pi\lambda dt}}e^{-\frac{(dB'(t))^2}{2\lambda dt}}\frac{\langle\phi, t|e^{2AdB'(t)-2A^2\lambda dt}|\phi, t\rangle}{\langle\phi, t|\phi, t\rangle} \\ &= \frac{1}{\sqrt{2\pi\lambda dt}}\frac{\langle\phi, t|e^{-\frac{(dB'(t)-2\lambda dtA)^2}{2\lambda dt}}|\phi, t\rangle}{\langle\phi, t|\phi, t\rangle} && (9.15a) \\ &= \frac{1}{\sqrt{2\pi\lambda dt}}e^{-\frac{(dB'(t))^2}{2\lambda dt}}\frac{\langle\phi, t+dt|\phi, t+dt\rangle}{\langle\phi, t|\phi, t\rangle} && (9.15b) \end{aligned}$$

has these two moments.

Define $N \equiv t/dt$. We multiply Eq. (9.15b) by itself $N+1$ times with successively smaller values of t, obtaining the joint probability of the independent increments dB at successive values of t:

$$\prod_{n=0}^{N} P\big[dB'(t-ndt)\big] = \prod_{n=0}^{N}\frac{1}{\sqrt{2\pi\lambda dt}}e^{-\frac{(dB'(t-ndt))^2}{2\lambda dt}} \cdot \frac{\langle\phi, t+dt|\phi, t+dt\rangle}{\langle\phi, 0|\phi, 0\rangle}. \tag{9.16}$$

Since $\langle\phi, 0|\phi, 0\rangle = 1$, and using Eq. (9.13) to write

$$\begin{aligned} \langle\phi, t+dt|\phi, t+dt\rangle &= \langle\phi, 0|e^{2AB'(t+dt)-2A^2\lambda(t+dt)}|\phi, 0\rangle \\ &= \langle\phi, 0|\prod_{n=0}^{N} e^{2A(B'(t+dt-ndt)-B'(t-ndt)-2A^2\lambda dt)}|\phi, 0\rangle \end{aligned} \tag{9.17}$$

(taking $B(0) = 0$), it follows from Eqs. (9.16), (9.17) that the joint probability of the values of B at successive values of t is

$$P\big[B'(t+dt), B'(t), \ldots, B'(dt)\big]$$

$$= \prod_{n=0}^{N} \frac{1}{\sqrt{2\pi\lambda dt}} \langle\phi, 0| e^{-\frac{(B'(t+dt-ndt)-B'(t-ndt)-2\lambda dtA)^2}{2\lambda dt}} |\phi, 0\rangle. \tag{9.18}$$

We have written $dB'(t - ndt) = B'(t + dt - ndt) - B'(t - ndt)$. We have also written P as the joint probability of $\{B'(t + dt), B'(t), \ldots, B'(dt)\}$, instead of the joint probability of $\{dB'(t), dB'(t - dt), \ldots dB'(0)\}$, which we can do since the Jacobian determinant for the change of variables is 1.

Now, we are interested in finding the probability $P[B'(t + dt)]$, regardless of what Brownian path leads to the value of $B'(t + dt)$. To obtain this, we integrate Eq. (9.18) over all B''s except $B'(t + dt)$. These integrals are easily done, since each $B'(mdt)$ appears in just two (adjacent) gaussians in the product, and

$$\int_{-\infty}^{\infty} dB \frac{1}{\sqrt{2\pi c_1}} e^{-\frac{(B-a_1)^2}{2c_1}} \frac{1}{\sqrt{2\pi c_2}} e^{-\frac{(B-a_2)^2}{2c_2}}$$
$$= \frac{1}{\sqrt{2\pi(c_1 + c_2)}} e^{-\frac{(a_1-a_2)^2}{2(c_1+c_2)}}.$$

The result for $P[B'(t + dt)]$ is

$$P = \frac{1}{\sqrt{2\pi\lambda(t + dt)}} \langle\phi, 0| e^{-\frac{(B'(t+dt)-2\lambda(t+dt)A)^2}{2\lambda dt}} |\phi, 0\rangle$$
$$= \frac{1}{\sqrt{2\pi\lambda(t + dt)}} e^{-\frac{(B'(t+dt))^2}{2\lambda dt}} \langle\phi, t + dt|\phi, t + dt\rangle. \tag{9.19}$$

9.3 CSL

We have arrived at CSL's two equations in the form of the Schrödinger equation (9.13) and the Probability Rule (9.19). It is useful to incorporate the exponential factor in (9.19) in the Schrödinger equation (and un-prime B, and replace $t + dt$ by t), so that its solution then becomes

$$|\phi, t\rangle = e^{-\frac{1}{4\lambda t}[B(t)-2\lambda tA]^2} |\phi, 0\rangle, \tag{9.20}$$

leaving the Probability Rule in the simple form

$$P\big[B(t)\big]dB(t) = \frac{dB(t)}{\sqrt{2\pi\lambda t}} \langle\phi, t|\phi, t\rangle. \tag{9.21}$$

Equations (9.20), (9.21) define CSL. Everything that follows is based upon these two equations.

9.3.1 Collapse Mechanism

Let's see how they contrive to collapse the state vector. Suppose the initial state vector is

$$|\phi, 0\rangle = \sum_n c_n |a_n\rangle. \tag{9.22}$$

$|\phi, 0\rangle$ is assumed normalized to 1, so $\sum_n |c_n|^2=1$. Equations (9.20), (9.21) then become

$$|\phi, t\rangle = \sum_n c_n |a_n\rangle e^{-\frac{1}{4\lambda t}[B(t)-2\lambda t a_n]^2} \tag{9.23}$$

and

$$P\big[B(t)\big]dB(t) = \frac{dB(t)}{\sqrt{2\pi\lambda t}} \sum_n |c_n|^2 e^{-\frac{1}{2\lambda t}[B(t)-2\lambda t a_n]^2}. \tag{9.24}$$

Now, assume all the a_n are unequal. Then, Eq. (9.24) describes a bunch of gaussians whose centers at $2\lambda t a_n$ drift further and further apart, while their widths $\sqrt{\lambda t}$ spread much more slowly. As t increases, the gaussians have less and less overlap.

Then, for the set of $B(t)$'s which lie within the mth gaussian, the state vector and probability are, to an excellent approximation (which becomes exact for $t \to \infty$),

$$|\phi, t\rangle \approx c_m |a_m\rangle e^{-\frac{1}{4\lambda t}[B(t)-2\lambda t a_m]^2} \tag{9.25}$$

and

$$P\big[B(t)\big]dB(t) \approx \frac{dB(t)}{\sqrt{2\pi\lambda t}} |c_m|^2 e^{-\frac{1}{2\lambda t}[B(t)-2\lambda t a_m]^2}. \tag{9.26}$$

For this set of $B(t)$'s, the integrated probability in (9.26) is $\approx |c_m|^2$, giving the Born Rule.

For any $B(t)$ in this set, the normalized state vector is

$$|\psi, t\rangle \equiv \frac{|\phi, t\rangle}{\sqrt{\langle\phi, t|\phi, t\rangle}} \approx |a_m\rangle \tag{9.27}$$

giving the collapsed state vector.

9.3.2 Refinements: Density Matrix

The density matrix, as usual, is constructed from the state vectors and their associated probabilities, here given by Eqs. (9.20), (9.21):

$$\rho(t) \equiv \int_{-\infty}^{\infty} P\big[B(t)\big]dB(t) \frac{|\phi, t\rangle\langle\phi, t|}{\langle\phi, t|\phi, t\rangle}$$

$$
\begin{aligned}
&= \int_{-\infty}^{\infty} \frac{dB(t)}{\sqrt{2\pi\lambda t}} \langle\phi, t|\phi, t\rangle \frac{|\phi, t\rangle\langle\phi, t|}{\langle\phi, t|\phi, t\rangle} \\
&= \int_{-\infty}^{\infty} \frac{dB(t)}{\sqrt{2\pi\lambda t}} e^{-\frac{1}{4\lambda t}[B(t)-2\lambda t A]^2} |\phi, 0\rangle\langle\phi, 0| e^{-\frac{1}{4\lambda t}[B(t)-2\lambda t A]^2} \\
&= e^{-\frac{\lambda t}{2}[A_L - A_R]^2} |\phi, 0\rangle\langle\phi, 0|. \qquad (9.28)
\end{aligned}
$$

In Eq. (9.28), A_L, A_R mean that these operators act on the left or the right respectively of the initial density matrix $|\phi, 0\rangle\langle\phi, 0|$.

With the initial density matrix (9.22), the matrix elements of the density matrix at time t is found from (9.28) to be

$$
\langle a_n|\rho(t)|a_m\rangle = c_n c_m^* e^{-\frac{\lambda t}{2}[a_n - a_m]^2}, \qquad (9.29)
$$

showing how the off-diagonal elements decay while the diagonal elements remain constant, the collapse rate increasing as the eigenvalue differences increase.

According to Eq. (9.28), the differential equation satisfied by the density matrix is

$$
\frac{d\rho(t)}{dt} = -\frac{\lambda}{2}\big[A, \big[A, \rho(t)\big]\big], \qquad (9.30)
$$

which is the simplest possible Lindblad equation [5–7].

9.3.3 Refinements: Hamiltonian

To add the Hamiltonian to the state vector dynamics, consider the evolution over an infinitesimal time interval: it and the probability rule become

$$
\begin{aligned}
|\phi, t\rangle &= e^{-iHdt - \frac{1}{4\lambda dt}[dB(t) - 2\lambda dt A]^2} |\phi, t - dt\rangle \\
&= e^{-dt[iH + \frac{1}{4\lambda}[w(t) - 2\lambda A]^2} |\phi, t - dt\rangle \qquad (9.31a)
\end{aligned}
$$

$$
P(w)dw = \frac{\langle\phi, t|\phi, t\rangle}{\langle\phi, t - dt|\phi, t - dt\rangle} \frac{dw(t)}{\sqrt{2\pi\lambda/dt}} \qquad (9.31b)
$$

where $w(t) \equiv dB(t)/dt$ is called white noise.

Over a finite time interval, Eqs. (9.31a), (9.31b) imply

$$
|\phi, t\rangle = \mathcal{T} e^{-\int_0^t dt'\{iH(t') + \frac{1}{4\lambda}[w(t') - 2\lambda A]^2\}} |\phi, 0\rangle \qquad (9.32a)
$$

$$
P(w)Dw = \langle\phi, t|\phi, t\rangle \prod_{t'=0}^{t-dt} \frac{dw(t')}{\sqrt{2\pi\lambda/dt}} \qquad (9.32b)
$$

where $\mathcal{T}$ is the time-ordering operator.

To summarize, for each white noise function $w(t)$ there is a corresponding state vector $|\phi, t\rangle$ given by Eq. (9.32a), one of which is supposed to be realized in nature with probability (9.32b). Generally, the hamiltonian evolution and the collapse-hamiltonian evolution compete against each other. This can give rise to effects which suggest experimental tests of the collapse theory vis-à-vis standard quantum theory/collapse postulate.

'White noise' was named after the sound which has all frequencies in equal amounts, in analogy to white light. It was named in a paper on the acoustics in airplanes [8], where the authors wrote:

> That white noise is annoying needs little argument. No one has been found who really enjoys it.

However, here it is enjoyed, in its role as the "chooser" of the collapsed state.

9.3.4 *Refinements: More Collapse-Generating Operators*

It is a straightforward generalization to describe collapse to a joint basis of operators A^α which commute, $[A^\alpha, A^\beta] = 0$. This requires one white noise function w^α for each A^α. The state vector evolution is

$$|\phi, t\rangle = \mathcal{T}e^{-\int_0^t dt'\{iH(t') + \frac{1}{4\lambda}\sum_\alpha [w^\alpha(t') - 2\lambda A^\alpha]^2\}}|\phi, 0\rangle \tag{9.33}$$

and the corresponding density matrix evolution is

$$\frac{d\rho(t)}{dt} = -i\big[H, \rho(t)\big] - \frac{\lambda}{2}\sum_\alpha \big[A^\alpha, \big[A^\alpha, \rho(t)\big]\big]. \tag{9.34}$$

The ensemble average of an operator $\mathcal{O}$ shall be denoted $\overline{\mathcal{O}}(t) \equiv \mathrm{Tr}\,\mathcal{O}\rho(t)$, where Tr is the trace operation. Then, Eq. (9.34) gives

$$\frac{d\overline{\mathcal{O}}(t)}{dt} = -i\overline{[\mathcal{O}, H]}(t) - \frac{\lambda}{2}\sum_\alpha \overline{\big[A^\alpha, \big[A^\alpha, \mathcal{O}\big]\big]}(t). \tag{9.35}$$

9.3.4.1 Nonlocality

We can illustrate, with two local operators A_L and A_R, where nonlocality enters the theory. This nonlocality, for example, is responsible for the violation of Bell's inequality, that is, responsible for giving the quantum theory result instead of the local realism result.

Consider an object, a cluster of N particles, in a superposition of two places, L and R. At L, there are two possible states, $|0\rangle_L$ which describes nothing at L and

$|N\rangle_L$ which describes the object at L. We define $A_L|0\rangle_L = 0$, $A_L|N\rangle_L = N|N\rangle_L$. Similar relations hold for R. Then, the initial state vector is

$$|\phi, 0\rangle = \frac{1}{\sqrt{2}}|0\rangle_L|N\rangle_R + \frac{1}{\sqrt{2}}|N\rangle_L|0\rangle_R.$$

The state vector and probability rule for this example are respectively

$$|\phi, t\rangle = \frac{1}{\sqrt{2}}e^{-\frac{1}{4\lambda t}\{B_L^2+[B_R-2\lambda tN]^2\}}|0\rangle_L|N\rangle_R + \frac{1}{\sqrt{2}}e^{-\frac{1}{4\lambda t}\{[B_L-2\lambda tN]^2+B_R^2\}}|N\rangle_L|0\rangle_R,$$

$$PdB_LdB_R = \frac{dB_LdB_R}{2\pi\lambda t}\frac{1}{2}\big[e^{-\frac{1}{2\lambda t}\{B_L^2+[B_R-2\lambda tN]^2\}} + e^{-\frac{1}{2\lambda t}\{[B_L-2\lambda tN]^2+B_R^2\}}\big].$$

We see that there are two probable ranges of B_L, B_R. If B_L lies within a number of standard deviations of 0 *and* B_R lies within a number of standard deviations of $2\lambda Nt$, then $|\phi, t\rangle \sim |0\rangle_L|N\rangle_R$, and the integrated probability of this occurring is 1/2: similarly for $L \leftrightarrow R$.

Thus, the Probability Rule demands of B_L, B_R that they *cooperate* in order to achieve a high probability of occurrence. Although they represent values of B at widely separated places, nonetheless they must have correlated values. So, the Schrödinger equation is local while all the nonlocality is vested in the Probability Rule.

It is easy to check from Eq. (9.34) that:

$$\frac{d}{dt}\,{}_R\langle 0|_L\langle N|\rho(t)|0\rangle_L|N\rangle_R = -\lambda N^2\,{}_R\langle 0|_L\langle N|\rho(t)|0\rangle_L|N\rangle_R,$$

so the collapse rate is λN^2.

9.4 Non-relativistic CSL

Finally, here is the CSL proposal to describe the non-relativistic world [3, 9]. The index α in Eq. (9.33) is changed to a continuum index $\mathbf{x}$, so the 'chooser' $w(\mathbf{x}, t)$, rather than being a set of random functions, is a random field:

$$|\phi, t\rangle = \mathcal{T}e^{-i\int_0^t dt' H(t') - \frac{1}{4\lambda}\int_0^t dt' \int d\mathbf{x}'[w(\mathbf{x}',t') - 2\lambda A(\mathbf{x}')]^2}|\phi, 0\rangle. \qquad (9.36)$$

The set of collapse-generating operators are mass-density operators, 'smeared' over a sphere of radius a:

$$A(\mathbf{x}) \equiv \sum_n \frac{m_n}{M}\frac{1}{(\pi a^2)^{3/4}}\int d\mathbf{z}e^{-\frac{1}{2a^2}[\mathbf{x}-\mathbf{z}]^2}\xi_n^\dagger(\mathbf{z})\xi_n(\mathbf{z}). \qquad (9.37)$$

Here, $\xi_n^\dagger(\mathbf{z})$ is the creation operator for a particle of type n at $\mathbf{z}$. m_n is the mass of this particle and M is the mass of a nucleon (say, the neutron). Thus, in ordinary matter, it is the nucleons which are mostly responsible for collapse. Experimental results [10–13] have dictated that the effective collapse rate in Eq. (9.36) be mass-proportional, $\sim \lambda m_n$.

Assuming the theory is correct, the parameter values of λ, a should be determined by experiment [14]. Until then, we shall provisionally adopt the parameter values chosen by Ghirardi, Rimini and Weber [15–17] in their instantaneous collapse theory, $\lambda \approx 10^{-16}$ sec^{-1}, $a \approx 10^{-5}$ cm. However, it should be mentioned that Adler [18] has given an argument for λ to be as large as $\approx 10^{-8}$ sec^{-1}.

One might very well extend this theory to include massless particles by replacing mass-density of $A(\mathbf{x})$ in Eq. (9.37) by energy-density/c^2. One might then regard it as holding in the co-moving frame of the universe [20], or as the limit of a relativistic CSL.[1]

The density matrix evolution equation (9.34) becomes, using (9.37),

$$\begin{aligned}\frac{d\rho(t)}{dt} &= -i\big[H, \rho(t)\big] - \frac{\lambda}{2}\sum_{k,n}\frac{m_k m_n}{M^2} \\ &\quad\cdot \frac{1}{(\pi a^2)^{3/2}}\int d\mathbf{x}\int d\mathbf{z}\int d\mathbf{z}' e^{-\frac{1}{2a^2}[\mathbf{x}-\mathbf{z}]^2} e^{-\frac{1}{2a^2}[\mathbf{x}-\mathbf{z}']^2} \\ &\quad\cdot \big[\xi_k^\dagger(\mathbf{z})\xi_k(\mathbf{z})\big[\xi_n^\dagger(\mathbf{z}')\xi_n(\mathbf{z}'), \rho(t)\big]\big] \\ &= -i\big[H, \rho(t)\big] - \frac{\lambda}{2}\sum_{k,n}\frac{m_k m_n}{M^2} \\ &\quad\cdot \int d\mathbf{z}\int d\mathbf{z}' e^{-\frac{1}{4a^2}[\mathbf{z}-\mathbf{z}']^2}\big[\xi_k^\dagger(\mathbf{z})\xi_k(\mathbf{z})\big[\xi_n^\dagger(\mathbf{z}')\xi_n(\mathbf{z}'), \rho(t)\big]\big]. \end{aligned} \tag{9.38}$$

9.5 Free Small Clump

In the rest of this paper, we shall illustrate CSL by discussing the force-free behavior of the center of mass (cm) of a small (dimensions $< a$) clump of ordinary matter.[2]

For simplicity we shall neglect the electrons, and take there to be N nucleons, regarded as a single type of particle, of mass $m = M$, which are very good

---

[1] Attempts to make a relativistic CSl have had a long and unsuccessful history: [21], [22], [23], [24] ...until the recent successful work by D.J. Bedingham [25].

[2] The internal excitation of the matter is not discussed here. Collapse narrows wave packets, resulting in atomic and nuclear 'anomalous' excitation (i.e., collapse-generated, not predicted by standard quantum theory) [10–13]. Experimental limits on such excitation strongly suggests the effective mass-proportionality of the collapse rate, as we have mentioned. Incidentally, it can be argued [19] that the increasing particle energy entails a concomitant decrease in the w-field energy, so total energy is conserved.

approximations for our calculations. Then, in the particle position basis, $|\mathbf{x}\rangle \equiv |\mathbf{x}_1,\ldots.\mathbf{x}_i,\ldots\mathbf{x}_N\rangle$, using $\xi_k^\dagger(\mathbf{z})\xi_k(\mathbf{z})|\mathbf{x}\rangle = \sum_{i=1}^N \delta(\mathbf{z}-\mathbf{x}_i)|\mathbf{x}\rangle$, Eq. (9.38) becomes

$$\frac{d\langle\mathbf{x}|\rho(t)|\mathbf{x}'\rangle}{dt} = -i\langle\mathbf{x}|\big[H,\rho(t)\big]|\mathbf{x}'\rangle - \langle\mathbf{x}|\rho(t)|\mathbf{x}'\rangle \cdot \frac{\lambda}{2}\sum_{i,j=1}^N \big[e^{-\frac{1}{4a^2}[\mathbf{x}_i-\mathbf{x}_j]^2} + e^{-\frac{1}{4a^2}[\mathbf{x}'_i-\mathbf{x}'_j]^2} - 2e^{-\frac{1}{4a^2}[\mathbf{x}_i-\mathbf{x}'_j]^2}\big]. \qquad (9.39)$$

Define the center of mass coordinate $\mathbf{X} \equiv N^{-1}\sum_{i=1}^N \mathbf{x}_i$ and the relative coordinates $\mathbf{y}_i \equiv \mathbf{x}_i - \mathbf{X}$. Because it is a 'small' clump, $\exp -[\mathbf{y}_i-\mathbf{y}_j]^2/4a^2 \approx 1$ and $\exp -[\mathbf{X}+\mathbf{y}_i-\mathbf{X}'-\mathbf{y}'_j]^2/4a^2 \approx \exp -[\mathbf{X}-\mathbf{X}']^2/4a^2$. With the density matrix assumed to have the form of the direct product of cm and internal coordinate density matrices, we can take the trace over the internal coordinates in Eq. (9.39) to obtain the equation for the evolution of $\langle\mathbf{X}|\rho(t)|\mathbf{X}'\rangle$.

It is useful to express this equation in operator form, writing the cm operator as $\hat{\mathbf{X}}$, and its conjugate cm momentum as $\hat{\mathbf{P}}$:

$$\frac{d\rho(t)}{dt} = -i\left[\frac{\hat{\mathbf{P}}^2}{2MN},\rho(t)\right] - \lambda N^2\big[1-e^{-\frac{1}{4a^2}[\hat{\mathbf{X}}_L-\hat{\mathbf{X}}_R]^2}\big]\rho(t). \qquad (9.40)$$

The associated state vector evolution equation (9.36) is

$$|\phi,t\rangle = \mathcal{T}e^{-i\int_0^t dt'\frac{1}{2MN}\hat{\mathbf{P}}^2 - \frac{1}{4\lambda}\int_0^t dt'\int d\mathbf{x}'[w(\mathbf{x}',t')-2\lambda NA(\mathbf{x}')]^2}|\phi,0\rangle, \qquad (9.41)$$

where

$$A(\mathbf{x}') \equiv \frac{1}{(\pi a^2)^{3/4}}e^{-\frac{1}{2a^2}[\mathbf{x}'-\hat{\mathbf{X}}]^2}. \qquad (9.42)$$

To illustrate the use of the collapse part of Eq. (9.40), consider the initial wave function

$$|\phi,0\rangle = \frac{1}{\sqrt{2}}[|L\rangle + |R\rangle], \qquad (9.43)$$

where the states describe the clump to the left or right, with the two wave packet cm's separated by the distance D. Then (ignoring the kinetic energy),

$$\frac{d\langle L|\rho(t)|R\rangle}{dt} = -\lambda N^2\big[1-e^{-\frac{D^2}{4a^2}}\big]\langle L|\rho(t)|R\rangle. \qquad (9.44)$$

Thus, for $D \gg a$, the collapse is described by exponential decay of the off-diagonal density matrix element with characteristic time $\lambda^{-1} = 10^{16}$ sec for a single nucleon. For a 10^{-5} cm cube of gold, where $N \approx 10^8$, the characteristic collapse time is $1/\lambda N^2 = 1$ sec.

9.6 Collapse of a Packet

We shall consider how a single wave packet undergoes collapse.

9.6.1 *Big Packet*

Consider a spread-out, real, positive, initial wave function such as

$$\langle \mathbf{X}|\phi, 0\rangle = \frac{1}{(2\pi D^2)^{3/4}} e^{-\frac{\mathbf{x}^2}{4D^2}}, \tag{9.45}$$

where $D \gg a$. We shall neglect the effect of the Hamiltonian. We shall see that the wave function collapses fairly rapidly to an approximately spherical wave function of size a, center location consistent with the Born Rule, and thereafter collapses more and more slowly to a smaller and smaller radius.

First we calculate the ensemble average of the operator $\mathcal{O} \equiv |\mathbf{X}\rangle\langle\mathbf{X}|$, so $\overline{\mathcal{O}}$ is the ensemble probability density at $\mathbf{X}$. For any density matrix, it follows from the collapse part of Eq. (9.40), using Eq. (9.35), that

$$\frac{d\overline{|\mathbf{X}\rangle\langle\mathbf{X}|}}{dt} = -\lambda N^2 \langle\mathbf{X}|\big[1 - e^{-\frac{1}{4a^2}[\hat{\mathbf{X}}_L - \hat{\mathbf{X}}_R]^2}\big]\rho(t)|\mathbf{X}\rangle = 0. \tag{9.46}$$

This, of course, doesn't say that collapse occurs but, if there is collapse, it says that the ensemble position probability distribution does not change from the initial distribution (Born Rule).

In order to see that there is indeed collapse, consider the ensemble average of the *modular momentum* [26] operator $\mathcal{O} \equiv \cos\hat{\mathbf{P}} \cdot \mathbf{n}L$, where $\mathbf{n}$ is a unit vector pointing in some direction. This is 1/2 the sum of two operators, one which translates the wave function by distance L in the $\mathbf{n}$ direction and the other in the $-\mathbf{n}$ direction. Thus, its expectation value gives the overlap of the wave function with itself (all the wave functions are real and positive) when so translated. For any density matrix, it follows from the collapse part of Eq. (9.40) that

$$\frac{d\overline{\cos\hat{\mathbf{P}} \cdot \mathbf{n}L}}{dt} = -\lambda N^2\big[1 - e^{-\frac{L^2}{4a^2}}\big]\overline{\cos\hat{\mathbf{P}} \cdot \mathbf{n}L} \tag{9.47}$$

Thus, for $L \gg a$, the ensemble average of the overlap rate of the collapsing wave functions decreases as $\approx \lambda N^2$ but then it slows, e.g., for $L = a$, the collapse rate is $\approx .2\lambda N^2$.

9.6.2 *Small Packet*

If the size of the wave function is less than a, one can utilize an approximate density matrix evolution equation obtained by expanding the exponential in Eq. (9.40), retaining only the leading term:

$$\frac{d\rho(t)}{dt} = -i\left[\frac{\hat{\mathbf{P}}^2}{2MN}, \rho(t)\right] - \frac{\lambda N^2}{4a^2}\sum_{1=1}^{3}[\hat{X}_i, [\hat{X}_i, \rho(t)]]. \tag{9.48}$$

The state vector evolution which yields this density matrix evolution is[3]

$$|\phi, t\rangle = \mathcal{T}e^{-i\int_0^t dt' \frac{1}{2MN}\hat{\mathbf{P}}^2} e^{-\frac{1}{4\lambda}\int_0^t dt' \sum_{i=1}^{3}[w_i(t') - \sqrt{2}\lambda N a^{-1}\hat{X}_i]^2}|\phi, 0\rangle. \tag{9.49}$$

When the initial wave function is a gaussian, such as Eq. (9.45) with $D < a$, since the Schrödinger equation is quadratic in $\hat{P}$ and $\hat{X}$, the solution is a gaussian. The exact solution to this problem can be found [27–29]. We shall arrive at it here using the formalism we have presented. It suffices to solve the one-dimensional problem since, with initial wave function (9.45), Eq. (9.49) is the product of three terms, one for each dimension.

We assume that the wave function at any time has the form

$$\psi(X, t) = e^{-A(t)X^2 + B(t)X + C(t)}, \tag{9.50}$$

and proceed to solve the Schrödinger equation which follows from the time derivative of (9.49):

$$\begin{aligned}\frac{\partial}{\partial t}\psi(X, t) &= \frac{i}{2m}\frac{\partial^2}{\partial X^2}\psi(X, t)\\ &\quad - \left[\frac{1}{4\lambda}w^2(t) - \frac{\tilde{\lambda}}{\lambda}w(t)X + \frac{\tilde{\lambda}^2}{\lambda}X^2\right]\psi(X, t),\end{aligned} \tag{9.51}$$

where $m \equiv NM$ and $\tilde{\lambda} \equiv \lambda N/\sqrt{2}a$. Inserting (9.50) into (9.51) we obtain

$$\dot{A} = \frac{-2i}{m}A^2 + \frac{\tilde{\lambda}^2}{\lambda}, \tag{9.52a}$$

$$\dot{B} = \frac{-2i}{m}AB + \frac{\tilde{\lambda}}{\lambda}w(t). \tag{9.52b}$$

Equation (9.52a) is a Ricatti equation, and can be solved by the ansatz $A = (m/2i)\dot{F}/F$. It follows from (9.52a) that $\ddot{F} = F(2i\tilde{\lambda}^2/m\lambda)$. Thus, $F = \exp \pm\alpha(1 + i)t$, where $\alpha \equiv \tilde{\lambda}/\sqrt{m\lambda}$ and

$$A = \frac{m\alpha(1-i)}{2}\frac{e^{\alpha(1+i)t} - Ke^{-\alpha(1+i)t}}{e^{\alpha(1+i)t} + Ke^{-\alpha(1+i)t}}, \tag{9.53}$$

---

[3]How can Eq. (9.49), where w is just a function of t, arise from Eq. (9.41), where w is a field, depending upon $\mathbf{x}$ as well as t? As far as I am aware, this has not been discussed before, so we treat it in Appendix B. More generally, it involves changing the collapse-generating operators A^α to a new, equivalent set, with concomitant change of white noise functions $w^\alpha(t)$ to a new, equivalent set.

where K is a constant depending upon D.

We see that A, which characterizes the squared standard deviation of X, is the same for all $w(t)$. Thus, the wave function *approaches an equilibrium size*, independently of its initial spread D. The equilibrium occurs because the Schrödinger evolution tends to spread the wave function while the collapse evolution tends to narrow it. This takes place in characteristic time $\alpha^{-1} \approx 5 \times 10^4/N$ sec. The equilibrium spread in X (its standard deviation) is $1/\sqrt{2(A+A^*)} = 1/\sqrt{2m\alpha} \approx 4/N^{1/2}$ cm.

We shall henceforth assume either that the collapse process starts at negative times so that equilibrium is reached at time 0, or that A initially has its equilibrium value. Putting that value into (9.52b) gives

$$\dot{B} = -\alpha(1+i)B + \frac{\tilde{\lambda}}{\lambda}w(t), \tag{9.54}$$

with solution

$$B(t) = \frac{\tilde{\lambda}}{\lambda}\int_0^t dt' w(t')e^{-\alpha(1+i)(t-t')}. \tag{9.55}$$

Knowing A and B, the expectation values of position and squared position can be found from (9.50):

$$\begin{aligned}\langle X\rangle &\equiv \frac{\langle\psi,t|X|\psi,t\rangle}{\langle\psi,t|\psi,t\rangle} = \frac{B+B^*}{2(A+A^*)}\\ &= \frac{\alpha}{\tilde{\lambda}}\int_0^t dt' w(t')e^{-\alpha(t-t')}\cos\alpha(t-t'), \end{aligned}\tag{9.56a}$$

$$\langle X^2\rangle \equiv \frac{\langle\psi,t|X^2|\psi,t\rangle}{\langle\psi,t|\psi,t\rangle} = \langle X\rangle^2 + \frac{1}{2(A+A^*)}. \tag{9.56b}$$

To complete the solution, we need to find $C(t)$ but, since it is used to find the probability density $\langle\psi,t|\psi,t\rangle$, it is best that we calculate that directly from the Schrödinger equation:

$$\begin{aligned}\frac{d}{dt}\langle\psi,t|\psi,t\rangle &= -\frac{w^2(t)}{2\lambda}\langle\psi,t|\psi,t\rangle + 2\frac{\tilde{\lambda}}{\lambda}w(t)\langle\psi,t|X|\psi,t\rangle - 2\frac{\tilde{\lambda}^2}{\lambda}\langle\psi,t|X^2|\psi,t\rangle\\ &= -\frac{1}{2\lambda}\big[w(t) - 2\tilde{\lambda}\langle X\rangle\big]^2\langle\psi,t|\psi,t\rangle - \frac{\tilde{\lambda}^2}{\lambda}\frac{1}{A+A^*}\langle\psi,t|\psi,t\rangle. \end{aligned}\tag{9.57}$$

Therefore, omitting the time-dependent factor arising from the last term of (9.57) (which is absorbed in the normalization of the probability), and defining a new set of white noise functions

$$v(t) \equiv w(t) - 2\tilde{\lambda}\langle X\rangle = w(t) - 2\alpha\int_0^t dt' w(t')e^{-\alpha(t-t')}\cos\alpha(t-t'), \tag{9.58}$$

the probability density is simply

$$\langle\psi,t|\psi,t\rangle = e^{-\frac{1}{2\lambda}\int_0^t dt' v^2(t')}. \tag{9.59}$$

We note that $Dw = Dv$, since it follows from (9.58) that the Jacobian of the transformation from w's to v's has 1's on the diagonal and 0's above the diagonal.

In order to use (9.59), it is necessary to obtain the inverse of the transformation (9.58). This can be done by taking the second derivative of (9.58), with the result

$$\frac{d^2w(t)}{dt^2} = \frac{d^2v(t)}{dt^2} + 2\alpha\frac{dv(t)}{dt} + 2\alpha v(t). \tag{9.60}$$

Defining $v(t)$'s Brownian motion $\tilde{B}(t)$ by $v(t) = d\tilde{B}(t)/dt$. It then follows from (9.60) that $w(t)$ can variously be written as

$$w(t) = v(t) + 2\alpha\int_0^t dt_1 v(t_1) + 2\alpha^2\int_0^t dt_1\int_0^{t_1} dt_2 v(t_2), \tag{9.61a}$$

$$= v(t) + 2\alpha\int_0^t dt_1 v(t_1)\big[1+\alpha(t-t_1)\big], \tag{9.61b}$$

$$= v(t) + 2\alpha\tilde{B}(t) + 2\alpha^2\int_0^t dt'\tilde{B}(t'). \tag{9.61c}$$

It then follows from the first equation in (9.58) that $\langle X\rangle$ can be written as

$$\langle X\rangle = \frac{1}{\sqrt{m\lambda}}\left[\tilde{B}(t) + \alpha\int_0^t dt'\tilde{B}(t')\right]. \tag{9.62}$$

One can then show, using (9.50), (9.55), (9.61a)–(9.61c) and (9.62), that

$$\langle P\rangle = 2iA\langle X\rangle - iB(t) = \frac{\tilde{\lambda}}{\lambda}\tilde{B}(t). \tag{9.63}$$

This problem is completely solved. We see from Eqs. (9.62), (9.63) that, after the equilibrium packet size is achieved, the momentum expectation value undergoes Brownian motion and the position expectation value undergoes a motion that can be described as Brownian+.

Any expectation value can be calculated, and any ensemble average expectation value can be calculated. For example, although it can readily be found using the density matrix, the ensemble average of the squared position expectation value can be found from Eq. (9.56b), using $\langle X\rangle = (1/2\tilde{\lambda})[w(t)-v(t)]$ (Eq. (9.58)), Eq. (9.61b), and $\overline{v(t)v(t')} = \lambda\delta(t-t')$ (which follows from (9.59)):

$$\overline{\langle X^2\rangle} = \frac{1}{2m\alpha} + \frac{1}{m\lambda}\overline{\left[\int_0^t dt' v(t')\big[1+\alpha(t-t')\big]\right]^2}$$

$$= \frac{1}{2m\alpha} + \frac{1}{m}\int_0^t dt'\big[1 + \alpha(t - t')\big]^2$$
$$= \frac{1}{2m\alpha} + \frac{1}{m}\big[t + \alpha t^2 + \alpha^2 t^3/3\big]. \tag{9.64}$$

$\overline{X}^2 \sim t$ behavior occurs for classical Brownian motion, modeled as a particle undergoing Newtonian dynamics with a random force and a viscous damping force. In this case, the average Brownian 'step' size is constant in time.

$\overline{X}^2 \sim t^3$ behavior occurs for classical Brownian motion when the viscous damping is removed. This is essentially because the average Brownian 'step' size increases with time.

This 'anomalous' $\sim t^3$ behavior is also the quantum behavior of an object larger than a and, since it grows larger than the $\sim t$ random walk predicted by standard quantum theory, it can be used to experimentally test CSL [28].

So, we have the picture of the final result of collapse, a wave packet of equilibrium size which undergoes classical random walk without viscous damping, with momentum generally increasing as it undergoes classical random walk.

9.7 Collapse of Interfering Packets

It follows from the density matrix evolution Eq. (9.40) that the interaction picture density matrix $\tilde{\rho}(t) \equiv U^\dagger(t)\rho(t)U(t)$ $(U(t) \equiv \exp(-i\hat{\mathbf{P}}^2/2m))$ satisfies

$$\frac{d\tilde{\rho}(t)}{dt} = -\lambda N^2 U^\dagger(t)\big[1 - e^{-\frac{1}{4a^2}[\hat{\mathbf{X}}_L - \hat{\mathbf{X}}_R]^2}\big]U(t)\tilde{\rho}(t), \tag{9.65}$$

with solution

$$\tilde{\rho}(t) = \mathcal{T}e^{-\lambda N^2 \int_0^t dt' U^\dagger(t')[1 - e^{-\frac{1}{4a^2}[\hat{\mathbf{X}}_L - \hat{\mathbf{X}}_R]^2}]U(t')}\rho(0) \tag{9.66}$$

or, going back to the density matrix $\rho(t)$,

$$\rho(t) = \mathcal{T}e^{-\lambda N^2 \int_0^t dt' U(t-t')[1 - e^{-\frac{1}{4a^2}[\hat{\mathbf{X}}_L - \hat{\mathbf{X}}_R]^2}]U^\dagger(t-t')}\rho_0(t), \tag{9.67}$$

where $\rho_0(t) \equiv \exp(-iHt)\rho(0)\exp(iHt)$ is the density matrix without collapse. In the position representation, Eq. (9.67) is

$$\langle\mathbf{X}|\rho(t)|\mathbf{X}'\rangle = \mathcal{T}e^{-\lambda N^2 \int_0^t dt'[1 - e^{-\frac{\mathbf{Z}^2(t-t')}{4a^2}}]}\langle\mathbf{X}|\rho_0(t)|\mathbf{X}'\rangle, \tag{9.68a}$$
$$\mathbf{Z}(t - t') \equiv \left(\mathbf{X} - \frac{t - t'}{mi}\nabla\right) - \left(\mathbf{X}' + \frac{t - t'}{mi}\nabla'\right). \tag{9.68b}$$

We now note that, because $[X_i - X_i', \nabla_j + \nabla_j'] = 0$, it follows that $[\mathbf{Z}(t - t'), \mathbf{Z}(t - t'')] = 0$, and so the time-ordering operation $\mathcal{T}$ may be removed from

Eq. (9.68a). Also because this commutator vanishes, any product of powers of Z's can be written in 'normal-ordered form,' by which we mean that the X's are to the left of the ∇'s. Denoting the normal ordered form by : :, Eq. (9.68a) becomes

$$\langle\mathbf{X}|\rho(t)|\mathbf{X}'\rangle = :e^{-\lambda N^2\int_0^t dt'[1-e^{-\frac{\mathbf{Z}^2(t-t')}{4a^2}}]}:\langle\mathbf{X}|\rho_0(t)|\mathbf{X}'\rangle \tag{9.69}$$

We shall apply Eq. (9.69) to the case where the uncollapsed density matrix $\rho_0(t)$ is constructed from a number of wave packets,

$$\langle\mathbf{X}|\rho_0(t)|\mathbf{X}'\rangle = \sum_{n,n'} c_n c_{n'}^* \phi_n(\mathbf{X},t)\phi_{n'}^*(\mathbf{X}',t). \tag{9.70}$$

The wave packets $\phi_n(\mathbf{X},t)$ are to have well-defined momenta $\mathbf{k}_n(\mathbf{X})$ at (almost) each point of the wave packet, which itself has dimensions large compared to the wavelength. Thus, a wave packet could be a laboratory 'plane wave,' a good approximation to an eigenstate of momentum $\mathbf{k}$. It could be a cylindrical wave packet or a spherical wave packet of momentum magnitude k such as might be obtained by putting the 'plane' wave packet through a slit or a circular hole.

An important feature of such a packet $\phi_j(\mathbf{X},t)$ is that

$$\langle\mathbf{X}|\hat{\mathbf{P}}|\phi_n,t\rangle = \frac{1}{i}\nabla\phi_n(\mathbf{X},t) \approx \mathbf{k}_n(\mathbf{X})\phi_j(\mathbf{X},t) \tag{9.71}$$

is a very good approximation. Another important feature of such a packet is that (almost) each point in each wave packet can be considered as moving on a straight-line trajectory with constant velocity $\mathbf{k}_n(\mathbf{X})/m$.

Putting together Eqs. (9.68b), (9.69), (9.70), we obtain for the ensemble's probability density at $\mathbf{X}$:

$$\langle\mathbf{X}|\rho(t)|\mathbf{X}\rangle = \sum_{n,n'} c_n c_{n'}^* \phi_n(\mathbf{X},t)\phi_{n'}^*(\mathbf{X},t)$$
$$\cdot e^{-\lambda N^2\int_0^t dt'[1-e^{-\frac{1}{4a^2}[\mathbf{X}_n(t-t')-\mathbf{X}_{n'}(t-t')]^2}]}, \tag{9.72}$$

where

$$\mathbf{X}_n(t-t') \equiv \mathbf{X} - \frac{\mathbf{k}_n(\mathbf{X})}{m}(t-t'). \tag{9.73}$$

That is, consider a point on the nth packet which is located at $\mathbf{X}$ at time t. Then, $\mathbf{X}_n(t-t')$ is the location that point had on the nth packet at the earlier time t'.

To summarize, we have seen in Eq. (9.40) or (9.44) that, when a clump is put into a superposition of two places with constant separation D, the two states play the gambler's ruin game, so that the off-diagonal elements of the density matrix decay at the rate $\lambda N^2[1-\exp-(D^2/4a^2)]$. Equation (9.72) says that, for a superposition of packets, the points on the packets, which end up at the same place $\mathbf{X}$ at time t, may be thought of as playing the gambler's ruin game with each other on the way to **X**, with the above-mentioned distance-determining rate now varying with time, governing the collapse all along the way.

Although it is not of concern here, we mention that, of course, the spatially separated points of a single packet, or of multiple packets, likewise mutually play the gambler's ruin game, and that description is obtained by considering the off-diagonal elements of the density matrix.

9.7.1 Mach-Zender Interference

As is well known, the Mach-Zender interferometer has a rectangular shape, say, with half-silvered beam-splitters at the lower left and upper right corners, and fully-silvered mirrors at the other two corners. An incoming wave packet splits into two equal packets at the first beam splitter. The packet going $\rightarrow$, $\uparrow$ has its sign reversed when it reflects at 90° from the front-surfaced mirror. The packet going $\uparrow$, $\rightarrow$, $\uparrow$ gets no net sign change: one sign change at the first, front-surfaced, beam splitter, one at the front-surfaced mirror, none at the second, back-surfaced, beam splitter. Thus, without collapse, there is no output in the $\uparrow$ direction.

Although there are certainly velocity changes of the packets, they take place over a relatively brief time interval, so Eq. (9.72) may be applied seriatim. Let t be the time interval separating emergence from the two beam splitters. At time 0, the two packets start off with $c_1 = c_2 = 1/\sqrt{2}$. Thereafter, $|\mathbf{X}_1(t-t') - \mathbf{X}_2(t-t')| \gg a$. Moreover, the collapse rate is unaffected if a packet changes sign. Finally, at time t, the second beam splitter has just made the amplitudes $c_1 = c_2 = 1/2$ and, if packet 1 is $\phi_1(\mathbf{X}, t)$, then packet 2 is $-\phi_1(\mathbf{X}, t)$.

Therefore, in Eq. (9.72), since $\exp -[\mathbf{X}_n - \mathbf{X}_n]^2/4a^2 = 1$, $\exp -[\mathbf{X}_1 - \mathbf{X}_2]^2/4a^2 \approx 0$, for a point $\mathbf{X}$ in the superposed wave packets,

$$\langle \mathbf{X}|\rho(t)|\mathbf{X}\rangle = \frac{1}{4}|\phi_1(\mathbf{X}, t)|^2 2\big[1 - e^{-\lambda N^2 t}\big], \tag{9.74}$$

and the probability that the clump emerges in the upward direction is

$$P_\uparrow = \int d\mathbf{X} \langle \mathbf{X}|\rho(t)|\mathbf{X}\rangle = \frac{1}{2}\big[1 - e^{-\lambda N^2 t}\big]. \tag{9.75}$$

We see that, as time spent in the interferometer increases, $P_\uparrow \rightarrow 1/2$ since, asymptotically, only one packet survives to hit the second beam splitter, and that packet has equal likelihood of going $\uparrow$ or $\rightarrow$.

9.7.2 Two Slit Interference

It should be clear that Eq. (9.72) can be applied to any interference or diffraction problem. Here we shall just consider the effect of collapse on the two-slit Fraunhofer interference pattern, with neglect of single-slit diffraction.

Take the two slits to be located at $x = \pm b$ and parallel to the z-axis. The pattern is observed at a point $\mathbf{X}$ on a screen located at $y = L \gg b$. The vectors from the slits to the point on the screen are $\mathbf{r}_{1,2} \equiv \mathbf{X} \pm \mathbf{i}b$. Defining θ as the angle $\mathbf{X}$ makes with the y-axis, and working only to first order in θ, then $\mathbf{X} \approx L[\mathbf{i}\theta + \mathbf{j}]$, and $r_{1,2} \approx |\mathbf{X}| \pm b\theta$.

The two cylindrical packets $\phi_{1,2}$ have wave number k, and are of equal amplitude $c_{1,2} = 1/\sqrt{2}$ when they emanate from the two slits. They reach the screen at point $\mathbf{X}$ at time t, where their amplitudes are $A\exp \pm ikb\theta$. We are to consider the parts of the two packets traveling with speed k/m on a straight line, each from its slit to the screen. Their separation at any time t' is $|\mathbf{X}_1(t-t') - \mathbf{X}_2(t-t')| = 2b[1-(t'/t)]$. Putting this into Eq. (9.72) yields

$$\begin{aligned}
\langle \mathbf{X}|\rho(t)|\mathbf{X}\rangle &= \frac{A^2}{2}\Big[2 + \big(e^{2ikb\theta} + e^{-2ikb\theta}\big)e^{-\lambda N^2[t - \int_0^t dt' e^{-(\frac{b}{a})^2(1-\frac{t'}{t})^2}]}\Big] \\
&= A^2\Big[1 + \cos(2kb\theta)e^{-\lambda t N^2[1 - \frac{a}{b}\int_0^{\frac{b}{a}} dv e^{-v^2}]}\Big] \\
&= 2A^2\cos^2(kb\theta)e^{-\lambda t N^2[1 - \frac{\sqrt{\pi}a}{2b}\chi(b/a)]} \\
&\quad + A^2\Big[1 - e^{-\lambda t N^2[1 - \frac{\sqrt{\pi}a}{2b}\chi(b/a)]}\Big]
\end{aligned} \tag{9.76}$$

where $\chi(b/a) \equiv \mathrm{erf}(b/a)$.

Since t is the time to reach the screen, then $t \approx Lm/k$. Thus, we see that the two-slit two-packet interference pattern decays while the single packet non-interference pattern builds up as the screen is put further and further away. For $b \gg a$, the packet separation is $\gg a$ for almost all the time of travel, and the collapse rate is λN^2, as in the previous section. For $b \ll a$, the collapse rate is $\lambda N^2 b^2/3a^2$.

This concludes our discussion of free particle collapse dynamics, and this paper.

Appendix A: Proof That *R* and *S* Must be Diagonal

We prove here that the real symmetric operators R and S in the Stratonovich Schrödinger equation for the un-normalized state vector,

$$d|\phi, t\rangle = \big[R dB' + S dt\big]|\phi, t\rangle \tag{9.77}$$

must be diagonal in the $|a_n\rangle$ basis. This is in order that Eq. (9.77) give rise to the Itô gambler's ruin condition Eq. (9.5),

$$dx_n(t) = b_n(\mathbf{x})dB(t). \tag{9.78}$$

After putting Eq. (9.4), $dB' = dB + f dt$, into Eq. (9.77), we convert that Stratonovich equation to an Itô equation, with the result

$$d|\phi, t\rangle = [R dB + V dt]|\phi, t\rangle \quad \text{where } V \equiv S + Rf + \frac{\lambda}{2}R^2. \tag{9.79}$$

We note that V is also a real symmetric operator and, if we show R and V must be diagonal, then S must also be diagonal.

Using the rules for manipulating Itô equations, it is straightforward to find

$$d|\phi,t\rangle\langle\phi,t| = \big\{[RdB + Vdt], |\phi,t\rangle\langle\phi,t|\big\} + \lambda dt R|\phi,t\rangle\langle\phi,t|R, \tag{9.80a}$$

$$d\langle\phi,t|\phi,t\rangle = 2[\langle\phi,t|R|\phi,t\rangle dB + \langle\phi,t|V|\phi,t\rangle dt] + \lambda dt\langle\phi,t|R^2|\phi,t\rangle, \tag{9.80b}$$

where $\{M,N\} \equiv MN + NM$. Defining the density matrix $\rho(t) \equiv |\phi,t\rangle\langle\phi,t|/\langle\phi,t|\phi,t\rangle$ and $\overline{M} \equiv \text{Trace}M\rho$, we obtain from Eqs. (9.80a)–(9.80b) and the Itô rules:

$$\begin{aligned} d\rho &= \big[\{R,\rho\} - 2\rho\overline{R}\big]dB + dt\big[\{V,\rho\} - 2\rho\overline{V}\big] \\ &\quad + \lambda dt\big[\big[R\rho R - \rho\overline{R^2}\big] - 2\overline{R}\big[\{R,\rho\} - 2\rho\overline{R}\big]\big]. \end{aligned} \tag{9.81}$$

Now, $x_n(t) = \langle a_n|\rho(t)|a_n\rangle$. Thus, in order that the diagonal elements of Eq. (9.81) agree with Eq. (9.78), we see that the diagonal elements of Eq. (9.81) which do not multiply dB must vanish for *arbitrary* ρ:

$$0 = \big[\{V,\rho\}_{nn} - 2\rho_{nn}\overline{V}\big] + \lambda\big[\big[(R\rho R)_{nn} - \rho_{nn}\overline{R^2}\big] - 2\overline{R}\big[\{R,\rho\}_{nn} - 2\rho_{nn}\overline{R}\big]\big], \tag{9.82}$$

where $M_{nm} \equiv \langle a_n|M|a_m\rangle$

First, suppose that $\rho_{mm} = 1$, where $m \neq n$, and all other matrix elements of ρ vanish. It follows from Eq. (9.82) that

$$0 = (R\rho R)_{nn} = (R_{nm})^2. \tag{9.83}$$

That is, all the off-diagonal elements of R vanish, so R is diagonal.

Second, choose a density matrix for which ρ_{nn}, $\rho_{mm} = 1 - \rho_{nn}$, ρ_{nm} do not vanish, but all other matrix elements of ρ do vanish. Then, using the diagonal nature of R, Eq. (9.82) may be written as

$$\begin{aligned} 0 &= 2V_{nm}\rho_{nm}[1 - 2\rho_{nn}] \\ &\quad + \rho_{nn}[1 - \rho_{nn}][2(V_{nn} - V_{mm}) \\ &\quad + \lambda(R_{nn} - R_{mm})\big[(R_{nn} + R_{mm}) \\ &\quad - 4\big(R_{nn}\rho_{nn} + R_{mm}(1 - \rho_{nn})\big)\big]. \end{aligned} \tag{9.84}$$

For fixed ρ_{nn}, a viable density matrix (non-negative eigenvalues which add up to 1) exists for $|\rho_{nm}| \leq \sqrt{\rho_{nn}(1 - \rho_{nn})}$. But, as ρ_{nm} is varied, the first term in Eq. (9.84) varies while the rest of the terms remain fixed. Thus, the first term must vanish, and this means that $V_{nm} = 0$ for $n \neq m$, i.e., V is diagonal as well as R.

Appendix B: Transformation of Operators and White Noise

Consider the general CSL form for the evolution of the state vector, Eq. (9.33)

$$|\phi,t\rangle=\mathcal{T}e^{-\int_0^t dt'\{iH(t')+\frac{1}{4\lambda}\sum_\alpha[w^\alpha(t')-2\lambda A^\alpha]^2\}}|\phi,0\rangle \tag{9.85}$$

We introduce a real orthonormal set of vectors u^α_β, i.e., $\sum_\alpha u^\alpha_\beta u^\alpha_{\beta'}=\delta_{\beta\beta'}$, $\sum_\beta u^\alpha_\beta u^{\alpha'}_\beta=\delta_{\alpha\alpha'}$. Defining white noise functions $v^\beta(t)$ and complete commuting set of operators Z^β by $w^\alpha(t)\equiv\sum_\beta u^\alpha_\beta v^\beta(t)$ and $A^\alpha(t)\equiv\sum_\beta u^\alpha_\beta Z^\beta(t)$, one readily sees that, in the exponent of Eq. (9.85),

$$\sum_\alpha\left[w^\alpha(t)-2\lambda A^\alpha\right]^2=\sum_\beta\left[v^\beta(t)-2\lambda Z^\beta\right]^2. \tag{9.86}$$

The Jacobian of the transformation from w's to v's is 1 so, in using the Probability Rule (9.32b), $Dw=Dv$.

We wish to apply such a transformation to Eqs. (9.41), (9.42) which, for simplicity, we limit to one-dimensional space:

$$|\phi,t\rangle=\mathcal{T}e^{-i\int_0^t dt'\frac{1}{2MN}\hat{P}^2}e^{-\frac{1}{4\lambda}\int_0^t dt'\int dx'[w(x',t')-2\lambda NA(x')]^2}|\phi,0\rangle, \tag{9.87a}$$

$$A(x')\equiv\frac{1}{(\pi a^2)^{1/4}}e^{-\frac{1}{2a^2}[x'-\hat{X}]^2}. \tag{9.87b}$$

We shall use as orthonormal functions the harmonic oscillator wave functions

$$u_n(x)\equiv C_nH_n(x/a)e^{-\frac{1}{2a^2}x^2}\quad\text{where } C_n\equiv\frac{1}{\sqrt{\pi^{1/2}2^nn!a}}. \tag{9.88}$$

With the definitions $v_n(t)\equiv\int dx\,w(x,t)u_n(x)$ and $\hat{Z}_n\equiv\int dx\,A(x)u_n(x)$, the exponent in Eq. (9.87a) may be written as

$$\begin{aligned}-\frac{1}{4\lambda}&\int_0^t dt'\int dx'\left[w(x',t')-2\lambda NA(x')\right]^2\\&=-\frac{1}{4\lambda}\sum_{n=0}^\infty\int_0^t dt'\left[v_n(t')-2\lambda N\hat{Z}_n\right]^2.\end{aligned} \tag{9.89}$$

Thus, we see how a white-noise field gets converted to an equivalent sum of white noise functions.

Using the identity $\exp(-t^2+2tz)=\sum_{n=0}^\infty t^nH_n(z)/n!$, with $t\equiv\hat{X}/2a$, $z\equiv x'/a$, we find

$$\hat{Z}_n=\int dx\frac{1}{(\pi a^2)^{1/4}}e^{-\frac{1}{2a^2}[x'-\hat{X}]^2}u_n(x)$$

$$= \frac{1}{C_n(\pi a^2)^{1/4}} e^{-\frac{1}{4a^2}\hat{X}^2} \sum_{m=0}^{\infty} \frac{(\hat{X}/2a)^m}{m!} \int dx u_m(x) u_n(x)$$

$$= e^{-\frac{1}{4a^2}\hat{X}^2} \frac{(\hat{X}/\sqrt{2}a)^n}{\sqrt{n!}}. \tag{9.90}$$

This leads to the density matrix evolution equation

$$\frac{d\rho(t)}{dt} = -i\left[\frac{\hat{P}^2}{2MN}, \rho(t)\right] - \frac{\lambda N^2}{2} \sum_{n=0}^{\infty} \left[e^{-\frac{1}{4a^2}\hat{X}^2} \frac{(\hat{X}/\sqrt{2}a)^n}{\sqrt{n!}}, \left[e^{-\frac{1}{4a^2}\hat{X}^2} \frac{(\hat{X}/\sqrt{2}a)^n}{\sqrt{n!}}, \rho(t)\right]\right]. \tag{9.91}$$

If we expand $\exp -\hat{X}^2/4a^2$, we see that the $n = 0$ term goes as $(\hat{X}/a)^4$ and the rest of the terms go as $(\hat{X}/a)^n$ to lowest order. Therefore, the lowest order term comes from $n = 1$. Upon neglect of the higher order terms, this gives the density matrix evolution equation

$$\frac{d\rho(t)}{dt} = -i\left[\frac{\hat{P}^2}{2MN}, \rho(t)\right] - \frac{\lambda N^2}{4a^2}\left[\hat{X}, \left[\hat{X}, \rho(t)\right]\right], \tag{9.92}$$

which is identical to the one-dimensional version of Eq. (9.48).

References

1. W. Feller, *An Introduction to Probability Theory and Its Applications* (Wiley, New York, 1950). Chap. 14
2. P. Pearle, Found. Phys. **12**, 249 (1982)
3. P. Pearle, Phys. Rev. A **39**, 2277 (1989)
4. P. Pearle, Phys. Rev. D **13**, 857 (1976)
5. G. Lindblad, Commun. Math. Phys. **48**, 119 (1976)
6. V. Gorini, A. Kossakowski, E.C.G. Sudarshan, J. Math. Phys. **17**, 821 (1976)
7. P. Pearle, Eur. J. Phys. **33**, 805 (2012)
8. Ł.D. Carson, W.R. Miles, S.S. Stevens, Vision, hearing and aeronautical design. Sci. Mon. **56**, 446 (1943)
9. G.C. Ghirardi, P. Pearle, A. Rimini, Phys. Rev. A **42**, 78 (1990)
10. P. Pearle, E. Squires, Phys. Rev. Lett. **73**, 1 (1994)
11. B. Collett, P. Pearle, F. Avignone, S. Nussinov, Found. Phys. **25**, 1399 (1995)
12. P. Pearle, J. Ring, J.I. Collar, F.T. Avignone, Found. Phys. **29**, 465 (1999)
13. G. Jones, P. Pearle, J. Ring, Found. Phys. **34**, 1467 (2004)
14. W. Feldmann, R. Tumulka, J. Phys. A, Math. Theor. **45**, 065304 (2012)
15. G.C. Ghirardi, A. Rimini, T. Weber, Phys. Rev. D **34**, 470 (1986)
16. G.C. Ghirardi, A. Rimini, T. Weber, Phys. Rev. D **36**, 3287 (1987)
17. G.C. Ghirardi, A. Rimini, T. Weber, Found. Phys. **18**, 1 (1988)

18. S.L. Adler, J. Phys. A **40**, 2935 (2007)
19. P. Pearle, Phys. Rev. A **72**, 022112 (2005)
20. P. Pearle, Phys. Rev. A **71**, 032101 (2005)
21. P. Pearle, in *Sixty-Two Years of Uncertainty*, ed. by A. Miller (Plenum, New York, 1990), p. 193
22. G.C. Ghirardi, R. Grassi, P. Pearle, Found. Phys. **20**, 1271 (1990)
23. P. Pearle, Phys. Rev. A **59**, 80 (1999)
24. O. Nicrosini, A. Rimini, Found. Phys. **33**, 1061 (2003)
25. D.J. Bedingham, Found. Phys. **41**, 686 (2010)
26. Y. Aharonov, D. Rohrlich, *Quantum Paradoxes* (Wiley, Weinheim, 2005). Chap. 5
27. L. Diosi, Phys. Rev. A **40**, 1165 (1989)
28. B. Collett, P. Pearle, Found. Phys. **33**, 1495 (2003). Appendix B, pp. 1524–1529
29. A. Bassi, J. Phys. A, Math. Gen. **38**, 3173 (2005)

Chapter 10
Many Worlds, the Born Rule, and Self-Locating Uncertainty

Sean M. Carroll and Charles T. Sebens

Abstract We provide a derivation of the Born Rule in the context of the Everett (Many-Worlds) approach to quantum mechanics. Our argument is based on the idea of self-locating uncertainty: in the period between the wave function branching via decoherence and an observer registering the outcome of the measurement, that observer can know the state of the universe precisely without knowing which branch they are on. We show that there is a uniquely rational way to apportion credence in such cases, which leads directly to the Born Rule. [*Editors note*: for a video of the talk given by Prof. Carroll at the Aharonov-80 conference in 2012 at Chapman University, see quantum.chapman.edu/talk-14.]

10.1 Introduction

A longstanding puzzle in the Many-Worlds or Everett approach to quantum mechanics[1] (EQM) is the origin of the Born Rule: the probability of finding a post-measurement system in an eigenstate $|a\rangle$ of an observable A, given that the system is prepared in state $|\psi\rangle$, is given by $|\langle a|\psi\rangle|^2$. Here we summarize and discuss the resolution of this problem that we recently developed [3], in which the Born Rule is argued to be the uniquely rational way of dealing with the self-locating uncertainty that inevitably accompanies branching of the wave function. A similar approach has been advocated by Vaidman [4]; our formal manipulations closely parallel those of Zurek [5].

Ours is certainly not the first attempt to derive the Born Rule within EQM. One approach is to show that, in the limit of many observations, branches that do not

S.M. Carroll (✉)
Physics Department, California Institute of Technology, Pasadena, CA, USA
e-mail: seancarroll@gmail.com

C.T. Sebens
Philosophy Department, University of Michigan, Ann Arbor, MI, USA
e-mail: csebens@gmail.com

[1]Everett's original paper is [1]. A comprehensive introduction to the theory in its modern version can be found in [2].

D.C. Struppa, J.M. Tollaksen (eds.), *Quantum Theory: A Two-Time Success Story*,
DOI 10.1007/978-88-470-5217-8_10,

obey the Born Rule have vanishing measure [6–8]. A more recent twist is to use decision theory to argue that a rational agent should act as if the Born Rule is true [9–12]. Another approach is to argue that the Born Rule is the only well-defined probability measure consistent with the symmetries of quantum mechanics [5, 13].

While all of these ideas have some degree of merit, they don't seem to have succeeded in convincing a majority of experts in the field. Our purpose here is not to criticize other approaches (there may be many valid ways to derive a correct answer), but to provide a simple and hopefully transparent alternative derivation that is physics-oriented while offering a clear answer to the question of how probabilities arise in EQM, a deterministic theory.

The main idea we use is that of *self-locating uncertainty* [14]: the condition of an observer who knows that the environment they experience occurs multiple times in the universe, but doesn't know which example they are actually experiencing. We argue that such a predicament inevitably occurs in EQM, during the "post-measurement/pre-observation" period between when the wave function branches and when the observer registers the affect of the branching. A naive analysis might indicate that, in such a situation, each branch should be given equal likelihood; here we demonstrate that a more careful treatment leads us inevitably to the Born Rule for probabilities.

10.2 Everettian Quantum Mechanics

In EQM, the quantum state is described by a vector $|\Psi\rangle$ in a Hilbert space $\mathcal{H}$, evolving under the influence of a self-adjoint Hamiltonian H according to Schrödinger's equation

$$H|\Psi\rangle = i\hbar\partial_t|\Psi\rangle. \tag{10.1}$$

This smooth unitary evolution is supposed to account for absolutely all the quantum dynamics; there is no separate rule governing "wave function collapse." Rather, we model the observer as well as the system as part of the quantum state, and unitary evolution causes the state of the universe to split into multiple non-interacting branches, each associated with a possible measurement outcome.

Consider an example in which the "system" is a single qubit initially in a state $|\psi\rangle = (1/\sqrt{2})(|\uparrow\rangle + |\downarrow\rangle)$, where the arrow denotes the value of the spin along the z-axis. The observer is initially uncorrelated with the spin, in a ready state $|O_0\rangle$. The measurement process is described via the following form of unitary evolution:

$$|\Psi\rangle = \frac{1}{\sqrt{2}}|O_0\rangle\,(|\uparrow\rangle + |\downarrow\rangle) \tag{10.2}$$

$$\rightarrow \frac{1}{\sqrt{2}}\left(|O_\uparrow\rangle|\uparrow\rangle + |O_\downarrow\rangle|\downarrow\rangle\right). \tag{10.3}$$

Here, $|O_\uparrow\rangle$ and $|O_\downarrow\rangle$ represent states in which the observer has measured spin-up and spin-down, respectively. The wave function has not collapsed, but the observer is now described by a superposition of different measurement outcomes.

The first challenge for such an approach is obvious: in the real world, it never *feels* like we are in a superposition of measurement outcomes. We see the spin up or down, Schrödinger's Cat alive or dead—never a superposition of different possibilities. Everett's insight was that, if a measurement of a spin that was originally in either of the eigenstates $|\uparrow\rangle$ or $|\downarrow\rangle$ leaves the observer with the impression of a definite measurement outcome, then the linearity of quantum mechanics implies that a superposition of such states should lead to two definite experiences. The wave function in Eq. (10.3) represents two agents seeing two different outcomes, not one agent somehow experiencing an indeterminate outcome.

This story becomes more plausible once decoherence is understood as a crucial part of quantum mechanics. In a realistic situation, the observer and system do not constitute the entire universe; there is also an environment, generally with many more degrees of freedom. Initially the environment, like the observer, is in a state $|\omega_0\rangle$ that is unentangled with the system under consideration. But if the system and the environment are allowed to interact—as is practically inevitable if the system is a macroscopic object like Schrödinger's Cat, constantly radiating and breathing (or failing to) and so forth—then entanglement with the environment quickly ensues (typically before entanglement with the observer):

$$\begin{aligned} |\Psi\rangle &= \frac{1}{\sqrt{2}}|O_0\rangle\,(|\uparrow\rangle + |\downarrow\rangle)\,|\omega_0\rangle \\ &\to \frac{1}{\sqrt{2}}|O_0\rangle\,\big(|\uparrow\rangle|\omega_\uparrow\rangle + |\downarrow\rangle|\omega_\downarrow\rangle\big)\,. \end{aligned} \tag{10.4}$$

In a generic situation, the entangled environment states will be nearly orthogonal: $\langle\omega_\uparrow|\omega_\downarrow\rangle \approx 0$. In that case, the component describing the up spin will no longer be able to interfere with the component describing the down spin. We say that decoherence has occurred, and the wave function has branched. Decoherence helps explain how EQM is a theory of distinct causally well-isolated "worlds."[2]

A popular objection to EQM is that it is ontologically extravagant—an incredible number of unobservable worlds are invoked to help explain observations within the single world to which we have access. This objection is misplaced. Any viable version of quantum mechanics involves a Hilbert space $\mathcal{H}$ of very high dimensionality. The holographic principle suggests that the dimensionality of the Hilbert space describing our observable universe is at least $\exp(10^{120})$ [15], and there is good reason to believe it is infinite [16]. In EQM, the size of Hilbert space remains fixed, but the state vector describes an increasing number of distinct worlds as it evolves. The potential for describing many worlds was always there; the objection that there are too many universes is really an objection that Hilbert space is too big, which would apply equally well to any approach to quantum mechanics which includes a state vector. A proper measure of ontological extravagance relies on the number of types of fundamental entities proposed by the theory (like the wave function) and laws

[2] EQM is time-symmetric, but branching occurs toward the future, and not toward the past, because the low-entropy early universe was relatively free of entanglements between subsystems.

that govern them, not the number of large scale structures (like quantum worlds) which emerge from them. EQM, which requires only a vector in a Hilbert space and a single evolution law, is ontologically quite restrained.

A more pressing concern is that the formalism of EQM offers little guidance to the preferred basis problem—why do we collapse onto certain states and not others? We do not address this question in this paper, but there has been significant progress in understanding the origin of "pointer states" which arise from decoherence and are robust under macroscopic perturbations [17, 18]. Philosophically, there has been progress made in understanding how the many worlds of quantum mechanics can be emergent, arising dynamically from unitary evolution and not requiring the addition of new laws to govern their creation [19].

Our concern here is with the origin of the Born Rule. In an equation such as (10.4), it is unclear what role the coefficients multiplying each branch should play for an observer living within the wave function. We will argue that they play a crucial role in justifying a probability calculus that leads us to the Born Rule. Along the way, we will see how probability can arise in a deterministic theory as agents evolve from perfect knowledge to self-locating uncertainty.

10.3 Self-Locating Uncertainty

Modern theories of cosmology often invoke "large universes"—ones in which any given local situation (such as a particular observer, in a particular macroscopic quantum state, with particular data about their surroundings) is likely to occur multiple times [20–22]. The setting could be something as dramatic as an inflationary multiverse, or as relatively pedestrian as an homogeneous cosmology with sufficiently large spatial sections. If the likely number of such duplicate observers is infinite, we face the cosmological measure problem. Even if it is finite, however, any one such observer finds themselves in a situation of self-locating uncertainty. They can know everything there is to know about the state of the universe and an arbitrary amount about their local environment, but still not be able to determine which such instantiation of that data they are experiencing.

This situation has been extensively studied in the philosophical literature (see *e.g.* [23–25]), often in the case of hypothetical exact duplications of existing persons rather than large universes. One intuitively obvious principle for assigning probabilities in the face of such uncertainty is "indifference," roughly: if all you know is that you are one of N occurrences of a particular set of observer data, you should assign equal credence (a.k.a. "degree of belief" or just "probability") $1/N$ to each possibility. Elga [26] has given convincing arguments in favor of indifference in the case of identical classical observers. Crucially, this result is not simply postulated as the simplest approach to the problem, but rather derived from seemingly innocuous principles of rational reasoning.

In EQM, self-locating uncertainty is inevitable: not with respect to different locations in space, but with respect to different branches of the wave function. Consider again the branching process described in Eq. (10.4), but now we include an explicit measuring apparatus A (which might represent an electron microscope, a Geiger counter, or other piece of experimental equipment). Imagine that we preserve quantum coherence in the system long enough to perform a measurement with the apparatus, which then (as a macroscopic object) rapidly becomes entangled with the environment and causes decoherence. Only then does the observer record the outcome of the measurement (normalizations have been omitted for convenience):

$$\begin{aligned} |\Psi\rangle &= |O_0\rangle\,(|\uparrow\rangle + |\downarrow\rangle)\,|A_0\rangle|\omega_0\rangle && (10.5)\\ &\to |O_0\rangle\,\big(|\uparrow\rangle|A_\uparrow\rangle + |\downarrow\rangle|A_\downarrow\rangle\big)\,|\omega_0\rangle && (10.6)\\ &\to |O_0\rangle\,\big(|\uparrow\rangle|A_\uparrow\rangle|\omega_\uparrow\rangle + |\downarrow\rangle|A_\downarrow\rangle|\omega_\downarrow\rangle\big) && (10.7)\\ &= |O_0\rangle|\uparrow\rangle|A_\uparrow\rangle|\omega_\uparrow\rangle + |O_0\rangle|\downarrow\rangle|A_\downarrow\rangle|\omega_\downarrow\rangle && (10.8)\\ &\to |O_\uparrow\rangle|\uparrow\rangle|A_\uparrow\rangle|\omega_\uparrow\rangle + |O_\downarrow\rangle|\downarrow\rangle|A_\downarrow\rangle|\omega_\downarrow\rangle. && (10.9) \end{aligned}$$

Each line represents unitary time evolution except for (10.8), in which we have merely distributed the observer state for clarity. That is the moment we describe as post-measurement/pre-observation. At that step, the wave function has branched—decoherence has occurred, as indicated by the different environment states. The observer is still described by a unique state $|O_0\rangle$, but there are two copies, one in each branch. Such an observer (who by construction doesn't yet know the outcome of the measurement) is in a state of self-locating uncertainty.

A particularly clear example of such uncertainty would be a real-world version of the Schrödinger's Cat experiment. An actual cat interacts strongly with its environment and would not persist in a coherent superposition of alive and dead for very long; the wave function would branch long before a human experimenter opens the box. But in fact such uncertainty is generic. The timescale for decoherence for a macroscopic apparatus is extremely short, generally much less than 10^{-20} sec. Even if we imagine an experimenter looking directly at a quantum system, the state of the experimenter's eyeballs would decohere that quickly. The timescale over which human perception occurs, however, is tens of milliseconds or longer. Even the most agile experimenter will experience *some* period of self-locating uncertainty in which they don't know which of several branches they are on, even if it is too brief for them to notice. Although the experimenter may not be quick-thinking enough to reason during this period, there are facts about what probabilities they ought to assign before they get the measurement data.

Naively, the combination of indifference over indistinguishable circumstances and self-locating uncertainty when wave functions branch is a disaster for EQM, rather than a way forward. Consider a case in which the amplitudes are unequal for two branches:

$$|\Psi\rangle = \sqrt{\frac{1}{3}}|O_0\rangle|\uparrow\rangle|\omega_\uparrow\rangle + \sqrt{\frac{2}{3}}|O_0\rangle|\downarrow\rangle|\omega_\downarrow\rangle. \qquad (10.10)$$

The conditions of the two observers would seem to be indistinguishable from the inside; there is no way they can "feel" the influence of the amplitudes multiplying their branches of the wave function. Therefore, one might be tempted to conclude that Elga's principle of indifference implies that probabilities in EQM should be calculated by *branch-counting* rather than by the Born Rule—every branch should be given equal weight, regardless of its amplitude. In this case, Eq. (10.10), that means assigning equal 50/50 probability to up and down even though the branch weights are unequal. This would be empirically disastrous, as real quantum measurements don't work that way. We will now proceed to show why such reasoning is incorrect, and in fact a proper treatment of self-locating uncertainty leads directly to the empirically desirable conclusion.[3]

10.4 The Epistemic Separability Principle

We base our derivation of the Born Rule on what we call the *Epistemic Separability Principle* (ESP), roughly: the outcome of experiments performed by an observer on a specific system shouldn't depend on the physical state of other parts of the universe (for a more careful discussion see [3]). If I set out to measure the z-component of a spin in my laboratory, the probability of a particular outcome should be independent of the quantum state of some other spin in a laboratory on an alien planet around a distant star in the Andromeda galaxy. An essentially equivalent assumption is made by Elga in his discussion of classical self-locating uncertainty [26].[4] The ESP applies in both quantum and classical contexts. In classical contexts, the ESP is compatible with Elga's indifference principle (see [3]). In quantum contexts, it mandates the Born rule. In EQM, the ESP amounts to the idea that the state of the environment shouldn't affect predictions that are purely about the observer/system Hilbert space.

Consider a Hilbert space that describes an observer, a system, and an environment:

$$\mathcal{H} = \mathcal{H}_O \otimes \mathcal{H}_S \otimes \mathcal{H}_E. \tag{10.11}$$

We consider general states of the universe, described by a state vector

$$|\Psi\rangle = \sum_{a,i,\mu} \Psi_{a,i,\mu} |\psi_a\rangle |\phi_i\rangle |\omega_\mu\rangle, \tag{10.12}$$

where $\{\psi_a\}$, $\{\phi_i\}$, and $\{\omega_\mu\}$ are bases for the observer, system, and environment respectively, all of which are orthonormal: $\langle\psi_a|\psi_b\rangle = \delta_{ab}$ etc. Consider unitary trans-

[3]Page has recently argued that the prospect of classical self-locating uncertainty in large universes poses a crisis for quantum mechanics, as the Born Rule becomes insufficient for calculating the probability of measurement outcomes [8, 27–30]. Our approach provides a unified treatment of classical and quantum self-locating uncertainties, defusing the would-be crisis.

[4]The ESP is implicit in Elga's discussion of his TOSS & DUPLICATION thought experiment, where he notes that the outcome of an additional coin toss should not affect the credence we assign to being either an original or a duplicated person with identical experiences.

formations that act only on the environment, which we can write as

$$T = \mathbf{I}_{OS} \otimes U_E, \tag{10.13}$$

where U_E is a unitary matrix that acts on $\mathcal{H}_E$. Then we can formulate the ESP in the context of EQM as the statement that probabilities in the observer/system subspace are unchanged by such transformations (here s is a possible outcome of a measurement of the system S):

$$P(O \text{ measures } s|\Psi) = P(O \text{ measures } s|T[\Psi]). \tag{10.14}$$

In [3] we also offer a version of this principle using density matrices rather than directly in terms of transformations on states; roughly, the outcome of an experiment depends only on the reduced density matrix of the observer/system subspace. The two formulations are equivalent if we are comparing states with identical Hilbert spaces for the environment.

A key motivation behind EQM is that no additional assumptions should be added to the basic structure of the Hilbert space and unitary evolution. While the ESP might seem like an additional assumption, we believe it simply reflects the structure of a quantum theory in which the space of states can be factorized. Note that we are not assuming the absence of interactions between the system and environment (which would make decoherence impossible); only that changing the environment without changing the observer or system should leave experimental predictions unaltered.

10.5 Deriving the Born Rule

Consider a specific example where we have an observer, a spin with equal amplitudes to be up or down, and an environment (again omitting the overall normalization):

$$|\Psi\rangle = |O\rangle|\uparrow\rangle|\omega_1\rangle + |O\rangle|\downarrow\rangle|\omega_2\rangle. \tag{10.15}$$

Like Eq. (10.8), this is a state in which the observer has yet to observe the outcome and thus ought to be uncertain which branch they are on. The environment states are assumed to be orthonormal (by decoherence), and without loss of generality we can take them to be the first two elements of an orthonormal basis $\{|\omega_\mu\rangle\}$.

We can write any environment unitary U_E in the form

$$U_E = \sum_\mu |\tilde{\omega}_\mu\rangle\langle\omega_\mu|, \tag{10.16}$$

where the states $|\tilde{\omega}_\mu\rangle$ are another set of orthonormal vectors. We decompose the environment into a tensor product of two subsystems, one of which we will label as a "coin" (although it could be of arbitrary dimension) and the other includes everything else:

$$\mathcal{H}_E = \mathcal{H}_C \otimes \mathcal{H}_{\widehat{E}}. \tag{10.17}$$

Then we can construct an orthonormal basis $\{|\tilde{\omega}_\mu\rangle\}$ for the environment $\mathcal{H}_E$ in which the first two basis vectors take the form

$$|\tilde{\omega}_1\rangle = |H\rangle \otimes |\Omega\rangle, \tag{10.18}$$

$$|\tilde{\omega}_2\rangle = |T\rangle \otimes |\Omega\rangle, \tag{10.19}$$

where $|H\rangle$ (heads) and $|T\rangle$ (tails) are two orthonormal vectors in $\mathcal{H}_C$, and $|\Omega\rangle \in \mathcal{H}_{\widehat{E}}$.

Now we can construct two specific environment unitaries:

$$U_E^{(1)} = \sum_\mu |\tilde{\omega}_\mu\rangle\langle\omega_\mu|, \tag{10.20}$$

$$U_E^{(2)} = |\tilde{\omega}_2\rangle\langle\omega_1| + |\tilde{\omega}_1\rangle\langle\omega_2| + \sum_{\mu>2} |\tilde{\omega}_\mu\rangle\langle\omega_\mu|. \tag{10.21}$$

Acting on our state (10.15) we get

$$\begin{aligned} |\Psi_1\rangle &\equiv \big(\mathbf{I}_{OS} \otimes U_E^{(1)}\big)|\Psi\rangle \\ &= |O\rangle|\uparrow\rangle|H\rangle|\Omega\rangle + |O\rangle|\downarrow\rangle|T\rangle|\Omega\rangle \end{aligned} \tag{10.22}$$

and

$$\begin{aligned} |\Psi_2\rangle &\equiv \big(\mathbf{I}_{OS} \otimes U_E^{(2)}\big)|\Psi\rangle \\ &= |O\rangle|\uparrow\rangle|T\rangle|\Omega\rangle + |O\rangle|\downarrow\rangle|H\rangle|\Omega\rangle. \end{aligned} \tag{10.23}$$

In the $|\Psi_1\rangle$, the spin and the "coin" have become entangled so that the coin is heads if the particle was spin up, in $|\Psi_2\rangle$ the coin is heads if the particle was spin *down*.

By the ESP, Eq. (10.14), the probability that the observer will measure spin up or spin down is equal in all of these states, since they are related by unitary transformations on the environment:

$$P(\uparrow|\Psi) = P(\uparrow|\Psi_1) = P(\uparrow|\Psi_2), \tag{10.24}$$

$$P(\downarrow|\Psi) = P(\downarrow|\Psi_1) = P(\downarrow|\Psi_2). \tag{10.25}$$

However, we can also consider the coin to be "the system," and the spin as part of the environment. In that case, the two environments are related by a unitary transformation on the spin:

$$U_S = |\uparrow\rangle\langle\downarrow| + |\downarrow\rangle\langle\uparrow|. \tag{10.26}$$

Therefore, by analogous logic, the probability of the observer measuring heads or tails is equal in the two states $|\Psi_1\rangle$ and $|\Psi_2\rangle$:

$$P(H|\Psi_1) = P(H|\Psi_2), \tag{10.27}$$

$$P(T|\Psi_1) = P(T|\Psi_2). \tag{10.28}$$

Looking at the specific states in (10.22) and (10.23), we notice that the branch of the wave function in which the coin is heads is the same as the one where the spin is up in $|\Psi_1\rangle$, but the one where the spin is down in $|\Psi_2\rangle$. So, in $|\Psi_1\rangle$ the particle is spin up if and only if the coin is heads, and in $|\Psi_2\rangle$ the particle is spin down if and only if the coin is heads. (See Fig. 10.1.) We therefore have

$$P(\uparrow|\Psi_1) = P(H|\Psi_1), \tag{10.29}$$

$$P(\downarrow|\Psi_2) = P(H|\Psi_2). \tag{10.30}$$

Comparing with (10.27) we immediately get

$$P(\uparrow|\Psi_1) = P(\downarrow|\Psi_2), \tag{10.31}$$

and comparing that with (10.24) and (10.25) reveals

$$P(\uparrow|\Psi) = P(\downarrow|\Psi) = 1/2. \tag{10.32}$$

This is, of course, the result we expect from the Born Rule: when the components of the wave function have equal amplitudes, they get assigned equal probabilities. This shouldn't be surprising, as it is also what we would expect from naive branch-counting. However, notice that the equality of the amplitudes was crucially important, rather than merely incidental; had they not been equal, we would have been unable to fruitfully compare results from different unitary transformations on the environment.

It is therefore crucial to consider branches with unequal amplitudes. Here our logic follows that of Zurek [5]. Start with a state where one branch has an amplitude greater than the other by a factor of $\sqrt{2}$:

$$|\Psi\rangle = |O\rangle|\uparrow\rangle|\omega_1\rangle + \sqrt{2}|O\rangle|\downarrow\rangle|\omega_2\rangle. \tag{10.33}$$

We can change to a new environment basis $\{|\widehat{\omega}_\mu\rangle\}$, defined by

$$\begin{aligned}
|\omega_1\rangle &= |\widehat{\omega}_1\rangle \\
|\omega_2\rangle &= \frac{1}{\sqrt{2}}|\widehat{\omega}_2\rangle + \frac{1}{\sqrt{2}}|\widehat{\omega}_3\rangle \\
|\omega_3\rangle &= \frac{1}{\sqrt{2}}|\widehat{\omega}_2\rangle - \frac{1}{\sqrt{2}}|\widehat{\omega}_3\rangle \\
|\omega_\mu\rangle &= |\widehat{\omega}_\mu\rangle, \quad \mu > 3.
\end{aligned} \tag{10.34}$$

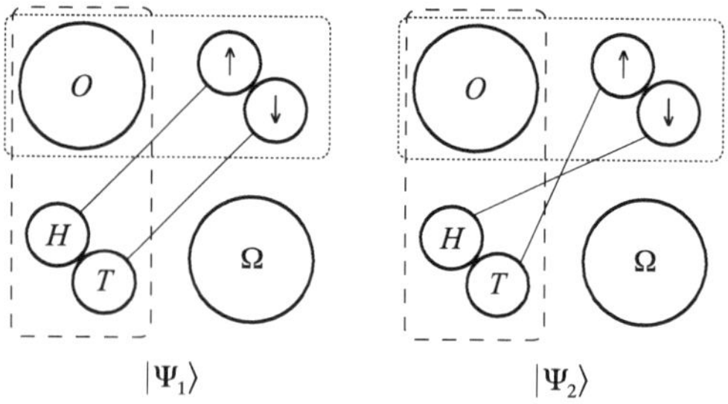

Fig. 10.1 A schematic representation of the setup behind our derivation of the Born Rule. The states $|\Psi_1\rangle$ and $|\Psi_2\rangle$ are *on the left and right*, respectively. Factors denote the observer, the spin, the coin, and the rest of the environment. Thin *diagonal lines* connecting the spin and coin represent entanglement within different branches of the wave function. The *horizontal/vertical boxes* made from *dotted/dashed lines* show two different ways of carving out the "Observer+System" subsystem from the "Environment." The ESP implies that the probability of the system being in a particular state is independent of the state of the environment. Applying that rule to both the spin and coin systems implies the Born Rule as the uniquely rational way of assigning credences

Then our state is

$$|\Psi\rangle = |O\rangle|\uparrow\rangle|\widehat{\omega}_1\rangle + |O\rangle|\downarrow\rangle|\widehat{\omega}_2\rangle + |O\rangle|\downarrow\rangle|\widehat{\omega}_3\rangle. \tag{10.35}$$

This reduces the problem of two branches with unequal amplitudes to that of three branches with equal amplitudes.

Following our previous logic, we construct a new orthonormal environment basis involving both a coin and a playing card, the latter of which has basis vectors $\{|\heartsuit\rangle, |\diamondsuit\rangle, |\spadesuit\rangle, |\clubsuit\rangle\}$. In terms of these we write a third set of environment basis vectors $\{|\tilde{\omega}_\mu\rangle\}$ as:

$$|\tilde{\omega}_1\rangle = |H\rangle \otimes |\heartsuit\rangle \otimes |\Omega\rangle, \tag{10.36}$$

$$|\tilde{\omega}_2\rangle = |T\rangle \otimes |\heartsuit\rangle \otimes |\Omega\rangle \tag{10.37}$$

$$|\tilde{\omega}_3\rangle = |H\rangle \otimes |\diamondsuit\rangle \otimes |\Omega\rangle, \tag{10.38}$$

$$|\tilde{\omega}_4\rangle = |T\rangle \otimes |\diamondsuit\rangle \otimes |\Omega\rangle \tag{10.39}$$

$$|\tilde{\omega}_5\rangle = |H\rangle \otimes |\clubsuit\rangle \otimes |\Omega\rangle \tag{10.40}$$

$$\cdots \tag{10.41}$$

Again we construct environment unitaries

$$U_E^{(1)} = \sum_\mu |\tilde{\omega}_\mu\rangle\langle\widehat{\omega}_\mu|, \tag{10.42}$$

$$U_E^{(2)} = |\tilde{\omega}_4\rangle\langle\widehat{\omega}_1| + |\tilde{\omega}_1\rangle\langle\widehat{\omega}_2| + |\tilde{\omega}_5\rangle\langle\widehat{\omega}_3| + |\tilde{\omega}_2\rangle\langle\widehat{\omega}_4| + |\tilde{\omega}_3\rangle\langle\widehat{\omega}_5| + \sum_{\mu>5} |\tilde{\omega}_\mu\rangle\langle\widehat{\omega}_\mu|. \tag{10.43}$$

Acting on our state (10.35) we get

$$|\Psi_1\rangle = |O\rangle|\uparrow\rangle|H\rangle|\heartsuit\rangle|\Omega\rangle + |O\rangle|\downarrow\rangle|T\rangle|\heartsuit\rangle|\Omega\rangle + |O\rangle|\downarrow\rangle|H\rangle|\diamondsuit\rangle|\Omega\rangle \qquad (10.44)$$

and

$$|\Psi_2\rangle = |O\rangle|\uparrow\rangle|T\rangle|\diamondsuit\rangle|\Omega\rangle + |O\rangle|\downarrow\rangle|H\rangle|\heartsuit\rangle|\Omega\rangle + |O\rangle|\downarrow\rangle|H\rangle|\clubsuit\rangle|\Omega\rangle \qquad (10.45)$$

From the form of $|\Psi_2\rangle$, in particular the first term in the superposition, it is easy to see that

$$P(\uparrow|\Psi_2) = P(T|\Psi_2) = P(\diamondsuit|\Psi_2). \qquad (10.46)$$

From treating different combinations of spin/coin/card as parts of the environment, we also derive

$$P(\uparrow|\Psi_1) = P(\uparrow|\Psi_2), \qquad (10.47)$$

$$P(T|\Psi_1) = P(T|\Psi_2), \qquad (10.48)$$

$$P(\diamondsuit|\Psi_1) = P(\diamondsuit|\Psi_2). \qquad (10.49)$$

From Eqs. (10.46)–(10.49), we can safely conclude that each of the three branches represented in (10.44) have equal probability, one-third each. Since $|\Psi_1\rangle$ is related to the original $|\Psi\rangle$ by a unitary transformation on the environment, the ESP implies

$$P(\uparrow|\Psi) = \frac{1}{2}P(\downarrow|\Psi) = \frac{1}{3}. \qquad (10.50)$$

This is precisely the Born Rule prediction for this particular case of unequal amplitudes. The spin-down component of the original state was greater than the spin-up component by a factor of $\sqrt{2}$, and ends up with twice the probability. Other possibilities follow by straightforward extension of the above method. Admittedly, this reasoning only strictly applies when the ratio of different amplitudes is the square root of a rational number; however, since this is a dense set, it seems reasonable to conclude that the Born Rule is established.

This route to the Born Rule has a simple physical interpretation. Take the wave function and write it as a sum over orthonormal basis vectors with equal amplitudes for each term in the sum (so that may terms may contribute to a single branch). Then the Born Rule is simply a matter of counting—every term in that sum contributes an equal probability.

10.6 Discussion

We have proposed that self-locating uncertainty is generic in the process of quantum measurement, and that a proper treatment of such uncertainty leads us directly to the Born Rule [3]. In spirit our approach is similar to that of Vaidman [4], although we have carried the program through in more explicit detail. The result has the virtue of being relatively physically transparent. The wave function of the universe branches, and initially you don't know which branch you are on; close investigation reveals that the only rational way to apportion credence to the different possibilities is to use the Born Rule.

Formally, our derivation bears a close resemblance to the envariance program of Zurek [5], although we believe there are some conceptual advantages. Most importantly, while envariance helps us understand why the Born Rule is a sensible prescription if one thinks of EQM as a probabilistic theory at all, our emphasis on self-locating uncertainty provides a direct explanation for how such probabilities can arise in a perfectly deterministic theory. In a fundamentally stochastic theory, one thinks of probability as the answer to a question of the form "how likely is it that this particular outcome will occur?" That philosophy fails in EQM, where it is clear that *every* outcome with nonvanishing support in the wave function will occur (in some branch) with probability one. The Born Rule does not tell you the probability that you will end up as "the observer who measures spin up" (for example); rather, you know with certainty that you will evolve into multiple observers with different eventual experiences. In our approach, the question is not about which observer you will end up as; it is how the various future selves into which you will evolve should apportion their credences. Since every one of them should use the Born Rule, it is justified to talk *as if* future measurement outcomes simply occur with the corresponding probability. It is the journey from perfect knowledge to inevitable self-locating uncertainty that is the basis of probability talk in quantum mechanics.

Another advantage of our approach is that it provides a unified framework in which to discuss classical and quantum self-locating uncertainty. This has become an important issue in modern cosmology, in which models of the universe very often predict "large" spacetimes with multiple copies of various observers. Our formalism only provides unambiguous guidance in cases where the number of classical observers is finite, so it does not directly address the cosmological measure problem as it appears in models of eternal inflation—but it seems reasonable that getting the finite case right is an important step towards understanding the infinite case.

Acknowledgements Sean Carroll feels that it has been an honor and a pleasure to take part in the celebration of Yakir Aharonov's 80th birthday and would like to thank Jeff Tollaksen and the organizers of a very stimulating meeting. His work was supported in part by the U.S. Department of Energy, the National Science Foundation, and the Gordon and Betty Moore Foundation. Charles Sebens's work was supported by the National Science Foundation Graduate Research Fellowship under Grant No. DGE 0718128.

References

1. H. Everett, "Relative state" formulation of quantum mechanics. Rev. Mod. Phys. **29**, 454–462 (1957)

2. D. Wallace, *The Emergent Multiverse: Quantum Theory According to the Everett Interpretation* (Oxford University Press, London, 2012)
3. C.T. Sebens, S.M. Carroll, *Self-Locating Uncertainty and the Origin of Probability in Everettian Quantum Mechanics* (2013)
4. L. Vaidman, Probability in the many-worlds interpretation of quantum mechanics, in *The Probable and the Improbable: Understanding Probability in Physics, Essays in Memory of Itamar Pitowsky*, ed. by Y. Ben-Menahem, M. Hemmo (Springer, Berlin, 2011)
5. W.H. Zurek, Probabilities from entanglement, Born's Rule $p_k = |\psi_k|^2$ from envariance. Phys. Rev. A **71**(5), 052105 (2005)
6. J.B. Hartle, Quantum mechanics of individual systems. Am. J. Phys. **36**, 704–712 (1968)
7. E. Farhi, J. Goldstone, S. Gutmann, How probability arises in quantum mechanics. Ann. Phys. **192**, 368 (1989)
8. A. Aguirre, M. Tegmark, Born in an infinite universe: a cosmological interpretation of quantum mechanics. Phys. Rev. D **84**, 105002 (2011)
9. D. Deutsch, Quantum theory of probability and decisions. Proc. R. Soc. Lond. Ser. A **458**, 3129–3137 (1999)
10. D. Wallace, Quantum probability and decision theory (2002). arXiv preprint quant-ph/0211104
11. H. Greaves, Understanding Deutsch's probability in a deterministic multiverse. Stud. Hist. Philos. Sci. Part B, Stud. Hist. Philos. Mod. Phys. **35**(3), 423–456 (2004)
12. D. Wallace, How to prove the Born Rule, in *Many Worlds?: Everett, Quantum Theory, & Reality*, ed. by S. Saunders, J. Barrett, A. Kent, D. Wallace (Oxford University Press, London, 2010), pp. 227–263
13. A.M. Gleason, Measures on the closed subspaces of a Hilbert space. J. Appl. Math. Mech. **6**, 885–894 (1957)
14. D. Lewis, Attitudes *de dicto* and *de se*. Philos. Rev. **88**(4), 513–543 (1979)
15. T. Banks, Cosmological breaking of supersymmetry? or Little lambda goes back to the future 2 (2000)
16. S.M. Carroll, *What If Time Really Exists?* (2008)
17. W.H. Zurek, Pointer basis of quantum apparatus: into what mixture does the wave packet collapse? Phys. Rev. D **24**, 1516–1525 (1981)
18. E. Joos, H.D. Zeh, The emergence of classical properties through interaction with the environment. Z. Phys. B, Condens. Matter **59**, 223–243 (1985)
19. D. Wallace, Decoherence and ontology, in *Many Worlds?: Everett, Quantum Theory, & Reality*, ed. by S. Saunders, J. Barrett, A. Kent, D. Wallace (Oxford University Press, London, 2010), pp. 53–72
20. J.B. Hartle, M. Srednicki, Are we typical? Phys. Rev. D **75**(12), 123523 (2007)
21. D.N. Page, Typicality defended (2007). arXiv:0707.4169
22. M. Srednicki, J. Hartle, Science in a very large universe. Phys. Rev. D **81**(12), 123524 (2010)
23. N. Bostrom, *Anthropic Bias: Observation Selection Effects in Science and Philosophy* (Routledge, London, 2002)
24. C.J.G. Meacham, Sleeping beauty and the dynamics of *de se* beliefs. Philos. Stud. **138**(2), 245–269 (2008)
25. D. Manley, *Self-Location, Existence, and Evidential Uniqueness* (2013)
26. A. Elga, Defeating Dr. Evil with self-locating belief. Philos. Phenomenol. Res. **69**(2), 383–396 (2004)
27. D.N. Page, The Born Rule dies. J. Cosmol. Astropart. Phys. **0907**, 008 (2009)
28. D.N. Page, *Born Again* (2009)
29. D.N. Page, *Born's Rule Is Insufficient in a Large Universe* (2010)
30. A. Albrecht, D. Phillips, *Origin of Probabilities and Their Application to the Multiverse* (2012)

Chapter 11
Physics and Narrative

David Albert

Abstract I present a very simple thought experiment—which has somehow been overlooked in the literature—that has surprising consequences about the Lorentz-transformation properties of the quantum states of multiple-particle systems. [*Editors note*: for a video of the talk given by Prof. Albert at the Aharonov-80 conference in 2012 at Chapman University, see quantum.chapman.edu/talk-29.]

Thirty years ago, when I was in my late twenties, and I had a post-doctoral position with Yakir in the Physics department at Tel Aviv University, I was asked to give a toast at a celebration for Yakir's 50th birthday. And what I remember saying is that I couldn't think of any of my contemporaries at the time—that I couldn't think of anybody (that is) in their twenties—who were even remotely as brave, or as open, or as creative, or as experimental, or as overcome with wonder, or as bursting with life, or as constantly and resolutely expecting the impossible, or (in brief) as young, as Yakir. But even I could not have imagined at the time that thirty years later he would turn out to be younger still.

I cannot begin to catalogue my debts to Yakir here. I owe him—to put it simply—everything. And it is a great honor and pleasure to be able to celebrate with him today. And I thought it might be fun to remind him of some thoughts that he and I had together thirty years or so ago—and to tell a little about what has become of them, in my own imagination, since.

Consider a system of four distinguishable quantum-mechanical spin-1/2 particles. Call it S. And suppose that the complete history of the motions of those particles in position-space—as viewed from the perspective of some particular Lorentz-frame K—is as follows: Particle 1 is permanently located in the vicinity of some particular spatial point, and particle 2 is permanently located in the vicinity of some *other* spatial point, and particles 3 and 4 both move with uniform velocity along

D. Albert (✉)
Department of Philosophy, Columbia University, 706 Philosophy HallMail Code: 4971, 1150 Amsterdam Avenue, New York, NY 10027, USA
e-mail: da5@columbia.edu

D.C. Struppa, J.M. Tollaksen (eds.), *Quantum Theory: A Two-Time Success Story*,
DOI 10.1007/978-88-470-5217-8_11, © Springer-Verlag Italia 2014

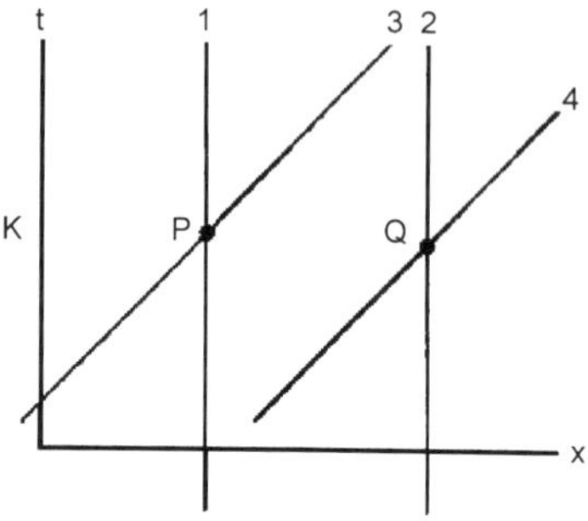

Fig. 11.1 Situation in Lorentz-inertial system K

parallel trajectories in space-time.[1] The trajectory of particle 3 intersects the trajectory of particle 1 at space-time point P—as in Fig. 11.1—and the trajectory of particle 4 intersects the trajectory of particle 2 at space-time point Q. And P and Q are simultaneous, from the perspective of K.

And suppose that the state of the *spin* degrees of freedom of S, at $t = -\infty$, is $|\varphi\rangle_{12}|\varphi\rangle_{34}$, where

$$|\varphi\rangle_{AB} = \frac{1}{\sqrt{2}}|\uparrow\rangle_A|\downarrow\rangle_B - \frac{1}{\sqrt{2}}|\downarrow\rangle_A|\uparrow\rangle_B \tag{11.1}$$

I want to compare the effects of two different possible *Hamiltonians* on this system. In one, S evolves freely throughout the interval from $t = -\infty$ to $t = +\infty$. The other includes an impulsive contact interaction term that *exchanges spins*—a term (that is) which is zero except when two of the particles occupy the same point, and which (when it *isn't* zero) generates precisely the following unitary evolution:

$$\begin{aligned}
|\uparrow\rangle_A|\downarrow\rangle_B &\rightarrow |\downarrow\rangle_A|\uparrow\rangle_B \\
|\downarrow\rangle_A|\uparrow\rangle_B &\rightarrow |\uparrow\rangle_A|\downarrow\rangle_B \\
|\uparrow\rangle_A|\uparrow\rangle_B &\rightarrow |\uparrow\rangle_A|\uparrow\rangle_B \\
|\downarrow\rangle_A|\downarrow\rangle_B &\rightarrow |\downarrow\rangle_A|\downarrow\rangle_B
\end{aligned} \tag{11.2}$$

A minute's reflection will show that the entire history of the quantum state of this system, from the perspective of K—the entire history (that is) of the quantum-mechanical *wave-function* of this system, even down to the overall phase, from the perspective of K—will be *identical* on these two scenarios. On both scenarios (that is) the state of S, from the perspective of K, throughout the interval from $t = -\infty$ to $t = +\infty$, will be precisely $|\varphi\rangle_{12}|\varphi\rangle_{34}$.

And what's interesting is that the situation is altogether *different* from the perspective of every *other* frame. On the first scenario—the scenario in which S evolves freely—the state of S is going to be precisely $|\varphi\rangle_{12}|\varphi\rangle_{34}$, in *every* frame, throughout

[1]This sort of permanent localization can be accomplished, say, by placing the particles in boxes, or by making their masses large.

the interval from $t' = -\infty$ to $t' = +\infty$.[2] But on the *second* scenario, when viewed from the perspective of frames other than K, the interactions at P and Q occur at *different times*. In those other frames, then, throughout the interval between P and Q, the state of S is going to be $|\varphi\rangle_{14}|\varphi\rangle_{23}$.

And it follows immediately that the complete history of the quantum state of S in frames *other* that K cannot be deduced, either by means of the application of a geometrical space-time point-transformation or *in any other way*, from the complete history of the quantum state of S in K—because transformation in question would need (per impossible!) to map *precisely the same* history in K into one of two *entirely distinct* histories in K′, depending on which one of the above two *Hamiltonians* obtains.

All of this is as easy as can be. And all of it has been taken note of, on a number of different occasions, in the literature of the foundations of quantum mechanics. It was pointed to in a 1984 [1] paper by Yakir Aharonov and myself—for example—and in a paper by Wayne Myrvold from 2002 [2], and it must at least have occurred in passing to a great many people.[3] But nobody seems to have been able to look it *straight in the face*, nobody seems to have entirely *taken it in*.[4]

Let's back up (then) and slow down, and see if we can figure out what it means.

---

[2] I am going to be supposing, throughout, that the velocities of these other frames with respect to K are small compared to the speed of light, so that the effects of Lorentz-transformations on the spins can be neglected. The effects of transforming to other frames that are going to interest us here can all be made as large as one likes, even at small relative velocities, by separating the two particles from one another by a great spatial distance.

[3] The example presented here, however, is a good deal cleaner and more perspicuous than either the one discussed by Aharonov and I in 1984 or the one discussed by Myrvold in 2002. The example cited in the paper by Aharonov and myself involves *measurement*-type interactions, and the one Myrvold presents involves an external field that violates Poincare-invariance. Neither of those sorts of distractions come up, however, in the example presented here.

[4] What Yakir and I had to say about it—in the 1984 paper—was that in so far as frame K is concerned, the interaction "disrupts (as it were) the *transformation* properties of the state and disrupts its *covariance*, without in any way disrupting the history of the state itself". But precisely how it is that the *transformation* properties of something can be disrupted without *in any way* disrupting *the history of the thing itself* I confess I can no longer imagine. It seems panicked—looking back on it now—and incoherent, and mad.

Professor Myrvold (on the other hand) thinks it shows that the Lorentz-transformation of quantum-mechanical wave-functions is not so much a *geometrical* or even a *kinematical* matter as it is a matter of *dynamics*, a matter of the *Hamiltonian* of the system whose wave-function is being transformed. According to Professor Myrvold, the business of performing a Lorentz-transformation on the complete temporal history of the wave-function of an isolated system is in general going to require that we know, and are able to solve, the system's *dynamical equations of motion*. But if we go *that* route, nothing whatever is going to remain of the intuition that carrying out such a transformation is merely a matter of looking at *precisely the same set of physical events* from two different *perspectives*, from two different *points of view*. Dynamics—after all—is not the business of changing one's perspective on already *existing* events, but of generating entirely *new* ones!

Call a world *narratable* if the entirety of what there is to say about it can be presented as a single *story*, if the entirety of what there is to say about it can be presented as a *single temporal sequence* of *instantaneous global physical situations*.

The possible worlds of Newtonian Mechanics can each be presented, in its entirety, by means of a specification of the local physical conditions at every point in a four-dimensional manifold. And there is a way of *slicing that manifold up* into a one-parameter collection of infinite three-dimensional hyperplanes such that the dynamical *significance* of the parameter in question—the dynamical *role* of the parameter in question—is precisely that of a *time*.[5] A Newtonian-Mechanical *instantaneous global physical situation*, then, is a specification of the local physical conditions at each one of the points on any particular one of those infinite three-dimensional hyperplanes. And since all of those instantaneous global Newtonian-Mechanical physical situations taken together amount—by construction—to a specification of the local physical conditions at every point in the manifold, the possible worlds of Newtonian Mechanics are invariably *narratable*. Moreover, they are *uniquely* narratable, in the sense that the number of different ways of slicing the manifold up in such a way as to *satisfy* the conditions described above—in a Newtonian-Mechanical world—is invariably, precisely, *one*.

The possible worlds of *Non-Relativistic Quantum Mechanics* can each be presented, in its entirety, by means of a specification of the values of a real two-component field—a specification, that is, of the quantum-mechanical *wave-function*—at every point in a $3N + 1$ dimensional manifold (where N is the number of particles in the world in question). And there is a way of *slicing that manifold up* into a one-parameter collection of infinite $3N$-dimensional hyperplanes such that the dynamical role of the parameter in question is precisely that of a time. A *Non-Relativistic Quantum-Mechanical* instantaneous global physical situation, then, is a specification of the local physical conditions at each one of the points on any particular one of those infinite $3N$-dimensional hyperplanes. And since all of those instantaneous global Non-Relativistic Quantum-Mechanical physical situations taken together amount to a specification of the local physical conditions at every point in the manifold, the possible worlds of Non-Relativistic Quantum-Mechanics are invariably narratable. And the narratability here is again *unique*, in the sense that the number of different ways of slicing the manifold up in such a way as to *satisfy* the conditions described above is invariably, precisely, *one*.

The possible worlds of *Classical Relativistic Maxwellian Electrodynamics*—just like those of Newtonian Mechanics—can each be presented, in its entirety, by

[5]It means a host of things, by the way, to speak of the parameter in question here as "playing the dynamical role of a time". It means (for example) that the trajectory of every particle in the world intersects every one of the three-dimensional hyperplanes in question here exactly once, and it means that the total energy on any one of these hypersurfaces is the same as the total energy on any *other* one of them, and it means (principally and fundamentally and in sum) that the equation

$$F = m(\partial 2x/\partial \rho 2),$$

where ρ is the parameter in question, is true.

means of a specification of the local physical conditions at every point in a four-dimensional manifold. And there is, again, a way of *slicing that manifold up* into a one-parameter collection of infinite three-dimensional hyperplanes such that the dynamical significance of the parameter in question is precisely that of a *time*. And so a Classical Relativistic Maxwellian instantaneous global physical situation is a specification of the local physical conditions at each one of the points on any particular one of those infinite three-dimensional hyperplanes. And since all of those instantaneous global Classical Maxwellian physical situations taken together amount to a specification of the local physical conditions at every point in the manifold, the possible worlds of Classical Relativistic Maxwellian Electrodynamics are narratable. But in *this* case the narratability is manifestly *not* unique—Classical Relativistic Maxwellian Electrodynamics is (rather) *multiply* narratable. In the case of Classical Relativistic Maxwellian Electrodynamics, each different *Lorentz-frame* is plainly going to correspond to a different way of slicing the manifold up in so as to satisfy the conditions described above.

But Relativistic *Quantum* Theories are an altogether different matter. In both the non-relativistic and the relativistic cases, an instantaneous Quantum-Mechanical *state of the world*—an instantaneous Quantum-Mechanical *global physical situation*—is a specification if the expectation-values of all of the local and non-local quantum-mechanical observables that refer exclusively to the time in question. And the lesson of the example we went through above is that the entirety of what there is to say about a Relativistic Quantum-Mechanical world can *not* be presented as a one-parameter family of situations like that. The lesson of the example we went through above (more particularly) is that any one-parameter family of situations like that is necessarily going to leave the expectation-values of non-local Quantum-Mechanical observables that refer to several *different* times—the expectation-values of non-local Quantum-Mechanical observables (that is) which are instantaneous from the perspective of *other* Lorentz-frames—unspecified. In order to present the entirety of what there is to say about a Relativistic Quantum-Mechanical world, we need to specify, *separately*, the Quantum-Mechanical state of the world associated with *every separate space-like hypersurface*. If the theory is to be relativistic in the sense of Einstein, in the sense of Minkowski, nothing less is going to do.

The relationship between the quantum-mechanical states of the world associated with any set of space-like hypersurfaces and the quantum-mechanical states of the world associated with any *other* set of space-like hypersurfaces is therefore, invariably, a matter of *dynamical evolution*—even (for example) if one of those sets happens to be the complete family of equal-time hyperplanes for K and the other one of those sets happens to be the complete family of equal-time hyperplanes for K′.[6]

[6]In this respect, then, Professor Myrvold (see note 4) is perfectly right. Where Myrvold goes wrong is in imagining that a relationship like that is consistent with the claim that an assignment of a quantum state of the system in question to every one of the equal-time hyperplanes of K can amount to a complete *history* of that system—where he goes wrong (that is) is in imagining that a relationship like that can leave the world *narratable*.

Fig. 11.2

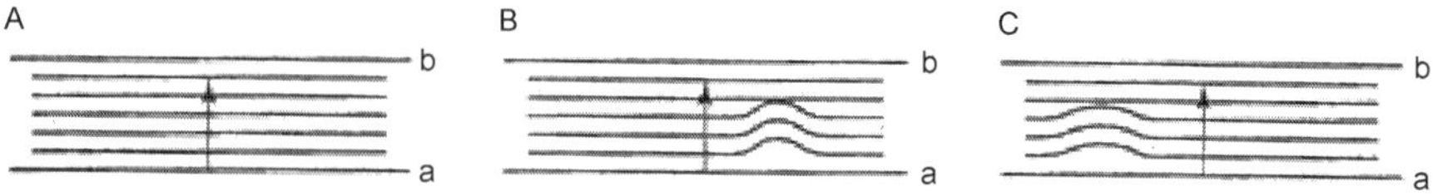

Fig. 11.3

Moreover, the elementary *unit* of dynamical evolution here is plainly not an infinitesimal *translation in time* (which is generated by the global Hamiltonian of the world H, as in Fig. 11.2a) but an arbitrary infinitesimal *deformation*, an arbitrary infinitesimal *undulation*, of the *space-like hypersurface* (which is generated by the *local Hamiltonian density* of the world δH, as in Fig. 11.2b).

The dynamical laws of the evolutions of relativistic quantum-mechanical systems therefore have a much richer mathematical structure than the laws of the evolutions of *non*-relativistic quantum-mechanical systems do. Suppose (for example) that we should like to calculate the wave-function of some particular isolated quantum-mechanical system on hypersurface b, given the wave-function of that system on some *other* hypersurface a—where a may be either in the past of b or in its future. In the *non*-relativistic case, which is depicted in Fig. 11.3a, there is always exactly *one* continuous one-parameter family of hypersurfaces—the continuous one-parameter family of *absolute simultaneities* between a and b—along which a calculation like that is going to have to *proceed*, along which the system in question can be pictured as *evolving*. In the relativistic case, on the other hand, there is invariably an *infinity* of continuous one-parameter families of space-like hypersurfaces along which such a calculation can proceed, and along which the system in question can be pictured as evolving. Two such families are displayed in Figs. 11.3b and 11.3c.

And one of the necessary conditions of the existence of a solution to the dynamical equations of motion of a theory like this, one of the necessary conditions of the internal consistency of the dynamical equations of motion of a theory like this, is that the calculation that proceeds along the route pictured in Fig. 11.3b and the calculation that proceeds along the route pictured in Fig. 11.3c, so long as they both *start out* with precisely the same wave-function at a, will both necessarily *produce* precisely the same wave-function at b. And while there can be no such thing as a Lorentz-transformation of the complete temporal sequence of the quantum states of any isolated system S in frame K into the complete temporal sequence of quantum states of that system in frame K′, there is nonetheless a perfectly clear and perfectly explicit idea of the Lorentz-transformation of any comprehensive summary of the world; there is a perfectly clear and perfectly explicit idea (that is) of the Lorentz transformation of any assignment of states to every space-like hypersurface. Given

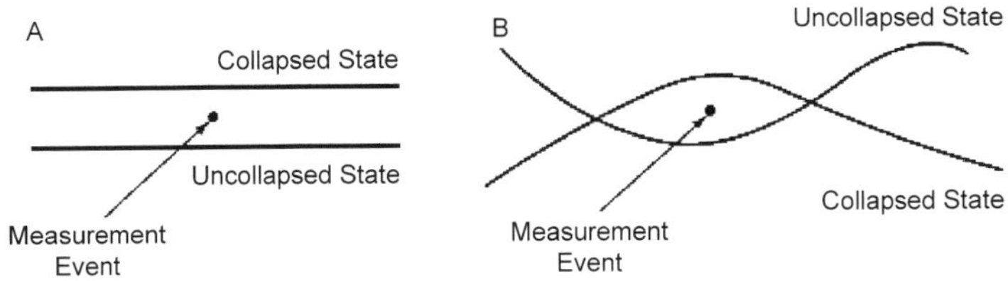

Fig. 11.4

any such assignment, the way to Lorentz-transform it is just to assign the same set of states to a Lorentz-transformed set of hypersurfaces, and we speak of a set of dynamical laws of the evolutions of wave-functions as Lorentz Invariant just in case any Lorentz-transformation of any comprehensive summary of the world which is in accord with those laws yields another comprehensive summary of the world which is in accord with those laws.

Here are two stories—which pull in very different directions—about where all this might leave us:

11.1 Story 1

This story is about the *collapse* of the wave-function. And it will work best—for the moment—to tell it in the language of the old-fashioned and idealized and unscientific and altogether outmoded postulate of collapse on which collapses are brought about by means of the intervention of localized, external, un-quantum-mechanical *measuring-devices*. On this picture, collapses involve a discontinuous and probabilistic projection of the wave-function of the *measured* system, the *quantum-mechanical* system, onto an eigenfunction of some particular one of its *local observables* (the observable, that is, which the external device in question is designed to measure) at some particular *space-time point* (the so-called 'measurement-event'—the point at which the measured system *interacts* with the measuring-device). The probability of a projection onto this or that particular eigenfunction of the measured observable is determined, in the familiar way, by the Born rule.

On the *non-relativistic* version of the collapse postulate (which is depicted in Fig. 11.4a) the collapse occurs as the 'now' sweeps forwards across the measurement-event—the collapse (that is) affects the wave-function of the system in question in the *future* of that event, but not in its *past*. And twenty or so years ago, I wrote a paper with Yakir Aharonov which proposed a manifestly Lorentz-invariant *relativistic* version of that postulate (which is depicted in Fig. 11.4b) on which the collapse occurs as an *undulating space-like hypersurface*, *any* undulating space-like hypersurface, *deforms* forwards across the measurement-event—on which (that is) the collapse affects the wave-function of the system in question on those space-like hypersurfaces that intersect the *future light-cone* of measurement-event, but not on those space-like hypersurfaces that intersect its *past light-cone*.

Fig. 11.5

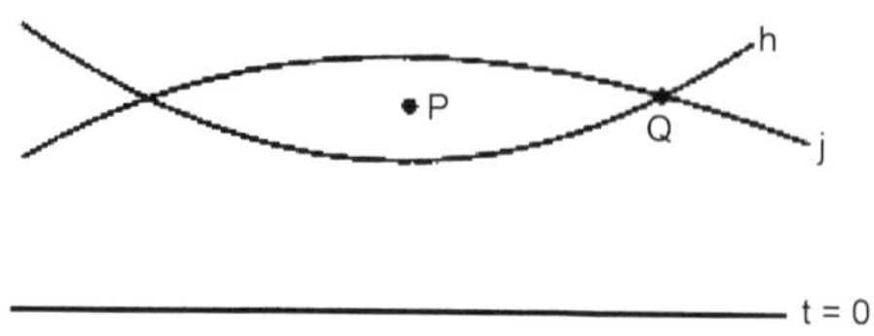

Suppose (then) that we are given the wave-function of some isolated relativistic quantum-mechanical system S along some space-like hypersurface a, and suppose that we are given the addresses of all of the space-time points in the future of a at which measurements of local observables of S are to be carried out, and suppose that we are told what particular local observable of S each particular one of those measurements is to be a measurement *of*. The relativistic postulate of collapse just described—together with the deterministic laws of the ordinary dynamical evolutions of the wave-functions of *isolated* relativistic quantum-mechanical systems under infinitesimal deformations of the space-like hypersurface—will assign a definite probability to any particular assignment of *outcomes* to those measurements, and it will assign a definite probability to any particular assignment of a quantum-mechanical *wave-function* to any particular space-like hypersurface b which is entirely in the *future* of a, and (moreover) it will do both of those things *uniquely*—completely independent (that is) of which one of the above-mentioned *routes* the calculation of those probabilities take.

The *trouble* here—or so I imagined until now—is that the possible worlds of this sort of a theory aren't going to be *narratable*. Suppose (for example) that the momentum of a free particle is measured along the hypersurface marked $t = 0$ in Fig. 11.5, and that later on a collapse leaves the particle localized at P. Then the projection-postulate that Aharonov and I proposed is going to stipulate (among other things) that the wave-function of the particle along hypersurface h is an eigenstate of momentum, and that the wave-function of the particle along hypersurface j is (very nearly) an eigenstate of *position*. And so the quantum-mechanical wave-functions associated with hypersurfaces h and j, in this example, are going to disagree with one another even about the expectation-values of *local* quantum-mechanical observables at points like Q, where they intersect.[7] And that (of course) puts narratability quite decisively out of the question.

[7]Note, however, that the expectation-values of all local observables at Q *given the state along* $t= 0$ will still be completely independent of the *route* by which one chooses to *calculate* from $t = 0$ to Q. On *certain* routes (for example) Q is going to come up as an element of h, and on certain *others* it will come up as an element of j. If Q comes up as an element of h, then the expectation-values of all local observables at Q, given the state along $t = 0$, will be determined—in the familiar way—by the state at h. But if Q comes up as an element of j, then the expectation-values of all local observables at Q, given the state along $t = 0$, will be determined by a *probability-distribution* over various different *possible* states at j—corresponding to the different possible outcomes of the measurement at P. The Lorentz-invariance of the dynamical equations of motion and the collapse-postulate, however, will guarantee that those two sets of expectation-values will invariably be identical.

But a case might be made that the example we went through at the outset of this paper sheds a very different light on all this. We can now see—it might be argued—that the narratability of Relativistic Quantum Theories is dead before the measurement problem ever even *comes up*, before the non-locality that Bell discovered ever even *enters the picture*. Adding a postulate of the collapse of the wave-function to a Relativistic Quantum Theory, on this view, solves the measurement problem, and costs *nothing*. The Lorentz-invariance of the theory is preserved perfectly intact, and as for the failure of *narratability*, *that* price turns out to have been *paid*, unbeknownst to us, long before the question of measurement ever arose.

If all this is right, then many-worlds and many-minds and many-histories theories have no advantage whatever—in so far as questions of Lorentz-invariance are concerned—over collapse theories. The Lorentz-invariance of the theories of many-worlds and many-minds and many-histories comes, after all, at the price of non-narratability—just as that of collapse theories does.

Moreover, there is now reason to hope that these considerations may turn out not to depend all that sensitively on the unrealistic idealizations of the measurement-process I described a few paragraphs back. A talented young German physicist named Roderich Tamulka has recently published a fully relativistic version of the GRW collapse-theory for non-interacting particles a theory (as it turns out) that fits around the schematic general principles that Yakir and I laid out twenty years ago like skin. It still remains and it may turn out to be a highly non-trivial business—to generalize Tamulka's theory to the case of interacting particles and to fields. We shall have to wait and see. But what Tamulka has already accomplished represents an immense and encouraging step in the right direction.

11.2 Story 2

This is a story about the linear, unitary, deterministic evolution of the wave-functions of quantum-mechanical systems, altogether un-adorned by any mechanism of collapse. Consider a Relativistic Quantum-Mechanical world W in which the free Hamiltonian of a certain pair of electrons is identically zero, and in which the *wave-function* of that pair, along every space-like hypersurface whatever, is precisely the wave-function $|\varphi\rangle_{12}$ of Eq. (11.1). And let $t' = \alpha$ be a flat space-like hypersurface all of whose points are simultaneous with respect to some particular Lorentzian frame of reference K′. And imagine an experiment designed to measure and record the total spin of that pair of electrons along $t' = \alpha$. The experiment involves two localized pieces of apparatus, which have previously been brought together, and prepared in a state in which certain of their internal variables are quantum-mechanically entangled with one another, and then separated in space. One of those pieces of apparatus then interacts with particle 1 at point L (in Fig. 11.6) and the other interacts with particle 2 at point Q. And the positions of the relevant *pointers* on those two pieces of apparatus, at the *conclusions* of those interactions, are measured, and the values of those positions are transmitted to F, and those values are mathematically combined

Fig. 11.6

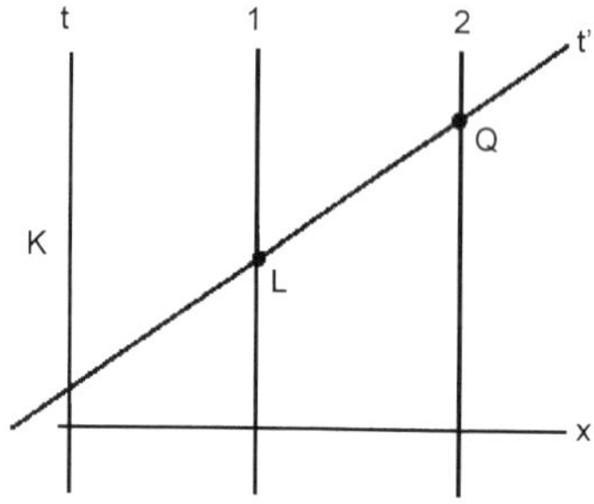

with one another in such a way as to determine the outcome of the measurement of the total spin of the pair of electrons along $t' = \alpha$, and (finally) that outcome is *recorded*, in ink (say), in English, on a piece of paper, at G.[8] No such experiment is actually *carried out* in W—mind you—but it is a fact about W that *if* such an experiment *were* to have been carried out, it would with certainty have been recorded at G that the total spin of that pair along $t' = \alpha$ was zero.

Now, the most obvious and most straightforward way of *accounting* for that fact, the most obvious and most straightforward way of *explaining* that fact, is to point out (1) that the state of the electron-pair, along the hypersurface $t' = \alpha$, is $|\varphi\rangle_{12}$, and (2) that $|\varphi\rangle_{12}$ is an eigenstate of the total spin of that pair, with eigenvalue zero, and (3) that a *measurement* of the total spin of that pair along $t' = \alpha$—if it had been carried out—would therefore, with certainty, have found that the total spin of that pair *is* zero. Note that this explanation depends only on the *state* of the pair of electrons at $t' = \alpha$, and not *at all* on the *dynamical laws* by which that state *evolves*.[9]

But another explanation—or rather a *continuous infinity* of other explanations—have plainly got to be available as well. If (for example) we trace out the development of the world exclusively along the continuous one-parameter family of hypersurfaces of simultaneity in K, the experiment in question is going to look not so much like an *instantaneous measurement* as an extended sequence of *dynamical interactions*. At $t = 0$, state of the electron-pair is $|\varphi\rangle_{12}$, and the pair of *apparatuses* are in the specially prepared quantum-mechanically entangled state—call it $|\Theta\rangle$-alluded to above. Then—at L—electron 1 interacts with one of the localized pieces of apparatus, and this interaction leaves the *electron-pair* quantum-mechanically *entangled* with the pair of *apparatuses*. Then—at Q—electron 2 interacts with the *other* localized piece of apparatus in precisely such a way as to *undo* that latter

[8]Detailed instructions for the construction and preparation of measuring-apparatuses like these—using only local interactions—can be found in an old paper of Yakir Aharonov's and mine [3].

[9]The account *does* depend on the dynamics of the two pieces of *measuring-apparatus*—of course—and on the dynamics of the mechanism whereby the positions of the relevant pointers on those two pieces of apparatus are *transmitted to F*, and on the dynamics of the mechanism whereby those position-values are mathematically combined with one another in such a way as to determine the *outcome* of the total-spin measurement, and (finally) on the dynamics of the mechanism whereby that outcome is *recorded* at *G*—but it doesn't depend *at all* on the dynamics of the pair of electrons *themselves*.

entanglement—leaving the electron-pair once again in the state $|\varphi\rangle_{12}$ and the pair of apparatuses once again in the state $|\Theta\rangle$. Thereafter, the various transmitters and receivers and compilers and recorders go to work, and the end-product of all this activity—the end-product (that is) which is *entailed* with *certainty* by the state of the world along $t = 0$ and the deterministic quantum-mechanical equations of motion, no matter *which* continuous one-parameter family of space-like hypersurfaces the intervening calculation traces through—is a sheet of paper at G, bearing the inscription "total spin equals Zero".

This particular experiment's having this particular outcome, then, can be given a *complete* and *satisfactory* and *deterministic* explanation which traces out the development of the world *exclusively* along the continuous one-parameter family of hypersurfaces of simultaneity in K, and which makes *no mention whatsoever* of the state of the pair of electrons—or of anything else—at $t' = \alpha$.

And we are plainly going to be able to produce very much the same sort of an explanation—very much the same *continuous infinity* of explanations—of the outcome any hypothetical experiment whatsoever.

And this suggests a way of picturing relativistic quantum-mechanical worlds—for a price—as *narratable*. All that needs to be given up is the Einsteinian insistence that the unfolding of the world in every separate Lorentz-frame and along every continuous one-parameter family of space-like hypersurfaces all be put on an *equal metaphysical footing*. Suppose—on the contrary—that it is stipulated that an assignment of a quantum state of the world to every one of the hypersurfaces of simultaneity of (say) K—and to no *other* space-like hypersurfaces—amounts in and of itself to a complete and exhaustive and unaugmentable account of the world's history. Then there would be no facts at all about the 'state of the world' along, say, $t' = \alpha$. And all *talk* of such 'facts' in the physical literature would need to be re-interpreted as shorthand for *counterfactual* talk about how this or that hypothetical *experiment*—if it were to be performed—would come out. The world would be narratable—and (moreover) *uniquely* so.

On this way of thinking, the impulse away from an *Einsteinian* understanding of special relativity—the impulse (that is) towards a *Lorentzian* understanding of special relativity—would arise not (in the first instance) from the *non-locality* of the *collapse*, but *earlier* and *farther down*, from *the geometry of the Hilbert space* and the demand for *narratability*. And the way would seem to be open to trying out new fundamental theories of the world which violate Lorentz-invariance—a little bit—even in their empirical predictions.

Acknowledgements I am grateful to my son, Ben Albert, for doing such a wonderful job with the diagrams.

References

1. Y. Aharonov, D. Albert, Is the usual notion of time evolution adequate for quantum-mechanical systems? II. Relativistic considerations. Phys. Rev. D **29**, 228–234 (1984)

2. W.C. Myrvold, On peaceful coexistence: is the collapse postulate incompatible with relativity? Stud. Hist. Philos. Mod. Phys. **33**, 435–466 (2002)
3. Y. Aharonov, D. Albert, Can we make sense out of the measurement process in relativistic quantum mechanics. Phys. Rev. D **24**, 359–364 (1981)

Part V
Nonlocality

Chapter 12
Quantum Correlations in Newtonian Space and Time:

Faster than Light Communication or Nonlocality

Nicolas Gisin

Abstract We investigate possible explanations of quantum correlations that satisfy the principle of continuity, which states that everything propagates gradually and continuously through space and time. In particular, following (Bancal et al. in Nat. Phys., 2012) we show that any combination of local common causes and direct causes satisfying this principle, i.e. propagating at any finite speed, leads to signalling. This is true even if the common and direct causes are allowed to propagate at a supraluminal-but-finite speed defined in a Newtonian-like privileged universal reference frame. Consequently, either there is supraluminal communication or the conclusion that Nature is nonlocal (i.e. discontinuous) is unavoidable. [*Editor's note*: for a video of the talk given by Prof. Gisin at the Aharonov-80 conference in 2012 at Chapman University, see quantum.chapman.edu/talk-28.]

It is an honor to dedicate this article to Yakir Aharonov, the master of quantum paradoxes.

12.1 Introduction

Correlations cry out for explanations [1]. This is true in all sciences, from correlations between measurement results in quantum physics to correlations between earthquakes and tsunamis in geophysics, and correlations between tides and the moon's positions in classical physics, to name but a few examples. Once a correlation has been identified, the next task of science consists in developing a theoretical model explaining the correlation. Such models take the form of a story supported by mathematical equations. Particularly challenging is the search of an explanation for quantum correlations when considering several measurements per party on two or more distant systems initially in an entangled state.

N. Gisin (✉)
Group of Applied Physics, University of Geneva, 22, Ch. de Pinchat, 1211 Geneva 4, Switzerland
e-mail: Nicolas.Gisin@unige.ch

D.C. Struppa, J.M. Tollaksen (eds.), *Quantum Theory: A Two-Time Success Story*,
DOI 10.1007/978-88-470-5217-8_12,

In all sciences besides quantum physics, all correlations are explained by a combination of only two basic mechanisms. Either a first system influences a second one, i.e. Direct Causation (DC), as for example the earthquake that causes the tsunami. Or the correlated events share a local Common Cause (CC) in their common past as two readers of this text whose readings are highly correlated. Sometimes the common or direct causes may be subtle and not easy to detect, as twins that look extraordinarily alike thanks to common genes (local variables, i.e. CC), or as one's yawning triggers others to yawn, thanks to delicate influences (i.e. DC).

Many correlations involve a combination of the two basic mechanisms, common and direct causes, like for instance the correlations between hockey players: they trained together, hence share common causes, and, during games, influence each other.

Formally a correlation between two parties A and B is a conditional probability distribution $p(a, b|x, y)$, where a, b denote the measurement results collected by A and B, and x, y the measurement settings freely (i.e. independently from each other and from all CC and DC) chosen by A and B, respectively. This generalizes straightforwardly to n parties. If A's marginal $p(a|x, y) \equiv \sum_b p(a, b|x, y)$ depends explicitly on B's choice y, then A can get information about B's choice by merely observing her local statistics. This is called *signalling*. The no-signalling principle states that A's marginal is independent of B's choice, $p(a|x, y) \equiv \sum_b p(a, b|x, y) = p(a|x)$, and B's marginal is independent of A's choice, $p(b|x, y) = p(b|y)$. Note that all physical communication should be carried by some physical object (atoms, photons, energy, waves, etc). Hence, assuming only local Common Causes carried by the (localized) physical systems in Alice and Bob's hands, signalling would be non-physical communication. But Direct Cause may allow signalling as discussed in Sect. 12.8.

This paper is organized as follows. In the next section, we present the intuition behind our result [2]. Next, in Sect. 12.3, we define formally v-causal models. Then, before presenting the main result in Sect. 12.5, we analyze the case of DC (without additional variables) in Sect. 12.4. Finally, we discuss experiments that could test our results in Sect. 12.6 and discuss the interpretation of our results.

12.2 Explanations of Correlations

First attempts at explaining correlations between distant quantum measurement results assumed that the source producing the entangled quantum systems produces additional variables, hidden to today's physics, which would locally (i.e. continuously) determine the probabilities of the measurement results. This would provide a local Common Cause explanation. Such local hidden variable models must obey the famous Bell inequalities. But quantum theory predicts and experiments confirm that Bell inequalities can be violated; hence all explanations based only on local common causes have been experimentally refuted.[1]

Direct Cause explanations of quantum correlations received relatively little attention, compared to CC explanations (up to some noticeable exceptions, in particular Eberhard who proposed an explicit model already in 1989 [3]). This is due to the fact that Bell inequality violations have been convincingly demonstrated between space-like separated measurements [4–6], hence a DC explanations would require influences that propagate faster than light.

The assumption of faster than light influences does not respect the spirit of relativity. However, the assumption of a universal privileged reference frame with respect to which a faster than light influence can be defined, is not in contradiction with relativity.[2] Think for example of the reference frame in which the micro-wave back ground radiation, residue of the big bang, is isotropic; our Earth propagates with respect to this universal frame at the well defined speed of 369 km/s in a direction known at each moment [7].

There is thus no definite reason not to investigate the possibility of explaining quantum correlation with a combination of DC and CC. Actually, many authors who thought seriously about quantum non-locality noticed that correlation between distant events strongly suggest that "something is going on behind the scene", using John Bell's words [8, 9]. David Bohm and Basil Hiley, for example, have been very explicit when writing "it is quite possible that quantum nonlocal connections might be propagated, not at infinite speeds, but at speeds very much greater than that of light. In this case, we could expect observable deviations from the predictions of current quantum theory (e.g. by means of a kind of extension of the Aspect-type experiment)" [10]. Let us also note that most (non relativistic) text books tell a story like "first measurement collapses the entire wavefunction, hence changes (influences) the state of all systems entangled with the measured system". Consequently, it is good scientific practice to study the assumption that quantum correlations are caused by faster than light influences propagating in a hypothetical universal privileged reference frame and analyze its consequences.

We call v-**causal** all explanations that combine local Common Causes and Direct Causes where the influence (describing the direct cause) propagates at a supraluminal-but-finite speed v defined in a hypothetical universal privileged reference frame: $c < v < \infty$. Note that such a universal privileged reference frame would be quite similar to Newton's space and time, but with a given fixed maximal velocity v. It is thus quite familiar to physicists,[3] see Fig. 12.1. When two events can

[1]Up to some combinations of loopholes that seem highly implausible; however, this being science, this logical possibility should be addressed experimentally.

[2]One could also consider the history-fiction case that quantum theory would have been developed before the discovery of relativity. In such a case, quantum nonlocality would have been equally surprising and fascinating and physicists would naturally have been led to search for explanations of these extraordinary correlations in terms of delicate influences yet to be discovered.

[3]Though this strongly contrasts with ideas in quantum gravity where space-time is sometimes thought of as an emergent concept, as e.g. in loop quantum gravity.

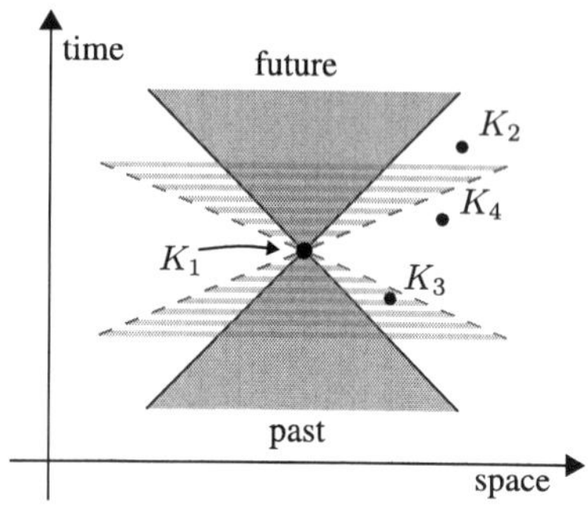

Fig. 12.1 Space-time diagram in the privileged reference frame. The *shaded light cone* is delimited by *solid lines. Points inside the v-cone (hatched)*, e.g. K2 and K3, are v-connected to K1; while *points outside the v-cone*, like K4, are not-v-connected to K1. (Taken with permission from Nature [2])

be connected by a hidden influence at speed v, we say that they are v-**connected**; otherwise we say that the events are **not-v-connected**.

The kind of experiment that Bohm and Hiley had in mind to test such a DC or v-causal explanation is quite intuitive: if the influence carrying the DC propagates at finite speed, it should be possible to arrange an experiment between distant quantum systems with good enough synchronization (in the universal privileged reference frame), so that the influence doesn't arrive on time to establish the correlation. Thus, in such situations, the measured correlation should necessarily be local, i.e. satisfy all Bell inequalities, even in cases where quantum theory predicts a violation of some Bell inequality. Hence, v-causal explanations can't reproduce all quantum predictions. Accordingly, they can be tested experimentally against quantum theory.

Such experiments face two intrinsic difficulties. First, since today we don't know the hypothetical privileged reference frame, it is not clear in which reference frame the synchronization should be optimized. Indeed, if two events are simultaneous in one frame, e.g. the laboratory frame, then, according to special relativity, they are not simultaneous with respect to others frames, e.g., to the cosmic microwave background radiation frame. Second, within an assumed privileged reference frame, perfect synchronization is impossible in practice; hence if nonlocal correlations are observed, this only sets a lower bound on the speed of the hypothetical hidden influence. Nevertheless, experiments have been carried out, setting stringent lower bounds of this speed, assuming the lab frame [11–14], the microwave background radiation frame [15] and even scanning all possible privileged reference frames [16, 17]. These experiments have excluded speeds up to about 50'000 times the speed of light.

At this point the case for a definite experimental test of DC explanation may seem quite hopeless: two-party experiments can only hope to increase the lower bound of the speed of the hypothetical hidden influence or to find the breakdown of quantum theory. But in 2002 Valerio Scarani and myself noticed that the situation changes dramatically when analyzing situations with more than two parties [18, 19]. The original scenario we considered involves 3 parties (see also Ryff [20] whose argument is recalled in Sect. 12.4). The general idea is the following. If two out of all parties measure simultaneously, e.g. Bob and Charlie are not-v-connected, then their correlation must be local. If moreover, the correlations between the other pairs of parties, those whose measurements are v-connected, allow one to guarantee that Bob and Charlie share nonlocal correlations, then one could infer a contradiction with any v-causal explanation without the need for any demanding synchronization. That one can infer the nonlocality between Bob and Charlie without ever measuring them in the same run of an experiment is quite counterintuitive, though it is known that sometimes one can infer a property of some quantum state or probability distribution from only the knowledge of some of their marginals [21–23].

The next step was made by Stefan Wolf and colleagues who introduced the concept of transitivity of nonlocality [24]. They showed that, assuming only no-signalling, there are examples of 3-party correlations, $p(a, b, c|x, y, z)$, such that if both marginals A-B and A-C are nonlocal, then the third marginal B-C is necessarily also nonlocal. This beautifully illustrates the idea Scarani and myself had in 2002. But unfortunately, Wolf and colleagues's example uses correlations that can't be achieved with measurements on quantum systems and, today, no quantum example of transitivity of nonlocality has been found. This is why the example we present in this paper doesn't use the concept of transitivity of nonlocality, but the theorem [2] recalled in Sect. 12.5.

To conclude this introduction let us consider some consequences of the assumption that v-causality is the explanation of all quantum correlations. As already mentioned, this would imply that some predictions of quantum theory are wrong: if two events are not-v-connected, then their correlation would be local even in cases where quantum theory predicts a violation of some Bell inequality. But could this departure from quantum predictions be used to communicate, in particular to communicate faster than light? In the 2-party case, Alice and Bob could arrange to be just at the border of being v-connected. So, if Bob makes his measurement early enough, the hidden influence doesn't arrive on time and they observe local correlations; but if Bob delays a little bit his measurement, then the influence arrives on time and they observe quantum correlations. This, however, can't be used by Alice and Bob to communicate. Indeed, their local statistics would be identical in both cases, whether the hidden influence arrives on time or not; it is only later, once they compare their data, that Alice and Bob can notice whether or not they violated some Bell inequality. Consequently, with only two parties, the hidden influence could remain hidden for ever: there would be a hidden layer at which faster than light hidden influences carry Direct Causes and thus establish correlations that appear nonlocal, but at our higher level nothing travels faster than light. In this paper, following [2], we prove that such a peaceful coexistence between relativity and faster than light hidden influences can't exist. But for this we'll need to consider more than two parties.

12.3 v-Causality

In this section we define formally local Common Cause, Direct Cause and v-causal explanations. Readers who feel they understand CC, DC and v-causality may like to jump to Sect. 12.4.

Consider a 2-party scenario, denoted Alice and Bob, with measurement settings x and y and measurement results a and b, respectively. The generalization to more parties is straightforward, as summarized at the end of this section. The conditional probability distribution, or in short correlation, $p(a, b|x, y)$, is the probability of results a, b when the settings x, y are chosen.

A pure **local Common Cause** explanation of $p(a, b|x, y)$ assumes additional variables, traditionally labeled λ, such that:

$$p(a, b|x, y) = \sum_{\lambda} \rho(\lambda) p(a|x, \lambda) p(b|y, \lambda) \tag{12.1}$$

where $\rho(\lambda)$ denotes the probability that the additional variable assumes the value λ (note that λ may include the quantum state ρ). For a justification see, e.g. [1, 25–28]. In a v-causal model, the information carried by the variable λ propagates gradually and continuously from some common v-past of Alice and Bob. If v would be the speed of light, this would merely be the usual intersection between the past light cones. But here the common v-past is the intersection of wider, more open, cones, see Fig. 12.1. Important in a common cause explanation is that $p(a|x, \lambda)$ doesn't depend on y and symmetrically $p(b|y, \lambda)$ is independent of x. Hence all correlations are due to the common local variable λ.

A pure **Direct Cause** explanation of $p(a, b|x, y)$ assumes that there is an absolute time ordering of the events at Alice and Bob (defined in the hypothetical universal privileged reference frame). For example, assume Alice is first to chose her measurement settings x and collect her result a. Direct cause[4] assumes that as soon as Alice performed her measurement, a signal—which we call a hidden influence—informs the rest of the universe, in particular Bob, of her measurement setting x and result a. In this case there are two possibilities:

1. The information reaches Bob's system before it produces the result b, i.e. Alice and Bob are v-connected. In this case:

$$p(a, b|x, y, v\text{-connected}) = p(a|x) p(b|y,\ x, a) \tag{12.2}$$

   For example, quantum correlations between v-connected events can be described as due to DC: $p(a|x) = Tr(A_a^x \rho_A)$ where ρ_A Alice's partial trace quantum state and A_a^x the projector representing her measurement, and $p(b|y,\ x, a) = Tr(B_b^y \rho_a^x)$ where $\rho_a^x = \frac{A_a^x \rho A_a^x}{Tr(A_a^x \rho)}$ is Bob's reduced state that depends on Alice's measurement setting x and result a. Note that in this case direct cause exactly reproduces the quantum prediction: $p(a, b|x, y, v\text{-connected}) = Tr(A_a^x \otimes B_b^y \cdot \rho)$.

[4] Standard text book descriptions of measurements collapsing the quantum state is an explicit example of a hidden influences explanation; however, in such descriptions the influence propagates at infinite speed. Hence it is more a direct action at a distance than an influence propagating in space and time. Note that because of the infinite speed, all parties are v-connected. Such descriptions also require a universal privileged reference frame.

2. Bob's system has to produce the result b before the information carried by the hidden influences arrives from Alice's system, i.e. Alice and Bob are not-v-connected:

$$p(a,b|x,y,\ \text{not-}v\text{-connected}) = p(a|x)p(b|y) \tag{12.3}$$

In the case that Bob's probability depends only on his local quantum state $\rho_B = Tr_A(\rho)$, one has:

$$p(a,b|x,y,\ \text{not-}v\text{-connected}) = Tr(A_a^x \cdot \rho_A) \cdot Tr(B_b^y \cdot \rho_B) \tag{12.4}$$

In general, for entangled states ρ, this prediction differs from the quantum prediction.

A v-**causal** explanation of $p(a,b|x,y)$ combines additional local variables and hidden influences,[5] all propagating at a speed v (or lower) defined in the universal privileged reference frame. This frame defines an absolute time ordering, as for direct cause explanations. Here again one has to distinguish two possibilities depending on whether Alice and Bob are v-connected or not:

1. The information reaches Bob's system before it produces the result b, i.e. Alice and Bob are v-connected. In this case:

$$p(a,b|x,y,v\text{-connected}) = \sum_\lambda \rho(\lambda)p(a|x\lambda)p(b|y,\lambda,\ x,a) \tag{12.5}$$

Since we look for an explanation of quantum correlations, one expects that, in the case of v-connected events, quantum correlations are reproduced.
2. Bob's system has to produce the result b before the information carried by the hidden influences arrives from Alice's system, i.e. Alice and Bob are not-v-connected:

$$p(a,b|x,y,\ \text{not-}v\text{-connected}) = \sum_\lambda \rho(\lambda)p(a|x,\lambda)p(b|y,\lambda) \tag{12.6}$$

where λ includes the quantum state ρ. Consequently, in any v-causal model, unconnected events must satisfy all Bell inequalities.

The generalization to an arbitrary number of parties should be straightforward: when a system undergoes a measurement it takes into account all the information it received, whether additional local variables or hidden influences, and sends out information about itself in all directions by hidden influences propagating at speed v.

---

[5]The De-Broglie-Bohm pilot wave model is an explicit example of a v-causal explanation; however, in this model the influence propagates at infinite speed. Hence it is more a direct action at a distance than an influence propagating in space and time. Note that because of the infinite speed, all parties are v-connected, hence Bohm's model recovers all quantum predictions. This model also requires a universal privileged reference frame.

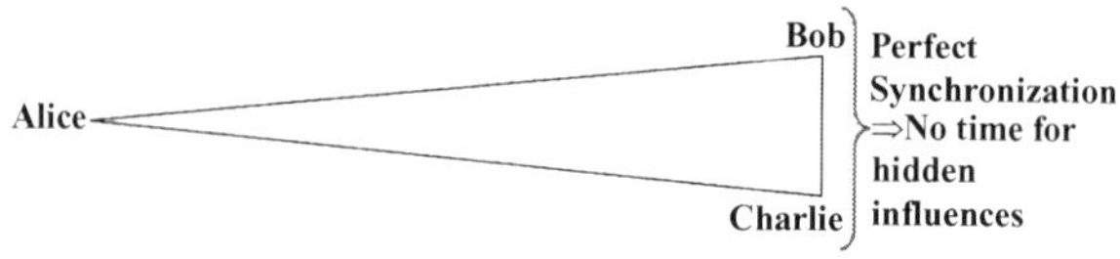

Fig. 12.2 Spacial configuration of the 3-party scenario discussed in Sect. 12.4 to show that pure Direct Cause leads to signalling

Since we are looking for an explanation of quantum correlations, one expects that, whenever possible, v-connected events reproduce quantum correlations. However, unconnected events necessarily produce local correlations, hence correlations that may differ from the quantum predictions. This constraint is what limits the power of v-causal explanations and makes experimental tests possible.

Note that the speed of light c doesn't appear in the definitions of local Common Cause and Direct Cause, nor v-causality.[6]

12.4 No Direct Cause Explanation

In this section we study the assumption that correlations between quantum measurement results are due to DC carried by hidden influences propagating at a finite but supraluminal speed v. More precisely, we consider hidden influence plus the usual quantum state, but no additional local variables. This section is greatly inspired by [20] (note that Eberhard published a related argument also involving 3 parties [3]).

Consider a 3-party scenario, Alice, Bob and Charlie, where Alice is far away from both Bob and Charlie. Bob and Charlie are relatively close to each other, but distant enough (in the hypothetical universal reference frame) so that they can synchronize their measurements well enough to be not-v-connected, see Fig. 12.2. Alice, Bob and Charlie know the relative positions of each other and at what time Bob and Charlie perform their measurements. Assume they share a GHZ state $\Psi = |0,0,0\rangle + |1,1,1\rangle$ and all measure σ_z. Quantum theory predicts that all three collect the same result: $a = b = c$. We shall see that if this correlation is due to some supraluminal hidden influence (without additional variables, i.e. pure DC), then Alice could communicate faster than light to Bob and Charlie (Bob and Charlie need to collaborate).

The argument runs as follows. First, if Alice chooses to communicate "yes", she performs her measurement early enough that the hidden influence arrives on time to Bob and Charlie. In this case the hidden influence tells Bob and Charlie's system which result a Alice obtained, hence Bob and Charlie's system produce that same

[6]Nor does c appear in the definition of "Bell locality" (12.1). Nevertheless, physicists have always been interested in tests of Bell inequalities between space-like separated events, i.e. between events not-c-connected. This illustrates that v-causal models were always in the back of the mind of those physicists, though with $v = c$.

result: $b = a$ and $c = a$. Next, if Alice chooses to communicate "no", she doesn't perform any measurement, or only too late for the hidden influence to arrive on time. In this case Bob and Charlie obtain random and independent results (recall that they are not-υ-connected, hence their result are produced independently of each other), whence half the time $b \neq c$. Consequently, once Bob and Charlie compared their results (which they can do in a time very short relative to the time light would take to propagate from Alice to them), they can infer with good probability Alice's message.

This is faster than light communication from Alice to Bob-Charlie. By elongating the triangle the speed of this communication gets arbitrarily close to the speed υ of the hidden influence. Hence, the hidden influence doesn't remain hidden, but can be activated.

This simple example shows that with 3 parties one can activate the hidden influence, something impossible with only 2 parties. However, this example also shows that there is a simple way around the argument. Indeed, the correlation is a simple and local one: $a = b = c$. Hence, one could merely supplement the DC explanation with a shared random bit r and assume that in the case the hidden influence doesn't arrive on time, all systems produce the result r. This motivates the investigation of υ-causality, where DC is combined with additional local variables as explained in Sect. 12.3 and analyzed in the next section.

12.5 No υ-Causal Explanation

At this stage of the search for an explanation of quantum correlations, local common causes and hidden influences are both individually excluded. The first one predicts Bell inequalities that have been violated, while the second one can't remain hidden as recalled in the previous section. Let us thus analyse the hypotheses that υ-causality, i.e. an arbitrary combination of Direct and Common Causes, is the explanation of all correlations. This might sound bizarre. But quantum correlations are bizarre and there is simply no other type of explanations that satisfy the principle of continuity (we discuss this principle in more detail in Sect. 12.7). This section is greatly inspired by [2].

Consider the 4-party configuration of Fig. 12.3, represented in the hypothetical privileged reference frame. Alice, Bob, Charlie and Dave have a choice between two measurement settings, labeled x, y, z and w and collect binary results $a, b, c, d \in \{-1, +1\}$, respectively. Alice measures first, hence is not influenced by any of the other parties. Next, Dave measures at a time such that the hypothetical influence from Alice arrives on time to Dave. Finally, Bob and Charlie measure quasi-simultaneously, i.e. Bob and Charlie are not-υ-connected, but such that the hypothetical influences from Alice and Dave arrive on time both to Bob and to Charlie.

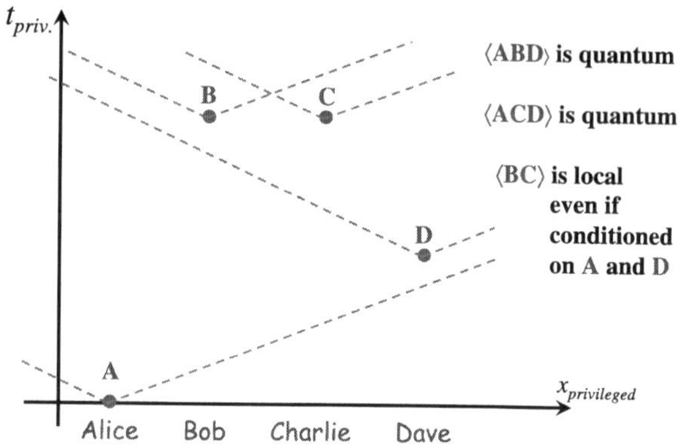

Fig. 12.3 Space-time configuration in the privileged reference frame of the 4-party scenario discussed in Sect. 12.5 to show that all v-causal models lead to signalling. (Taken with permission from Nature from [2])

If we were considering only DC, the joined probability would read:

$$p(a,b,c,d|x,y,z,w) = p(a|x)\cdot p(d|w,x,a)\cdot p(b|y,x,a,w,d) \cdot p(c|z,x,a,w,d) \tag{12.7}$$

It is not difficult to see that the correlation (12.7) leads to signalling from Alice to Bob-Charlie-Dave (who need to cooperate). But in this configuration, contrary to the triangular configuration of the previous section, we can exclude the possibility that additional variables allow one to avoid the activation of the hypothetical hidden influence.

The idea is to find an inequality satisfied by all no-signalling correlations where the not-v-connected parties are local with the following two properties:

1. all terms in the inequality involve only v-connected parties (hence, to evaluate the inequality one never has to measure in a same round of the experiment not-v-connected parties, one thus avoids the synchronization difficulty),
2. the n-party correlation can be violated by quantum correlations (i.e. quantum theory predicts a violation of the inequality).

The technical difficulty of this strategy is that, first one has to study the intersection of the n-party no-signalling polytope with the local polytope of the not-v-connected parties. Next, one has to project this intersection polytope on the subspace of correlations containing only terms corresponding to v-connected parties.

This strategy can obviously not work with only two parties (both would either be v-connected or both not-v-connected). Hence, with my co-authors of [2] we spent a long time searching for an example involving 3 parties, one pair being not-v-connected and two pairs v-connected. But no example has been found, though the search continues, varying the number of inputs (measurements settings) and outcome for each party [29]. The breakthrough came when Jean-Daniel Bancal and Stefano Pironio had the courage to consider 4 parties in the configuration of Fig. 12.3. After heavy numerical search they found the following [2].

Theorem *Let $p(a, b, c, d|x, y, z, w)$ be a correlation, i.e. a conditional probability distribution, with binary inputs $x, y, z, w \in \{0, 1\}$ and outcomes $a, b, c, d \in \{-1, +1\}$.*

If

1. *The correlation $p(a, b, c, d|x, y, z, w)$ is non-signalling, and*
2. *$p(b, c|y, z, \ a, x, d, w)$ is local*[7] *for all a, x, d, w,*

then $S \leq 7$, where

$$\begin{aligned} S = &-3\langle A_0\rangle - \langle B_0\rangle - \langle B_1\rangle - \langle C_0\rangle - 3\langle D_0\rangle \\ &- \langle A_1B_0\rangle - \langle A_1B_1\rangle + \langle A_0C_0\rangle \\ &+ 2\langle A_1C_0\rangle + \langle A_0D_0\rangle + \langle B_0D_1\rangle \\ &- \langle B_1D_1\rangle - \langle C_0D_0\rangle - 2\langle C_1D_1\rangle \\ &+ \langle A_0B_0D_0\rangle + \langle A_0B_0D_1\rangle + \langle A_0B_1D_0\rangle \\ &- \langle A_0B_1D_1\rangle - \langle A_1B_0D_0\rangle - \langle A_1B_1D_0\rangle \\ &+ \langle A_0C_0D_0\rangle + 2\langle A_1C_0D_0\rangle - 2\langle A_0C_1D_1\rangle \end{aligned} \tag{12.8}$$

In (12.8) $\langle A_1B_0\rangle$ *denotes the average of the product of Alice and Bob's outcomes when Alice chooses $x = 1$ and Bob $y = 0$ and similarly for the other terms.*

The above inequality S is remarkable because none of its 23 terms involves both Bob and Charlies, hence it can be evaluated without ever measuring Bob and Charlie in the same run of an experiment. Nevertheless,

1. Assuming no-signalling, a violation implies that Bob and Charlie share nonlocal correlations, i.e. correlations that can't be explained by Common Causes, and
2. Assuming that Bob and Charlie are local, as they are in any v-causal model, a violation implies that $p(a, b, c, d|x, y, z, w)$ is signalling.

It is not difficult to check that the inequality $S \leq 7$ can be violated by the following 4 qubit state [2]

$$\begin{aligned} |\Psi\rangle = &\frac{17}{60}|0000\rangle + \frac{1}{3}|0011\rangle - \frac{1}{\sqrt{8}}|0101\rangle + \frac{1}{10}|0110\rangle \\ &+ \frac{1}{4}|1000\rangle - \frac{1}{2}|1011\rangle - \frac{1}{3}|1101\rangle + \frac{1}{2}|1110\rangle \end{aligned} \tag{12.9}$$

with the measurements

$$\hat{A}_0 = -U\sigma_x U^\dagger \qquad \hat{A}_1 = U\sigma_z U^\dagger \tag{12.10}$$

---

[7]I.e. satisfy the Clauser-Horn inequality: $p(b = c = 0|0, 0, \ a, x, d, w) + p(b = c = 0|0, 1, \ a, x, d, w) + p(b = c = 0|1, 0, \ a, x, d, w) - p(b = c = 0|1, 1, \ a, x, d, w) - p(b = 0|y = 0, \ a, x, d, w) - p(c = 0|z = 0, \ a, x, d, w) \leq 0$ and all its symmetric forms obtained by permuting the inputs and outcomes.

$$\hat{B}_0 = H \qquad \hat{B}_1 = -\sigma_x H \sigma_x \tag{12.11}$$

$$\hat{C}_0 = -\hat{D}_0 = \sigma_z \qquad \hat{C}_1 = \hat{D}_1 = -\sigma_x \tag{12.12}$$

where $U = \cos(\frac{4\pi}{5})\sigma_z - \sin(\frac{4\pi}{5})\sigma_x$, the σ's denote the Pauli matrices and H the Hadamard matrix. Quantum theory predicts for these state and measurement settings $S \approx 7.2$.

Accordingly, the supraluminal hidden influence in any v-causal model can be activated. Indeed, in any v-causal model Bob and Charlie are local, hence, one can deduce from the quantum prediction that the 4-party correlation is signalling (recall that the 4-party correlation is not quantum, because Bob and Charlie are not-v-connected, only the 3-party marginals A-B-D and A-C-D are quantum, but this suffices to evaluate S).

Consequently, at least one of the four 3-party marginals depends on the fourth's input. Consider first the A-B-D 3-party marginal; since A-B-D are all v-connected, $p(a, b, d|x, y, w)$ is quantum and thus non-signalling (it doesn't depend on Charlie's input z). Moreover, the A-B-D correlation can't depend on Charlie's input z, because z is chosen outside of A-B-D past v-cones. Similarly for the A-C-D 3-party marginal. Consequently, it must be either A-B-C that depends on Dave's input w or B-C-D that depends on Alice's input x (or both). Both cases are similar; let us thus consider the case that $p(b, c, d|y, z, w, x)$ depends explicitly on x. This is signalling from Alice to Bob-Charlie-Dave. Moreover, this can be used for faster than light communication: it suffices that Bob and Charlie send (at the speed of light) their inputs y, z and outcomes b, c to Dave so that Dave can evaluate their 3-party marginal B-C-D. Since this marginal depends on Alice's measurement setting choice x, Alice can communicate to Dave. Figure 12.4 shows that this communication can be faster than light. By moving B-C-D away from A, but such that the hidden influence from Alice still arrives on time to all of them, one can make this faster than light communication tend to the speed v of the hidden influence.

In summary, the hidden influence of any v-causal explanation of quantum correlation can never remain hidden: it necessarily allows for faster than light communication. We'll come back to this remarkable conclusion in Sect. 12.8.

12.6 Experiment

In this section we consider how experiments could test the contradiction we have established between quantum theory, v-causality and no faster than light communication. At first, one may wonder whether such an experiment is necessary at all. Indeed, quantum correlations have been measured abundantly. Hence, it seems highly likely that the state (12.9) and the quantum measurements (12.10)–(12.12) can be realized with good enough approximation to violate inequality (12.8). Moreover, the very assumption that quantum correlations are explained by v-causality implies that the ABD and the ACD correlation predicted by any v-causal model are identical to the quantum prediction, hence that any v-causal model violates the inequality. If not,

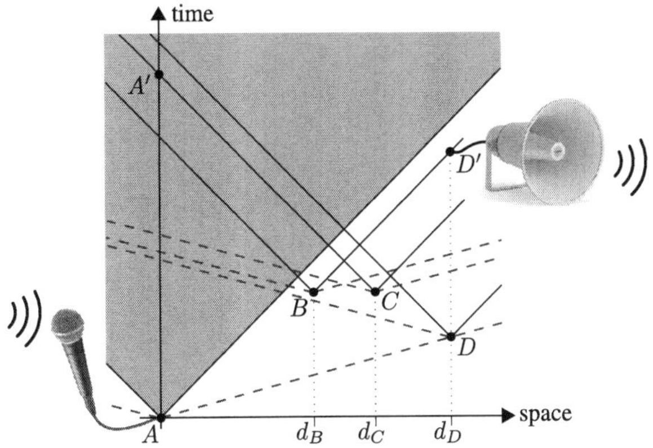

Fig. 12.4 In the 4-party scenario, signalling leads to faster than light communication. Here we illustrate the case where the signalling goes from A to BCD; the case D→ABC is similar (the other cases don't happen, because they are quantum, see text). (Taken with permission from Nature from [2])

the v-causal model would not be an explanation for the quantum correlation.[8] Furthermore, if it would turn out impossible to violate inequality (12.8), then quantum theory would fail even in cases were the events are v-connected. This would be very difficult to explain and v-causality might not be of much help. This is in sharp contrast to Bell's inequality: had it turn out impossible to violate Bell's inequality, local CC would have been vindicated. Hence, whether or not one eventually observes a violation of the inequality (12.8), in both cases explanations based on v-causality seem difficult to maintain!

But, physics being an experimental science, one should check that correlations violating the inequality we used in the previous section to derive our conclusion can indeed be realized.

So, imagine a source producing a state close to the 4 qubit state (12.9) and distributing each qubit to Alice, Bob, Charlie and Dave. Alice is first to choose her measurement setting x, measure her qubit and collect her outcome a. Alice and her qubit may be aware of the locations of her partners, who may perform some measurement, or are measuring quasi simultaneously, such that the corresponding hidden influence did not reach her yet. However, Alice can't know when such possible measurements will be performed by her partners, or whether they will be performed at all (the so-called "free-will" assumption). Accordingly, Alice (more precisely her qubit) has to send out her hidden influence at speed v independently of when Bob, Charlie and Dave may measure (or not measure) their qubits.

Dave should be second to measure, but not too early, so as to make sure that Alice's hidden influence reaches him on time. This can be guaranteed by merely letting Dave measure his qubit at a time such that even light would arrive on time. Since the hidden influence propagates faster than light, it will necessarily also arrive on time, irrespectively of which reference frame is the privileged one. Here we assume that Alice's hidden influence always propagates at the same speed v, independently of the protocol of the experiment. This is a standard assumption in science: one never

[8]Though, if quantum theory is falsified, then one would no longer be looking for an explanations of all quantum correlations.

assumes that the experimental protocol changes the laws one is testing. In summary, it is easy to guarantee that Dave measures far enough in the future to respect the time ordering of Fig. 12.3.

This just leaves Bob and Charlie. They should both measure in the future of Dave so that its hidden influence arrives on time. This can again be achieved by setting Bob and Charlie in the future light cone of Dave (and thus also of Alice). However, according to the configuration depicted in Fig. 12.3, Bob and Charlie should be well enough synchronized to guarantee that no hidden influence from one can reach the other. This is impossible without knowing an upper bound on the speed of the hidden influence and the privileged frame. This difficulty is circumvented, as already explained in the previous section, by the observation that in the inequality (12.8), no term involves both Bob and Charlie. Hence, one doesn't need to ever measure them in a same round of the experiment. It suffices that, after Dave measured his qubit, a random choice is made by the experimentalist to measure either Bob or Charlie's qubit. In each case, another, fourth (independent) random choice is made to select the measurement setting. Both these choices are made in the future light cone of Dave. Again, we assume that the qubit chosen to be measured, whether it is Bob's or Charlie's, produces a result that is independent of the protocol. In other words, in case Bob's qubit is chosen to be measured, the probability of the result b is the same as if Charlie's qubit would be measured simultaneously: Bob's result probability can't depend on when Charlie's qubit is measured as long as Charlie qubit can't influence Bob's. And if Charlie's qubit is not measured at all, Bob's result probability can't depend on "when Charlie's qubit is not measured".

In summary, a first random bit decides Alice's measurement setting x, next in the absolute future a second random bit chooses Dave's setting w, finally, again in the absolute future, a third random bit decides whether Bob's or Charlie's qubit is measured and a fourth random bit decides the measurement setting y or z. Note that all these 4 random bits must be independent of the hypothetical additional variables and hidden influences, as in Bell inequality analysis (this is sometimes called the "free will" or the "measurement independence" assumption [30, 31]).

In this way, the experimental test of quantum predictions for the configuration depicted in Fig. 12.3 can be realized. If a violation is observed, as one expects from quantum theory, then one has to conclude that

1. either the hypothetical hidden influence can't remain hidden, but necessarily leads to signalling and to faster than light communication,
2. or, all v-causal explanations are ruled out, i.e. no combination of Direct Cause and local Common Cause can explain the experimental result.

Both these alternatives are fascinating and will be discussed in the conclusion section.

One might be surprised that the proposed experiment doesn't involve any space-like separated measurements. But, as mentioned at the end of Sect. 12.3, the speed of light doesn't appear in the definition of v-causality. Hence, according to v-causality, one doesn't expect any difference when measurements are time-like or space-like separated. Furthermore, signalling between time-like separated events would be about as bizarre as between space-like separated events. Indeed, imagine that Alice is located in a safe, e.g. in the basement of the Swiss national bank. One expects that this would not affect the correlations between her measurements and those of her partners, wherever they are located. In particular they could be in the future light-cone of Alice, somewhere outside of the bank. But then, signalling from Alice to BCD, as v-causality and the violation of (12.8) predict, implies that Alice could communicate to her partners, whatever physical security measures and isolation one imposes on Alice[9]!

12.7 Newton and the Principle of Continuity

It is not the first time in history that physics is confronted with nonlocality. Actually, physics almost always presented a nonlocal world-view of nature, first with Newton's theory of universal gravitation, then with quantum nonlocality. Only during a short time window of about 10 years did physics present a local world-view.

Newton was very concerned by the nonlocal predictions of his theory of universal gravitation. Indeed, he noticed that his theory predicts that any change in the local configuration of matter would have an immediate effect on the entire universe. Hence, by moving to the left or to the right a stone on the moon,[10] one could, in principle, signal at arbitrary speed to Earth and to any place in the universe. Let us read how the great man described the situation [32]:

> That Gravity should be innate, inherent and essential to Matter, so that one Body may act upon another at a Distance through a Vacuum, without the mediation of any thing else, by and through which their Action and Force may be conveyed from one to another, is to me so great an Absurdity, that I believe no Man who has in philosophical Matters a competent Faculty of thinking, can ever fall into it. Gravity must be caused by an Agent acting constantly according to certain Laws, but whether this Agent be material or immaterial, I have left to the Consideration of my Readers.

Accordingly, "no action at a distance" is not a principle of relativity nor of Einstein, but is part of Newtonian space-time. Let us emphasize that "no action at a distance" implies that nothing propagates at infinite speed, in particular there are no infinite speed influences nor ∞-causality.

Usually, quantum correlations are seen as being in tension with (special) relativity, remember Shimony's statement about the peaceful coexistence of quantum theory and relativity. But it is natural to go beyond these tensions and investigate the consequences of assuming that the correct interpretation of Lorentz transformation is not mere geometry of space-time, but real Fitzgerald contractions of lengths and Larmor dilation of time intervals, as Lorentz and Poincaré themselves thought

---

[9]This would be similar to signalling using gravitation—no way to prevent it—but at the speed v.

[10]To move the stone one shouldn't take support on the moon, as this would not move the center of mass of the moon-&-stone, but use a small rocket.

and as John Bell considered [9, 33]. Hence, the interest for studying quantum correlations in Newtonian space-time, or, equivalently for that matter, in a universal privileged reference frame.

Notice that all v-causal explanations of correlations satisfy a **principle of continuity** that states that everything (mass, energy and information) propagates gradually and continuously through space as time passes, i.e. nothing jumps instantaneously from here to there. In other worlds, there is no action at a distance. Reciprocally, all explanations of correlations that satisfy the principle of continuity are v-causal. Hence, Newton and Einstein would have bet on a v-causal explanation of all correlations, including quantum correlations.

An experimental violation of the inequality $S \leq 7$ either implies a violation of the principle of continuity or implies faster than light communication.

12.8 No-signalling in v-Causality

No-signalling is generally considered as a fundamental principle that has to hold in any meaningful physical theory. However, if the correlations between some events are due to hidden influences, then there is no reason to assume that the influences don't allow one to signal (at the speed of the hidden influence or slower). This is for example the case with gravity. Had someone before Einstein had the technology to check the correlation between the displacement of a stone on the moon and the weight of some mass on Earth, even when displaced and measured simultaneously, he would have observed a null correlation (at least for good enough synchronization) and thus have falsified Newton's theory of universal gravitation. He could also have observed that the correlation establishes when the weight measurement is performed about a second after the displacement of the mass on the moon. This would have allowed him to signal at the speed of what was then a hidden influence, i.e. the speed of gravitons that, according to general relativity, carry the cause of the change in the weight of the mass on Earth. This is a typical Direct Cause explanation. Note that one could have used this hidden influence to signal even without knowing the theory of general relativity.

Similarly, if the speed of the hidden influence that explains quantum correlations propagates faster than the speed of light, then the corresponding signalling would equally be faster than the speed of light. Consequently, there are only two possibilities:

1. either the hidden influence remains hidden for ever, i.e. is intrinsically hidden,[11] hence doesn't allow for signalling, or
2. the hidden influence can be used to communicate at a speed equal or lower than the speed of the hidden influence, i.e. the hidden influence doesn't remain hidden.

[11] I am quite suspicious of explanations relying on intrinsically hidden stuff, hence I dislike this part of the alternative.

In Sect. 12.5 we demonstrated that the hypothetical hidden influence of all v-causal model can't remain hidden, but on the contrary leads to faster than light communication at the level of the classical measurement settings and results. Hence, the first of the above two alternative is excluded.

12.9 Conclusion

The main conclusion of this paper is that an experimental violation of the inequality $S \leq 7$ would imply

1. either a violation of the principle of continuity (that states that everything propagates gradually and continuously through space as time passes as discussed, in Sect. 12.7), i.e. the falsification of all v-causal models, or
2. the possibility of faster than light communication at the level of the classical measurement settings and results.

It is unlikely that many physicists will contemplate seriously the second alternative.[12] However, one should realize that the first alternative is about as difficult to swallow as the second one. A violation of the principle of continuity implies that the world is truly and definitively not local, i.e. Nature is not continuous, but nonlocal. This conclusion has already been claimed by many physicists (including this author), though only based on the violation of Bell inequality between space-like separated events. These physicists made the (admittedly highly plausible) assumption that space-time is described by relativity. In this paper we have extended the conclusion: even if one is willing to consider a Newtonian-type privileged reference frame, but without faster than light communication, the conclusion that Nature is nonlocal is unavoidable.

Should then Physicists give up the great Enterprize of explaining how Nature does it [37]? Certainly not! But physicists have only two options:

1. Pursue the search for the speed of v-causal explanations by improving the "Salart-type" experiments [16, 17]. Note that the finding of such a speed would falsify both quantum theory and relativity, a result not many physicists are willing to envisage. However, the tension between these two pillars of today's physics may well dissolve not merely by saving one of them at the cost of the other, but

---

[12] One recent exception is B. Cocciaro [34]. In this paper the author also recalls that faster than light communication in one universal global privileged reference frame, as consider in this paper, doesn't lead to the "grand father" time paradox. Indeed, for time paradoxes one should communicate to one's own past; this requires a go-&-return communication. But if both the go and the return signal are defined in the same reference frame and at the same—possibly supraluminal—speed v, then the "return" signal will necessarily arrive in the absolute future of the start of the "go" signal. It is straightforward to see this in the privileged frame. But then, the start of the "go" and the arrival of the "return" signals are necessarily also time-like in all other reference frames, hence the impossibility to communicate to one's own past. This is not new and was emphasized, e.g., in [26, 34–36]. Consequently, supraluminal communication might not have said it's last word.

by finding the limits of both theories. Accordingly, "Salart-type" experiments are still needed, but the result of [2]—recalled in this paper—shows that a positive result would definitively lead to faster than light communication, hence it would not only falsify quantum theory but also falsify relativity.

2. Accept quantum nonlocality and enlarge our story tool-box by inventing new tools—necessarily nonlocal—to tell explanatory stories. Possibly something like "one random event can manifest itself at several locations".

Acknowledgements This article greatly profited from numerous exchanges with my co-authors of [2] and from comment by Rob Thew and many colleagues over the years. This work has been supported by the ERC-AG grant QORE, the CHIST-ERA DIQIP project, and by the Swiss NCCR *Quantum Science and Technology*—QSIT.

References

1. J.S. Bell, *Speakable and Unspeakable in Quantum Mechanics: Collected Papers on Quantum Philosophy* (Cambridge University Press, Cambridge, 2004)
2. J.D. Bancal, S. Pironio, A. Acin, Y.C. Liang, V. Scarani, N. Gisin, Nat. Phys. **8**, 867–870 (2012)
3. Ph.H. Eberhard, A realistic model for Quantum Theory with a locality property, in *Quantum Theory and Pictures of Reality*, ed. by W. Schommers (Springer, New York, 1989), pp. 169–215
4. A. Aspect, J. Dalibard, G. Roger, Phys. Rev. Lett. **49**, 1804 (1982)
5. W. Tittel, J. Brendel, N. Gisin, H. Zbinden, Phys. Rev. Lett. **81**, 3563–3566 (1998)
6. G. Weihs, T. Jennewein, C. Simon, H. Weinfurter, A. Zeilinger, Phys. Rev. Lett. **81**, 5039 (1998)
7. C. Lineweaver et al., Astrophys. J. **470**, 38 (1996). http://pdg.lbl.gov
8. J.S. Bell, La nouvelle cuisine, in [1]
9. J.S. Bell, in *The Ghost in the Atom*, ed. by P.C.W. Davies, J.R. Brown (Cambridge University Press, London, 1993), pp. 45–57
10. D. Bohm, B.J. Hiley, *The Undivided Universe* (Routledge, London, 1993) (p. 347 of the paperback edition)
11. N. Gisin, V. Scarani, W. Tittel, H. Zbinden, 100 years of Q theory. Proc. Ann. Phys. **9**, 831–842 (2000)
12. N. Gisin, V. Scarani, W. Tittel, H. Zbinden. quant-ph/0009055
13. A. Stefanov, H. Zbinden, N. Gisin, Phys. Rev. Lett. **88**, 120404 (2002)
14. N. Gisin, Sundays in a quantum engineer's life, in *Proceedings of the Conference in Commemoration of John S. Bell*, Vienna, 10–14 November (2000)
15. V. Scarani, W. Tittel, H. Zbinden, N. Gisin, Phys. Lett. A **276**, 1–7 (2000)
16. D. Salart, A. Baas, C. Branciard, N. Gisin, H. Zbinden, Nature **454**, 861 (2008)
17. B. Cocciaro, S. Faetti, L. Fronzoni, Phys. Lett. A **375**, 379 (2011)
18. V. Scarani, N. Gisin, Phys. Lett. A **295**, 167 (2002)
19. V. Scarani, N. Gisin, Braz. J. Phys. **35**, 2A (2005)
20. L.C. Ryff. arXiv:0903.1076
21. N. Linden, S. Popescu, W.K. Wootters, Phys. Rev. Lett. **89**, 207901 (2002)
22. N. Linden, W.K. Wootters, Phys. Rev. Lett. **89**, 277906 (2002)
23. L.E. Würflinger, J.D. Bancal, A. Acin, N. Gisin, T. Vertesi, Phys. Rev. A **86**, 032117 (2012)
24. S. Coretti, E. Hänggi, S. Wolf, Phys. Rev. Lett. **107**, 100402 (2011)
25. D. Mermin, Am. J. Phys. **49**, 940 (1981)

26. T. Maudlin, *Quantum Non-locality and Relativity: Metaphysical Intimations of Modern Physics* (Blackwell Publishers, Oxford, 2002)
27. T. Norsen, John S. Bell's concept of local causality. Am. J. Phys. **79**, 1261–1275 (2012)
28. N. Gisin, Found. Phys. **42**, 80–85 (2012)
29. T. Barnea Master thesis, University of Geneva (2012). arXiv:1304.1812
30. M.J.W. Hall, Phys. Rev. Lett. **105**, 250404 (2010)
31. J. Barrett, N. Gisin, Phys. Rev. Lett. **106**, 100406 (2011)
32. I. Newton, *Papers & Letters on Natural Philosophy and Related Documents* (Harvard University Press, Cambridge, 1958), p. 302. Edited, with a general introduction, by Bernard Cohen, assisted by Robert E. Schofield
33. J.S. Bell, How to teach special relativity, in [1]
34. B. Cocciaro. arXiv:1209.3685
35. F. Reuse, Ann. Phys. **154**, 161 (1984)
36. P. Caban, J. Rembielinski, Phys. Rev. A **59**, 4187 (1999)
37. N. Gisin, Quantum nonlocality: how does nature do it? Science **326**, 1357–1358 (2009)

Chapter 13
PR-Box Correlations Have No Classical Limit

Daniel Rohrlich

Abstract One of Yakir Aharonov's endlessly captivating physics ideas is the conjecture that two axioms, namely relativistic causality ("no superluminal signalling") and nonlocality, so nearly contradict each other that a unique theory—quantum mechanics—reconciles them. But superquantum (or "PR-box") correlations imply that quantum mechanics is not the most nonlocal theory (in the sense of nonlocal correlations) consistent with relativistic causality. Let us consider supplementing these two axioms with a minimal third axiom: there exists a classical limit in which macroscopic observables commute. That is, just as quantum mechanics has a classical limit, so must any generalization of quantum mechanics. In this classical limit, PR-box correlations *violate* relativistic causality. Generalized to all stronger-than-quantum bipartite correlations, this result is a derivation of Tsirelson's bound without assuming quantum mechanics. [*Editors note*: for a video of the talk given by Dr. Rohrlich at the Aharonov-80 conference in 2012 at Chapman University, see quantum.chapman.edu/talk-10.]

I first met Yakir Aharonov when I was a post-doc at Tel Aviv University, more than two decades ago. I discovered that he could answer questions about quantum mechanics that no one had answered to my satisfaction, not even Nobel Prize-winning physicists. I wanted to understand what he understood about quantum mechanics, and gladly accepted his offer to write a book together [1]. Around that time, he told me an idea that has fascinated me ever since. What follows is the story of that idea.

Of course quantum mechanics is baffling, he said. Look at its axioms: "Physical states are normalized vectors in Hilbert space"; "Measurable physical quantities correspond to self-adjoint operators"; the Born rule; etc. Are these statements about the physical world, or statements about applied mathematics? By contrast, special relativity has an exemplary logical structure: two axioms, each with a clear physical meaning—a fundamental physical constant (the speed of light) and a fundamental invariance (the equivalence of inertial reference frames)—are so nearly incompatible that a unique kinematics reconciles them. Suppose we tried to derive special

D. Rohrlich (✉)
Department of Physics, Ben Gurion University of the Negev, Beersheba 84105, Israel
e-mail: rohrlich@bgu.ac.il

D.C. Struppa, J.M. Tollaksen (eds.), *Quantum Theory: A Two-Time Success Story*, DOI 10.1007/978-88-470-5217-8_13, © Springer-Verlag Italia 2014 relativity from the wrong axioms, e.g. "Fast objects contract in the direction of their motion" and "Moving clocks slow down." How could we understand the theory? If the analogy seems far-fetched, note that as late as 1909, Henri Poincaré gave a lecture entitled "La méchanique nouvelle" ("The new mechanics"). The lecture, after mentioning the two axioms of special relativity, goes on to mention the FitzGerald contraction: "One needs to make still a third hypothesis, much more surprising, much more difficult to accept, one which is of much hindrance to what we are currently used to. A body in translational motion suffers a deformation in the direction in which it is displaced [2]." The phrase "of much hindrance"—wouldn't it apply to any of the axioms of quantum mechanics?! The point here is that for Poincaré, the FitzGerald contraction was an axiom and therefore hard to accept; for us, the FitzGerald contraction is a logical consequence of clear physical axioms and therefore not so hard to accept.

But, Aharonov continued, quantum mechanics, too, reconciles two nearly incompatible physical statements. On the one hand, quantum mechanics has to obey *relativistic causality*, the constraint that no signal can travel faster than light. On the other hand, quantum mechanics is nonlocal, in at least two different ways. There are the Aharonov-Bohm [3] and related effects, in which the motions of particles depend on nonlocal relative phases; and there are nonlocal quantum correlations that violate a Bell inequality [4, 5]. Quantum nonlocality is action at a distance: a cause *here* has an immediate effect *there*. How can action at a distance be compatible with relativistic causality? The only way to reconcile them is via *uncertainty*. If the effect *there* of a cause *here* is uncertain, it may not lead to superluminal signalling. Thus we can obtain uncertainty as a logical consequence of axioms of relativistic causality and nonlocality. It is hard to accept uncertainty as an axiom (the Born rule); it leaves us asking, "Why does God play dice?" But it is not so hard to accept uncertainty as a logical consequence of relativistic causality and nonlocality; we can say, "God plays dice because it is the only way for nonlocality and relativistic causality to coexist."

Is this derivation of quantum uncertainty only qualitative, or is it also quantitative? To answer this question, we must specify what we mean by quantum nonlocality. Aharonov defined "modular" quantum variables that are nonlocal in space or time. The nonlocality of modular variables arises from nonlocal relative phases. He showed that there is always just enough uncertainty in measurements of modular variables to prevent their use for noncausal signalling [6, 7]. Independently, Shimony noted that quantum mechanics is remarkable in reconciling relativistic causality with the nonlocality implicit in nonlocal quantum correlations. He gave this coexistence the apt name "passion at a distance" [8, 9]. But can we say anything quantitative about the nonlocality of quantum correlations?

Let two physicists, Alice and Bob, share many identical pairs of particles with a common origin. Alice measures a or a' on her particles and Bob measures b or b' on his, where a, a', b and b' are observables taking values ± 1. Their respective measurements on any given pair are spacelike separated. At some point they pool their data and calculate the correlation functions $C(a,b)$, $C(a,b')$, $C(a',b)$ and $C(a',b')$. By definition,

$$C(a,b) = \mathrm{p}_{ab}(1,1) + \mathrm{p}_{ab}(-1,-1) - \mathrm{p}_{ab}(1,-1) - \mathrm{p}_{ab}(-1,1), \tag{13.1}$$

where $\mathrm{p}_{ab}(i,j)$ is the probability that measurements of a and b on a given pair yield $a = i$ and $b = j$. If the correlations are local, then a certain linear combination of correlations,

$$S_{CHSH} = C(a,b) + C(a,b') + C(a',b) - C(a',b'), \tag{13.2}$$

is bounded by 2 in absolute value [5]. Quantum mechanics, however, obeys—and saturates—"Tsirelson's bound" [10], namely $|S_{CHSH}| \le 2\sqrt{2}$. The quantum correlations that saturate this bound are $C(a,b) = C(a,b') = C(a',b) = \sqrt{2}/2 = -C(a',b')$, and they are nonlocal. But it is easy to define correlations that yield $S_{CHSH} = 4$: let $C(a,b) = C(a,b') = C(a',b) = 1 = -C(a',b')$. Since these correlations violate Tsirelson's bound, which is a theorem of quantum mechanics, they are not quantum correlations; we call them "superquantum" or "PR-box" correlations. Now the ideas of Aharonov and Shimoney inspire a conjecture: quantum correlations obey relativistic causality, while superquantum correlations do not. It sounds like a plausible conjecture; if Alice measures a and obtains $a = 1$, quantum correlations do not tell her the result of Bob's measurement on his paired particle; but superquantum correlations tell her that Bob's result was 1 whether he measured b or b'. Apparently, superquantum correlations give *too much* information about a spacelike-separated event (Bob's measurement) to be consistent with relativistic causality. Alas, the conjecture fails, and fails straightforwardly. Suppose that whether Alice measures a or a', the results ± 1 are equally probable (whatever Bob measures); and whether Bob measures b or b', the results ± 1 are equally probable (whatever Alice measures). Then Bob and Alice cannot send each other superluminal signals, or even subluminal signals, because Bob's only choice—what to measure, b or b', on his paired particle—does not affect the statistics of Alice's results, and vice versa. It is straightforward to check that these local probabilities (for Alice and Bob on their own) are compatible with both nonlocal quantum correlations and superquantum correlations, and therefore relativistic causality is compatible with both [11]. What a disappointment! It should not be so easy to disprove such a lovely conjecture!

We might therefore ask, "If quantum correlations are nonlocal, why aren't they *more* nonlocal than they are?" Over the years, others have shown, remarkably, that an additional axiom of communication complexity [12–19] is sufficient to rule out superquantum correlations, and comes close to ruling out all stronger-than-quantum correlations. So does a stronger axiom of relativistic causality called "information causality" [20]. However, the physical meaning of communication complexity and information causality is unclear.

But let us take a closer look at superquantum correlations. We see that if Alice measures a on her particle in a given pair, she knows that Bob, whether he measures b or b' on his particle in the same pair, gets the same result she does; if she measures a' on her particle, she knows that Bob gets the same result she does if he measures

b and the opposite result if he measures b'. She can even prepare an "uncertainty-free" ensemble for Bob, as follows. Let them share a large number of pairs, and let Alice measure a on all her particles. If she gets -1, she tells Bob to throw out his particle from the same pair; if she gets 1, she tells Bob to keep his particle. When Alice is done, Bob is left with an ensemble of particles corresponding to $a = 1$, and Alice already knows that if Bob measures b on any particle in the ensemble, he will obtain 1, while if he measures b' on any particle in the ensemble, he will obtain 1. Thus—at least from Alice's point of view—the observables b and b' are not subject to any uncertainty principle on this ensemble.

Now that is strange. Complementarity—the quantum obligation to choose from among incompatible measurements—is intimately tied to the uncertainty principle [21]. If superquantum correlations are not subject to any uncertainty principle, why can't Bob measure both b *and* b' on each of his particles? The inevitable answer is that if Bob could measure both b and b' on even one of his particles, then Alice could certainly send him a superluminal message. For if Alice chooses to measure a, then $b = b'$. If Alice chooses to measure a', then $b = -b'$. Thus Bob could read a (one-bit) signal from Alice if he could measure both b and b'. We conclude that, whereas complementarity is intrinsic to quantum mechanics, here in the context of superquantum correlations it is nothing more than a fig leaf—an extraneous item tacked onto the model to prevent us from seeing Nature's pudenda.

It is not only strange, it is rotten. No respectable theory should contain such an artificial, cheap fix. The PR box is rotten. Unfortunately, by proving it rotten, we have not thereby eliminated it. What if the universe is rotten? It is logically possible.

Yet quantum mechanics has a classical limit; in the classical limit, all observables commute. Let us define macroscopic observables B and B':

$$B = \frac{b_1 + b_2 + \ldots + b_N}{N}, \qquad B' = \frac{b'_1 + b'_2 + \ldots + b'_N}{N}, \tag{13.3}$$

where b_m and b'_m represent b and b', respectively, on the m-th pair. Alice already knows the values of B and B' and, for large N, there must be "weak" measurements [22] that Bob can make to obtain partial information about both B *and* B'; for, in the classical limit, there can be no complementarity between B and B'. The classical limit is an inherent constraint, a kind of boundary condition, on quantum mechanics, and we should apply this constraint to any generalization of quantum mechanics. We should not claim that superquantum correlations are more nonlocal than quantum correlations unless they are subject to the same constraints. Perhaps quantum mechanics would be more nonlocal than it is—even maximally nonlocal—if it were not constrained to have a classical limit. In this sense, the axiom of a classical limit is minimal: if it applies to quantum correlations, it should apply also to proposed generalizations of quantum correlations, including superquantum correlations. Thus superquantum correlations, too, must have a classical limit.

We therefore assume that for large enough N, there must be some measurements Bob can make to obtain partial information about *both* B *and* B'. And now the game changes. It is true that $a = 1$ and $a = -1$ are equally likely, and so the average values of B and B' vanish, whether Alice measures a or a'. But if she measures

a on each pair, then typical values of B and B' will be $\pm 1/\sqrt{N}$ (but possibly as large as ± 1) and correlated. If she measures a' on each pair, then typical values of B and B' will be $\pm 1/\sqrt{N}$ (but possibly as large as ± 1) and *anti*-correlated. Thus Alice can signal to Bob by consistently choosing whether to measure a or a'. This claim is delicate because the large-N limit in which B and B' commute is also the limit that suppresses the fluctuations of B and B'. To insure that Bob has a good chance of measuring B and B' accurately enough to determine whether they are correlated or anti-correlated, N may have to be large and therefore the fluctuations in B and B' will be small. However, Alice and Bob can repeat this experiment (on N pairs at a time) as many times as it takes to give Bob a good chance of catching and measuring large enough fluctuations. They can repeat the experiment exponentially many times. Alice and Bob's expenses and exertions are not our concern. Relativistic causality does not forbid superluminal signalling only when it is cheap and reliable. Relativistic causality forbids superluminal signalling altogether.

It might be that the errors and uncertainties in Bob's measurements are so large that some, or even most, of the (anti-)correlations between B and B' that he detects are erroneous. We cannot specify exactly how the approach to the classical limit depends on N. But this is no objection. What matters is only that when Bob detects a correlation, it is more likely that Alice measured a than when he detects an anti-correlation. If it were not more likely, it would mean that Bob's measurements yield zero information about B or about B', contradicting the fact that there is a classical limit in which B and B' are jointly measurable.

As a concrete example, let us suppose Bob considers only those sets of N pairs in which $B = \pm 1$ and $B' = \pm 1$. The probability of obtaining $B = 1$ is 2^{-N}. But if Alice is measuring a consistently, the probability of obtaining $B = 1$ *and* $B' = 1$ is also 2^{-N}, and not 2^{-2N}, while the probability of obtaining $B = 1$ and $B' = -1$ vanishes. If Alice is measuring a' consistently, the probabilities are reversed. Thus with unlimited resources, Alice can send a (superluminal) signal to Bob. Superquantum (PR-box) correlations are *not* consistent with relativistic causality in the classical limit.

We have ruled out superquantum correlations. To recover quantum correlations, however, we have to rule out all stronger-than-quantum correlations, i.e. we have to derive Tsirelson's bound from the three axioms of nonlocality, relativistic causality, and the existence of a classical limit. The proof [23] is technical, but here are the main points.

Consider sets of N pairs exhibiting superquantum correlations. No matter what Alice measures, the averages $\langle B \rangle$, $\langle B' \rangle$ and therefore also $\langle B + B' \rangle$ tend towards 0. But if Alice measures a' on all the pairs, then $B + B' = 0$ identically for each set of N pairs, since a' and b are perfectly correlated while a' and b' are perfectly anti-correlated. If Alice measures a on all the pairs, then the values of $B + B'$ on successive sets of N pairs fall in a binomial distribution centered at 0. As long as Bob has *some* information about $B + B'$ and its variance, he will ultimately be able to read Alice's one bit of information, i.e. whether Alice measures a or a' on all her pairs. Now suppose that the (anti-)correlations are not perfect but near-perfect. Then the variance of $B + B'$ will not vanish if Alice measures a' on all her pairs, but will still be significantly smaller than if Alice measures a on all her pairs, and Bob will still ultimately be able to read Alice's one bit of information. As the (anti-)correlations get weaker, however, the variances of $B + B'$ and $B - B'$ will approach each other, until, at some critical correlation, Bob will not be able to read Alice's one bit. A natural guess is that this critical correlation corresponds to the maximum quantum correlation $\sqrt{2}/2$, the correlation that saturates Tsirelson's bound. Surprisingly, this guess fails: the critical correlation is 1/2, the correlation that saturates the CHSH inequality! How can it be?

We arrive at a paradox: quantum correlations conform to relativistic causality, yet the calculation of the maximal correlation consistent with relativistic causality passes right by quantum correlations and arrives at local correlations!

The root of this paradox is that the PR box, and more generally any nonlocal box, yields $\langle B \rangle$ and $\langle B' \rangle$ and therefore also $\langle B \pm B' \rangle$, and yields the variances of B and B'; but it does not yield the variances of $B \pm B'$ unless we tacitly assume that the b_m and b'_m combine in a certain way. When we calculate the variances of $B \pm B'$ *without* this assumption—that is, when we think "outside the box"—we do indeed obtain Tsirelson's bound. Indeed, we obtain more than Tsirelson's bound. If we examine the tacit assumption that led to the paradox, we see that it forces b_m and b'_m to add as *scalars*. If b_m and b'_m are allowed to add as vectors, they *can* saturate Tsirelson's bound. So this derivation of Tsirelson's bound shows that Hilbert space is implicit in quantum correlations.

Acknowledgements For over two decades I have had the great good fortune to work with Professor Yakir Aharonov, learning from his penetrating questions, his mastery of quantum and statistical fluctuations, his subtle formulations such as weak measurement and weak values, and his countless other insights.

References

1. Y. Aharonov, D. Rohrlich, *Quantum Paradoxes: Quantum Theory for the Perplexed* (Wiley-VCH, Weinheim, 2005)
2. H. Poincaré, *Sechs Vorträge aus der Reinen Mathematik und Mathematischen Physik* (Teubner, Leipzig, 1910). (Transl. and cited in A. Pais, *'Subtle is the Lord...': the Science and Life of Albert Einstein* (Oxford University Press, New York, 1982), pp. 167–168.)
3. Y. Aharonov, D. Bohm, Phys. Rev. **115**, 485 (1959)
4. J.S. Bell, Physics **1**, 195 (1964)
5. J.F. Clauser, M.A. Horne, A. Shimony, R.A. Holt, Phys. Rev. Lett. **23**, 880 (1969)
6. Y. Aharonov, H. Pendleton, A. Petersen, Int. J. Theor. Phys. **2**, 213 (1969)
7. Y. Aharonov, in *Proc. of the Int. Symp. on the Foundations of Quantum Mechanics*, Tokyo (1983), p. 10. See also Aharonov, Y. and Rohrlich, D., op. cit., Chaps. 5, 6 and 13
8. A. Shimony, in *Foundations of Quantum Mechanics in Light of the New Technology*, ed. by S. Kamefuchi et al. (Japan Physical Society, Tokyo, 1983), p. 225
9. A. Shimony, in *Quantum Concepts of Space and Time*, ed. by R. Penrose, C. Isham (Clarendon Press, Oxford, 1986), p. 182
10. B.S. Tsirelson (Cirel'son), Lett. Math. Phys. **4**, 93 (1980)
11. S. Popescu, D. Rohrlich, Found. Phys. **24**, 379 (1994). (See also D. Rohrlich, The Frontiers Collection, in *Probability in Physics*, ed. by Y. Ben-Menahem, M. Hemmo (Springer, Berlin, 2012), pp. 187–200.)

12. W. van Dam, Nonlocality & communication complexity. Ph.D. thesis, Oxford University (2000)
13. W. van Dam, Preprint (2005). quant-ph/0501159
14. D. Dieks, Phys. Rev. A **66**, 062104 (2002)
15. H. Buhrman, S. Massar, Phys. Rev. A **72**, 052103 (2005)
16. J. Barrett, S. Pironio, Phys. Rev. Lett. **95**, 140401 (2005)
17. G. Brassard, H. Buhrman, N. Linden, A.A. Méthot, A. Tapp, F. Unger, Phys. Rev. Lett. **96**, 250401 (2006)
18. J. Barrett, Phys. Rev. A **75**, 032304 (2007)
19. D. Gross, M. Müller, R. Colbeck, O.C.O. Dahlsten, Phys. Rev. Lett. **104**, 080402 (2010)
20. M. Pawłowski et al., Nature **461**, 1101 (2009)
21. N. Bohr, in *Albert Einstein: Philosopher–Scientist*, ed. by P.A. Schilpp (Tudor Publ. Co., New York, 1951), pp. 201–241
22. Y. Aharonov, D.Z. Albert, L. Vaidman, Phys. Rev. Lett. **60**, 1351 (1988). (See also, Y. Aharonov, D. Rohrlich, op. cit., Chaps. 16–17)
23. D. Rohrlich, Stronger-than-quantum bipartite correlations violate relativistic causality in the classical limit. Phys. Rev. Lett., submitted

Chapter 14
A Gravitational Aharonov-Bohm Effect, and Its Connection to Parametric Oscillators and Gravitational Radiation

Raymond Y. Chiao, Robert W. Haun, Nader A. Inan, Bong-Soo Kang, Luis A. Martinez, Stephen J. Minter, Gerardo A. Munoz, and Douglas A. Singleton

Abstract A thought experiment is proposed to demonstrate the existence of a gravitational, vector Aharonov-Bohm effect. We begin the analysis starting from four Maxwell-like equations for weak gravitational fields interacting with slowly moving matter. A connection is made between the gravitational, vector Aharonov-Bohm effect and the principle of local gauge invariance for nonrelativistic quantum matter interacting with weak gravitational fields. The compensating vector fields that are necessitated by this local gauge principle are shown to be incorporated by the DeWitt minimal coupling rule. The nonrelativistic Hamiltonian for weak, time-independent fields interacting with quantum matter is then extended to time-dependent fields, and applied to the problem of the interaction of radiation with macroscopically coherent quantum systems, including the problem of gravitational radiation interacting with superconductors. But first we examine the interaction of EM radiation with superconductors in a parametric oscillator consisting of a superconducting wire placed at the center of a high Q superconducting cavity driven by pump microwaves. Some room-temperature data will be presented demonstrating the splitting of a single microwave cavity resonance into a spectral doublet due to the insertion of a central

R.Y. Chiao (✉)
Schools of Natural Sciences and Engineering, University of California, Merced, Merced, CA 95344, USA
e-mail: raymond.chiao@gmail.com

R.W. Haun · N.A. Inan · B.-S. Kang · L.A. Martinez
School of Natural Sciences, University of California, Merced, P.O. Box 2039, Merced, CA 95344, USA

S.J. Minter
Vienna Center for Quantum Science and Technology, Faculty of Physics, University of Vienna, Boltzmanngasse 5, 1090 Vienna, Austria

G.A. Munoz
California State University, Fresno, Fresno, CA 93740, USA

D.A. Singleton
Department of Physics, Institut Teknologi Bandung, Bandung, Indonesia

D.C. Struppa, J.M. Tollaksen (eds.), *Quantum Theory: A Two-Time Success Story*,
DOI 10.1007/978-88-470-5217-8_14,

wire. This would represent an *unseparated* kind of parametric oscillator, in which the signal and idler waves would occupy the same volume of space. We then propose a *separated* parametric oscillator experiment, in which the signal and idler waves are generated in two disjoint regions of space, which are separated from each other by means of an impermeable superconducting membrane. We find that the threshold for parametric oscillation for EM microwave generation is much lower for the separated configuration than the unseparated one, which then leads to an observable dynamical Casimir effect. We speculate that a separated parametric oscillator for generating coherent GR microwaves could also be built. [*Editor's note*: for a video of the talk given by Prof. Chiao at the Aharonov-80 conference in 2012 at Chapman University, see quantum.chapman.edu/talk-20.]

14.1 Introduction

In this paper in honor of Yakir Aharonov's 80th birthday we discuss three apparently distinct phenomena: The gravitational Aharonov-Bohm effect, the dynamical Casimir effect arising from parametric oscillations, and gravitational waves. The first of these phenomena is simply the gravitational version of the electromagnetic Aharonov-Bohm effect. There has been recent interest in the gravitational version of the *scalar* Aharonov-Bohm effect [1]. Here we will discuss the gravitational version of the *vector* Aharonov-Bohm effect. In the second phenomenon, i.e., the dynamical Casimir effect, we propose a possible experiment in which photons could be "pumped out of the vacuum" via a vibrating superconducting (SC) "membrane" considered as a parametric oscillator. Finally, we discuss gravitational waves. We speculate that a gravitational version of "pumping gravitons out of the vacuum" via parametric amplification, and above threshold, parametric oscillation, might be possible. The analog of a laser for gravitational waves could thus be constructed. The concept that links these three phenomena together is the use of the DeWitt minimal coupling rule, whereby particles are coupled to both the electromagnetic and the gravitational vector potential.

14.2 Gravitational Aharonov-Bohm Effect

A gravitational analog of the vector Aharonov-Bohm effect is depicted in Fig. 14.1 [1–5]. Aharonov-Bohm (AB) interference can occur when an incoming single-electron wavepacket is split at point A by means of a beam splitter into two partial waves traveling along paths 1 and 2, respectively, that go around the outside of a "solenoid" which contains circulating *mass* currents (indicated by the blue arrows).

For instance, such mass currents in a cylindrical superconducting (SC) mass shell (indicated in black in Fig. 14.1), could be produced by rotating the shell at a constant angular frequency around its cylindrical axis. The two partial waves could then be recombined into an outgoing single-electron wavepacket at point B by means of

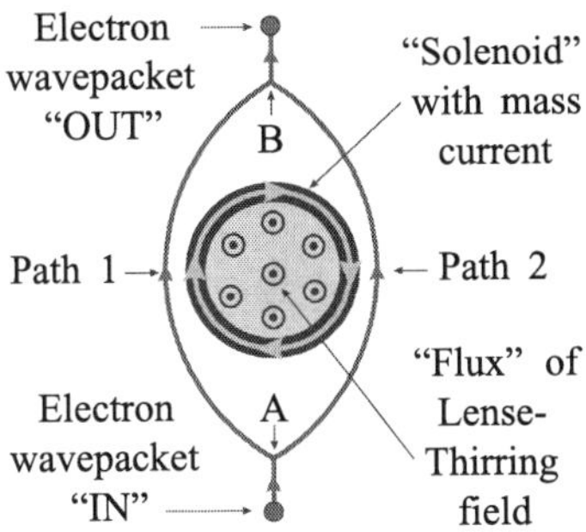

Fig. 14.1 (Color online) Sketch of a gravitational Aharonov-Bohm effect. A "solenoid" with circulating mass currents (*in blue*), produces "flux" (*black dots*) of a certain "gravito-magnetic" field (the Lense-Thirring field) in its interior (*in yellow*). In its exterior (*in white*), this field is zero. Nevertheless, an electron wave packet (*in red*), which is split at point A to go around the "solenoid" via paths 1 and 2, and then recombined at point B, will exhibit an Aharonov-Bohm fringe shift

another beam splitter. Just like in the purely electromagnetic AB effect, the "flux" of a certain gravito-magnetic field (i.e., the Lense-Thirring field) would be confined entirely to the interior region (indicated in yellow in Fig. 14.1) of the "solenoid," and would vanish at all points in its exterior (indicated in white). There results a quantum mechanical AB fringe shift due to the "flux" that is observable at point B, which cannot be explained classically.

To understand the thought experiment pictured in Fig. 14.1, we begin from Einstein's field equations, which, in the limit of weak gravitational fields near slowly moving matter (i.e., in the vicinity of nonrelativistic masses), become the following set of four Maxwell-like equations [11–15]:[1]

$$\nabla \cdot \mathbf{E}_g = -\frac{\rho_g}{\varepsilon_g} \tag{14.1}$$

[1]In the gravitational version of Maxwell's equations 14.1)–(14.4) and the gravitational Lorentz force (14.7), we are following the conventions of [10]. It should be noted that in the literature, there are several different versions of the gravitational Maxwell equations and gravitational Lorentz equation that differ by factors of 2 in various places (see, for example, [16] and [17]). It is not clear that the various formulations given in [11–17] are entirely consistent with one another. However, for the regime considered here, where we will use 14.1)–(14.4) with the quasi-static approximation in which the time derivatives of the gravito-electric and gravito-magnetic vector fields in 14.2) and (14.4) vanish, the various conventions of [10–17] agree.

One should also point out that writing (14.1)–(14.4) in terms of vector fields is questionable, since while electromagnetism consists of vector interactions, gravity consists of tensor interactions. Thus, instead of vector gravito-electric and gravito-magnetic fields, it would be better to write out the gravitational version of Maxwell's equations and the gravitational Lorentz force equation in terms of tensor fields. This approach has been advocated recently [6–8] in terms of tendex and vortex fields. A recent review [9] formulates the gravitational version of Maxwell's equations and the gravitational version of the Lorentz force equation by contracting the tensor fields, which do not vanish in any reference frame, to vector fields, which can transformed away by gauge choice. However, as before, for the present goals of this paper and in the regimes under which we will use the vector gravitational Maxwell equations, this distinction will not be necessary.

$$\nabla \times \mathbf{E}_g = -\frac{\partial \mathbf{B}_g}{\partial t} \tag{14.2}$$

$$\nabla \cdot \mathbf{B}_g = 0 \tag{14.3}$$

$$\nabla \times \mathbf{B}_g = \mu_g \left(-\mathbf{j}_g + \varepsilon_g \frac{\partial \mathbf{E}_g}{\partial t} \right) \tag{14.4}$$

where the gravitational analog ε_g of the electric permittivity ε_0 of free space is[2]

$$\varepsilon_g = \frac{1}{4\pi G} = 1.19 \times 10^9 \text{ SI units} \tag{14.5}$$

and where the gravitational analog μ_g of the magnetic permeability μ_0 of free space is[3]

$$\mu_g = \frac{4\pi G}{c^2} = 9.31 \times 10^{-27} \text{ SI units} \tag{14.6}$$

Here $G = 6.67 \times 10^{-11}$ in SI units is Newton's constant, and $c = 3.00 \times 10^8 \text{ m s}^{-1}$ in SI units is the vacuum speed of light.

In the four Maxwell-like equations, (14.1)–(14.4), the electric-like field $\mathbf{E}_g$ is the *gravito-electric* field (i.e., the local acceleration $\mathbf{g}$ of a freely falling test particle), which could be produced by the mass density ρ_g of nearby matter, via (14.1). Likewise, the magnetic-like field $\mathbf{B}_g$ is the *gravito-magnetic* field, which could be produced by the mass current density $\mathbf{j}_g$ of nearby nonrelativistically moving matter, and also by the gravitational analog of the Maxwell displacement current density $\varepsilon_g \partial \mathbf{E}_g / \partial t$, via (14.4). In the case of nearby *stationary* nonrelativistic mass currents, $\mathbf{B}_g$ can be identified with the Lense-Thirring field [18] that is generated by these currents.

A nonrelativistic test particle with a mass m moves in the presence of the weak fields $\mathbf{E}_g$ and $\mathbf{B}_g$ in accordance with the Lorentz-like force law [11–15]

$$\mathbf{F} = m\frac{d\mathbf{v}}{dt} = m(\mathbf{E}_g + 4\mathbf{v} \times \mathbf{B}_g) \tag{14.7}$$

where m is the mass of the test particle and $\mathbf{v}$ is its velocity (with $v \ll c$).

To understand the experiment pictured in Fig. 14.1, we only need the *stationary* version of (14.4), i.e., the gravitational analog of Ampere's law

$$\nabla \times \mathbf{B}_g = -\mu_g \mathbf{j}_g \tag{14.8}$$

---

[2]The constant ε_g is determined by the Newtonian limit of (14.1), which must yield Newton's law of gravitation, with Newton's constant G determined by Cavendish's experiment.

[3]The constant μ_g is defined in this way in order to obtain the gravitational version of Ampere's law given in (14.8).

where $\mathbf{j}_g$ and $\mathbf{B}_g$ do not depend on time. This implies the following gravitational analog of Ampere's circuital law:

$$\oint_C \mathbf{B}_g \cdot d\mathbf{l} = -\mu_g (I_g)_{\text{enc}} \tag{14.9}$$

where $d\mathbf{l}$ is a line element of an arbitrary closed curve C, and $(I_g)_{\text{enc}}$ is the mass current which is enclosed by C. Applying this Ampere's circuital law to the "solenoid" of Fig. 14.1, which could be a uniformly rotating SC cylindrical mass shell, and using an appropriately chosen closed curve C, one concludes that the $\mathbf{B}_g$ field in the interior to the "solenoid" is a uniform field pointing along the cylindrical axis, and that it has a constant magnitude

$$B_g = \mu_g I'_g \tag{14.10}$$

where I'_g is the mass current per unit length of the "solenoid" flowing around the circumference of the rotating cylindrical mass shell. Furthermore, by another appropriate choice of the closed curve C, one concludes that everywhere outside the "solenoid," it is the case that

$$B_g = 0 \tag{14.11}$$

i.e., that the Lense-Thirring field vanishes everywhere exterior to the "solenoid." This is analogous to the fact that the magnetic field vanishes at all points outside of an electromagnetic solenoid. Hence it follows from the Lorentz-like force law (14.7) that although the electron in Fig. 14.1 experiences a *radial* classical gravitational force due to the mass of the "solenoid," it could never have experienced any *azimuthal* classical gravitational force on its way from point A to point B via either path 1 or path 2 that could have caused the AB phase shift.

Put differently, if one thinks of the "solenoid" as a rotating SC cylindrical mass shell, the experiment has two independent parameters, namely, the linear mass density, and the angular velocity of the shell. That means one could shift the interference fringes by changing the angular velocity. Since the gravito-electric field from the mass of the shell does not depend on the angular velocity, a fringe shift will happen despite the fact that the classical force has not changed. Hence the AB fringe shift in the gravitational case could not have had a classical origin.

Now from the Maxwell-like equation (14.3), and from the vector identity

$$\nabla \cdot (\nabla \times \mathbf{h}) = 0 \tag{14.12}$$

it follows that it is always possible to express the magnetic-like field $\mathbf{B}_g$ as

$$\mathbf{B}_g = \nabla \times \mathbf{h} \tag{14.13}$$

for some vector field $\mathbf{h}$. The relationship (14.13) is formally identical to the relationship in electromagnetism between the magnetic field $\mathbf{B}$ and the electromagnetic vector potential $\mathbf{A}$

$$\mathbf{B} = \nabla \times \mathbf{A} \tag{14.14}$$

which follows from the Maxwell equation $\nabla \cdot \mathbf{B} = 0$. Therefore we shall call $\mathbf{h}$ the "gravitational vector potential."

In the gravitational case, just as in the electromagnetic case, the gravitational vector potential $\mathbf{h}$ possesses the gauge freedom

$$\mathbf{h} \to \mathbf{h} + \nabla\mu \tag{14.15}$$

where μ can be any arbitrary scalar function of position. This follows from the vector identity $\nabla \times \nabla\mu = 0$, and is formally identical to the case of electromagnetism, in which the vector potential $\mathbf{A}$ possesses the gauge freedom

$$\mathbf{A} \to \mathbf{A} + \nabla\lambda \tag{14.16}$$

where λ can be any arbitrary scalar function of position. Again, this follows from the vector identity $\nabla \times \nabla\lambda = 0$.

Now the principle of *local* gauge invariance in nonrelativistic quantum mechanics states that the phase of the time-independent wavefunction $\Psi(\mathbf{r})$ of any quantum system must always be able to be *locally* transformed without affecting the physics of the system. In other words, the transformation [19]

$$\Psi(\mathbf{r}) \to \Psi(\mathbf{r}) \exp\big(i\phi(\mathbf{r})\big) \tag{14.17}$$

where the phase $\phi(\mathbf{r})$ can be any *arbitrary* real scalar function of position $\mathbf{r}$, can neither change the properties of the quantum system, nor the physical laws governing the system and its interactions with its environment. In particular, this *local* transformation of the phase of the wavefunction cannot change the probability distribution of the system, since

$$\left|\Psi(\mathbf{r})\right|^2 \to \left|\Psi(\mathbf{r}) \exp\big(i\phi(\mathbf{r})\big)\right|^2 = \left|\Psi(\mathbf{r})\right|^2 \tag{14.18}$$

and therefore the Born probability interpretation of the wavefunction is unaffected by this transformation.

However, *gradients* of $\Psi(\mathbf{r})$ will be changed by the introduction of an *arbitrary* scalar function $\phi(\mathbf{r})$, and therefore will alter the momentum of the system. If so, one could *arbitrarily* alter the physical laws governing the system, including altering the conservation of momentum of a particle in the usual $\exp(i\mathbf{p} \cdot \mathbf{r}/\hbar)$ plane wave state of an electron within a force-free region of space, where one knows that $\mathbf{p}$ must be a constant. This obviously cannot be the case. Therefore the principle of local gauge invariance *necessitates* the existence of some compensating vector field (or fields), such as the $\mathbf{A}$ and $\mathbf{h}$ fields in the DeWitt minimal coupling rule [20–22]

$$\frac{\hbar}{i}\nabla \to \frac{\hbar}{i}\nabla - q\mathbf{A} - m\mathbf{h} \tag{14.19}$$

$$\mathbf{p} \to \mathbf{p} - q\mathbf{A} - m\mathbf{h} \tag{14.20}$$

where $\mathbf{p}_{\text{op}} = \frac{\hbar}{i}\nabla$ is the momentum operator, q is the charge, and m is the mass of the nonrelativistic quantum system under consideration. Here $\mathbf{A}$ and $\mathbf{h}$ are, respectively, the vector potentials for electromagnetism and for weak gravitation, which are being viewed here as being the requisite "compensating vector fields," whose existence is *necessitated* by the principle of local gauge invariance. Since the vector fields $\mathbf{A}$ and $\mathbf{h}$ have the gauge freedoms $\mathbf{A}(\mathbf{r}) \to \mathbf{A}(\mathbf{r}) + \nabla\lambda(\mathbf{r})$ and $\mathbf{h}(\mathbf{r}) \to \mathbf{h}(\mathbf{r}) + \nabla\mu(\mathbf{r})$, these freedoms can then be used to compensate for the gauge freedom in the transformation $\Psi(\mathbf{r}) \to \Psi(\mathbf{r})\exp(i\phi(\mathbf{r}))$, in just such a way that the quantum system can once again satisfy the principle of local gauge invariance.

Thus invoking the DeWitt minimal coupling rule (14.20), we demand that the nonrelativistic Hamiltonian of any quantum system in the presence of $\mathbf{A}$ and $\mathbf{h}$ fields must always have the following form [20–22]:

$$H = \frac{1}{2m}(\mathbf{p} - q\mathbf{A} - m\mathbf{h})^2 + V \tag{14.21}$$

where V is the potential energy of the system. Here, in the present context of the SC quantum systems that we are interested in, such as that of the SC "solenoid" pictured in Fig. 14.1, $q = 2e$ is the charge of a Cooper pair, and $m = 2m_e$ is its mass.

Although one can always *arbitrarily* choose a gauge locally so that both vector fields $\mathbf{A}$ and $\mathbf{h}$ are set identically equal to zero at each point exterior to the solenoid, nevertheless the *fluxes* interior to the solenoid

$$\Phi = \oint_C \mathbf{A} \cdot d\mathbf{l} \tag{14.22}$$

$$\Phi_g = \oint_C \mathbf{h} \cdot d\mathbf{l} \tag{14.23}$$

where C is a closed curve enclosing the solenoid, cannot be arbitrarily set equal to zero, but must instead be gauge-invariant, nonzero, globally *measurable* quantities. Hence the fluxes Φ and Φ_g must be *physical* quantities.

The gravitational Aharonov-Bohm effect depicted in Fig. 14.1 is closely related to the time holonomy which arises from the off-diagonal time-space components of the metric tensor g_{0i}.[4] It can be shown [23] that this time holonomy Δt can be expressed as follows:

$$\Delta t = -\frac{1}{c}\oint_C \frac{g_{0i}}{g_{00}} dx^i \tag{14.24}$$

where C is an arbitrary closed curve in space (such as the one enclosing the "solenoid" in Fig. 14.1), and dx^i is a spatial line element of this closed curve. In light of the time holonomy given by (14.24), it is impossible in general relativity to define a global time coordinate for an entire physical system, such as the topologically nontrivial superconductor in Fig. 14.1.

[4] We follow the notation and sign conventions of MTW [44], i.e., Greek indices denote space-time indices running from 0 to 3; Latin indices denote spatial indices running from 1 to 3; repeated indices are summed; the signature of the Minkowski metric is diag$(-1, +1, +1, +1)$.

In the weak field, slow matter approximation, in which

$$g_{0i} \approx h_{0i} \tag{14.25}$$

where h_{0i} are the time-space components of the small-deviation metric $h_{\mu\nu}$ from the Minkowski metric $\eta_{\mu\nu}$, and in which the time-time component of the metric can be approximated by

$$g_{00} \approx -1 \tag{14.26}$$

it follows that the time holonomy (14.24) becomes approximately

$$\Delta t \approx \frac{1}{c} \oint_C h_{0i} dx^i \tag{14.27}$$

For electron waves traveling around a closed curve C enclosing a "solenoid" such as that in Fig. 14.1, the *time* holonomy (14.27) becomes the *phase* holonomy

$$\Delta\phi = \omega_{\text{Compton}} \Delta t \approx \frac{m_e c^2}{\hbar} \frac{1}{c} \oint_C h_{0i} dx^i \neq 0 \tag{14.28}$$

where $\omega_{\text{Compton}} = m_e c^2/\hbar$ is the Compton frequency of the electron [1]. The phase shift (14.28) , which is nonvanishing for the "solenoid" configuration of Fig. 14.1, is the gravitational AB phase shift. It is closely related to Berry's phase [24], since both phases have a common origin in non-Euclidean geometry.

Since physically counting the number of fringes in a shift of an Aharonov-Bohm interference pattern, such that in Fig. 14.1, must yield the same result for all observers, independent of their state of motion under a restricted set of (Galilean) coordinate transformations, it is sufficient for the purposes of this paper to say that the flux Φ_G is a Galilean invariant, and therefore a measurable, *physical* quantity. Moreover, the closed-path integral of $h_{0i} dx^i$ in (14.27) is an intrinsic time holonomy which cannot vanish due to an arbitrary gauge transformation.

However, if one were to *arbitrarily* make the global gauge choice

$$h_{0i} = 0 \quad \text{everywhere} \tag{14.29}$$

as is done, for example, in the transverse-traceless gauge, then it follows that the phase holonomy must vanish identically, i.e.,

$$\Delta\phi \approx \frac{mc^2}{\hbar} \frac{1}{c} \oint_C h_{0i} dx^i = 0 \quad \text{by setting } h_{0i} = 0 \text{ everywhere} \tag{14.30}$$

and the gravitational Aharonov-Bohm phase shift predicted for the electron interference pattern in Fig. 14.1 would disappear.

However, just as in the case of the electromagnetic Aharonov-Bohm effect where

$$\Delta\phi = \frac{q}{\hbar} \oint_C A_i dx^i = 0 \quad \text{by setting } A_i = 0 \text{ everywhere} \tag{14.31}$$

the results (14.30) and (14.31) are both unphysical whenever the closed curve C encloses either a solenoid with a nonvanishing electromagnetic flux $\Phi \neq 0$, or a "solenoid" with a nonvanishing Lense-Thirring flux $\Phi_g \neq 0$, since both of these fluxes are gauge-invariant, measurable, physical quantities that cannot be arbitrarily set equal to zero. Hence the transverse-traceless gauge choice (14.29) is unphysical in situations that involve the gravitational Aharonov-Bohm effect depicted in Fig. 14.1 where $\Phi_g \neq 0$, and also, by extension, in time-varying situations that involve $\Phi_g(t) \neq 0$, for example, in situations where gravitational radiation is interacting with superconducting systems.

The usual Aharonov-Bohm phase shift follows from DeWitt's minimal coupling rule (14.20) when one sets $\Phi_g = 0$, for then the phase shift arising from (14.20) in the configuration pictured in Fig. 14.1, reduces down to the usual expression

$$\Delta\phi = \frac{q}{\hbar}\oint_C \mathbf{A}\cdot d\mathbf{l} \neq 0 \tag{14.32}$$

and we recover the standard form for the AB phase shift. However, if $\Phi_g \neq 0$, then the total AB phase, upon integration over any closed curve C enclosing the "solenoid," becomes

$$\Delta\phi_{\text{tot}} = \frac{q}{\hbar}\oint_C \mathbf{A}\cdot d\mathbf{l} + \frac{m}{\hbar}\oint_C \mathbf{h}\cdot d\mathbf{l} = \frac{q\Phi}{\hbar} + \frac{m\Phi_g}{\hbar} \tag{14.33}$$

The vector potential **h** in (14.33) can arise either from a Lense-Thirring field, or from rotations of a quantum system, such as from a rotating SC ring. Since experiments with rotating SC systems are much easier to perform than experiments involving Lense-Thirring fields, we shall confine our attention for now to these much easier experiments.

One can check experimentally the expression given by (14.33) for a rotating SC ring, in which case the single-valuedness of the macroscopic wavefunction of the Cooper pairs demands that $\Delta\phi_{\text{tot}} = 2\pi n$, where n is an integer. It follows from this that a magnetic field must be generated by rotating a superconducting ring, i.e., that a "London moment" must accompany this rotation. Precision measurements of the London moment of a rotating SC ring thus provide a test for the correctness of the expression for the total AB phase in (14.33). Cabrera and co-workers [25, 26] performed these measurements to 100 parts per million. Thus the formula in (14.33) for the total AB phase has been experimentally verified. In this way, the expression in (14.20) for the DeWitt minimal coupling rule has been experimentally tested to high precision.

So far we have been considering only the case of stationary, time-independent, charge and mass currents, such as those in a SC magnet, or in a steadily rotating SC ring. These quantum currents can be the quantum mechanical sources of time-independent **A** and **h** fields that give rise to the AB effect.

14.3 The Dynamical Casimir Effect Via Parametric Oscillations

In this section, we begin by discussing the case when the vector potentials **A** and **h** are time dependent, i.e. $\mathbf{A}(\mathbf{r},t)$ and $\mathbf{h}(\mathbf{r},t)$. From this we will propose a version of the dynamical Casimir effect for the electromagnetic vector potential **A**, in which photons are "pumped" out of the vacuum via parametric oscillations of a SC membrane.

It is natural to extend the time-independent Hamiltonian (14.21) to the following time-dependent one:[5]

$$H = \frac{1}{2m}\left(\mathbf{p} - q\mathbf{A}(\mathbf{r},t) - m\mathbf{h}(\mathbf{r},t)\right)^2 + V \tag{14.34}$$

in which the fields $\mathbf{A}(\mathbf{r},t)$ and $\mathbf{h}(\mathbf{r},t)$ are to be first treated as classical fields, but the matter (e.g., the vibrating SC wire in Fig. 14.2) is to be treated quantum mechanically, in the so-called "semi-classical approximation." Expanding the square in (14.34), one obtains the following interaction Hamiltonian terms:

$$H_{\mathbf{p}\cdot\mathbf{A}} = -\frac{q}{m}\mathbf{p}\cdot\mathbf{A}(\mathbf{r},t) \tag{14.35}$$

which leads to the interaction of the quantum system with electromagnetic (EM) radiation, such as in the stimulated emission and absorption of EM waves by the quantum matter, and

$$H_{\mathbf{p}\cdot\mathbf{h}} = -\mathbf{p}\cdot\mathbf{h}(\mathbf{r},t) \tag{14.36}$$

which leads to the interaction of the quantum system with gravitational (GR) radiation, such as in the stimulated emission and absorption of GR waves by the quantum matter, and

$$H_{\mathbf{A}\cdot\mathbf{h}} = +q\mathbf{A}(\mathbf{r},t)\cdot\mathbf{h}(\mathbf{r},t) \tag{14.37}$$

which leads to the interaction between EM and GR radiation fields mediated by the quantum system, such as in the transduction of EM waves into GR waves mediated by the quantum matter [27], and

$$H_{\mathbf{A}\cdot\mathbf{A}} = +\frac{q^2}{2m}\mathbf{A}(\mathbf{r},t)\cdot\mathbf{A}(\mathbf{r},t) \tag{14.38}$$

which leads to Landau-diamagnetism type of interactions of the quantum system with EM radiation, such as in the parametric amplification of EM waves by

[5] It should be noted that the use of the time-dependent gauge in which $(\mathbf{h}(\mathbf{r},t))_i = ch_{0i}(\mathbf{r},t) \neq 0$ in this context is only considered for fields within matter, and not in vacuum, where the time-dependent transverse-traceless gauge would be more appropriate.

a strongly driven, i.e., "pumped," quantum system (see, for example, Figs. 14.2 and 14.5), and

$$H_{\mathbf{h}\cdot\mathbf{h}} = +\frac{m}{2}\mathbf{h}(\mathbf{r},t)\cdot\mathbf{h}(\mathbf{r},t) \tag{14.39}$$

which leads to *gravitational* Landau-diamagnetism type of interactions of the quantum system with GR radiation, such as in the parametric amplification of GR waves, again by a strongly driven, i.e., "pumped," quantum system (again, see, for example, Figs. 14.2 and 14.5).

All of the above interaction terms will be treated as small perturbations of the unperturbed Hamiltonian

$$H_0 = \frac{\mathbf{p}^2}{2m} + V \tag{14.40}$$

and can thus be treated using standard perturbation theory.

At a fully quantum mechanical level of description, both the matter and the radiation fields **A** and **h** would have to be quantized. The radiation fields could be quantized by invoking the commutation relations

$$[a, a^\dagger] = 1 \tag{14.41}$$

$$[b, b^\dagger] = 1 \tag{14.42}$$

where a and $a^\dagger$ are, respectively, the annihilation and creation operators for a quantum of a single mode of the EM radiation field, and where b and $b^\dagger$ are, respectively, the annihilation and creation operators for a quantum of a single mode of the GR radiation field.

For now, let us focus solely on the interactions of quantized EM radiation with matter. The second quantized form for the EM vector potential operator $A_{\mathrm{op}}(\mathbf{r})$, when summed over all the modes of a cavity enumerated by the index κ, is [28]

$$A_{\mathrm{op}}(\mathbf{r}) = \sum_{\kappa}\sqrt{\frac{\hbar}{2\varepsilon_0\omega_\kappa}}(a_\kappa + a_\kappa^\dagger)\mathcal{E}_\kappa(\mathbf{r}) \tag{14.43}$$

where ω_κ is the frequency of mode κ, and $\mathcal{E}_\kappa(\mathbf{r})$ is one Cartesian component of the classical electric field distribution associated with this mode. Therefore, when there is a single dominant mode in the problem, the vector potential operator simplifies to the expression

$$A_{\mathrm{op}}(\mathbf{r}) \propto \left(a + a^\dagger\right) \tag{14.44}$$

where the mode index κ and the proportionality constant have been suppressed.

The expansion of the square in the interaction Hamiltonian $H_{\mathbf{A}\cdot\mathbf{A}} \propto (a + a^\dagger)^2$ in (14.38) will therefore contain the term [29]

$$K_{\mathrm{op}} \propto a^\dagger a^\dagger + \text{hermitian adjoint} \tag{14.45}$$

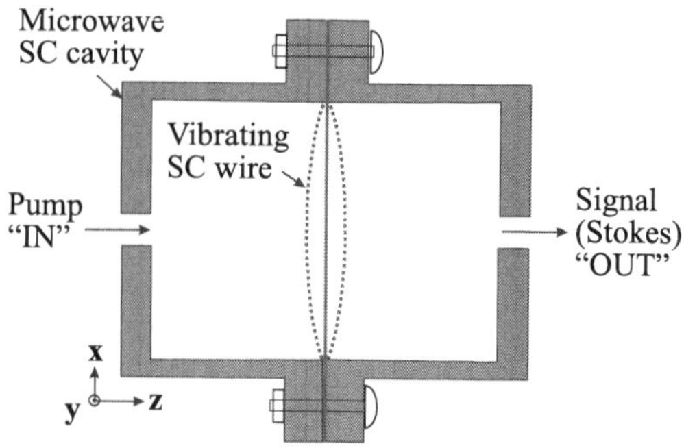

Fig. 14.2 (Color online) A parametric amplifier or oscillator, whose active element is the vibrating SC wire (*in red*) placed in the middle of a microwave SC cavity (*in grey*). The moving wire can be viewed as if it were an oscillating "semipermeable membrane," which does work upon some "seed" radiation initially present in the cavity, thus amplifying this radiation in a reciprocating, piston-like action. Photons incident upon this moving "membrane" experience a Doppler shift that changes their energy. Thus when pump microwaves enter through the *left hole*, an amplified signal (Stokes) wave will exit through the *right hole*

The $a^\dagger a^\dagger$ term of the operator K_{op} corresponds to the process of photon *pair creation* in the parametric amplification arising from the pumping action of some strong "pump" wave upon a quantum system. It can be shown that (14.45) has the form of an infinitesimal generator of a squeezed state of light [29].

Instead of enumerating all the possible resulting second quantized forms of the above interaction Hamiltonian terms, let us just focus on one such term, namely, $H_{\mathbf{A}\cdot\mathbf{A}}$ in (14.38), which is associated with parametric amplification, such as that in the setup depicted in Fig. 14.2. This Figure represents an "opto-mechanical" parametric amplifier, which becomes, above threshold, a parametric oscillator, whose active element is the central vibrating SC wire (indicated in red), placed across the middle of an extremely high-Q SC microwave cavity. Here, instead of using optical cavities, as is usual in ongoing opto-mechanical experiments [30], we shall be using SC microwave cavities. The reason for this is that the quality factor for SC microwave cavities has already been demonstrated by Haroche and co-workers [31] to be on the order of

$$Q \sim 10^{10} \tag{14.46}$$

which can be much higher than that of typical optical cavities.

The motion of the SC wire in the middle of the microwave cavity will modulate the "pump" microwaves coming through the "IN" port so as to produce radiation at new sideband frequencies via the Doppler effect. The "seed" radiation initially in one of these sidebands, namely, the first "Stokes" sideband, can then become the exponentially amplified. Macroscopic, easily detectable radiation in the form of a strong Stokes wave emitted by the parametric amplifier can then leave the cavity through the "OUT" port. This kind of strong Stokes emission would be similar to the Stokes emission observed in the stimulated Raman effect in nonlinear optics [32].

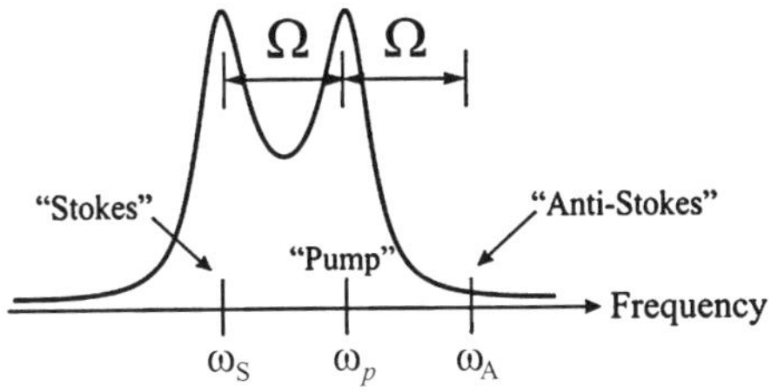

Fig. 14.3 The excitation spectrum of a microwave cavity with a wire placed at its center (see Fig. 14.2). The spectrum consists of a doublet of resonances at the "signal" mode at the "Stokes" frequency ω_S and the "pump" mode at the "pump" frequency ω_p. The difference between "pump" and "Stokes" frequencies is resonant with the frequency Ω of the vibrating wire (i.e., $\omega_p - \omega_S = \Omega$). The "anti-Stokes" frequency ω_A is off resonance with respect to the doublet, and hence is suppressed. (Cf. the stimulated Raman effect in [32])

However, here we shall first analyze classically the parametric amplification process in Fig. 14.2, in order to answer the following questions: What is the threshold for parametric oscillation in Fig. 14.2? Is this experiment feasible to perform? The key concept that we shall use in this classical analysis is that of the *work done* by the moving wire, viewed as if the wire were a moving "piston" acting on some "seed" radiation initially present in the cavity. This "piston-like" action of the moving wire can also be viewed as if the wire were a partially reflecting "moving mirror," and will lead to the *exponential* amplification of the "seed" radiation at the Stokes frequency above the threshold of parametric oscillation.

The reason for using a vibrating wire instead of a vibrating membrane is that the wire is one dimensional, whereas the membrane is two dimensional. The mass of a thin wire can be made much smaller than the mass of a thin membrane, and therefore a wire can be driven more easily into motion.

However, here we shall model the vibrating wire in Fig. 14.2 as if the wire were a vibrating "semi-permeable membrane," for which the pressure acting on the membrane can be converted into an easily calculable force. The justification for this "membrane" model is that the scattering cross section of a thin conducting wire, when placed symmetrically across the mouth of a waveguide, can be comparable in size to the cross-sectional area of the waveguide, because the wire tends to "short out" the electric field of the TE mode of a waveguide. Thus the reflection coefficient of the wire placed across the middle of a cavity (as in Fig. 14.2), can be made quite high. There results a splitting of a microwave cavity mode into a spectral doublet as illustrated in Fig. 14.3 [33, 34], due to the presence of the central wire. The splitting frequency of the doublet can typically be on the order of 1 GHz for a microwave mode frequency of around 10 GHz.

We have observed the splitting of a microwave cavity resonance into a spectral doublet (see Fig. 14.4). An RF cylindrical copper cavity of length $L = 1.284''$ and diameter $D = 1.02''$ with two parallel conducting end plates was constructed to support a TE 112 mode at 11.42 GHz [35]. (The mode indices l, m, n are chosen to correspond to the number of half wavelengths along their respective axes; angular, radial, and axial respectively.) The input coupler, a short straight wire placed

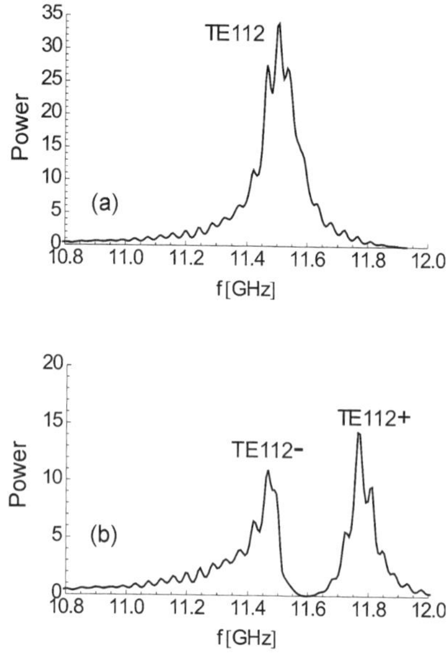

Fig. 14.4 Splitting of the TE 112 mode in cavity with a bisecting copper wire at its midpoint perpendicular to the axial direction and parallel to the input coupler. The splitting is on the order of 400 MHz. S21 transmission measurements (**a**) of a copper cavity with TE112 resonant frequency at 11.5 GHz and (**b**) splitting due to the placement of copper wire placed at the center. The vertical axes use the same arbitrary power reference in the conversion from logarithmic to linear scale

perpendicular to the axial direction, was placed on the cylinder at approximately a quarter wavelength from one end plate. The output coupler is a small loop placed at the other end plate of the cavity.

An off-diagonal scattering matrix element S21 transmission measurement, preformed with a HP 8720C network analyzer, shows the resonance of the TE 112 mode at approximately 11.50 GHz; see Fig. 14.4(a). A splitting is observed by placing a 22AWG copper wire at the midpoint of the cavity, perpendicular to the axial direction and parallel to the input coupler. The splitting is approximately 400 MHz; see Fig. 14.4(b).

In Fig. 14.3, the pump frequency ω_p of the parametric amplifier is assumed to be tuned to coincide with the upper member of the spectral doublet, and the signal frequency ω_S is assumed to be tuned to coincide with the lower member of this doublet, which we shall call the "Stokes frequency," in analogy with the stimulated Raman effect [32]. The idler frequency ω_i, i.e., the frequency of the mechanical motion of the central wire in Fig. 14.2, is the beat frequency $\Omega = \omega_p - \omega_S$ between the pump and signal frequencies. Note that the parasitic, Doppler upshifted "anti-Stokes frequency" ω_A is automatically suppressed by this spectral doublet.

To calculate the force acting on the central membrane (as a model of the force acting on the central wire) in the middle of the microwave cavity of Fig. 14.2, we begin from the Maxwell stress tensor [36, 37]

$$T_{ij} = \varepsilon_0\left(E_i E_j - \frac{1}{2}\delta_{ij} E^2\right) + \frac{1}{\mu_0}\left(B_i B_j - \frac{1}{2}\delta_{ij} B^2\right) \tag{14.47}$$

Now starting from the electromagnetic force exerted on charges and currents given by the Lorentz force law

$$\mathbf{F} = q(\mathbf{E} + \mathbf{v} \times \mathbf{B}) \tag{14.48}$$

it can be shown that there results the following relationship between the force $\mathbf{F}$ and the total Maxwell stress tensor T_{ij} and the Poynting vector $\mathbf{S}$ [36, 37]:

$$(\mathbf{F})_i = \oint_{S(V)} T_{ij} \cdot (d\mathbf{a})^j - \varepsilon_0 \mu_0 \frac{d}{dt} \int_V (\mathbf{S})_i dV \quad (\text{where } i = 1, 2, 3) \tag{14.49}$$

where T_{ij} is the stress tensor evaluated at $d\mathbf{a}$, an infinitesimal area element of an arbitrary surface $S(V)$ which encloses the volume V, dV is an infinitesimal volume element of the matter inside the volume V enclosed by the surface $S(V)$, and $\mathbf{S} = \mathbf{E} \times \mathbf{H}$ is the Poynting vector evaluated at dV inside V.

Now the tangential electric field must vanish at the boundary of any conductor. Hence, for all transverse electric modes of the microwave cavity pictured in Fig. 14.2, if one chooses the surface $S(V)$ to be that of a small pillbox straddling a patch of the surface of an equivalent SC membrane, the contribution to the force (14.49) from the Poynting vector term evaluated at the pillbox enclosing the patch of the surface, must vanish. In the case of transverse magnetic modes of the cavity, the Poynting vector $\mathbf{S} = \mathbf{E} \times \mathbf{H}$ does not vanish at the surface, since there will be a longitudinal component of the electric field at the surface along with a tangential component of the magnetic field. Hence there will arise a *tangential* component of $\mathbf{S}$ at the surface of the SC membrane, but this $\mathbf{S}$ cannot contribute to any *normal* force acting on the conducting surface.

Therefore the only contribution to the normal force acting on the equivalent SC membrane arises solely from the Maxwell stress tensor term of (14.49), which, for the case of transverse electric modes evaluated at the membrane, reduces down to

$$(T_{ij}) = \frac{1}{2\mu_0} \begin{pmatrix} -B_y^2 & 0 & 0 \\ 0 & +B_y^2 & 0 \\ 0 & 0 & -B_y^2 \end{pmatrix} \tag{14.50}$$

because the magnetic field of the transverse electric mode (e.g., the TE112 mode), whose electric field is pointing in the x direction (in the Cartesian coordinate system of Fig. 14.2), will be pointing in the y direction at the surface of the equivalent membrane.[6]

If one therefore replaces the SC wire by an equivalent SC membrane, then the diagonal terms of (14.50) can be interpreted as a "field pressure" acting on the mem-

[6]Note that the application of a *magnetic field* to any system of particles with a given charge-to-mass ratio is equivalent, by Larmor's theorem, to *rotating* the system at the Larmor angular frequency. Thus there exists a natural connection between the $\mathbf{A}$ and $\mathbf{h}$ fields in the proposed experiments in Figs. 14.2 and 14.5.

brane with a maximum amplitude of

$$P_{\max} = \frac{1}{2\mu_0}\left(B^2\right)_{\max} = (u_B)_{\max} \tag{14.51}$$

where $(B^2)_{\max}$ is the maximum of the square of the magnetic field, and $(u_B)_{\max}$ is the maximum magnetic energy density, evaluated at the surface of the membrane.

Because the stress tensor depends *quadratically* on the field, there results a pressure being exerted on the membrane at a beat note frequency due to the beating between the fields at pump frequency ω_p and the fields at the Stokes frequency ω_S in the spectral doublet of Fig. 14.3. This beat note can drive the membrane (or the wire) at the beat frequency $\Omega = \omega_p - \omega_S$, i.e., at the splitting of the upper and lower members of the doublet. The force in the z direction acting on the membrane (or wire) at the beat frequency Ω will therefore have the form

$$\begin{aligned} F_\Omega &= \frac{1}{\mu_0}\mathcal{B}_p\mathcal{B}_S^* \exp\bigl(-i(\omega_p - \omega_s)t\bigr) \cdot \mathcal{A}_{\text{eff}} + \text{c.c.} \\ &\propto \exp(-i\Omega t) + \text{c.c.} \end{aligned} \tag{14.52}$$

where $\mathcal{A}_{\text{eff}}$ is an effective area of the membrane (or, equivalently, the effective scattering cross-section of the wire), and where

$$B_p = \mathcal{B}_p \exp(-i\omega_p t) + \text{c.c.} \tag{14.53}$$

is the pump waveform, with $\mathcal{B}_p$ being the complex amplitude for the pump magnetic field waveform, and where

$$B_S = \mathcal{B}_S \exp(-i\omega_S t) + \text{c.c.} \tag{14.54}$$

is the Stokes waveform of some small amount of "seed" radiation already present inside the cavity, with $\mathcal{B}_S$ being the complex amplitude for the Stokes magnetic field waveform. (Note that such "seed" radiation could in principle be vacuum fluctuations of the EM field inside the cavity.) In the expression (14.52) for the force F_Ω at the beat frequency Ω, we have assumed that the pump wave (14.53) is always much stronger that the "seed" Stokes wave (14.54), i.e., $|\mathcal{B}_p| \gg |\mathcal{B}_S|$.

Since the driving force F_Ω at the beat frequency Ω can be made resonant with the acoustical resonance frequency of the membrane, we shall model the resulting motion of the membrane as that of a simple harmonic oscillator with a resonance frequency of Ω. Using Newton's equation of motion for a damped simple harmonic oscillator moving in the z direction, viz.,

$$m\bigl(\ddot{z} + \gamma\dot{z} + \Omega^2 z\bigr) = F_\Omega \tag{14.55}$$

where m is the mass of the membrane, and γ is its damping coefficient, and using an ansatz of the form

$$z = z_{\max} \exp(-i\Omega t) + \text{c.c.} \tag{14.56}$$

for the displacement of the membrane in the z direction, one finds that its maximum, on-resonance complex displacement amplitude is

$$z_{\max} = i\frac{\mathcal{B}_p\mathcal{B}_S^*}{\mu_0 m\gamma\Omega}\mathcal{A}_{\text{eff}} \tag{14.57}$$

The velocity of the membrane in the z direction will then have the form

$$v = v_{\max}\exp(-i\Omega t) + \text{c.c.} \tag{14.58}$$

where the complex velocity amplitude $v_{\max} = -i\Omega z_{\max}$ becomes, on resonance,

$$v_{\max} = \frac{\mathcal{B}_p\mathcal{B}_S^*}{\mu_0 m\gamma}\mathcal{A}_{\text{eff}} \tag{14.59}$$

There results a Doppler effect arising from the velocity of the membrane moving in the z direction in Fig. 14.2, giving rise to upper and lower Doppler sidebands. However, only the lower Doppler sideband will be excited, since only the lower sideband will be resonant with lower member of the doublet in Fig. 14.3.

The maximum, on-resonance time-averaged power $\langle \mathrm{P}\rangle$ being delivered from the pump wave into the simple-harmonic membrane motion at the beat frequency Ω, and therefore into the lower Doppler sideband, i.e., into the Stokes wave (14.54), is

$$\langle \mathrm{P}_{\max}\rangle = \langle F_\Omega \cdot v\rangle_{\max} = 2|\mathcal{B}_p|^2|\mathcal{B}_S|^2\frac{1}{\mu_0^2 m\gamma}\mathcal{A}_{\text{eff}}^2 \tag{14.60}$$

Now invoking the conservation of energy, we find that, if we for the moment neglect all losses, the Stokes wave will be amplified by this power transfer, such that the power $\langle \mathrm{P}_{\max}\rangle$ being transferred into the Stokes wave must equal the rate of growth of the time-averaged Stokes energy inside the cavity

$$\langle U_S\rangle = \frac{1}{\mu_0}|\mathcal{B}_S|^2 V_{\text{eff}} = \frac{1}{\mu_0}|\mathcal{B}_S|^2\mathcal{A}_{\text{eff}}\mathcal{L}_{\text{eff}} \tag{14.61}$$

where $\mathcal{L}_{\text{eff}}$ is an effective length of the cavity (i.e., $V_{\text{eff}} = \mathcal{A}_{\text{eff}}\mathcal{L}_{\text{eff}}$ is an effective volume of the cavity). In other words, from energy conservation it follows that

$$\langle \mathrm{P}_{\max}\rangle = \frac{d}{dt}\langle U_S\rangle \tag{14.62}$$

Substituting in from (14.60) and (14.61), one infers that

$$2|\mathcal{B}_p|^2|\mathcal{B}_S|^2\frac{1}{\mu_0^2 m\gamma}\mathcal{A}_{\text{eff}}^2 = \frac{d}{dt}\left(\frac{1}{\mu_0}|\mathcal{B}_S|^2\mathcal{A}_{\text{eff}}\mathcal{L}_{\text{eff}}\right) = \kappa_S\left(\frac{1}{\mu_0}|\mathcal{B}_S|^2\mathcal{A}_{\text{eff}}\mathcal{L}_{\text{eff}}\right) \tag{14.63}$$

where κ_S is the exponential gain coefficient for parametric amplification of the Stokes wave. Thus we arrive at an exponential-growth ODE for the energy stored (i.e. U_S) in the cavity at the Stokes frequency

$$\frac{d}{dt}\langle U_S\rangle = \kappa_S\langle U_S\rangle \tag{14.64}$$

Solving for the gain coefficient κ_S from (14.63), we conclude that

$$\kappa_S = \frac{2|\mathcal{B}_p|^2\mathcal{A}_{\text{eff}}}{\mu_0 m\gamma\mathcal{L}_{\text{eff}}} \tag{14.65}$$

Thus the gain of the Stokes wave is directly proportional to the pump power stored in the cavity, just like in the stimulated Raman effect [32].

Like in a laser, the threshold for oscillation [38, 39] occurs when

$$\text{Gain} = \text{Loss} \tag{14.66}$$

Above threshold, i.e., when the gain exceeds the loss, macroscopic amounts of coherent radiation can be produced inside the cavity by the *exponential* amplification of the "seed" radiation, i.e., of vacuum fluctuations. Thus even if one were to start off only with vacuum fluctuations as the "seed," one can produce macroscopic amounts of coherent radiation by the stimulated emission of radiation, just like in a laser. In principle, this should also apply to GR radiation, as well as to EM radiation. Thus *generators* of microwave frequency GR radiation should *in principle* be possible to construct, as well as amplifiers and detectors for this kind of radiation. The only remaining question is whether such devices are feasible *in practice*.

The loss of the Stokes wave (i.e., the "signal") from the cavity depicted in Fig. 14.2 can result from emission of radiation through an outcoupling hole into the environment, or from remnant ohmic losses in the components of the cavity. We shall call the resulting quality factor of the cavity at the Stokes frequency under working conditions (i.e., including internal losses and the outcoupling into the environment) the "loaded Q", which is defined as follows:

$$Q_S = \omega_S\tau_S \tag{14.67}$$

where "S" stands for "Stokes wave" whose frequency is ω_S, and whose stored energy inside the cavity decays away with a time scale τ_S (the so-called "cavity ring-down time") after the pump has been turned off.

Moreover, the loss coefficient γ in the motion of the simple harmonic oscillator leads a decay time $\tau_\Omega = 1/\gamma$ for the energy stored in the oscillator. This leads to a mechanical oscillator quality factor Q_Ω, which is defined as follows:

$$Q_\Omega = \Omega\tau_\Omega \tag{14.68}$$

It follows from the gain-equals-loss condition (14.66) that at threshold

$$(\kappa_S)_{\text{threshold}} = \frac{2|\mathcal{B}_p|^2_{\text{threshold}}\mathcal{A}_{\text{eff}}}{\mu_0 m\gamma\mathcal{L}_{\text{eff}}} = \frac{2|\mathcal{B}_p|^2_{\text{threshold}}Q_\Omega\mathcal{A}_{\text{eff}}}{\mu_0 m\Omega\mathcal{L}_{\text{eff}}} = \frac{1}{\tau_S} = \frac{\omega_S}{Q_S} \tag{14.69}$$

Since the time-averaged stored energy in the pump wave inside the cavity depicted in Fig. 14.2 is

$$\langle U_p \rangle = \frac{1}{\mu_0} |\mathcal{B}_p|^2 \mathcal{A}_{\text{eff}} \mathcal{L}_{\text{eff}} \tag{14.70}$$

we conclude from (14.69) that the threshold pump power needed for parametric oscillation is

$$\langle U_p \rangle_{\text{threshold}} = \frac{1}{2} \frac{m\Omega \omega_{\text{S}} \mathcal{L}_{\text{eff}}^2}{Q_{\text{S}} Q_{\Omega}} \tag{14.71}$$

This result is to be compared with the threshold pump power needed for parametric oscillation for exciting an elastic mode of a mirror of a Fabry-Perot resonator obtained by Braginsky and co-workers [38]

$$\langle U_p \rangle_{\text{Braginsky}} = \frac{1}{2} \frac{m\omega_s^2 L^2}{Q_\iota Q_s} \tag{14.72}$$

where m is the mass of the mirror, ω_s is the frequency of the elastic mode, L is the length of the Fabry-Perot resonator, Q_i is the quality factor of a down-shifted "idler" optical mode of the resonator, and Q_s is the quality factor of the elastic mode. By inspection of (14.71) and (14.72), we see that these thresholds are quite similar.

However, the electrodynamic Q factor of SC microwave cavities is typically on the order of 10^{10} [31], whereas the typical mechanical Q factor for the best opto-mechanical oscillators, which are composed of non-SC materials in the ongoing opto-mechanical experiments, is at most on the order of 10^5 [30]. Therefore the question naturally arises whether it is possible to replace these low-Q, non-SC mechanical oscillators, with high-Q SC mechanical oscillators, in which their mechanical Q can approach the typical electrodynamic $Q \sim 10^{10}$ of SC microwave cavities.

One possible answer to this question is the "triple SC microwave Fabry-Perot resonator" shown in Fig. 14.5 [33], in which a *charged* SC membrane is extremely tightly coupled via its electrostatic charge to the longitudinal electric fields of a transverse magnetic mode of a high Q SC microwave cavity. The idea here is that when the charge on the SC membrane is sufficiently large, then the mechanical dynamics of the membrane will be "slaved" to follow closely the electromagnetic dynamics of the high-Q microwave SC mode. Calculations[7] show that one only needs a charge of pico-Coulombs for this to happen. Note that as a result of the "slaved"

[7] If the electrostatic charge $+q$ on the left side of the membrane were to be sufficiently large (i.e., greater than around 20 picocoulombs; see Appendix B of [33]), so that the membrane becomes extremely tightly coupled to the "single" Fabry-Perot cavity mode on the left side of the membrane, which would happen if this cavity were to be excited in an appropriate transverse magnetic mode, then the mechanical dynamics of the membrane would be "slaved" to the electromagnetic dynamics of this "signal" frequency cavity. For then the displacement of the membrane would be completely determined by Newton's equation of motion (14.55), so that there results a fixed relationship between the *instantaneous* longitudinal electric field and the *instantaneous* membrane

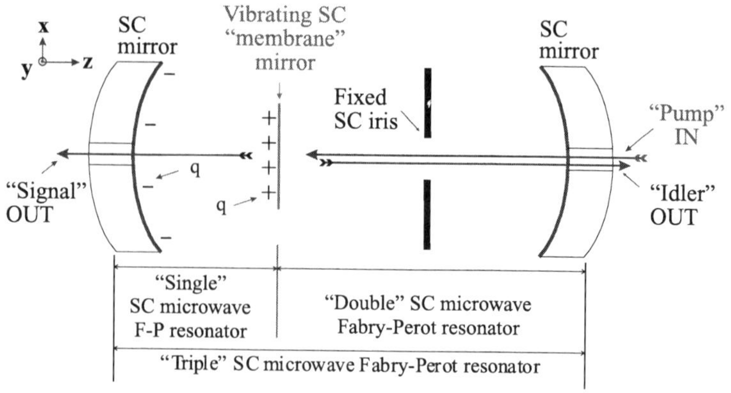

Fig. 14.5 (Color online) A "triple" SC microwave Fabry-Perot cavity [33] consists of a "single" SC cavity separated by a vibrating SC "membrane" *(in red)* from a "double" SC cavity with a fixed SC iris *(in black)* at its center. This separating membrane is totally *impermeable* to all microwaves. The membrane is electrostatically charged on it left surface with a charge $+q$, and the left SC mirror is charged on its right surface with a charge $-q$. "Pump" microwaves *(in blue)* enter into the system through the right hole, and "signal" waves leave the system through the left hole, but "idler" waves leave the system through the right hole

dynamics, the SC membrane will be moving at microwave, and not at acoustical, frequencies. This means that the motion of the membrane will be essentially that of a "free" mass, which is being driven solely by Maxwell's stress tensor. Therefore this microwave-frequency motion will be independent of the elastic and dissipative mechanical properties of the membrane.

Another important feature of the configuration shown in Fig. 14.5 is that the signal and the idler waves are *spatially separated* into two disjoint, high Q SC cavities, which are separated from each other by a common, vibrating SC membrane. It turns out that this leads to two *separate* Q factors in its denominator of the threshold for parametric oscillation, which arises due to this separation. We shall therefore call the parametric oscillator configuration of Fig. 14.5 a "separated parametric oscillator," in contrast to that in Fig. 14.2, which we shall call an "unseparated parametric oscillator." The threshold for the separated parametric oscillator of Fig. 14.5 will

---

displacement, which is given by

$$(\varepsilon_\Omega)_s = -\frac{q(\mathcal{E}_z)_s}{m\omega_s^2} \tag{14.73}$$

where $(\mathcal{E}_z)_s$ is the complex amplitude of the longitudinal electric field of the transverse magnetic mode at the "signal" frequency ω_s of the "single" Fabry-Perot cavity and $(\varepsilon_\Omega)_s$ is the complex amplitude of the displacement of the membrane, which oscillates at the same frequency ω_s. In this way, the mechanical vibration frequency Ω of the membrane would be forced to become identical to the microwave frequency ω_s of this cavity, and the kinetic energy of the mechanical vibrational motion would be forced to become identical to the electromagnetic energy stored inside this cavity, because these two degrees of freedom would be so tightly coupled to each other that they would no longer be independent degrees of freedom.

turn out to be at least a factor of 10^5 lower than that of the unseparated parametric oscillator of Fig. 14.2.

We start the analysis of the separated parametric amplifier depicted in Fig. 14.5 by examining the work done by the "pump" wave on the moving SC "membrane," when it produces a displacement by an amount Δz of the membrane to the left along the axis of the "double" Fabry-Perot resonator on the right side of the membrane. The work done during this displacement is

$$\Delta W = \left(\frac{1}{2\mu_0}B^2\right) \cdot \mathcal{A}_{\text{eff}}\Delta z \tag{14.74}$$

where $\mathcal{A}_{\text{eff}}$ is the effective hemiconfocal spot size at the membrane, Δz is the displacement of this membrane, and

$$u_B = \frac{1}{2\mu_0}B^2 \tag{14.75}$$

is the energy density of the magnetic field evaluated at the right surface of the "membrane," which is the pressure arising from the Maxwell stress tensor (14.50), i.e., a pressure being exerted upon the membrane that can cause a change of the volume inside the "double" Fabry-Perot resonator on the right side of the membrane

$$\Delta V = \mathcal{A}_{\text{eff}}\Delta z \tag{14.76}$$

where $\mathcal{A}_{\text{eff}}$ is the effective area of the membrane, which is determined by the hemiconfocal spot size of the mode on the right side of the membrane. For simplicity, we shall assume here that the hemiconfocal spot size of the mode on the left side of the membrane also has the same $\mathcal{A}_{\text{eff}}$.

The instantaneous mechanical work ΔW in (14.74) done by the "pump" upon the fields of the resonator can be rewritten in the form

$$\Delta W = P\Delta V \tag{14.77}$$

where the instantaneous pressure P on the "membrane" is

$$P = \frac{1}{2\mu_0}B^2 \tag{14.78}$$

which is equal to the instantaneous energy density u_B in (14.75) evaluated at the right surface of the membrane. It is clear from the expression for the work in (14.77) that ΔW can be interpreted as if it were the work being done by a moving piston acting on a thermodynamic system, here, the radiation fields inside a cavity.

Now we shall presently see that if energy were to be continually supplied to the "double Fabry-Perot" resonator on the right side of the membrane by some continuous-wave, external microwave pump waveform oscillating at a frequency ω_p (i.e., the "pump" frequency) entering through the right hole of Fig. 14.5, then the exponential amplification of some seed "signal" waveform at a frequency of ω_s

within the "single" Fabry-Perot on the left side of the membrane, simultaneously with the exponential amplification of some seed "idler" waveform at a frequency of ω_i within the "double" Fabry-Perot on the right side of the membrane, can occur. This amplification effect can arise from the mutual reinforcement of the signal and idler waves at the expense of the pump wave, in which the pump wave beats with the idler wave via the Maxwell stress tensor to produce more of the signal wave, and the signal wave modulates the pump wave via the Doppler effect to produce more of the idler wave, etc. The mutual reinforcement of the two "seed" waves will lead to an instability above a certain threshold, i.e., to the parametric oscillation of both the signal and idler waves that produces macroscopic amounts of both kinds of waves, which then leave the system in opposite directions via the left hole and the right hole of Fig. 14.5, respectively, just like in a laser.

For parametric amplification to occur, the frequency-matching condition

$$\omega_p = \omega_s + \omega_i \tag{14.79}$$

must be satisfied. The meaning of the relationship can be most easily seen by multiplying it by the Planck's constant $\hbar$ so that one obtains the relationship

$$\hbar\omega_p = \hbar\omega_s + \hbar\omega_i \tag{14.80}$$

In other words, in the parametric amplification process, one signal photon $\hbar\omega_s$ is simultaneously created along with one idler photon $\hbar\omega_i$ at the expense of one pump photon $\hbar\omega_p$, which is annihilated during this "photon pair-creation process." In this process, an entangled pair of signal and idler photons, with the signal photon appearing on the left side, and the idler photon appearing on the right side of the membrane, respectively, will be produced in a correlated emission event inside the "triple" Fabry-Perot resonator depicted in Fig. 14.5. This photon pair-creation process is described by the interaction Hamiltonian

$$H_{\text{int}} \propto a_p a_s^{\dagger} a_i^{\dagger} + \text{hermitian adjoint} \tag{14.81}$$

which is a generator of a *two-mode* squeezed state [28].

Let the microwave pump magnetic field just outside of the right surface of the membrane have the form

$$B_p = \mathcal{B}_p \exp(-i\omega_p t) + \text{c.c.} \tag{14.82}$$

where the pump magnetic field vector points transversely to the membrane immediately outside of its right surface, with $\mathcal{B}_p$ being the complex amplitude of the pump magnetic field.

Similarly, let the "seed" idler magnetic field just outside of the right surface of the membrane have the form

$$B_i = \mathcal{B}_i \exp(-i\omega_i t) + \text{c.c.} \tag{14.83}$$

which is a vector parallel to the magnetic field vector of the pump wave immediately outside of its right surface, with $\mathcal{B}_i$ being the complex amplitude of the idler magnetic field. We shall assume that the pump is tuned to be on resonance with the upper member of the spectral doublet of the "double" Fabry-Perot (see Fig. 14.3), and that the idler is tuned to be on resonance with the lower member of this doublet. However, the off-resonance, parasitic "anti-Stokes" (i.e., the Doppler up-shifted) frequency component arising from the motion of the membrane will be suppressed, and hence neglected.

To calculate the coefficient of parametric amplification, let us assume that the pump wave is much stronger than both a very weak "seed" idler wave and a very weak "seed" signal wave, so that $|\mathcal{B}_p| \gg |\mathcal{B}_i|$ and $|\mathcal{B}_p| \gg |\mathcal{B}_s|$. It follows from (14.79) and (14.83) that the square of the total magnetic field evaluated at the right surface of the "membrane" will have the form

$$B^2 = (B_p + B_i)^2 = \big(\mathcal{B}_p \exp(-i\omega_p t) + \mathcal{B}_i \exp(-i\omega_i t) + \text{c.c.}\big)^2 \tag{14.84}$$

If we define the "beat frequency" as

$$\Omega = \omega_p - \omega_i \tag{14.85}$$

then we see that there will arise cross terms in the square of the magnetic field (14.84) which will contain terms that vary at the beat frequency Ω, viz.,

$$\begin{aligned}\big(B^2\big)_\Omega &= \mathcal{B}_p \exp(-i\omega_p t)\mathcal{B}_i^* \exp(+i\omega_i t) + \text{c.c.}\\ &= \mathcal{B}_p\mathcal{B}_i^* \exp(-i\Omega t) + \text{c.c.}\end{aligned} \tag{14.86}$$

Therefore there will exist a pressure being exerted on the membrane that varies at the beat frequency Ω of the form

$$(P)_\Omega = \frac{1}{2\mu_0}\big(B^2\big)_\Omega = \frac{1}{2\mu_0}\big(\mathcal{B}_p\mathcal{B}_i^* \exp(-i\Omega t) + \text{c.c.}\big) \tag{14.87}$$

If we define the complex pressure amplitude $\mathcal{P}_\Omega$ as follows:

$$\mathcal{P}_\Omega = \frac{1}{2\mu_0}\mathcal{B}_p\mathcal{B}_i^* \tag{14.88}$$

then the pressure exerted on the membrane which varies at the beat frequency Ω will have the form

$$(P)_\Omega = \mathcal{P}_\Omega \exp(-i\Omega t) + \text{c.c.} \tag{14.89}$$

But the beat frequency Ω will be assumed to be tuned into resonance with the signal frequency, i.e.,

$$\Omega = \omega_p - \omega_i = \omega_s \tag{14.90}$$

so that the membrane can be driven at the resonance frequency ω_s of the "single" Fabry-Perot resonator to the left of the membrane. Thus power from the right side of the membrane can be fed resonantly by the motion of the membrane into the signal "seed" waveform on the left side.

Due to the presence of the electrostatic charge $+q$ on the left surface of the membrane, and the relationship

$$F_z(t) = qE_z(t) \tag{14.91}$$

where $F_z(t)$ is the Coulomb force on the membrane exerted on the charge $+q$ by the longitudinal electric field $E_z(t)$ of a transverse magnetic mode of the "single" Fabry-Perot resonator on the left side of the membrane [33], it follows that the motion of the membrane in the longitudinal z direction will be "slaved" through $E_z(t)$ to the dynamics of this transverse magnetic mode [33, Appendix B, where it was shown that one only needs $q \approx 20$ pC for the Coulomb force to dominate the dynamics of the membrane]. This is due to the tight coupling between $\Delta z(t)$ and $E_z(t)$ which arises from the electrostatic charge $+q$ (see footnote 7).

Hence let us introduce an ansatz that the displacement of the membrane has the form

$$\big(\Delta z(t)\big)_\Omega = \varepsilon_\Omega(t)\exp(-i\Omega t) + \text{c.c.} \tag{14.92}$$

where $\varepsilon_\Omega(t)$ is some *slowly-varying* complex displacement amplitude of the membrane, which is the slowly-varying envelope of the *fast* beat frequency phase factor $\exp(-i\Omega t)$.

Hence the velocity of the membrane will have the form

$$(v)_\Omega = \frac{d(\Delta z)_\Omega}{dt} \approx v_\Omega(t)\exp(-i\Omega t) + \text{c.c.} \tag{14.93}$$

where the *slowly-varying* complex velocity amplitude of the moving membrane is

$$v_\Omega(t) \approx -i\Omega\varepsilon_\Omega(t) \tag{14.94}$$

within the "slowly varying envelope approximation" [40], and the acceleration of the membrane will have the form

$$(a)_\Omega = \left(\frac{dv}{dt}\right)_\Omega \approx \alpha_\Omega(t)\exp(-i\Omega t) + \text{c.c.} \tag{14.95}$$

where the *slowly-varying* complex acceleration amplitude of the moving membrane is

$$\alpha_\Omega(t) \approx -\Omega^2\varepsilon_\Omega(t) \tag{14.96}$$

also within the slowly varying envelope approximation.

The force due to the pressure (14.87) being exerted on the membrane will have the form

$$(F)_\Omega = \mathcal{F}_\Omega(t)\exp(-i\Omega t) + \text{c.c.} \tag{14.97}$$

where the *slowly-varying* complex force amplitude acting on the membrane is

$$\mathcal{F}_\Omega(t) = \frac{1}{\mu_0}\mathcal{B}_p\mathcal{B}_i^*(t)\mathcal{A}_{\text{eff}} \tag{14.98}$$

where $\mathcal{A}_{\text{eff}}$ is the effective area of the membrane, and where, in the "undepleted pump" approximation [32], we have assumed that the pump amplitude $\mathcal{B}_p$ is independent of time, but that the idler amplitude $\mathcal{B}_i^*(t)$ can be a slowly varying function of time due to its amplification.

Then the time-averaged mechanical power fed into the membrane's motion from the radiation, and hence into the signal wave of the "single" Fabry-Perot cavity of Fig. 14.5, using (14.94) and (14.98), is

$$\begin{aligned}\left\langle\frac{dW}{dt}\right\rangle_{\text{signal}} &= \langle F\cdot v\rangle \\ &= \langle \mathcal{F}_\Omega \cdot v_\Omega^* + \text{c.c.}\rangle = \langle \mathcal{F}_\Omega \cdot i\Omega\varepsilon_\Omega + \text{c.c.}\rangle \\ &= -2\text{Im}(\mathcal{F}_\Omega\Omega\varepsilon_\Omega^*) = -2\text{Im}\left(\frac{1}{\mu_0}\mathcal{B}_p\mathcal{B}_i^*(t)\mathcal{A}_{\text{eff}}\cdot\Omega\varepsilon_\Omega^*(t)\right)\end{aligned} \tag{14.99}$$

Let the complex amplitudes of the pump, idler, and signal waveforms have the complex polar forms

$$\mathcal{B}_p = |\mathcal{B}_p|\exp(i\phi_p) \tag{14.100}$$

$$\mathcal{B}_i = |\mathcal{B}_i|\exp(i\phi_i) \tag{14.101}$$

$$\varepsilon_\Omega = |\varepsilon_\Omega|\exp(i\phi_s) \tag{14.102}$$

By inspection of (14.99), we see that the maximum power transfer from the radiation fields into the membrane's motion occurs when the phases of pump, signal, and idler waveforms (i.e., (14.100), (14.101), and (14.102)) are adjusted so as to satisfy the condition

$$\phi_p - \phi_i - \phi_s = -\frac{\pi}{2} \tag{14.103}$$

whereupon the maximum mechanical power fed into the membrane becomes

$$\left\langle\frac{dW}{dt}\right\rangle_{\text{max},\text{signal}} = +\frac{2\mathcal{A}_{\text{eff}}\Omega}{\mu_0}|\mathcal{B}_p|\big|\mathcal{B}_i(t)\big|\big|\varepsilon_\Omega(t)\big| \tag{14.104}$$

Neglecting for the moment all dissipative losses, the kinetic energy of the membrane must *grow* due to this *positive* mechanical power being fed into it. Hence, invoking the principle of the conservation of energy, we get the equation

$$\left\langle\frac{dW}{dt}\right\rangle_{\text{max},\text{signal}} = \frac{d}{dt}\left(\frac{1}{2}m\langle v^2\rangle\right) = \frac{d}{dt}\left(m\Omega^2\big|\varepsilon_\Omega(t)\big|^2\right) \tag{14.105}$$

where we have used (14.94) and the fact that $\langle v^2\rangle = 2|v_\Omega|^2$. Therefore

$$\frac{d}{dt}\left(m\Omega^2|\varepsilon_\Omega|^2\right) = 2m\Omega^2|\varepsilon_\Omega|\frac{d|\varepsilon_\Omega|}{dt} = \frac{2\mathcal{A}_{\text{eff}}\Omega}{\mu_0}|\mathcal{B}_p||\mathcal{B}_i||\varepsilon_\Omega| \quad (14.106)$$

We thus arrive at an ODE for the rate of growth of the magnitude $|\varepsilon_\Omega|$ of the displacement of the membrane

$$\frac{d|\varepsilon_\Omega|}{dt} = \frac{\mathcal{A}_{\text{eff}}}{\mu_0 m\Omega}|\mathcal{B}_p||\mathcal{B}_i| \quad (14.107)$$

Next, we shall obtain a similar ODE for the rate of growth of the magnitude $|\mathcal{B}_i|$ of the idler wave. We start from the motional EMF created by the motion of the vibrating SC membrane, which leads to the generation of the motional electric field

$$\mathbf{E} = \mathbf{v}\times\mathbf{B} \quad (14.108)$$

This relationship implies that the sinusoidal, back-and-forth motion of the mirror at a frequency Ω will be modulating the magnetic field oscillating at the pump frequency ω_p, such that an idler electric field at the surface of the mirror oscillating at the idler frequency ω_i will be generated, i.e.,

$$(\mathbf{E})_i = (\mathbf{v})_\Omega \times (\mathbf{B})_p \quad (14.109)$$

This is a manifestation of the Doppler effect, in which the sinusoidal motion of the mirror will produce Stokes and anti-Stokes sidebands around the pump frequency ω_p. However, due to the doublet spectrum depicted in Fig. 14.3 in the "double" Fabry-Perot resonator, only the down-shifted, first-order Stokes sideband will be resonant with the resonator. Therefore we shall neglect henceforth the anti-Stokes sideband, and all the other higher order Doppler sidebands.

Note that when the membrane is moving towards the left in Fig. 14.5, which is the direction in which the magnetic pressure due to the pump wave is pushing, this pressure will deliver power into the motion of the moving membrane, and *simultaneously*, will deliver power into the red-shifted, first-order Doppler sideband, i.e., the idler wave, via the relationship (14.109). This will lead to parametric amplification of the membrane's motion.

Using the Cartesian coordinate system shown in Fig. 14.5, we shall assume that the instantaneous velocity of the mirror is pointing in the $-z$ direction, and that the instantaneous pump magnetic field vector $\mathbf{B}_p$ is pointing in $+x$ direction, so that the instantaneous motional $\mathbf{E}$ field will be pointing in the $+y$ direction. Thus, in terms of the complex amplitudes, (14.109) reduces down to

$$\mathcal{E}_i = v_\Omega^* \mathcal{B}_p \quad (14.110)$$

Now the time-averaged power delivered into the idler wave by the motional $\mathbf{E}$ field acting on the current density $\mathbf{j}$ induced by the idler wave at the right surface of the SC mirror, is

$$\left\langle \frac{dW}{dt} \right\rangle_{\text{idler}} = \int \langle \mathbf{j} \cdot \mathbf{E} \rangle dV = \left(j_i^* \mathcal{E}_i \right) \mathcal{A}_{\text{eff}} \delta + \text{c.c.} \tag{14.111}$$

where j_i^* is the complex conjugate of the supercurrent density amplitude flowing on the surface of the mirror at the idler frequency ω_i, $\mathcal{E}_i$ is the complex motional electric field amplitude at ω_i, $\mathcal{A}_{\text{eff}}$ is the effective focal area of the membrane, and δ is the London penetration depth, within which the supercurrents j_i^* will be flowing near the surface of the membrane. To calculate j_i^* in (14.111), we use Ampere's circuital law and a rectangular loop straddling the surface of the SC mirror, to get

$$\mathcal{B}_i^* = \mu_0 j_i^* \delta \tag{14.112}$$

where δ is London's penetration depth. Solving for j_i^* from (14.112), and substituting it into (14.111) using (14.110), we get

$$\begin{aligned} \left\langle \frac{dW}{dt} \right\rangle_{\text{idler}} &= \frac{1}{\mu_0} \left(v_\Omega^* \mathcal{B}_p \mathcal{B}_i^* \right) \mathcal{A}_{\text{eff}} + \text{c.c.} \\ &= \frac{1}{\mu_0} \left((i \Omega \varepsilon_\Omega^*) \mathcal{B}_p \mathcal{B}_i^* \right) \mathcal{A}_{\text{eff}} + \text{c.c.} \end{aligned} \tag{14.113}$$

where in the last step we used (14.94) for the complex conjugate of the complex velocity amplitude v_Ω^* of the membrane.

To maximize the power transferred to the idler, we again choose the phase condition (14.103) between the pump, idler and signal complex amplitudes, and find

$$\left\langle \frac{dW}{dt} \right\rangle_{\text{max,idler}} = +\frac{2}{\mu_0} \Omega |\varepsilon_\Omega| |\mathcal{B}_p| |\mathcal{B}_i| \mathcal{A}_{\text{eff}} \tag{14.114}$$

Assuming the absence of all dissipation, and invoking once again the principle of the conservation of energy, but this time for the idler wave, we obtain

$$\begin{aligned} \left\langle \frac{dW}{dt} \right\rangle_{\text{max,idler}} &= \frac{d}{dt} \left(\frac{1}{2\mu_0} \langle B_i^2 \rangle \right) \mathcal{A}_{\text{eff}} \mathcal{L}_{\text{eff}} = \frac{d}{dt} \left(\frac{1}{\mu_0} |\mathcal{B}_i|^2 \right) \mathcal{A}_{\text{eff}} \mathcal{L}_{\text{eff}} \\ &= +\frac{2}{\mu_0} \Omega |\varepsilon_\Omega| |\mathcal{B}_p| |\mathcal{B}_i| \mathcal{A}_{\text{eff}} \end{aligned} \tag{14.115}$$

where $\mathcal{L}_{\text{eff}}$ is the effective length of the "double" Fabry-Perot resonator. We thus arrive at an ODE for the rate of growth of the idler wave

$$\frac{d}{dt} |\mathcal{B}_i| = \frac{\Omega}{\mathcal{L}_{\text{eff}}} |\varepsilon_\Omega| |\mathcal{B}_p| = K_2 |\varepsilon_\Omega| \tag{14.116}$$

where the constant of proportionality K_2 is

$$K_2 = \frac{\Omega}{\mathcal{L}_{\text{eff}}} |\mathcal{B}_p| \tag{14.117}$$

Note that this is of the same form as the ODE for the rate of growth of the signal wave (see footnote 7) obtained earlier in (14.107), viz.,

$$\frac{d}{dt}|\varepsilon_\Omega| = \frac{\mathcal{A}_{\text{eff}}}{\mu_0 m\Omega}|\mathcal{B}_p||\mathcal{B}_i| = K_1|\mathcal{B}_i| \tag{14.118}$$

where the constant of proportionality K_1 is

$$K_1 = \frac{\mathcal{A}_{\text{eff}}}{\mu_0 m\Omega}|\mathcal{B}_p| \tag{14.119}$$

This implies that there exists a mutual enhancement of the signal and idler waves that leads to their *exponential* growth, i.e., to the *parametric amplification* of both waves. To see this, let us rewrite the two equations (14.116) and (14.118) in the following 2×2 matrix form:

$$\frac{d}{dt}\begin{pmatrix}|\varepsilon_\Omega|\\|\mathcal{B}_i|\end{pmatrix} = \begin{pmatrix}0 & K_1\\K_2 & 0\end{pmatrix}\begin{pmatrix}|\varepsilon_\Omega|\\|\mathcal{B}_i|\end{pmatrix} = \Lambda\begin{pmatrix}|\varepsilon_\Omega|\\|\mathcal{B}_i|\end{pmatrix} \tag{14.120}$$

where Λ is the eigenvalue of the 2×2 matrix, viz.,

$$\Lambda = \pm\sqrt{K_1K_2} = \pm\sqrt{\frac{\mathcal{A}_{\text{eff}}|\mathcal{B}_p|^2}{\mu_0 m\mathcal{L}_{\text{eff}}}} \tag{14.121}$$

The solution of (14.120) is

$$\begin{pmatrix}|\varepsilon_\Omega|\\|\mathcal{B}_i|\end{pmatrix} = \begin{pmatrix}|\varepsilon_\Omega|\\|\mathcal{B}_i|\end{pmatrix}_{t=0}\exp(\Lambda t) \tag{14.122}$$

The meaning of the *positive* root for Λ is that it represents the rate of exponential *growth* of the amplitudes of the coupled signal and idler waves, when the phase condition (14.103), $\phi_p - \phi_i - \phi_s = -\pi/2$, for maximum power *delivery* to these coupled waves, is satisfied, whereas the *negative* root for Λ is that it represents the rate of exponential *decay* of the amplitudes of the coupled signal and idler waves, when the anti-phase condition, $\phi_p - \phi_i - \phi_s = +\pi/2$, for maximum power *extraction* from these coupled waves, is satisfied. Whether one gets exponential growth or exponential decay of the waves thus depends on the choices of the initial phases of the pump, signal, and idler waves. This kind of *phase-dependent* amplification is the signature of the production of a *squeezed state* of the vacuum.

Next, let us introduce a dissipative loss phenomenologically into the ODE for the idler (14.116) as follows:

$$\frac{d}{dt}|\mathcal{B}_i| - \frac{1}{2\tau_i}|\mathcal{B}_i| = K_2|\varepsilon_\Omega| \tag{14.123}$$

where τ_i is the "cavity ring-down time" for the energy stored in the idler cavity mode on the right side of the membrane after the pump wave has been suddenly shut off, and, similarly, into the ODE for the signal (14.118) as follows:

$$\frac{d}{dt}|\varepsilon_\Omega| - \frac{1}{2\tau_s}|\varepsilon_\Omega| = K_1|\mathcal{B}_i| \tag{14.124}$$

where τ_s is the "cavity ring-down time" for the energy stored in the signal cavity mode on the left side of the membrane after the pump wave has been suddenly shut off. The relationships between the cavity ring-down times τ_i and τ_s of the two cavity modes, and their loaded quality factors Q_i and Q_s, are

$$Q_i = \omega_i \tau_i \tag{14.125}$$

$$Q_s = \omega_s \tau_s \tag{14.126}$$

At the threshold of parametric oscillation, there is a balance between gain and loss such that there arises a steady-state situation in which

$$\frac{d}{dt}|\mathcal{B}_i| = \frac{d}{dt}|\varepsilon_\Omega| = 0 \tag{14.127}$$

Therefore, at threshold, the two ODE's (14.123) and (14.124) reduce down to the two algebraic equations

$$-\frac{1}{2\tau_i}|\mathcal{B}_i| = K_2|\varepsilon_\Omega| \tag{14.128}$$

$$-\frac{1}{2\tau_s}|\varepsilon_\Omega| = K_1|\mathcal{B}_i| \tag{14.129}$$

Multiplying the left sides and the right sides of these two equations together, we get

$$\frac{1}{4\tau_i \tau_s} = K_2 K_1 \tag{14.130}$$

By using the relationships (14.117), (14.119), (14.125), and (14.126), we get from (14.130) the threshold condition

$$\frac{\omega_i \omega_s}{4Q_i Q_s} = \frac{\mathcal{A}_{\text{eff}}|\mathcal{B}_p|^2}{\mu_0 m \mathcal{L}_{\text{eff}}} = \frac{|\mathcal{B}_p|^2 V_{\text{eff}}}{\mu_0 m \mathcal{L}_{\text{eff}}^2} \tag{14.131}$$

Since the time-averaged stored energy stored in the pump cavity mode is

$$\langle U_p \rangle = \frac{1}{2\mu_0}\langle B_p^2 \rangle V_{\text{eff}} = \frac{1}{\mu_0}|\mathcal{B}_p|^2 V_{\text{eff}} \tag{14.132}$$

we arrive from (14.131) at the conclusion that the threshold condition is

$$\langle U_p \rangle_{\text{threshold}} = \frac{m\omega_i \omega_s \mathcal{L}_{\text{eff}}^2}{4Q_i Q_s} \tag{14.133}$$

This is to be compared with Braginski's threshold condition (14.72) [38]

$$\langle U_p \rangle^{\text{Braginski}}_{\text{threshold}} = \frac{1}{2} \frac{m\omega_s^2 L^2}{Q_i Q_s} \tag{14.134}$$

The above two expressions agree as to an order-of-magnitude estimate for the threshold of parametric oscillation for the "triple" Fabry-Perot resonator configuration of Fig. 14.5.

The required threshold input pump power $\langle \mathrm{P}_p \rangle_{\text{threshold}}$ for parametric oscillation due to pump microwaves entering in through the right hole of the "triple" cavity configuration of Fig. 14.5, can be found via the steady-state condition

$$\langle \mathrm{P}_p \rangle_{\text{threshold}} = \frac{1}{\tau_p} \langle U_p \rangle_{\text{threshold}} \tag{14.135}$$

where the quality factor for the pump cavity mode Q_p is related to the pump cavity ring-down time τ_p by

$$Q_p = \omega_p \tau_p \tag{14.136}$$

Finally, putting this together with (14.133), we conclude that for parametric oscillation to occur in the configuration of Fig. 14.5, we need to inject a microwave pump power into the "triple" Fabry-Perot cavity the minimum amount of

$$\langle \mathrm{P}_p \rangle_{\text{threshold}} = \frac{m\omega_p \omega_i \omega_s \mathcal{L}_{\text{eff}}^2}{4 Q_p Q_i Q_s} \tag{14.137}$$

Numerically, if we assume that[8]

$$m = 2 \text{ mg} \tag{14.138}$$

$$\omega_p = 2\pi \times 20 \text{ GHz} \tag{14.139}$$

$$\omega_i \approx \omega_s \approx 2\pi \times 10 \text{ GHz} \tag{14.140}$$

$$\mathcal{L}_{\text{eff}} \approx \lambda_i/2 \approx \lambda_s/2 \approx 3 \text{ cm} \tag{14.141}$$

$$Q_p \approx Q_i \approx Q_s \approx 10^{10} \tag{14.142}$$

then we conclude that we would require a microwave pump power at a frequency of 20 GHz to be injected through the right hole of the "triple" Fabry-Perot cavity of at least

$$\langle \mathrm{P}_p \rangle_{\text{threshold}} \approx 0.2 \text{ microwatts} \tag{14.143}$$

[8]Instead of a solid SC membrane, one could use a grid of fine SC wires, for example, four fine SC wires arranged in a pattern similar to the number sign "#", in order to reduce the mass, and thus the threshold.

Above this minimum power level, there would result a parametric oscillation effect in which a macroscopic amount of signal and idler microwaves centered around 10 GHz, with powers on the order of microwatts (i.e., with powers comparable to the pump threshold power), would be emitted in opposite directions through the left and the right holes, respectively, of the "triple" Fabry-Perot cavity, like in a laser. If the Q is lowered by opening the outcoupling holes, the threshold will go up, but so will the output power of the parametric oscillator. For example, by lowering all the Q's to 10^9 instead of 10^{10}, the threshold will be increased to 0.2 milliwatts, but the output power of the dynamical Casimir effect will also increase by a few milliwatts.

It should be stressed that the leftmost cavity, that is, the "single" Fabry-Perot resonator of Fig. 14.5, is initially devoid of any radiation (i.e., it is initially an *empty* cavity), so that the emission of a macroscopic amount of signal microwaves through the left hole from the left side of the apparatus, would be a dramatic manifestation of the dynamical Casimir effect, in which the observed signal output must have built up exponentially starting solely from vacuum fluctuations inside this initially empty resonator. Since the dynamical Casimir effect is closely related to Hawking radiation according to [41], an observation of parametric oscillation resulting from the moving SC membrane in Fig. 14.5 would be a very interesting result from the point of view of quantum field theory.

14.4 The Gravitational Dynamical Casimir Effect, and the Generation of Coherent Gravitational Radiation

In this final section, we speculate that the above ideas can be extended to include the case of gravitational radiation. The physical concept that ties all these ideas together is the crucial use of the DeWitt minimal coupling rule in all of them.

In particular, we briefly comment on the possibility of extending the "separated parametric oscillator" idea for generating EM microwaves by means of the vibrating SC membrane placed inside the extremely high Q "triple" SC cavity, as depicted in Fig. 14.5, to the much more speculative idea of generating GR microwaves using the same vibrating SC membrane inside the same "triple" SC cavity. This extension is based on the fact that the interaction Hamiltonian $H_{\mathbf{h}\cdot\mathbf{h}}$ in (14.39) is mathematically identical to that of the interaction Hamiltonian $H_{\mathbf{A}\cdot\mathbf{A}}$ in (14.38). Furthermore, we are assuming that it is permissible for gravitational radiation fields to be second quantized (see (14.42)).

However, for this extension of the parametric oscillator idea to work, it is crucial that the walls SC cavity, including the surfaces of the moving SC membrane, reflect GR microwaves with as high a reflectivity as in the case of EM microwaves. In the paper "Do mirrors for gravitational waves exist?" [42], it was predicted that even *thin* SC films are highly reflective mirrors for GR plane waves. This surprising prediction was based on the DeWitt minimal coupling rule (14.20) applied to the Ginzburg-Landau theory of superconductivity. The "off-diagonal long-range order" (ODLRO) [43] nature of the coherent Cooper pairs causes these pairs to behave *differently* from the ions in the ionic lattice, for which ODLRO does not exist. As a result, inside the SC thin film, the coherent Cooper pairs, which exhibit constructive AB interference, *do not* undergo geodesic motion, in contrast to the incoherent ions, which *do* undergo geodesic motion, in response to incident GR radiation. This *difference* in the internal motions of the Cooper pairs and of the ions inside the SC in the presence of GR radiation, leads to a charge separation effect induced by an incoming GR plane wave, such that a huge back-action of the SC film on the GR wave that causes its reflection, results.

If such SC mirrors for GR waves were indeed to exist in Nature, then moving SC mirrors would not only be able to do work like a piston on these waves, but would also simultaneously lead to a Doppler effect that leads to the exponential amplification of these waves above the threshold for parametric oscillation, as explained above. Thus, a laser-like generation of coherent GR waves starting from vacuum fluctuations should become possible. If so, a Hertz-like experiment for GR radiation at microwave frequencies [27] would become feasible to perform.

Acknowledgements D.A.S. acknowledges the support of a 2012–2013 Fulbright Senior Scholar Grant. We thank Jay Sharping for his help in our experiments.

References

1. M.A. Hohensee, B. Estey, P. Hamilton, A. Zeilinger, H. Müller, Force-free gravitational redshift: proposed gravitational Aharonov-Bohm experiment. Phys. Rev. Lett. **108**, 230404 (2012). Here we consider a thought experiment to see the *vector* gravitational AB effect instead of the *scalar* gravitational AB effect.
2. Y. Aharonov, G. Carmi, Quantum aspects of the equivalence principle. Found. Phys. **3**, 493 (1973)
3. E.G. Harris, The gravitational Aharonov-Bohm effect with photons. Am. J. Phys. **64**, 378 (1996). Here we consider the AB effect with electrons rather than photons. The AB effect with photons could be understood entirely classically in terms of classical EM waves diffracting around a "solenoid." However, no such classical explanation would exist for the electron interference experiment described in Fig. 14.1
4. J.M. Cohen, B. Mashhoon, Standard clocks, interferometry, and gravitomagnetism. Phys. Lett. A **181**, 353 (1993)
5. A. Tartaglia, Gravitational Aharonov-Bohm effect and gravitational lensing. gr-qc/0003030
6. R. Owen et al., Frame-dragging vortexes and tidal tendexes attached to colliding black holes: visualizing the curvature of spacetime. Phys. Rev. Lett. **106**, 151101 (2011). arXiv:1012.4869
7. D.A. Nichols et al., Visualizing spacetime curvature via frame-drag vortexes and tidal tendexes: general theory and weak-gravity applications. Phys. Rev. D **84**, 124014 (2011). arXiv: 1108.5486
8. A. Zimmerman, D.A. Nichols, F. Zhang, Classifying the isolated zeros of asymptotic gravitational radiation by tendex and vortex lines. Phys. Rev. D **84**, 044037 (2011). arXiv:1107.2959
9. R.H. Price, J.W. Belcher, D.A. Nichols, Comparison of electromagnetic and gravitational radiation; what we can learn about each from the other. arXiv:1212.4730
10. M. Thorsbud, Post-Newtonian methods and the gravito-electromagnetic analogy. Master's Thesis, Department of Physics, University of Oslo (2010), p. 56
11. V.B. Braginsky, C.M. Caves, K.S. Thorne, Laboratory experiments to test relativistic gravity. Phys. Rev. D **15**, 2047 (1977)
12. R.L. Forward, General relativity for the experimentalist. Proc. IRE **49**, 892 (1961)

13. A. Tartaglia, M.L. Ruggiero, Gravito-electromagnetism versus electromagnetism. Eur. J. Phys. **25**, 203 (2004)
14. Section 4.4 in [18]
15. M. Agop, C.Gh. Buzea, P. Nica, Local gravitoelectromagnetic effects on a superconductor. Physica C **339**, 130 (2000)
16. B. Mashhoon, F. Gronwald, H. Lichtenegger, Gravitomagnetism and the clock effect. Lect. Notes Phys. **562**, 83 (2001)
17. A. Tartaglia, M.L. Ruggiero, Gravitoelectromagnetism versus electromagnetism. Eur. J. Phys. **25**, 203 (2004), and Sect. 4.4 of [18]
18. R.M. Wald, *General Relativity* (University of Chicago Press, Chicago, 1984)
19. H. Weyl, *The Theory of Groups and Quantum Mechanics* (Dover, New York, 1950). (The meaning of the arrow in "$a \to b$" is that "a is to be replaced by b in all the following equations.")
20. B.S. In, DeWitt's paper, "Superconductors and gravitational drag". Phys. Rev. Lett. **16**, 1092 (1966), the minimal coupling rule (14.20) was derived from the principle of general covariance which is behind all metric theories of gravity. This principle was applied to a classical, relativistic, spinless point particle, with the rule (14.20) emerging in the limit of low velocities and weak fields
21. G. Papini, A test of general relativity by means of superconductors. Phys. Lett. **23**, 418 (1966)
22. G. Papini, Detection of inertial effects using superconducting interferometers. Phys. Lett. **24A**, 32 (1967)
23. L.D. Landau, E.M. Lifshitz, *The Classical Theory of Fields*, 4th edn., vol. 2 (Butterworth-Heinemann, Stoneham, 2000)
24. M.V. Berry, Quantal phase factors accompanying adiabatic changes. Proc. R. Soc. Lond. A **392**, 45 (1984)
25. S.B. Felch, J. Tate, B. Cabrera, J.T. Anderson, Precise determination of h/m_e using a rotating, superconducting ring. Phys. Rev. B **31**, 7006 (1985). There remain some small, unexplained discrepancies for the inferred electron mass, which are probably due to some unknown systematic errors in the experiment.
26. M.D. Semon, Experimental verification of an Aharonov-Bohm effect in rotating reference frames. Found. Phys. **7**, 49 (1982)
27. R.Y. Chiao, New directions for gravitational-wave physics via 'Millikan oil drops', in *Visions of Discovery*, ed. by R.Y. Chiao, M.L. Cohen, A.J. Leggett, W.D. Phillips, C.L. Harper Jr. (Cambridge University Press, London, 2011), p. 348
28. J.C. Garrison, R.Y. Chiao, *Quantum Optics* (Oxford University Press, Oxford, 2008). Equation (2.103)
29. J.C. Garrison, R.Y. Chiao, *Quantum Optics* (Oxford University Press, Oxford, 2008). Equation (15.30)
30. M. Aspelmeyer, P. Meystre, K. Schwab, Quantum optomechanics. Phys. Today **65**, 29 (2012)
31. S. Kuhr, S. Gleyzes, C. Guerlin, J. Bernu, U.B. Hoff, S. Deléglise, S. Osnaghi, M. Brune, J.M. Raimond, S. Haroche, E. Jacques, P. Bosland, B. Visentin, Ultrahigh finesse Fabry-Perot superconducting resonator. Appl. Phys. Lett. **90**, 164101 (2007)
32. R.W. Boyd, *Nonlinear Optics* (Academic Press, San Diego, 2003)
33. R.Y. Chiao, L.A. Martinez, S.J. Minter, A. Trubarov, Parametric oscillation of a moving mirror driven by radiation pressure in a superconducting Fabry–Perot resonator system. Phys. Scr. T **151**, 014073 (2012). arXiv:1207.6885
34. M. Philipp, P. von Brentano, G. Pascovici, A. Richter, Frequency and width crossing of two interacting resonances in a microwave cavity. Phys. Rev. E **62**, 1922 (2000)
35. I.G. Wilson, C.W. Schramm, J.P. Kinzer, High Q resonant cavities for microwave testing. Bell Syst. Tech. J. **25**(3), 408–434 (1946)
36. D.J. Griffiths, *Introduction to Electrodynamics*, 3rd edn. (Prentice Hall, New York, 1999), p. 351
37. J.D. Jackson, *Classical Electrodynamics*, 3rd edn. (Wiley, New York, 1998), p. 261

38. V.B. Braginsky, S.E. Strigin, S.P. Vyatchanin, Parametric oscillatory instability in Fabry-Perot interferometer. Phys. Lett. A **287**, 331 (2001)
39. R.Y. Chiao, Analysis and estimation of the threshold for a microwave 'pellicle mirror' parametric oscillator, via energy conservation. arXiv:1211.3519
40. J.C. Garrison, R.Y. Chiao, *Quantum Optics* (Oxford University Press, Oxford, 2008), p. 89
41. P.D. Nation, J.R. Johansson, M.P. Blencowe, F. Nori, Colloquium: stimulating uncertainty: amplifying the quantum vacuum with superconducting circuits. Rev. Mod. Phys. **84**, 1 (2012)
42. S.J. Minter, K. Wegter-McNelly, R.Y. Chiao, Do mirrors for gravitational waves exist? Physica E **42**, 234 (2010). arXiv:0903.0661
43. C.N. Yang, Concept of off-diagonal long-range order and the quantum phases of liquid He and of superconductors. Rev. Mod. Phys. **34**, 694 (1962)
44. C.W. Misner, K.S. Thorne, J.A. Wheeler, *Gravitation* (Freeman, San Francisco, 1972)

Chapter 15
Paradoxes of the Aharonov-Bohm and the Aharonov-Casher Effects

Lev Vaidman

Abstract For a believer in locality of Nature, the Aharonov-Bohm effect and the Aharonov-Casher effect are paradoxes. I discuss these and other Aharonov's paradoxes and propose a local explanation of these effects. If the solenoid in the Aharonov-Bohm effect is treated quantum mechanically, the effect can be explained via local interaction between the field of the electron and the solenoid. I argue that the core of the Aharonov-Bohm and the Aharonov-Casher effects is that of quantum entanglement: the quantum wave function describes all systems together. [*Editor's note*: for a video of the talk given by Prof. Vaidman at the Aharonov-80 conference in 2012 at Chapman University, see quantum.chapman.edu/talk-21.]

15.1 Introduction

Thirty years ago, after calculating integrals for the QCD sum rules in the M.Sc. studies at Weitzmann Institute I went to Tel Aviv looking for a more intuitive physics research. There I met Yakir Aharonov who suggested to look at his recent papers on nonlocal measurements [1, 2]. I found a very different approach. Assume we can measure a certain nonlocal variable, say a product of two variables related to spatially separated regions. Then we can send signals faster than light: a paradox! A paradox that led us to find measurable nonlocal variables, define new types of measurements, construct a new formalism of nonlocal measurements [3, 4].

For Yakir, paradoxes are the main tool for developing new physics. Yakir is certain that we will have a new revolution in physics and the paradoxes we find lead us toward it. During last thirty years I learned from Yakir how to use paradoxes to do research. However, our philosophical approaches became different. I want to believe that apart from some (important) details we understand Nature today. Yakir

L. Vaidman (✉)
Raymond and Beverly Sackler School of Physics and Astronomy, Tel-Aviv University, Tel-Aviv 69978, Israel
e-mail: vaidman@post.tau.ac.il

L. Vaidman
Institute for Quantum Studies, Chapman University, Orange 92866, CA, USA

D.C. Struppa, J.M. Tollaksen (eds.), *Quantum Theory: A Two-Time Success Story*, DOI 10.1007/978-88-470-5217-8_15, taught me to use paradoxes as a powerful research tool, but instead of leading to new directions, it leads me to correct, improve, deepen and clarify our current physical theories.

At the beginning of the twentieth century in spite of serious paradoxes of physical theories the view that physics is "finished" was shared by many, but the theory of relativity and the quantum theory proved them all wrong. Today, when physics explains almost everything we can see, physicists rarely claim that physics is close to the final theory of the Universe. Quantum theory brings two elements which make it very difficult to believe that we completely understand Nature: randomness and nonlocality. Yakir accepts randomness and nonlocality: "God does plays dice to avoid contradiction with nonlocality". For me, accepting randomness is admitting limits to physics. I refuse to do it. I have to pay a big price for this: the only consistent way to avoid randomness in outcomes of quantum experiments is to accept that all outcomes take place in Nature. Thus, I have to accept existence of numerous parallel worlds corresponding to all possible outcomes of quantum experiments and adopt the many-worlds interpretation of quantum mechanics (MWI) [5, 6].

The MWI avoids, together with randomness, the nonlocality of the Bell-type correlations. In the picture of the whole Universe which incorporates all the worlds measurement of an entangled particle causes no action at a distance. When we perform measurement on one of the entangled particles, we change the quantum state of the other particle in each of the created worlds, but change nothing in local description of the other particle in the whole universe. It was a mixed state before the measurement and remains the same mixture after it. In a particular world, created by a quantum measurement, we experience an illusion of randomness and can observe nonlocal correlations.

There is one type of nonlocality which the MWI does not remove, the nonlocality of the Aharonov-Bohm (AB) effect [7]. The AB effect does not lead to "action at a distance", but it prevents a local explanation of the dynamics of charged particles in particular setups. This nonlocality is what I want to analyze here.

15.2 Mach-Zehnder Interferometer

In classical physics we can explain the behavior of particles in the following local way: Particles create fields (propagating with velocity of light or slower) and the other particles accelerate due to local action of these fields. Wave packets of classical electromagnetic field (which are not waves of some media made of particles) also change their propagation due to local interaction. The interference of overlapping wave packets is a local phenomenon too.

Consider a Mach-Zehnder interferometer (MZI), Fig. 15.1a. If the interferometer is properly tuned, the wave packets split inside, but invariably reunite toward detector A. Small shift of one of the mirrors increasing the length of one arm by a half a wave length leads to a classical lag between the wave packets which changes the interference after the final beam splitter such that the wave packet ends at detector B, Fig. 15.1b. All this behaviour is perfectly understood by local interaction

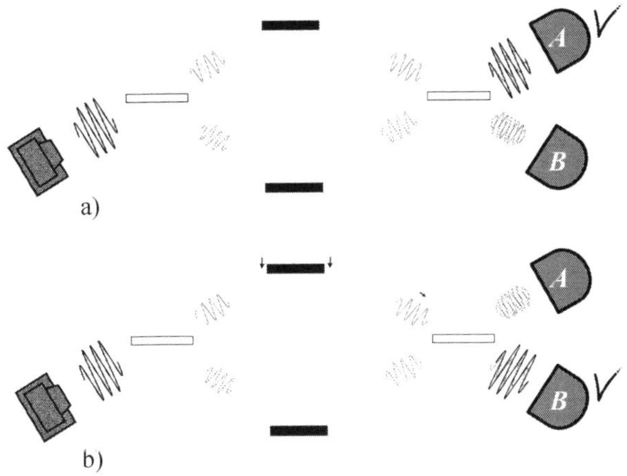

Fig. 15.1 *Mach-Zehnder interferometer.* The wave packet entering the interferometer splits in two beam splitters creating a pair of wave packets moving toward detector A and another pair moving toward detector B. (**a**) The MZI is tuned in such a way that the pair of wave packets moving toward A interfere constructively and the pair moving toward B interfere destructively. (**b**) A small shift of the upper mirror causes small shifts of one of the wave packet moving toward A and one moving toward B. The shifts by the half of the wavelength result in destructive interference toward A and constructive interference toward B

of the wave packets with beam splitters and mirrors and finally local interference of the wave packets moving toward detectors.

More than thirty years ago the interference experiment with MZI has been performed with single photons and, not surprisingly, showed the same results [8]. It is natural to assume that it can be explained in the same way as above with the replacement of a classical wave by the quantum wave of the photon. However, while the evolution of the wave packets of classical electromagnetic fields in the MZI setup can be observed locally, the evolution of the quantum wave is unobservable. The relative phase of the wave packets of the quantum wave which controls the interference when they finally overlap cannot be observed locally when the wave packets are still separated. (There is a possibility of measuring the relative phase of the wave packets using local coupling to parts of a composite measuring device [9], but then the result depends on the definition of the relative phase of the measuring device, so it does not provide an unambiguous measurement of the relative phase of the photon's wave packets.) Although the phase of each wave packet could not be measured during the propagation of the packet inside the interferometer, until the discovery of the Aharonov-Bohm effect [7], there was no reason to suspect that it changes in a conceptually different way from the change of phase of a classical wave packet.

To observe the AB effect, we introduce a solenoid with a flux Φ of the magnetic field inside an electron MZI, Fig. 15.2. The interferometer is tuned in such a way that when the flux vanishes, the electron ends at detector A with certainty. The solenoid leads to a relative phase $\phi_{AB} = \frac{e\Phi}{c\hbar}$ and choosing the flux such that $\phi_{AB} = \pi$ causes the electron to change the interference and to end at detector B.

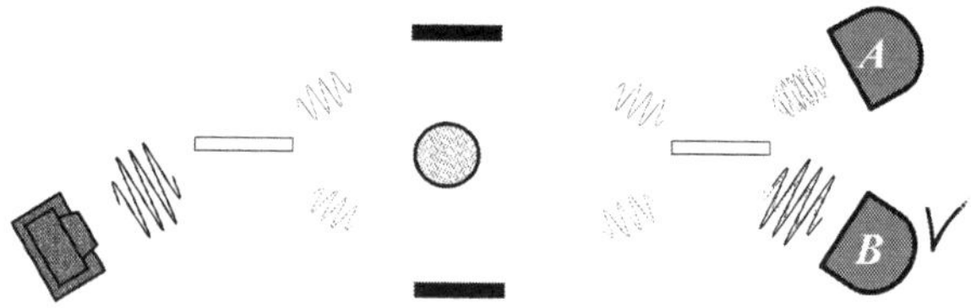

Fig. 15.2 *The Aharonov-Bohm effect.* The electron MZI with a solenoid inside the interferometer exhibits the AB effect. The wave packets are not shifted, but the solenoid leads to a relative phase between the lower and the upper wave packets which causes the change in the interference picture from constructive to destructive interference toward detector A, and destructive to constructive interference toward detector B

Contrary to the case where we moved a mirror and thus locally changed the evolution of a wave packet in one of the arms, we do not have a local explanation of the change in the evolution of the electron wave packets in the present case: The electromagnetic field of the solenoid vanishes at the trajectories of the wave packets of the electron and the vector potential can be made to vanish at any point along the trajectories by using the gauge freedom. The line integral of the vector potential which equals the enclosed magnetic flux is gauge invariant and proportional to the AB phase. However, this provides a topological rather than a local explanation as we had before.

15.3 Aharonov-Casher Effect

In the following, I present an attempt to provide a local explanation of the AB effect. However, I discuss first the Aharonov-Casher (AC) effect [10] which is dual to the AB effect, and show that Boyer's local explanation [11] of the AC effect fails. Aharonov and Casher noticed that there is a symmetry in the interaction between a polarized neutron and an electron. A vertical line of vertically polarized neutrons is equivalent to a solenoid of the AB effect. The symmetry of the interaction suggests that the replacement electron $\leftrightarrow$ neutron for all particles will transform the AB effect in the electron MZI with a solenoid to an analogous AC effect in the neutron MZI with a line of charges. In the AC effect, in contrast to the AB effect, the neutron wave packets do not move in the field-free region. The magnetic field is zero, so naively the neutron does not experience an electromagnetic force, but Boyer correctly realized that a commonly used current-loop model of the neutron leads to a non-vanishing electric dipole moment for a moving neutron and thus it experiences the electromagnetic force due to the electric field of the line of charges. The neutron accelerates while approaching the line of charges in one arm of the interferometer and decelerates in the other arm, then decelerates (accelerates) to the original velocity until it leaves the vicinity of the charged line. The classical lag between the two wave packets provides a local explanation of the AC effect.

Boyer's paper was published when Yakir, I, and Philip Pearle were together in South Carolina. Aharonov's intuition was that Boyer cannot be right. He came with

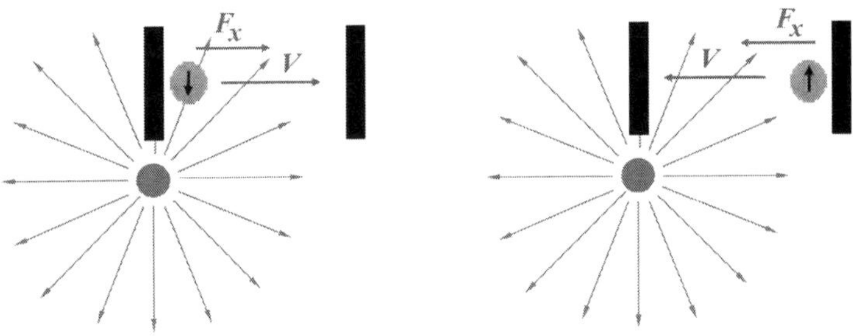

Fig. 15.3 *Free energy source.* According to Boyer, a polarized neutron bouncing between two mirrors in the presence of the line of charge will experience force in the direction of its motion. Thus it can be used as a never ending source of energy

the following paradox: Let us consider elastic mirrors for the neutron approaching the charged line, see Fig. 15.3. Since the induced electric dipole moment $\mathbf{d} = \frac{\mathbf{V}\times\mu}{c}$ changes its direction together with the change of the direction of the velocity, the Boyer's force $(\mathbf{d}\cdot\nabla)\mathbf{E}$ always accelerates the neutron in our setup. But nothing else is changed in the system, so the Boyer's force is a free source of energy!

After some time we found the resolution of the paradox [12, 13]. Boyer provides the correct expression for the force, but in this case the Second Law of Newton is more subtle than just $\mathbf{F} = m\mathbf{a}$. A current loop in an electric field has a "hidden" mechanical momentum $\frac{\mu\times\mathbf{E}}{c}$ and the Boyer's force just provides the time derivative of this momentum. There is no acceleration, so (unfortunately) there is no free source of energy. There is no classical lag between the wave packets and, therefore, the AC effect exhibits the same paradoxical nonlocal feature as the AB effect. In fact, later we learned that this hidden momentum is the core of the paradox discussed by Shockley and James [14] almost a half century ago. (Note that apparently the hidden momentum is not well enough understood until today, see the erroneous conclusion published just a few months ago [15].)

15.4 Local Mechanism for the Aharonov-Bohm Effect

So, we have now two effects which have no local explanation. The wave packets of the electron move in an identical way with or without the solenoid, or for the neutron, with or without the charged line, but, nevertheless, the interference depends on the electromagnetic sources. It seems that there is only global explanation of these effects. The final interference depends on the integral on a closed trajectory and it is apparently meaningless to ask in which part of the trajectory the influence of the solenoid (line of charges) took place.

For me this is a paradox. At every place on the paths of the wave packets of the particle there is no observable effect of any kind, but nevertheless, a relative phase is generated. In an attempt to find local explanation for everything I can see, I identify a weak point in the current descriptions of the AB and the AC effects: the fields with

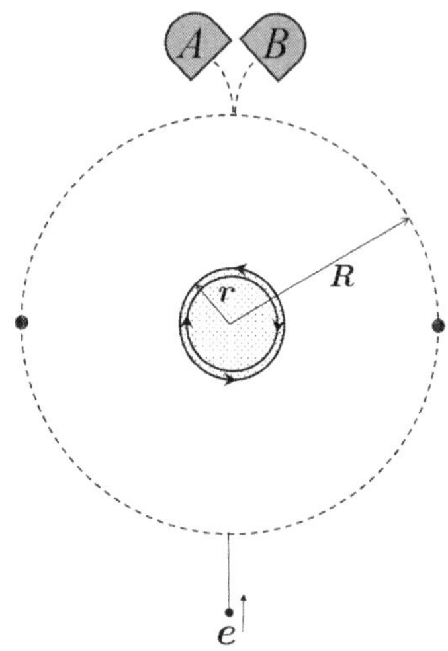

Fig. 15.4 *Solvabe model of the magnetic AB experiment.* The electron wave packet coming directly toward a circle splits into a superposition of two wave packets and after encircling the solenoid in the center interfere on the beam splitter toward detectors A and B

which the quantum particles interact are considered to be classical. I believe that everything is quantum. And indeed, considering the solenoid as a quantum object I have found a local explanation for the AB effect [16]. I will show that the AB effect arises from different shifts of the wave packets of the source of the magnetic flux which experience different local electric fields created by the two wave packets of the electron inside the MZI.

Consider the following setup. The solenoid consists of two cylinders of radius r, mass M, large length L and charges Q and $-Q$ homogenously spread on their surfaces. The cylinders rotate in opposite directions with surface velocity v. The electron encircles the solenoid with velocity u in superposition of being in the left and in the right sides of the circular trajectory of radius R, see Fig. 15.4.

The flux in the solenoid due to two cylinders is $2\pi r^2 \frac{4\pi}{c}\frac{Qv}{2\pi r L} = \frac{4\pi Qvr}{cL}$, and the AB phase, i.e., the change in the relative phase between left and right wave packets due to electromagnetic interaction is:

$$\phi_{AB} = \frac{4\pi e Q v r}{c^2 L \hbar}. \tag{15.1}$$

When the electron moves on the circular trajectory, it creates magnetic flux through a cross section of the solenoid seen at angle θ, see Fig. 15.5,

$$\Phi(\theta) = \pi r^2 \frac{ue\cos\theta}{c(\frac{R}{\cos\theta})^2} = \frac{\pi r^2 eu\cos^3\theta}{cR^2}. \tag{15.2}$$

Before the electron entered the circle, it provided no flux through this section. By entering one arm of the circle, the electron produces change in the the magnetic flux and causes an electromotive force on charged solenoids which changes their velocity. The change in the velocity is:

$$\delta v = \frac{1}{M}\int \frac{\pi r^2 eu\cos^3\theta}{c^2R^2}\frac{1}{2\pi r}\frac{R}{\cos^2\theta}2\pi r\frac{Q}{2\pi r L}d\theta = \frac{uQer}{c^2MRL}. \tag{15.3}$$

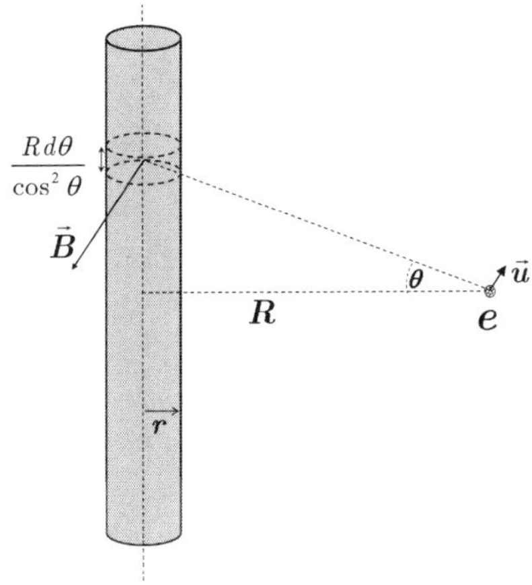

Fig. 15.5 *Magnetic field created by the electron.* The field at the cross section of the cylinder when the right wave packet is seen at angle θ

The shift of the wave packet of the cylinders due to this velocity change during the motion of the electron wave packet is

$$\delta x = \delta v \frac{\pi R}{u} = \frac{\pi Q e r}{c^2 M L}. \tag{15.4}$$

We consider here motion on the circle of radius r as a linear motion. The relevant wavelength of de Broglie wave of each cylinder is $\lambda = \frac{h}{Mv}$. For calculating the AB phase we should take into account that both cylinders are shifted and that they shifted (in opposite directions) in both branches. This leads to factor 4 and provides correct expression for the AB phase:

$$4\frac{2\pi\delta x}{\lambda} = \frac{8\pi^2 Q v e r}{h c^2 L} = \phi_{AB}. \tag{15.5}$$

The explanation of the AB effect is as follows. The electron in superposition of two arms of the interferometer creates different fields at the location of the source of the potential and thus the wave packet of the source is shifted differently. The change in the wave function of the source is, essentially, the change in the relative phase only. In the AB effect, due to topology of the circle, the envelope of the wave packet has no any change whatsoever. Since in quantum mechanics the wave function is for all parts of the system together, the change in the wave function of the source leads to an observable effect in the interference experiment with the electron.

15.5 Discussion

Another manifestation of the wholeness of quantum mechanical description is entanglement. In fact, during the process of observation of the AB effect, the shift in the momentum of the source might create entanglement with the electron; it disappears before the end of the experiment. But this entanglement is not necessary: the initial uncertainty in momentum of the source might be much larger than the intermediate shift in the momentum due to the interaction with the electron. Thus, although for the explanation it is crucial to understand that the quantum wave function is for all particles, the effect can appear even if the source particles and the electron at all times are described arbitrarily well by a product state.

The pictorial explanation with spatial wave packets shifted by fields will disappear when we go beyond physics of moving charges. A model for a solenoid can be a line of polarized neutrons. If the model of each neutron is a current loop, then we will have the picture as above, but if we describe it as a quantum spin, we will not have a picture of charged particles kicked by the field of the electron, we should say that the magnetic field of the electron changes the phase of the neutrons directly. (Note, however, that viewing the polarized spin state of the neutron as a superposition of two spin states polarized in perpendicular direction, we can restore the story of electron fields causing rotation of the spin around the axes of the solenoid.) This is also an explanation of the AC effect: the local field acting on the neutron is responsible for appearance of the AC phase.

One of the most revolutionary aspects of the AB effect is that contrary to classical mechanics, in quantum mechanics we cannot explain everything by action of local fields. Potentials, which were auxiliary objects in classical physics, have direct physical meaning in quantum mechanics. Explanation of the AB and the AC effects in local terms allows us to entertain the idea that potentials are auxiliary concepts after all. However, today all versions of quantum theory, and the Schrödinger equation in particular, are based on the concept of potential. There is no quantum analog of the Second Law of Newton. This work might open the challenge of developing a theory which will tell us how quantum particles evolve when the interaction between them is described by local fields which have no potential.

Even before developing of such a local theory, my assertion provides one useful corollary: If the fields vanish at locations of all particles then these fields yield no observable effect. In the magnetic AB effect I see no variant for which all systems move in a free-field area, but I can devise such an example for the electric AB effect.

Consider the following configuration of three charged particles on a straight line: In the center we place an electron with charge $-e$. On both sides at equal distances we place two charges $4e$. Immediate calculation shows that electric field created by any two particles at the location of the third particle vanishes. The electric potential at location of the electron due to other charges does not vanish, but according to my corollary it cannot cause any effect. The effect of potential might naively be expected when we consider an electron MZI in which we bring the two charges to the electron wave packet moving in one arm of the interferometer keeping all the time the configuration described above, see details in reference [16]. The subtlety here is that we have to bring the charges toward one arm of the interferometer only in the "branch" in which the electron wave packet is there. The charges are in the mixed quantum state of being near and far from the interferometer. The charges do not provide classical field and this explains why the standard approach to the AB effect fails.

15.6 Aharonov-Bohm Effect with a Superconducting Screening

Classical mechanics has local formulation with Newton's laws and global formulation via Lagrangian and Hamiltonian. Quantum mechanics has only global formulations: Hamiltonian, Feynman's path integrals etc. In the past I also learned from the AB effect that the potentials have direct observable effects in quantum mechanics and thus quantum mechanics cannot have a local formulation [17]. But the local explanation of the AB and the AC effects presented above give me hope for a local quantum mechanics. I know that Yakir's intuition is against local theory I am looking for (see, however, a discussion of similar ideas in Aharonov's recent publication [18]). As a reply to my proposal, Yakir, as usual, presented me a challenge in the form of a paradox: "Consider everything quantum, as you do, and explain in your local terms the AB effect experiment by Tonomura [19] in which the solenoid was screened by a superconductor and the AB phase (of the value of π) was observed in a very convincing way." The superconductor apparently screens the field of the electron, so my proposed mechanism for the AB phase via the motion of the charges in the solenoid fails. I guess that the phase appears due to a local action of the electron on charges in the superconductor, but I still cannot provide an explanation. Whatever the resolution of this paradox is, it will deepen our understanding of quantum mechanics as many other Aharonov's paradoxes already did in a very profound way.

Acknowledgements I thank Shmuel Nussinov for useful discussions. This work has been supported in part by the Binational Science Foundation Grant No. 32/08 and the Israel Science Foundation Grant No. 1125/10,

References

1. Y. Aharonov, D. Albert, Phys. Rev. D **21**, 3316 (1980)
2. Y. Aharonov, D. Albert, Phys. Rev. D **24**, 359 (1981)
3. Y. Aharonov, D. Albert, L. Vaidman, Phys. Rev. D **34**, 1805 (1986)
4. L. Vaidman, Phys. Rev. Lett. **90**, 010402 (2003)
5. H. Everett III, Rev. Mod. Phys. **29**, 454 (1957)
6. L. Vaidman, Many-worlds interpretation of quantum mechanics, in *Stan. Enc. Phil.*, ed. by E.N. Zalta (2002). http://plato.stanford.edu/entries/qm-manyworlds/
7. Y. Aharonov, D. Bohm, Phys. Rev. **115**, 485 (1959)
8. P. Grangier, G. Roger, A. Aspect, Europhys. Lett. **1**, 173 (1986)
9. Y. Aharonov, L. Vaidman, Phys. Rev. A **61**, 052108 (2000)
10. Y. Aharonov, A. Casher, Phys. Rev. Lett. **53**, 319 (1984)
11. T. Boyer, Phys. Rev. A **36**, 5083 (1987)
12. Y. Aharonov, P. Pearle, L. Vaidman, Phys. Rev. A **37**, 4052 (1988)
13. L. Vaidman, Am. J. Phys. **58**, 978 (1990)
14. W. Shockley, R.P. James, Phys. Rev. Lett. **18**, 876 (1967)
15. M. Mansuripur, Phys. Rev. Lett. **108**, 193901 (2012)
16. L. Vaidman, Phys. Rev. A **86**, 040101 (2012)
17. L. Vaidman, Aharonov-bohm effect, in *Hebrew Encyclopedia* (1994), supplementary volume III (in Hebrew)
18. Y. Aharonov, T. Kaufherr, S. Nussinov, J. Phys. Conf. Ser. **173**, 012020 (2009)
19. A. Tonomura et al., Phys. Rev. Lett. **56**, 792 (1986)

Part VI
Weak Values

Chapter 16
Weak Values: The Progression from Quantum Foundations to Tool

Andrew N. Jordan and Jeff Tollaksen

Abstract Since its introduction in 1988, the weak value has made a remarkable progression within the scientific world. This article will delineate each step of this progression, displayed in four courses. The first course is the notion of a weak value in itself, the second is the first experiments together with theoretical challenges, the third is acceptance as a phenomena and further experiments, and the fourth is the use of weak values as a tool for both the further understanding of quantum puzzles and for precision measurements. We will discuss recent developments in the field and argue that the notion of contextual values as generalized eigenvalues of an observable, contextualized to the measurement being done, is the next conceptual step beyond the weak value. [*Editor's note*: for a video of the talk given by Prof. Jordan at the YA80 conference at Chapman, see quantum.chapman.edu/talk-8 and for Prof. Howell, see quantum.chapman.edu/talk-9.]

16.1 Introduction

This proceedings article is dedicated to Yakir Aharonov, on the occasion of his becoming an octogenarian. And 2013 is the 25th anniversary of the first weak value paper. As such, it is organized as a testimonial to his work on weak values over the years. We begin with a history of the field, starting from the two-time reformulation of quantum mechanics, and moving on through the introduction of weak values, the theoretical disputes that followed and the first experiments. We will then discuss recent developments using the weak value as an amplification mechanism, first to detect novel physical effects, and then for the purposes of metrology. Since

A.N. Jordan
Department of Physics and Astronomy, University of Rochester, Rochester, NY 14627, USA

A.N. Jordan
Institute for Quantum Studies, Chapman University, Orange, CA 92866, USA

J. Tollaksen (✉)
Institute for Quantum Studies, Schmid College of Science and Technology, Chapman University, Orange, CA 92866, USA
e-mail: tollakse@chapman.edu

D.C. Struppa, J.M. Tollaksen (eds.), *Quantum Theory: A Two-Time Success Story*,
DOI 10.1007/978-88-470-5217-8_16, © Springer-Verlag Italia 2014

the birthday conference coincided with the creation of the Institute for Quantum Studies, directed by Yakir Aharonov and one of the coauthors, the article concludes with a discussion of some recent developments in weak values and a consideration of some ways forward for the field, both fundamental and applied.

16.2 History

The story of the weak value begins with the advent of the two-time reformulation of quantum mechanics, by Aharonov, Bergmann and Lebowitz (ABL) in 1964 [1]. The usual formulation of quantum mechanics is given in terms of an initial wavefunction or quantum state, which is then propagated forward in time according to the Schrödinger equation. Outcomes of experiments then occur randomly upon measurement with the probabilities given in terms of this forward evolved wavefunction. Thus, while the Schrödinger equation is time reversal symmetric, the introduction of measurements and actual recorded events spoils this feature. This time asymmetric view of quantum mechanics can be made symmetric by realizing that the process of preparation is actually a kind of filtering of results: only one state of many possible states is chosen to begin with. By introducing the concept of post-selection, filtering the final results by a selection criterion (just as one does in a preparation) the theory can be made once more time-symmetric. Once two boundary conditions are supplied, one in the past and one in the future, one can think of the past state moving forward in time, or equivalently the future state moving backwards in time (or both). The predictions of this reformulation are the same as in conventional quantum theory, as they must be in order to be the same theory. Any advantage to this reformulation would come from deeper insight into interpretation issues [2] or in the uncovering of new features and effects that were missed before (such as the weak value), or in leading to new mathematics [3], stimulating discoveries in other fields [4], or, ultimately, in suggesting generalizations of quantum mechanics [5].

While the two-time approach to quantum physics can be applied to any situation that conventional quantum physics can, perhaps its most famous flower is the *weak value*. This idea was floated in a 1988 Physical Review Letter, written by Aharonov, Albert, and Vaidman (AAV), with the provocative title, "How the result of a measurement of a component of the spin of a spin-1/2 particle can turn out to be 100" [6]. The idea is to take a pre- and post-selected average of the weak measurement results of an operator. Here, a weak measurement is simply weakly coupling a meter to the system, usually taken to be an impulsive interaction with the meter prepared in a Gaussian state. Without post-selection, the meter would be shifted either up or down by a small amount, depending on which eigenstate the system is prepared in. However, AAV show that with system post-selection, the meter can be deflected by an amount much larger (in principle arbitrarily larger) than the shift without post-selection. Without post-selection, the meter is shifted by the expectation value of a quantum operator $\hat{\boldsymbol{A}}$. If on the other hand, a system is pre-selected in an initial quantum state $|\psi_i\rangle$, and post-selected on a final state $|\psi_f\rangle$, then the result of weakly measuring the operator $\hat{\mathbf{A}}$ is not its expectation value, but rather its *weak value* (WV):

$$A_w = \frac{\langle \psi_f | \hat{\mathbf{A}} | \psi_i \rangle}{\langle \psi_f | \psi_i \rangle}. \tag{16.1}$$

This object has surprising properties, such as the fact that for a bounded operator the weak value can wildly exceed the eigenvalue range, and unlike the expectation value, the weak value can be complex.

16.2.1 Interpretation Issues

There are several interesting aspects of this topic to explore. First, and most importantly is the interpretation of the weak value. Judging from the title of the AAV paper, the authors interpret the meter shift as telling you something about the system's spin. Indeed, further papers by the Aharonov school developing these ideas interpret the weak value as the actual value the spin takes on in a pre and post-selected situation. (See Refs. [7–9] for reviews.) This notion was promptly taken to task by some of the leading physicists of the day who questioned whether this said anything at all about the particle's spin [10, 11]. Thus began the weak value interpretation controversy that continues to this day.

An important point of contention is that if you take literally the idea that the weak value tells you about the value the spin takes during the weak measurement, and if you want to stick with the eigenvalue-eigenstate paradigm, then you are forced to view the weak value as a weighted average of the eigenvalues with probabilities that are negative or exceed 1. While some physicists accept this consequence, in Sect. 16.5 of this paper, we describe another way to solve this puzzle: Rather than insist that the weak value is a weighted average of the eigenvalues of the operator in question, we argue that when an operator is measured weakly, you should not describe it with its eigenvalues, but rather a new set of values that depends on the measurement context. Each possible outcome of the detector is assigned a value that can then be averaged with the detector result probabilities in a straightforward way. It turns out that generically these values have an expanded range, and therefore if the weighted average of these values extends beyond the range of the operator's eigenvalues (but not beyond the contextual value range), it is no longer mysterious. Only when the measurement is projective do the contextual values take on the operator's eigenvalues in the special case (see also [12]).

One point that has come up frequently in the philosophical literature is whether the weak value (as well as the ABL rule) implicitly uses a counter-factual interpretation being used. Kastner, for example, argues that in general, the ABL rule [1] cannot be used to calculate the probabilities of possible outcomes of observables that have not actually been measured [13]. On the other end of the spectrum, Aharonov and collaborators claim we should take the time-symmetry of the two-state vector formalism seriously, even to the point of permitting retro-causation, where future events affect the past [14, 15]!

Despite these controversies, many puzzles in the foundations of quantum mechanics can be "explained" or rendered at least equally as strange as the weak value. For example, weak values can be used as a fundamental test of quantum mechanics by ruling out a class of macrorealistic hidden variable theories and are equivalent to the violation of generalized Leggett-Garg inequalities [16]. Weak values have been useful to help resolve paradoxes that a rise in quantum mechanics such as Hardy's paradox [17, 18], apparent superluminal travel [19], and other quantum problems such as the three-box problem [7].

16.2.2 Counterfactuals

There has been a widespread tendency to "resolve" paradoxes of quantum mechanics by pointing out that there is often an element of counter-factual reasoning in the paradox: the contradictions arise only because inferences are made that do not refer to actual experiments. Had the experiment actually been performed, then standard measurement theory predicts that the system would have been disrupted so that no paradoxical implications arises.

We have proven [18] that one shouldn't be so quick in throwing away counter-factual reasoning; though indeed counter-factual statements have no observational meaning, such reasoning is actually a very good pointer towards interesting physical situations. *Without invoking counter-factual reasoning*, we have shown, in general, that the apparently paradoxical reality implied counter-factually has new, *experimentally accessible* consequences. These observable consequences become evident in terms of *weak measurements*, which allow us to test—to some extent—assertions that have been otherwise regarded as counter-factual. For illustrative purposes, we will consider 2 examples.

16.2.2.1 Counterfactuals and Hardy's Paradox

Hardy's paradox consists of two "superposed" Mach-Zehnder interferometers (MZI) (Fig. 16.1), one with a positron and one with an electron.[1] Consider a single interferometer for the positron (labeled by +). By adjusting the arm lengths, it is possible to arrange specific relative phases in the propagation amplitudes for paths between the beam-splitters $BS1^+$ and $BS2^+$ so that the positron can only emerge towards the detector C^+. However, the phase difference can be altered by the presence of an object, for instance in the lower arm, in which case detector D^+ may be triggered.

[1] Adapted from [18, 20, 21].

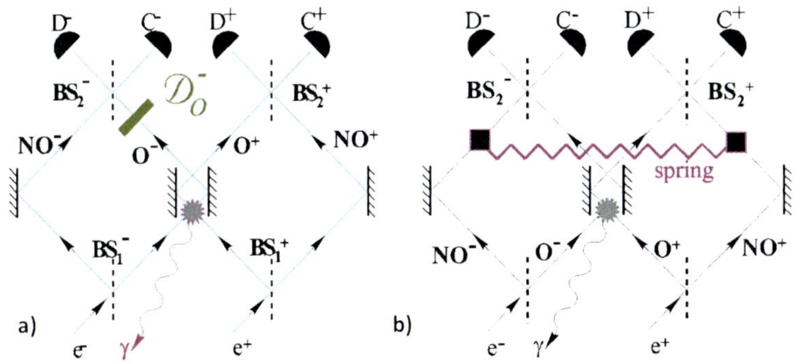

Fig. 16.1 (**a**) Counterfactual resolution: $\mathcal{D}_O^-$ disturbs the electron and the electron could end up in the D^- detector even if no positron were present in the overlapping arm, (**b**) $N^{+,-}_{\mathrm{NO,NOw}} = -1$, spring repulses

In the double-MZI [18], things are arranged so that there is now a region where the two particles overlap, so there is also the possibility that they will annihilate each other. We assume that this occurs with unit probability if both particles happen to be in this region. This overlapping region allows for a situation in which detectors D^- and D^+ may click in coincidence (in which case, obviously, there is no annihilation).

Suppose D^- and D^+ do click. (Given the overlap of the pre-selected state with this final possibility, we can that it is indeed allowed with probability $1/12$.) Trying to "intuitively" understand this situation leads to a paradox. For example, we should infer from the clicking of D^- that the positron must have gone through the overlapping arm; otherwise nothing would have disturbed the electron, and the electron couldn't have ended in D^-. Conversely, the same logic can be applied starting from the clicking of D^+, in which case we deduce that the electron must have also gone through the overlapping arm. But then they should have annihilated, and couldn't have reached the detectors. Hence the paradox.

These statements, however, are counter-factual, i.e. we haven't actually measured the positions. Suppose we actually measured the position of the electron by inserting a detector $\mathcal{D}_O^-$ in the overlapping arm of the electron-MZI. Indeed, the electron is always in the overlapping arm. But, we can no longer infer from a click at D^- that a positron should have traveled through the overlapping arm of the positron MZI in order to disturb the electron (Fig. 16.1.a). The paradox disappears.

Weak measurements produce only limited disturbance [22] and therefore can be performed simultaneously, allowing us to **experimentally** test such counter-factual statements. Therefore we would like to test [18] questions such as "Which way does the electron go?", "Which way does the positron go?", "Which way does the positron go when the electron goes through the overlapping arm?" etc. In other words, we would like to measure the single-particle "occupation" operators

$$\begin{aligned} \hat{N}^{+}_{\mathrm{NO}} &= |\mathrm{NO}\rangle_p\langle \mathrm{NO}|_p, & \hat{N}^{+}_{\mathrm{O}} &= |\mathrm{O}\rangle_p\langle \mathrm{O}|_p \\ \hat{N}^{-}_{\mathrm{NO}} &= |\mathrm{NO}\rangle_e\langle \mathrm{NO}|_e, & \hat{N}^{-}_{\mathrm{O}} &= |\mathrm{O}\rangle_e\langle \mathrm{O}|_e \end{aligned} \tag{16.2}$$

which tell us separately about the electron and the positron (and where NO refers to the non-overlapping arm and O refers to the overlapping arm). We note a most important fact, which is essential in what follows: the weak-value of a product of observables is *not* equal to the product of their weak-values. Hence, we have to measure the single-particle occupation-numbers independently from the pair occupation-operators:

$$\begin{aligned} \hat{N}^{+,-}_{\mathrm{NO,O}} &= \hat{N}^{+}_{\mathrm{NO}}\hat{N}^{-}_{\mathrm{O}}, & \hat{N}^{+,-}_{\mathrm{O,NO}} &= \hat{N}^{+}_{\mathrm{O}}\hat{N}^{-}_{\mathrm{NO}} \\ \hat{N}^{+,-}_{\mathrm{O,O}} &= \hat{N}^{+}_{\mathrm{O}}\hat{N}^{-}_{\mathrm{O}}, & \hat{N}^{+,-}_{\mathrm{NO,NO}} &= \hat{N}^{+}_{\mathrm{NO}}\hat{N}^{-}_{\mathrm{NO}}. \end{aligned} \tag{16.3}$$

These tell us about the simultaneous locations of the electron and positron. The results of all our weak-measurements on the above quantities, echo, to some extent, the counter-factual statements, but go far beyond that. They are now true observational statements in the sense of reading the meter position on a calibrated detector of repeated weak (post-selected) measurements (and such experiments have successfully verified these results [23–25]). In addition, weak-values obey an intuitive logic of their own which allows us to deduce them directly. While discussion of this intuition is left to the published articles [18], we discuss the essence of the paradox which is defined by three counterfactual statements:

- The electron is always in the overlapping arm.
- The positron is always in the overlapping arm.
- The electron and the positron are never both in the overlapping arms.

To these counterfactual statements correspond the following *observational* facts. In the cases when the electron and positron end up at D^- and D^+ respectively and if we perform a single ideal-measurement of:

- $\hat{N}^{-}_{O}$, we always find $\hat{N}^{-}_{O} = 1$.
- $\hat{N}^{+}_{O}$, we always find $\hat{N}^{+}_{O} = 1$.
- $\hat{N}^{+,-}_{O,O}$, we always find $\hat{N}^{+,-}_{O,O} = 0$.

The above statements seem paradoxical but, of course, they are valid only if we perform the measurements separately (i.e. they describe the results of separate experiments); they do not hold if the measurements are made simultaneously. However, a theorem [18] says that when measured weakly all these results remain true simultaneously:

$$N^{-}_{Ow} = 1, \qquad N^{+}_{Ow} = 1. \tag{16.4}$$

The subscript 'w' indicates weak value (16.1). All other weak-values can be trivially deduced [18]:

$$\begin{aligned} N^{-}_{NOw} &= 0, \qquad N^{+}_{NOw} = 0, \qquad N^{+,-}_{O,Ow} = 0 \\ N^{+,-}_{O,NOw} &= 1, \qquad N^{+,-}_{NO,Ow} = 1, \qquad N^{+,-}_{NO,NOw} = -1. \end{aligned} \tag{16.5}$$

What do all these results tell us?

First of all, the single-particle occupation numbers (16.4) are consistent with the intuitive statements that "the positron must have been in the overlapping arm otherwise the electron couldn't have ended at D^-" and also that "the electron must have been in the overlapping arm otherwise the positron couldn't have ended at D^+". But then what happened to the fact that they could not be both in the overlapping arms since this will lead to annihilation? We obtain a consistent result for this as well: the pair occupation number $N^{+,-}_{\mathrm{O,Ow}} = 0$ shows that there are zero electron-positron pairs in the overlapping arms!

We also feel intuitively that "the positron must have been in the overlapping arm otherwise the electron couldn't have ended at D^-, and furthermore, the electron must have gone through the non-overlapping arm since there was no annihilation". This is confirmed by $N^{+,-}_{\mathrm{O,NO}} = 1$. But we also have the statement "the electron must have been in the overlapping arm otherwise the positron couldn't have ended at D^- and furthermore the positron must have gone through the non-overlapping arm since there was no annihilation". This is confirmed too, $N^{+,-}_{\mathrm{NO,Ow}} = 1$. But these two statements together are at odds with the fact that there is in fact just one electron-positron pair in the interferometer. The paradox is resolved in a remarkable way—it tells us that $N^{+,-}_{\mathrm{NO,NOw}} = -1$. The original authors of [18] suggested that this result can be interpreted as *minus* one electron-positron pair in the non-overlapping arms which brings the total down to a single pair (Fig. 16.1.b)! However, this interpretation of -1 has indeed been controversial. Nevertheless, experiments confirmed our predictions [23–25]. A meter calibrated to give 0 for no particle and 1 for a particle of averaged weak measurements will show -1, for this post-selected average. (This issue is discussed in greater detail in Sect. 16.5.)

16.2.2.2 Counterfactuals and Contextuality

Our 2nd example of counterfactuals involves contextuality. The importance of contextuality in a variety of applications, such as quantum information has been growing. Of relevance to these endeavours are new experimentally accessible implications of contextuality through the use of weak measurements [20, 21]. We have analyzed contextuality in terms of pre- and post-selection, and have shown that it is possible to assign definite values to observables in new and surprising ways. We presented new physical reasons for restrictions on these assignments. Weak measurements suggest that novel *experimental* aspects of contextuality can be empirically demonstrated. We also proved that every Logical-pre-and-post-selection-paradox directly implies "quantum contextuality" which is introduced as the analog of contextuality at the level of quantum mechanics rather than at the level of hidden-variable theories. The proof utilizes our suggestion [18] to use weak measurements to probe assertions that were previously regarded as counterfactual. As Mermin emphasized: "... what follows is not idle theorizing about 'hidden variables'. It is a rock solid quantum mechanical effort to answer a perfectly legitimate quantum mechanical question" [26, 27]. Finally, we argued that certain results of these measurements (e.g. eccentric weak values outside the eigenvalue spectrum), cannot be explained by a "classical-like" hidden-variable theory unless disturbance of the hidden-variables is allowed [16, 28]. However, one of us has also argued [20, 21], that even in the limit of weak interaction, if invasive measurements are permitted in the assumed classical hidden variables, the subsequent post-selection can give rise to a post-selected average of measurement data that falls outside of the calibrated eigenvalue bound (in contrast to the non-postselected case) [16].

Although the outcomes of the weak measurements suggest a story which appears to be even stranger than the original one, the situation is in fact far better. Weak values obey a simple, intuitive, and, most important, *self-consistent* logic. This is in stark contrast with the logic of the original counter-factual statements which is not internally self-consistent and leads into paradoxes. Strangeness by itself is not a problem; self-consistency is the real issue. In this sense the logic of the weak values is similar to the logic of special relativity: That light has the same velocity in all reference frames is certainly highly unusual, but everything works in a self consistent way, and because of this special relativity is rather easy to understand. We are convinced that, due to its self-consistency, weak measurement logic will lead to a deeper understanding of the nature of quantum mechanics.

What is contextuality? Bell-Kochen-Specker (BKS) proved that one cannot assign unique answers (i.e. a Hidden-Variable-Theory, HVT) to yes-no questions in such a way that one can think that measurement simply reveals the answer as a pre-existing property that was intrinsic solely to the quantum system itself. BKS assumed that the specification of the HVT, which we call $V_\psi(\hat{A})$, should satisfy: $V_\psi(F\{\hat{A}\}) = F\{V_\psi(\hat{A})\}$, i.e. F represents any functional relation of an operator that is a member of a commuting subset of observables must also be satisfied if one substitutes the values for the observables into the functional relations.

By way of example, we consider Mermin's version of BKS with a set of 9 observables modeled by 2 spin-1/2 particles. It is intuitive [29] to represent all the "functional relationships between mutually commuting subsets of the observables," i.e. $V_\psi(F\{\hat{A}\}) = F\{V_\psi(\hat{A})\}$, by drawing them in Fig. 16.2.a and arranging them so that all the observables in each row (and column) commute with all the other observables in the same row (or column).

$V_\psi(F\{\hat{A}\}) = F\{V_\psi(\hat{A})\}$ requires that the value assigned to the product of all three observables in any row or column must obey the same identities that the observables themselves satisfy, i.e. the product of the values assigned to the observables in each oval yields a result of $+1$ except in the last column which gives -1. It is easy to see that the 9 numbers V_ψ cannot satisfy all the operator constraints because multiplying all 9 observables together gives 2 different results, a $+\mathbf{1}$ when it is done row by row and a $-\mathbf{1}$ when it is done column by column. There obviously

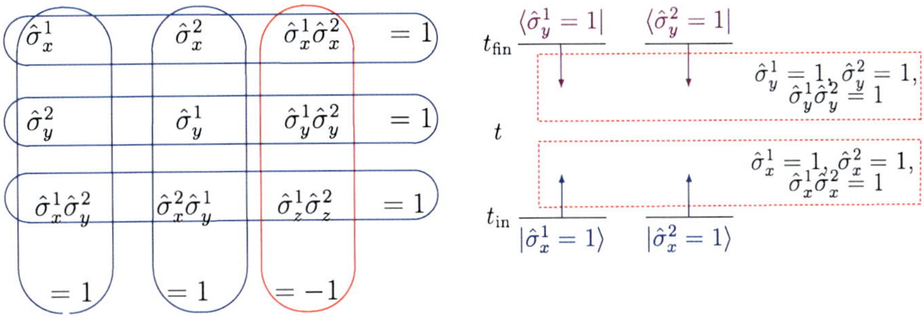

Fig. 16.2 (**a**) 4-D BKS example by Mermin, (**b**) some of the pre- and post-selected states for Mermin example

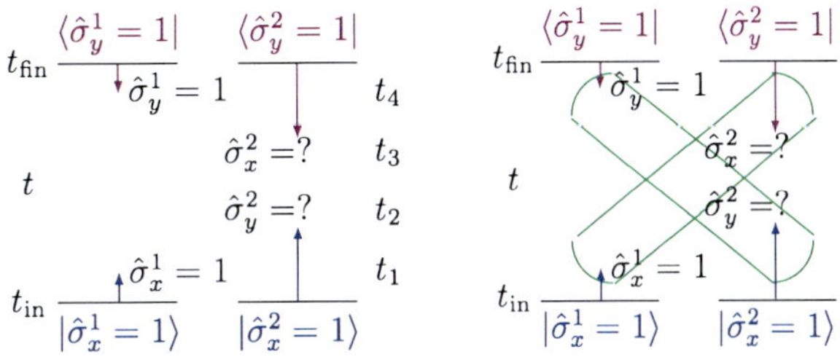

Fig. 16.3 (**a**) Time sequence of pre-and-post-selection measurements for Mermin example, (**b**) Measurement of $\hat{\sigma}_x^1\hat{\sigma}_y^2\hat{\sigma}_x^2\hat{\sigma}_y^1$ is diagonal

is no way to do this with numbers, since they are the same 9 numbers, simply ordered differently. Therefore the values assigned to the observables cannot obey the same identities that the observables themselves obey, $V_\psi(F\{\hat{A}\}) \neq F\{V_\psi(\hat{A})\}$, and an HVT would have to assign values to observables in a way that depended on the choice of which of 2 mutually commuting sets of observables that were also chosen to associate the observable with, i.e. the values assigned are contextual. For example, the assignment $\hat{\sigma}_z^1\hat{\sigma}_z^2 = \pm 1$ depends on whether we associate $\hat{\sigma}_z^1\hat{\sigma}_z^2$ with row-3 or with column-3.

However, it is possible to do this using pre- and post-selection, a result that Mermin describes as "intriguing" [26, 27]. The discovery of such situations was made possible through the use of "unique" multiple-time states which are natural in the 2-vector language [30] (see Fig. 16.4.b).

It may be fruitful to consider an aspect of this analysis: the outcome for the product of the first two observables in column 3 of Fig. 16.2.a with the pre-and-post-selection of Fig. 16.2.b is $\sigma_x^1\sigma_x^2\sigma_y^1\sigma_y^2 = +1$. However, if we measure the operators corresponding to the first 2 observables of row 3 in Fig. 16.2.a, i.e. $\hat{\sigma}_x^1\hat{\sigma}_y^2\hat{\sigma}_x^2\hat{\sigma}_y^1$, given this particular pre-and-post-selection shown in Fig. 16.2.b, then the sequence of measurements interfere with each other (as represented by the slanted ovals in Fig. 16.3.b). To see this, consider that $\hat{\sigma}_x^1\hat{\sigma}_y^2\hat{\sigma}_x^2\hat{\sigma}_y^1$ corresponds to the sequence of measurements represented in Fig. 16.3.a. While the pre-selection of particle 2 is $\hat{\sigma}_x^2 = 1$ at t_{in}, the next measurement after the pre-selection at t_2 is for $\hat{\sigma}_y^2$ and only *after* that a measurement of $\hat{\sigma}_x^2$ is performed at t_3. Thus, there is no guarantee that the $\hat{\sigma}_x^2$ measurement at t_3 will give the same value as the pre-selected state

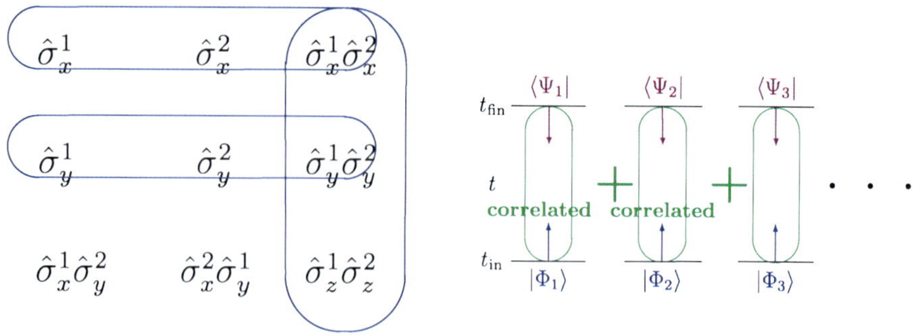

Fig. 16.4 (**a**) Products of observables that are not disturbed given the pre- and post-selection of Fig. 16.2.b, (**b**) generalized "multiple-time" state: superpositions of 2-vectors

of $\hat{\sigma}_x^2 = 1$ or that the $\hat{\sigma}_y^2$ measurement will give the same value as the post-selected state of $\hat{\sigma}_y^2 = 1$. From our picture, this is due to the disturbance of the 2-vector boundary conditions: the initial pre-selected vector $\hat{\sigma}_x^2 = 1$ from t_{in} is "destroyed" when the $\hat{\sigma}_y^2$ measurement at time t_2 is performed and therefore cannot inform the later $\hat{\sigma}_x^2$ measurement at time t_3. In other words, with the particular pre-and-post-selection given in Figs. 16.2.b and 16.3.a, the operator, $\hat{\sigma}_x^1\hat{\sigma}_y^2\hat{\sigma}_x^2\hat{\sigma}_y^1$ depends on information from both the pre-selected vector $\hat{\sigma}_x^1 = 1$, $\hat{\sigma}_x^2 = 1$ and the post-selected vector $\hat{\sigma}_y^1 = 1$, $\hat{\sigma}_y^2 = 1$ in a "diagonal pre- and post-selection" sense. Given the pre- and post-selection of Fig. 16.2.b, the subset of observables circled in Fig. 16.4.a (and the products of those circled observables) can be assigned eigenvalues in a way that satisfies the function relation requirement. But, the product of the other observables (e.g. $\hat{\sigma}_x^1\hat{\sigma}_y^2$ and $\hat{\sigma}_x^2\hat{\sigma}_y^1$) can only be ascertained (given this particular pre- and post-selection) using information from *both* the pre- and post-selected vector in a diagonal sense, and will thus violate the product rule. Our picture has thus given a physical answer for Mermin's "intriguing" question: there will always be a diagonal situation for any 2 observables. The observable $\hat{\sigma}_z^1\hat{\sigma}_z^2$ is assigned different values in different pre-and-post-selections. It is precisely because of this connection between particular pre-and-post-selections and different values for $\hat{\sigma}_z^1\hat{\sigma}_z^2$ that the issue of contextuality arises when we consider products of these observables. In other words, the contextuality here is manifested by the fact that $\hat{\sigma}_x^1\hat{\sigma}_y^2\hat{\sigma}_x^2\hat{\sigma}_y^1 = -1$ (given the pre-and-post-selection of Fig. 16.3.a) even though separately $\hat{\sigma}_x^1\hat{\sigma}_y^2 = +1$ and $\hat{\sigma}_x^2\hat{\sigma}_y^1 = +1$. But these 3 outcomes can be measured weakly without contradiction because the product of weak-values is not equal to the weak-value of the product. Therefore, instead of contextuality being an aspect of a hypothetical replacement for quantum mechanics (the HVT), we have shown that contextuality is directly part of quantum mechanics [20, 21].

In summary, as we understand better how contextuality can be used as a resource for quantum information tasks, the types of intuition and new empirical implications delivered by weak values will prove to be a useful tool.

16.2.3 First Experiments

The first experiment to investigate the weak value was an optical one by Ritchie, Story, and Hulet [31]. The system considered was the optical polarization of a beam, and the detector used to measure it was the translation of the laser beam perpendicular to the propagation direction. They used a birefringent crystal to separate the two polarization components of a laser beam by a distance small compared to the laser-beam waist (this is the weak measurement). It was followed by a strong measurement with a polarizer, which translated the centroid of the beam by a distance far larger than the birefringence-induced separation. While this experiment was done with a classical light beam, the interference of the coherent beam gave the same results as predicted by the quantum theory. This points to the fact that the large deflection of the beam can be understood as a wave interference effect.

Following this and related experiments, Pryde *et al.* performed the first single photon weak value experiments [32]. They used a photon's polarization, and a weak measurement that employed a two-photon entangling operation, and postselection. This measurement technique is quite different than the one just discussed, and ultimately gives the measurement result as a "yes" or "no" answer. The weak values cannot be explained by a semiclassical wave theory, due to the two-photon entanglement. They showed a weak value of a Stokes parameter of more than an order of magnitude outside of the operators spectrum for the smallest measurement strengths.

16.3 Precision Measurement

In the last line of the AAV paper, it was pointed out that for the Stern-Gerlach apparatus, the anomalously large deflection of the beam was controlled by the overlap between pre- and post-selected states, as well as the size of the magnetic field. Consequently, a small change in the magnetic field would result in a huge change in the beam displacement. Therefore, this effect could also be used to measure small changes in magnetic field, transforming an esoteric effect in wave interference into a novel metrological technique.

Although early optical experiments demonstrated the reality of the anomalously large shift of the beam [31], the first use of the weak value for metrology came in 2008 in an experiment by Hosten and Kwiat [33]. They were not interested in weak measurements or values, but rather in the optical spin Hall effect, the effect where different polarization states of light are shifted spatially in different, polarization-dependent directions when the beam is incident on a glass interface. However, they faced a severe problem in the fact that the effect theoretically corresponded to a spatial shift by an Angstrom, which is much smaller than the width of the optical beam. In order to measure this shift, they realized that this situation is exactly the one considered by AAV, and that by pre- and post-selecting the polarization of the beam by polarizers, they could then measure the amplified deflection on a position sensitive detector, with the cost of a drastically reduced intensity.

This experiment was followed by the Rochester group [34, 35], who were also primarily interested in the use of the weak value as a mechanism for amplification. However, rather than being concerned with learning about new physics, they were interested in the amplification mechanism itself. If Kwiat could use this technique to measure a small shift resulting from the optical spin Hall effect, why couldn't the Rochester group not use something similar to measure any small shift, and in so doing, introduce a new metrological technique?

One difficulty in the polarization based experiments is the fact that the source of deflection must be polarization dependent. The Rochester group were able to generalize this by switching over to an interferometer based system, where the different paths corresponded to different directions of deflection. The preselection, and post-selection corresponded to the optical beam entering and leaving different ports of the interferometer, while the weak measurement corresponded to a moving mirror slightly misaligning the interferometer. A piezo activated mirror moved the mirror slightly back and forth a known amount, and the test was to see how small the interferometric weak value technique could measure it. With an hour of integration time, the group reported 500 frad resolution, and later found a signal to noise ratio at the standard quantum limit. This was done with milliwatts of power in an open air experiment.

One of the main factors that improved the performance of this detector is the improved signal to noise ratio. Indeed, it is astonishing that by using only a fraction of the light, you can do better than using all of the light. How can this be? The answer is the fact that if you examine the signal to noise ratio—the quantity that tells you how confident you are that the average you measure is the actual value—the amplification of the deflection and the loss of photons in the signal compensate each other, so the signal to noise ratio is fundamentally the same. However, there are a number of practical issues that come into play through either technical noise or the feasibility of implementing noise reducing techniques where weak value techniques give an advantage over standard techniques. For example, Kedem points out that technical noise in the average shift of the position of the pointer does not appear in the momentum basis, and that noise in the average momentum shift increases the signal-to-noise ratio of the measured parameter [36]. Feizpour, Xing, and Steinberg showed that for additional technical noise that has a long correlation time, postselection can help boost the signal-to-noise ratio as well [37]. These results give a huge advantage to experiments using the signal-to-noise ratio to estimate the small parameter.

Subsequent experiments extended this technique to measuring phase, which turns out to be dual to the deflection measurement, and involves the inverse weak value [38–40]. Optical frequency differences could also be sensitively measured by putting a prism or other dispersive element into the interferometer, translating a change in frequency into a deflection. Optical frequency differences of down to $129 \pm 7\ \text{kHz}/\sqrt{\text{Hz}}$ were also measured.

16.4 New Directions in Precision Measurement

With these advances in applying weak value techniques to precision metrology, it is natural to ask what the next steps are. These experiments are essentially proof of principle type experiments, that could be further developed into commercial products for precision detectors. However, we should ask if 1) we can do better, and 2) in what new directions can these ideas be taken.

In terms of improving the sensitivity of the deflection detectors, we are pursuing several directions. The most promising is the simple realization that in a postselection based measurement technique, most of the light is simply thrown away. We saw above that the postselection combined with the amplified deflection compensated in the signal to noise ratio to give us the shot noise limited value. However, if we could somehow take that extra light we are throwing away and squeeze more measurement data out of it, we could do better.

One way of doing this is to recycle the light. In this way, the light falling on the detector would not be a small fraction of the postselected light, it would be all of the light, so that in principle, every photon would be postselected. In order to recycle, one could send a pulse of light into the interferometer, with N photons in it. A fraction pN of those photons would go to the detector with an amplified deflection, leaving $(1-p)N$ emerging from the other port of the interferometer. One could then send those photons back in the interferometer (for example by switching on a Pockels cell) and make a second pass. This would then repeated until all the photons are gone, or a new pulse is sent in (see [41]).

With this technique, one would then boost the signal to noise ratio by the square root of the number of passes, until losses kick in, doing significantly better than the single pass approach. The primary advantage is simply increasing the power incident on the detector. The only difficulty with this technique is that if all of the light eventually falls in the detector, because of the coherent addition of each pass, rather than each pass being amplified by the same amount, the gradual subtraction of photons in the detection process from one side of the beam will move the remaining beam in the opposite direction, leading to a walk-off effect as the recycle number increases. Eventually, when the last photon is detected, the integrated signal in the detector is exactly the same beam profile you started with, so the net signal (deflection) is exactly zero! To prevent this from happening, one can stabilize and recenter the beam on each pass with the quantum Zeno effect, or allow the beam to expand slightly [42].

Another direction to increase sensitivity is to combine this technique with other quantum optical techniques for precision measurement, such as the use of squeezed states. This promises to be challenging, however, since the advantages of squeezing are very sensitive to loss, so postselection is naively a very bad idea. However, we could mix the squeezing beam with the already postselected light as one possibility, or another possibility is to use the recycling scheme to collect all the correlated photons. Finally, we must address the question of what new application of these techniques can be found, whether commercial, industrial, or military.

For these kinds of applications, it is therefore interesting to see if the techniques can be applied to situations where there are cases of great intrinsic loss—such as light propagation in the atmosphere over large distances. Gisin *et al.* [43] point out that the everyday physics of fiber optics telecom networks naturally exhibits pre- and post-selected weak values through polarization-mode dispersion and polarization-dependent losses. In order for the loss of light to aid the measurement of a system parameter, it must somehow depend on the state of the system that is being measured; for example, the loss is greater for vertical than horizontal polarization.

Some open questions relate to what is the resolution gain that can be obtained:

- To what extent can weak value techniques help with overcoming the diffraction limit? In fact, a broad generalization of the weak value amplification paradigm [9] can be applied to a variety of classical and quantum applications.
- Doppler shift measurements—in comparing the frequencies on incoming and outgoing waves, is there a new weak values based technique?
- Single parameter versus many pixels imaging: If each weak value measurement requires only a small fraction of the light, perhaps by staging the measurements, we can obtain enhanced resolution of the overall signal or picture. Can compressive sensing help?
- The Rochester interference based measurement scheme has been adapted to measure small shifts in optical frequency [44]. By using pulses instead of continuous wave light, time-resolved measurements could also be explored. One possibility here would be to improve the sensitivity of accelerometers that typically use the Sagnac interferometer.
- While existing radar schemes do not exploit which-path interference, we propose to examine schemes where there is an intrinsic loop path in the system—such as bouncing a beam off of a concave surface. There, there will be two interfering paths that can be post-selected after a beam-splitter. This would be a single source/detector configuration. We can also explore a situation where there is a spatially separated source and detector, although it will be much more difficult to align such an interferometer.
- One of the authors has proposed using "robust" weak measurements to get even better precision out of weak value measurements [45]. (For example, "robust weak measurements" can yield accurate weak values that are outside the eigenvalue spectrum without requiring an exponentially rare ensemble.) However, this technique also requires the measurement of the *total* momentum of all of the particles, which is technically challenging. One possible laboratory solution would be to detect the photon-induced deflection of a mirror to realize this idea.

Here, we have summarized a number of open research questions related to the interplay between weak value amplification and quantum sensing. We view the most promising ideas to still be incremental advances to improve the sensitivity of existing experiments, and their scaling, but we also tried to give a sense of more daring ideas that would be of direct interest to commercial or military parties.

16.5 Contextual Values

In addition to the application of the weak value amplification schemes, there has been continuing work to better understand what weak values are and how to best interpret them. The introduction of contextual values permits a much broader tool in order to link information taken from a meter variable to the quantum system it is measuring. Contextual values is a new approach to generalized measurements that subsumed the weak value approach as a special case of a broader question of reconstructing or targeting averages or correlation functions of a system operator based on data provided to the observer from an ancillary meter or pointer variable that is correlated with the system in question.

The basic idea is very simple and can be explained with a classical example of detection physics that will be generalized simply to the quantum case. Consider being locked alone in a room, and on a table in front of you is a jar of red and blue marbles. Your job is to figure out the relative fraction of red versus blue marbles. Now, ordinarily, you could just count all of the red marbles, all of the blue marbles, and give the relative proportions. However, this task is complicated by the fact that you are color blind. Color blind! Then this is not possible without consulting another person or a color detector. However, you are not completely color blind, but only mostly color blind. You know from prior experience that if a red marble is held in front of you, you guess it is red 51 % of the time. If a blue marble is held in front of you, you guess it is blue 51 % of the time. With this knowledge, is it possible to figure out the fraction of red versus blue marbles in the jar?

Being a quantitative person, you assign numbers, or values, to the colors, red→ 1, blue→ −1. However, your eyes only guess the right answer with some probability, so you will record "thumbs up" (*tu*) if you think the marble is red, and "thumbs down" (*td*) if you think it is blue. Now, you assign *different* numbers, a and b, to the outcomes tu and td. This is done so you can still construct the correct color average from your mostly colorblind eyes. The color average is defined as

$$\langle \text{color} \rangle = p(\text{red})1 + p(\text{blue})(-1), \tag{16.6}$$

which is a weighted average of the numbers assigned with the probabilities of the colors in the jar. Since $p(\text{blue}) = 1 - p(\text{red})$, knowledge of the color average is enough to answer our question. We desire to pick the correct values of a and b, so the color average is the same,

$$\langle \text{color} \rangle = p(tu)a + p(td)b, \tag{16.7}$$

where we now use the data that is accessible to our eyes, the relative fraction of "thumbs up" versus "thumbs down" data taken on the marble jar.

We know how to relate the thumb probabilities with the color probabilities from your prior experience, or the "calibration" of your eyes,

$$p(tu) = .51\, p(\text{red}) + .49\, p(\text{blue}), \qquad p(td) = .49\, p(\text{red}) + .51\, p(\text{blue}). \tag{16.8}$$

Putting these relations into the color average, we find

$$p(\text{red})1 + p(\text{blue})(-1) = (.51\,a + .49\,b)p(\text{red}) + (.49\,a + .51\,b)p(\text{blue}). \quad (16.9)$$

Now, the critical point is that we desire to choose a and b, so that this relation is true no matter what the marble probabilities are, otherwise this choice would work only for those particular probabilities, and would be useless as a detector for a jar of marbles you have no *a priori* information about. Consequently, we have two equations for a and b, written in matrix form as,

$$\begin{pmatrix} 1 \\ -1 \end{pmatrix} = \begin{pmatrix} .51 & .49 \\ .49 & .51 \end{pmatrix} \begin{pmatrix} a \\ b \end{pmatrix}.$$

This is easily solved to give the result $a = 50, b = -50$. Thus, we see we must assign larger values to the thumb data to compensate for the ambiguous detection. This choice will produce the correct color average, no matter what the distribution of marbles is in the jar. Note that since the color average is just a weighted combination of $1, -1$, it must be between these "color eigenvalues". In this case, we call the precise degree of color blindness of the measurement the *context* of the measurement, while we call the values we assign to the measurement outcomes, a and b in this case, the *contextual values*.

We can now take the next step, and ask about conditioned averages. For example, as you are taking data, you notice that some of the marbles are spherical, while others are slightly ellipsoidal. One could then ask a question, *what is the color average of the jar, given that I only record data on the spherical marbles?* Clearly, this can drastically change the answer from the unconditioned average, such as the case where there are an equal number of red and blue marbles, but where the red marbles are all spherical and the blue marbles are all ellipsoidal.

We can calculate the conditioned average in the way described above, with the simple change of replacing probabilities with conditional probabilities. While the conditioning procedure can drastically change the answer from the unconditioned average, it is still bounded between -1 and 1 since

$${}_s\langle \text{color} \rangle = 1p(r|s) + (-1)p(b|s) = ap(tu|s) + bp(td|s). \quad (16.10)$$

This last step is true, provided that there is no *disturbance* of one property by the act of measuring the other. For example, if in the act of recording the color of the marble, you squeeze the marble from spherical to ellipsoidal in shape, then you will in general change the (conditioned) average of the color since the process of detection disturbed one of the properties you were measuring. In that case, Eq. (16.8) will not hold for the conditional probabilities.

Once we understand the contextual value approach on imperfect classical detectors, it is easy to extend it to the quantum case. In exactly the same way, we calibrate the detector by asking how the detector response is correlated with known states of the system. We identify an observable $\mathbf{A}$ we would like to measure (typically its average or expectation value), and we expand that observable in its eigendecomposition, as well as in terms of the probabilities of detector outcomes times the contextual values. As before, we want it to work for any quantum state or density matrix now, so now we get an operator equation to solve,

$$\mathbf{A} = \sum_j \alpha_j \mathbf{E}_j, \tag{16.11}$$

Where $\mathbf{E}_j$ are the POVM elements of the measurement, that together with the initial state ρ_i give the probabilities of the measurement outcomes, $P_j = \mathrm{Tr}[\rho \mathbf{E}_j]$. This operator equation is then solved for the contextual values $\{\alpha_j\}$, once we know about the context of the measurement, $\{\mathbf{E_j}\}$.

Conditioned averages of the operator are now defined as averages of the contextual values with the conditional probabilities of the detector, given a subsequent event (usually a second measurement on the system in a new basis). In the weak limit, the operators E_j are almost the identity up to a scale factor (since when the measurement strength vanishes, the state is undisturbed), we find under reasonable conditions a general expression for the conditioned average of an operator of

$${}_f\langle \mathbf{A}\rangle_i = \mathrm{Re}\,(\mathrm{Tr}[\mathbf{P}_f \mathbf{A} \rho_i]/\mathrm{Tr}[\mathbf{P}_f \rho_i]), \tag{16.12}$$

where $\mathbf{P}_f$ is the positive operator on the state, indexed by the final result f that is being post-selected on. In the case where $\rho_i = |\psi_i\rangle\langle\psi_i|$ and $\mathbf{P}_f = |\psi_f\rangle\langle\psi_f|$, the result (16.12) reduces to the *real part* of the weak value formula (16.1). We note that since we are averaging real values assigned to detector outcomes with the usual conditional probabilities, the final conditioned average must be real, which resolves the ambiguity of which part of the weak value comes from the conditioned average.

16.5.1 Imaginary Part of Weak Value

As we saw above, the contextual values approach to conditioned averages gives always the real part of the weak value in the weak limit, for pure pre and post selected states. This reflects its status as an operational formalism, always giving an answer involving detector probabilities and assigned real values. It is therefore natural to ask what the imaginary part of the weak value means from an operational perspective. We have come to the conclusion that, in terms of providing information about the conditioned average of its operator, it means nothing [28]. This is not to say it does not enter into physical expressions, say the momentum shift of a Gaussian detector variable that is coupled to the system via a Von Neumann interaction.

We can, however, provide another interpretation of the imaginary part of the weak value in terms of state disturbance [28, 46]. It corresponds to the rate of change of the postselection probability, in the direction of the operator A, in its role as generator of unitary transformations. This can be expressed concisely with the following formula (written for pure states for simplicity),

$$2\mathrm{Im}A_w = \lim_{\epsilon\to 0} \partial_\epsilon \log |\langle\psi_f| \exp[i\epsilon\mathbf{A}]|\psi_i\rangle|^2. \tag{16.13}$$

A simple illustration of this is the case of momentum, $\mathbf{p}$. In the case where the initial state is the wavefunction $\psi(x)$, and the postselection is for particular position x, the weak value is

$$p_w = \frac{\langle x|\mathbf{p}|\psi\rangle}{\langle x|\psi\rangle} = -i\frac{\partial_x \psi(x)}{\psi(x)} = \partial_x S - i\partial_x \log r, \tag{16.14}$$

where we have taken the polar decomposition, $\psi(x) = r\exp(iS)$. The real part of the weak value corresponds to the momentum in Madelung formulation of quantum mechanics [47], often called Bohmian mechanics today, while the imaginary part corresponds to the rate of change of the particle density at position x, $\rho(x) = r(x)^2$ (or postselection probability) also known as the Osmotic velocity in Nelson's quantum mechanics [48].

16.5.2 Generalizations to Wimpy Measurements

While the general expression for the conditioned average does not simplify away from the weak measurement regime, if we restrict our attention to the case of Von Neumann coupled detectors, we can make exact generalizations of the weak value to any measurement strength, not weak but not yet strong: wimpy. In this way, we will see that there is a surprising universality of the weak value, even when the weak value is no longer weak! The universality does not refer to the particular value the conditioned average takes, but rather to the fairly simple formula that captures a wide class of conditioned averages.

The price we pay in doing this is the fact that the state does get finitely disturbed away from the weak limit, however, the equations account for this in a very natural and simple way: the state that appears for the system is simply the dephased one after the interaction with the meter. Indeed, the weak value generalization (16.12) holds with the small change $\rho \to \rho'$, where ρ' is the post-interaction state, dephased after the interaction with the meter. This is true even for non-Gaussian detectors as well as arbitrary strength measurements [49].

16.6 Conclusions

The weak value, introduced by Yakir Aharonov in 1988 together with his collaborators, has shown itself to be a remarkably fruitful idea. We have tracked the scientific progress of this idea, from its first conception, its initial controversy, and the first proof of principle experiments. This was followed by gradual acceptance of the effect, even if the interpretation controversies still go on to the present day. The next step in this progression was the use of the weak value as a stepping stone to discover new physical effects that could not be otherwise detected, using the weak value not as a topic to untangle quantum mysteries, but as an amplifier. This role was soon stripped away from finding new physics to being used strictly as a metrological tool to measure a wide variety of physical quantities in optical systems.

Weak values continue to attract researchers to explore its secrets, and the lively current day activity in understanding the fundamental aspects shows no sign of stopping. We have every confidence that this area of metrological research with weak value amplification will continue to develop and further increase in precision as further applications for metrological measurements are sought in academic, industrial and military settings.

Acknowledgements This work was supported by the NSF grant No. DMR-0844899, Army Research Office "Advanced Quantum Sensing" grant W911NF-12-1-0237 and grant W911NF-09-1-0417. We thank Michael Berry for pointing out Madelung's work on the hydrodynamical formulation of quantum mechanics to us [47].

References

1. Y. Aharonov, P.G. Bergmann, J.L. Lebowitz, Phys. Rev. B **134**, 1410 (1964)
2. Y. Aharonov, S. Popescu, J. Tollaksen, in *Quantum Theory: A Two-Time Success Story. Yakir Aharonov Festschrift*, ed. by D. Struppa, J. Tollaksen (Springer, Milan, 2013). doi:10.1007/978-88-470-5217-8_3
3. Y. Aharonov, F. Colombo, I. Sabadini, D.C. Struppa, J. Tollaksen, J. Phys. A, Math. Theor. **44**, 365304 (2011)
4. S. Nussinov, J. Tollaksen, Color transparency in QCD and post-selection in quantum mechanics. Phys. Rev. D **78**, 036007 (2008)
5. Y. Aharonov, S. Popescu, J. Tollaksen, Each moment of time is a new universe, this volume
6. Y. Aharonov, D.Z. Albert, L. Vaidman, Phys. Rev. Lett. **60**, 1351 (1988)
7. Y. Aharonov, L. Vaidman, in *Time in Quantum Mechanics*, ed. by J.G. Muga, R. Sala Mayato, I.L. Egusquiza (Springer, Berlin, 2002). arXiv:quant-ph/0105101v2
8. Y. Aharonov, S. Popescu, J. Tollaksen, A time-symmetric formulation of quantum mechanics. Phys. Today **63**(11), 27–32 (2010)
9. Y. Aharonov, J. Tollaksen, New insights on time-symmetry in quantum mechanics, in *Visions of Discovery*, ed. by R.Y. Chiao, M.L. Cohen, A.J. Leggett, W.D. Phillips, C.L. Harper Jr. (Cambridge University Press, Cambridge, 2009)
10. A.J. Leggett, Phys. Rev. Lett. **62**, 2325 (1989)
11. A. Peres, Phys. Rev. Lett. **62**, 2326 (1989)
12. Y. Aharonov, A. Botero, Phys. Rev. A **72**, 052111 (2005)
13. R.E. Kastner, Found. Phys. **29**, 851 (1999)
14. Y. Aharonov, E. Cohen, D. Grossman, A.C. Elitzur. arXiv:1206.6224
15. Y. Aharonov, S. Popescu, J. Tollaksen, Phys. Today **63**, 27–32 (2011)
16. N.S. Williams, A.N. Jordan, Phys. Rev. Lett. **100**, 026804 (2008)
17. L. Hardy, Phys. Rev. Lett. **68**, 2981 (1992)
18. Y. Aharonov, A. Botero, S. Popescu, B. Reznik, J. Tollaksen, Phys. Lett. A **301**, 130 (2002)
19. D. Rohrlich, Y. Aharonov, Phys. Rev. A **66**, 042102 (2002)
20. J. Tollaksen, Pre- and post-selection, weak values, and contextuality. J. Phys. A, Math. Gen. **40**, 9033–9066 (2007). quant-ph/0602226
21. J. Tollaksen, Probing contextuality with pre-and-post-selection. J. Phys. Conf. Ser. **70**, 012014 (2007)
22. J. Tollaksen, Non-statistical weak measurements, in *Quantum Information and Computation V*, ed. by E. Donkor, A. Pirich, H. Brandt. Proc of SPIE, vol. 6573 (SPIE, Bellingham, 2007). CID 6573-33

23. J.S. Lundeen, K.J. Resch, A.M. Steinberg, Phys. Rev. A **72**(1), 016101 (2005)
24. S.E. Ahnert, M.C. Payne, Phys. Rev. A **70**, 042102 (2004)
25. K. Yokota, T. Yamamoto, M. Koashi, N. Imoto, New J. Phys. **11**, 033011 (2009)
26. N.D. Mermin, Phys. Rev. Lett. **74**, 831 (1995)
27. N.D. Mermin, in *Potentiality, Entanglement and Passion-at-a-Distance*, ed. by R.S. Cohen et al. (Kluwer Academic, Norwell, 1997), pp. 149–157
28. J. Dressel, A.N. Jordan, Phys. Rev. A **85**, 012107 (2012)
29. N.D. Mermin, Rev. Mod. Phys. **65**, 803 (1993)
30. Y. Aharonov, S. Popescu, J. Tollaksen, L. Vaidman, Phys. Rev. A **79**, 052110 (2009)
31. N.W.M. Ritchie, J.G. Story, R.G. Hulet, Phys. Rev. Lett. **66**, 1107 (1991)
32. G.J. Pryde, J.L. O'Brien, A.G. White, T.C. Ralph, H.M. Wiseman, Phys. Rev. Lett. **94**, 220405 (2005)
33. O. Hosten, P. Kwiat, Science **319**, 787 (2008)
34. P.B. Dixon, D.J. Starling, A.N. Jordan, J.C. Howell, Phys. Rev. Lett. **102**, 173601 (2009)
35. P.B. Dixon, D.J. Starling, A.N. Jordan, J.C. Howell, Phys. Rev. A **80**, 041803(R) (2009)
36. Y. Kedem, Phys. Rev. A **85**, 060102 (2012)
37. A. Feizpour, X. Xingxing, A.M. Steinberg, Phys. Rev. Lett. **107**, 133603 (2011)
38. D.J. Starling, P.B. Dixon, N.S. Williams, A.N. Jordan, J.C. Howell, Phys. Rev. A **82**, 011802(R) (2010). arXiv:0912.2357
39. A.K. Pan, A. Matzkin, Phys. Rev. A **85**, 022122 (2012)
40. A.G. Kofman, S. Ashhab, F. Nori, Phys. Rep. **520** (2012)
41. P. Kwiat et al., Increase of signal-to-noise ratio in weak value measurements, in *Quantum Theory: A Two-Time Success Story. Yakir Aharonov Festschrift*, ed. by D. Struppa, J. Tollaksen (Springer, Milan, 2013). doi:10.1007/978-88-470-5217-8_26
42. J. Dressel et al., Unpublished
43. N. Brunner, A. Acin, D. Collins, N. Gisin, V. Scarani, Phys. Rev. Lett. **91**, 180402 (2003)
44. D.J. Starling, P.B. Dixon, A.N. Jordan, J.C. Howell, Phys. Rev. A **82**, 063822 (2010)
45. J. Tollaksen, Robust weak measurements on finite samples. J. Phys. Conf. Ser. **70**, 012015 (2007). quant-ph/0703038
46. H.F. Hofmann, Phys. Rev. A **83**, 022106 (2011)
47. E. Madelung, Z. Phys. **40**, 322–326 (1926)
48. E. Nelson, Phys. Rev. **150**, 1079 (1966)
49. J. Dressel, A.N. Jordan, Phys. Rev. Lett. **109**, 230402 (2012)

Chapter 17
Entanglement and Weak Values: A Quantum Miracle Cookbook

Alonso Botero

Abstract The concept of the weak value has proved to be a powerful and operationally grounded framework for the assignment of physical properties to a quantum system at any given time. More importantly, this framework has allowed us to identify a whole range of surprising quantum effects, or "miracles", which are readily testable but which lie buried "under the noise" when the results of measurements are not post-selected. In all cases, these miracles have to do with the fact that weak values can take values lying outside the conventional ranges of quantum expectation values. We explore the extent to which such miracles are possible within the weak value framework. As we show, given appropriate initial and final states, it is generally possible to produce any set of weak values that is consistent with the linearity of weak values, provided that the states are entangled states of the system with some external ancillary system. Through a simple constructive proof, we obtain a recipe for arbitrary quantum miracles, and give examples of some interesting applications. In particular, we show how the classical description of an infinitely-localized point in phase-space is contained in the weak-value framework augmented by quantum entanglement. [*Editor's note*: for a video of the talk given by Prof. Botero at the Aharonov-80 conference in 2012 at Chapman University, see quantum.chapman.edu/talk-27.]

17.1 Introduction

The Two-Vector Formulation of quantum mechanics, proposed by Yakir Aharonov and his collaborators [1, 2], is a description of a quantum system at a given time between two complete measurements. The "Elements of Reality" are the weak values of quantum-Mechanical observables:

$$A_w \equiv \frac{\langle\psi_f|A|\psi_i\rangle}{\langle\psi_f|\psi_i\rangle}. \tag{17.1}$$

A. Botero (✉)
Physics Department, Universidad de los Andes, Cra 1 No. 18A-12, Bogotá, Colombia
e-mail: abotero@uniandes.edu.co

D.C. Struppa, J.M. Tollaksen (eds.), *Quantum Theory: A Two-Time Success Story*,
DOI 10.1007/978-88-470-5217-8_17,

One appealing aspect in this description is its non-contextual nature. Namely, from the non-disturbance of weak measurements, weak values describe the results of weak measurements even when the observables are jointly measured (as long as they're measured weakly). Therefore, the joint assignment of weak values to a set of generally incompatible observables is operationally consistent in a sense that is not possible in the context of strong measurements.

Our main question is then: What are the constraints on the possible joint weak values of a set of observables $\{A^{(1)}, A^{(2)}, \dots A^{(m)}\}$? More generally, can one find initial and final states $|\psi_i\rangle$ and $|\psi_i\rangle$ such that a set of arbitrary complex numbers $\{z^{(1)}, z^{(2)}, \dots z^{(m)}\}$ can be interpreted as the weak values

$$z^{(k)} \equiv \frac{\langle\psi_f|A^{(k)}|\psi_i\rangle}{\langle\psi_f|\psi_i\rangle} \tag{17.2}$$

of the corresponding observables $A^{(1)}, A^{(2)}, \dots A^{(m)}$?

There are two rather obvious constraints that one can identify: First, we note that weak values are *linear*, so that $C = \sum_k c_k A^{(k)}$ implies $C_w = \sum_k c_k A_w^{(k)}$. We also note that the weak value of the identity operator is unity. Therefore, one must assume that for any set of observables $\{A^{(1)}, A^{(2)}, \dots A^{(m)}\}$ one wishes to assign arbitrary weak values to, we must demand that $\{\text{Id}, A^{(1)}, A^{(2)}, \dots A^{(m)}\}$ are linearly independent operators.

Let us start with the trivial example of assigning weak values to a single observable A. In this case, $A_w \equiv \frac{\langle\psi_f|A|\psi_i\rangle}{\langle\psi_f|\psi_i\rangle}$ can indeed take any complex value with appropriate initial and final states in the system Hilbert space. For example, suppose a and a' are different eigenvalues of $\hat{A}$ and $|a\rangle$ and $|a'\rangle$ corresponding eigenstates. Then, for the initial and final (unnormalized) states

$$|\psi_f\rangle = |a\rangle + |a'\rangle \tag{17.3}$$

$$|\psi_i\rangle = (a' - z)|a\rangle + (z - a)|a'\rangle, \tag{17.4}$$

where z can be any complex number, it is easily verified that

$$\frac{\langle\psi_f|A|\psi_i\rangle}{\langle\psi_f|\psi_i\rangle} = z. \tag{17.5}$$

For the case of two or more observables, to assign weak values to m linearly-independent observables,

$$z^{(k)} \equiv \frac{\langle\psi_f|A^{(k)}|\psi_i\rangle}{\langle\psi_f|\psi_i\rangle}, \quad k = 1, 2, \dots m, \tag{17.6}$$

it is convenient to use an identity due to Aharonov [1], namely that

$$A^{(k)}|\psi_i\rangle = \langle A^{(k)}\rangle_i|\psi_i\rangle + (\Delta A^{(k)})_i|\phi^{(k)}\rangle, \tag{17.7}$$

where $\langle\phi^{(k)}|\psi_i\rangle = 0$ and $(\Delta A^{(k)})_i$ is the uncertainty of $A^{(k)}$ for the state $|\psi_i\rangle$. Replacing in (17.6), we have

$$z^{(k)} = \langle A^{(k)} \rangle_i + (\Delta A^{(k)})_i \frac{\langle \psi_f | \phi^{(k)} \rangle}{\langle \psi_f | \psi_i \rangle}. \tag{17.8}$$

Thus, if the set of orthogonal states $\{|\phi^{(1)}\rangle \dots |\phi^{(m)}\rangle\}$ is linearly independent, their overlaps with $|\psi_f\rangle$ are fixed:

$$\langle \psi_f | \phi^{(k)} \rangle = \frac{z_k - \langle A^{(k)} \rangle_i}{(\Delta A^{(k)})_i} \langle \psi_f | \psi_i \rangle. \tag{17.9}$$

If the Hilbert space of the system is of dimension d, then we can at most fix $d-1$ of these overlaps. And thus we conclude that for a system with a d-dimensional Hilbert space, weak values can be arbitrarily fixed for at most $d-1$ linearly-independent observables (besides the identity) using initial and final vectors in the system Hilbert space.

It follows from the foregoing that the possible weak values that can be generated with initial and final states in the system Hilbert space are quite constrained. Consider, for instance the weak value of the spin operators for a spin-1/2 particle. We can define the weak spin vector between two strong spin measurements

$$S_\mu \equiv \frac{\langle \hat{n} | \sigma_\mu | \hat{m} \rangle}{\langle \hat{n} | \hat{m} \rangle}, \tag{17.10}$$

where the unit vectors $\hat{m}$ and $\hat{n}$ refer to the polarizations of the initial and final states respectively. It is not hard to work out an expression for $\mathbf{S}$, namely,

$$\mathbf{S} = \frac{\hat{n} + \hat{m}}{1 + \hat{n} \cdot \hat{m}} + i \frac{\hat{n} \times \hat{m}}{1 + \hat{n} \cdot \hat{m}}. \tag{17.11}$$

The possible weak spin vectors that can be generated by varying over all $\hat{m}$ and $\hat{n}$ are then constrained to have the form

$$\mathbf{S} = x\hat{u} + iy\hat{v}, \quad \text{with } x^2 - y^2 = 1,\ \hat{u} \cdot \hat{v} = 0. \tag{17.12}$$

These constraints become rather more complicated for larger dimensional Hilbert spaces. We can think of weak values as a linear map between the Lie algebra of $SU(d)$, where d is the dimension of the system Hilbert space, and $\mathbb{C}^n$. For a d-dimensional Hilbert space, we can define a hermitian traceless basis $E_1, E_2, \dots E_{d^2-1}$ of the $SU(d)$ Lie algebra with an inner product

$$\mathrm{Tr}(E_i E_j) = \delta_{ij}. \tag{17.13}$$

The generators obey the following algebraic relations under multiplication [3]

$$E_i E_j = \frac{\delta_{ij}}{d} + d_{ijk} E_k + i f_{ijk} E_k, \tag{17.14}$$

where the f_{ijk} are the $SU(d)$ structure constants and the d_{ijk} can be chosen to be symmetric in the three indices. Let us then define the operator

$$W = \frac{|\psi_i\rangle\langle\psi_f|}{\langle\psi_f|\psi_i\rangle} \tag{17.15}$$

so that weak values can be obtained in analogy with expectation values given a density matrix: $A_w = \mathrm{Tr}(AW)$. When expanded in the Lie algebra basis, as

$$W = \frac{1}{d} + \sum_{i=1}^{d^2-1} E_k z_k, \tag{17.16}$$

the coefficients z_k are then given by the weak values of the Lie algebra basis elements:

$$z_k \equiv \frac{\langle\psi_f|E_k|\psi_i\rangle}{\langle\psi_f|\psi_i\rangle}. \tag{17.17}$$

But now notice that W is idempotent. Using the expansion (17.16), the identity $W^2 = W$ yields two constraints: a quadratic constraint similar to the one for spin-1/2:

$$\sum_i z_i^2 = 1 - \frac{1}{d}, \tag{17.18}$$

together with a set of cubic constraints involving the symmetric d-tensor:

$$z_i = \sum_{j,k} d_{ijk} z_j z_k. \tag{17.19}$$

These constraints define a complicated d-dimensional submanifold of $\mathbb{C}^n$ in which all the weak values must reside for pure initial and final states in the system Hilbert space.

17.2 Entangled Two-Vectors

The are, however, more general sets of pure initial and final states that can be attached to the system, if one considers it as part of a larger system $S + S'$, where S' is some ancillary system. Such conditions define what is known as a *generalized Two-Vector* [2] for the system. Given initial and final states $|\psi_i\rangle$ and $|\psi_i\rangle$ on the combined Hilbert space $H_S \otimes H_{S'}$, weak values for the system of interest can be obtained from an operator

$$W_S = \mathrm{Tr}_{S'}\left(\frac{|\psi_i\rangle\langle\psi_f|}{\langle\psi_f|\psi_i\rangle}\right), \tag{17.20}$$

where $\mathrm{Tr}_{S'}$ denotes the partial trace over the ancilla. Agaian, we can expand W_S as

$$W = \frac{1}{d} + \sum_{i=1}^{d^2-1} E_k z_k, \tag{17.21}$$

where $z_k \equiv \frac{\langle\psi_f|E_k|\psi_i\rangle}{\langle\psi_f|\psi_i\rangle}$ as before. The difference with the previous situation is that $W_S^2 = W_S$ is no longer a constraint. In fact, an independent assignment of weak values to each of the basis operators is possible, if the Hilbert space dimension of S' is sufficiently large.

To see this, suppose that S' has Hilbert space dimension d or larger. Now take the initial and final unnormalized entangled states

$$|\psi_i\rangle = \sum_i |i\rangle|i\rangle \tag{17.22}$$

$$|\psi_f\rangle = \left(1 + d\sum_k z_k^* E_k\right)|\psi_i\rangle, \tag{17.23}$$

where the z_k are arbitrary complex numbers. Then, taking the partial trace with respect to the ancilla, we find

$$W_S = \frac{1}{d} + \sum_{i=1}^{d^2-1} E_k z_k. \tag{17.24}$$

Thus we have shown that there always exist combinations of initial and final states on a d^2 dimensional Hilbert space that realize any arbitrary assignment of the $d^2 - 1$ complex weak values of the Lie algebra basis elements.

How much entanglement is required? To define two-vector entanglement, we note that W_S admits the singular value decomposition:

$$W_S = \sum_j \lambda_j |\chi_j\rangle\langle\eta_j|, \tag{17.25}$$

where the λ_j are positive singular values and $\langle\eta_i|\eta_j\rangle = \langle\chi_i|\chi_j\rangle = \delta_{ij}$. From the condition $\mathrm{Tr}(W_S) = 1$, we have

$$\sum_j \lambda_j \langle\eta_j|\chi_j\rangle = 1. \tag{17.26}$$

The operator W_S can then be realized with a canonical pair of equally-entangled states:

$$|\psi_i\rangle = \sum_j \sqrt{\alpha_j}\,|\chi_j\rangle|j\rangle \tag{17.27}$$

$$|\psi_f\rangle = \sum_j \sqrt{\alpha_j}\,|\eta_j\rangle|j\rangle, \tag{17.28}$$

where

$$\alpha_j = \frac{\lambda_j}{\sum_j \lambda_j}. \tag{17.29}$$

We can then define the entanglement of W_S as the entanglement entropy of either of the canonical states, namely:

$$E(W_S) \equiv H[\{\alpha_i\}], \tag{17.30}$$

where $H[\{\alpha_i\}]$ is the Shannon entropy of the α_i coefficients.

Let us work out the entanglement in the general spin-1/2, which we can now describe by a weak spin vector

$$\mathbf{S} = x\hat{u} + iy\hat{v} \tag{17.31}$$

with arbitrary x, y, $\hat{u}$ and $\hat{v}$. In this case, the singular values are of the form

$$\lambda_\pm = \left[1 + x^2 + y^2 \pm 2x\sqrt{1 + y^2 \sin^2\theta}\right]^{1/2}, \tag{17.32}$$

where θ is the angle between the directions $\hat{u}$ and $\hat{v}$; we see that when $x^2 - y^2 = 1$ and $\theta = \pi/2$, one of the singular values vanishes, in which case no entanglement is needed. For simplicity, let us further concentrate on the case $y = 0$. Then, the required entanglement is easily found to be

$$E(W_S) = \begin{cases} H_2\left(\frac{1+x}{2}\right) & x \le 1, \\ H_2\left(\frac{1+x^{-1}}{2}\right) & x > 1, \end{cases} \tag{17.33}$$

where H_2 is the binary entropy. The canonical states for $x \le 1$ and $x \ge 1$ are easily obtained. For $x \le 1$, they are indeed the same states:

$$|\psi_i\rangle = |\psi_f\rangle = \sqrt{1+x}\,|\hat{u}+\rangle|+\rangle + \sqrt{1-x}\,|\hat{u}-\rangle|-\rangle, \tag{17.34}$$

whereas in the case $x > 1$,

$$|\psi_i\rangle = \sqrt{1+x^{-1}}\,|\hat{u}+\rangle|+\rangle + \sqrt{1-x^{-1}}\,|\hat{u}-\rangle|-\rangle, \tag{17.35}$$

$$|\psi_f\rangle = \sqrt{1+x^{-1}}\,|\hat{u}+\rangle|+\rangle - \sqrt{1-x^{-1}}\,|\hat{u}-\rangle|-\rangle. \tag{17.36}$$

Note that a simple 180° relative phase in the Schmidt components of one of the states is all that is needed to map x to $1/x$.

17.3 Applications

We have thus seen that with the use of entangled initial and final states, it is possible to "cook up" any set of joint weak values consistent with the linearity constraints. Let us then look at some of the quantum miracles that can be understood in terms of these weak value assignments using entangled two-vectors.

17.3.1 Sharper Exotic Weak Values

In a weak measurement of some observable A, the wave function describing the conditional outcome of the measuring device is given by the Fourier transform

$$\phi_m(p) \propto \int dq\,\langle\psi_f|e^{iAq}|\psi_i\rangle\phi_i(q)e^{ipq}, \tag{17.37}$$

where p is the pointer variable, q its conjugate, and $\phi_i(q)$ the initial device wave function in its conjugate representation. Roughly speaking, a weak measurement occurs when $\phi_i(q)$ is sufficiently narrow around $q = 0$ that one can approximate the exponential by its first order expansion in q. More generally, the spread in the outcome distribution of a weak measurement, around the weak value, is determined by the width of the interval in q, around $q = 0$, in which the approximation

$$\langle\psi_f|e^{iAq}|\psi_i\rangle \simeq \langle\psi_f|\psi_i\rangle \times e^{iA_wq} \tag{17.38}$$

holds. If the interval is of order Δq, we can then expect a spread of order $1/\Delta q$ in the pointer wave function around the weak value.

One situation in which this interval can be relatively wide is when for all powers of the observable A up to some maximum power m are linearly independent. Then, from our previous results, we know that there exist initial and final states such that

$$(A^k)_w = (A_w)^k, \quad \forall k = 1, 2, \ldots m, \tag{17.39}$$

thus ensuring that the approximation (17.38) holds to order q^m, and hence for a much wider range in q around $q = 0$, allowing for narrower pointer-variable states. This property could be useful in enhancing the precision of experiments that exploit the amplification properties of weak measurements [4–6]. The linear independence conditions are satisfied by the J_z operator in a spin j representation, for all powers of J_z up to $2j$, so the accuracy of the above approximation can be quite good for high values of j.

17.3.2 Mean King's Problem

In the so-called "Mean King's problem", originally proposed by Aharonov, Albert, and Vaidman [7], the goal is to ascertain the outcome of one of three possible projective measurements of the three orthogonal spin components, that could be performed on a spin-1/2 particle, supposing one is granted access to the particle both before and after the measurement. The solution to this problem can be recast as the solution to a problem of simultaneous weak value assignment. To see this note that the conditional probability of obtaining the outcomes $\pm$ in a spin component measurement along the direction $\hat{e}$, with given initial and final states, is

$$p(\pm) \propto \left|\langle\psi_f\left|(1 \pm \hat{e}\cdot\sigma)\right|\psi_i\rangle\right|^2. \tag{17.40}$$

This probability can be re-written as

$$p(\pm) \propto \left|1 \pm \hat{e}\cdot\mathbf{S}\right|^2, \tag{17.41}$$

where $\mathbf{S}$ is the vector of weak values introduced earlier. Definite conditional outcomes, that is, probabilities $p(\pm)$ that are either one or zero, are obtained when each of the three components of $\mathbf{S}$ is either $+1$ or -1. Given our earlier construction of states yielding specific weak values, we can suppose that the particle and some ancillary particle are prepared in the state $|\psi_i\rangle = |+\rangle|+\rangle + |-\rangle|-\rangle$, and the problem then is to find four vectors $\mathbf{S}_\alpha$ (with $\alpha = 1, 2, 3, 4$), with components that are either $+1$ or -1, and such that the corresponding final states

$$|\psi_\alpha\rangle = (1 + \mathbf{S}_\alpha \cdot \sigma)|\psi_i\rangle \tag{17.42}$$

are mutually orthogonal, so that they can be interpreted as the outcomes of a projective measurement of the particle and the ancilla. The orthogonality condition is met when the vectors $\mathbf{S}_\alpha$ lie at 120° from each other. Explicitly,

$$\mathbf{S}_1 = \hat{x} + \hat{y} + \hat{z} \tag{17.43}$$

$$\mathbf{S}_2 = -\hat{x} - \hat{y} + \hat{z} \tag{17.44}$$

$$\mathbf{S}_3 = -\hat{x} + \hat{y} - \hat{z} \tag{17.45}$$

$$\mathbf{S}_4 = \hat{x} - \hat{y} - \hat{z}, \tag{17.46}$$

where $\hat{x}$, $\hat{y}$ and $\hat{z}$ are unit vectors. Note that these spin vectors are real and of length $\sqrt{3}$, and can therefore only be realized with entangled states, according to our previous discussion.

17.3.3 Joint, Conditionally Sharp Outcomes of Canonical Variables

In the "Mean king's problem", the task is to ascertain the outcome of one of three possible intermediate projective measurements that can be performed at the exclusion of the other two. Surprisingly, one can also cook up situations where one can ascertain the simultaneous outcomes of joint strong measurements of incompatible observables [8]. To see this, one must first define what is meant by a joint measurement of two observables A_1 and A_2. In analogy with our previous discussion, one can imagine coupling the system impulsively to two measuring devices, with pointer variables p_1 and p_2 and canonical conjugates q_1 and q_2. The coupling may be described by the effective evolution operator

$$U_{meas} = \exp\left[i(q_1 A_1 + q_2 A_2)\right], \tag{17.47}$$

which is the natural generalization for the von Neumann scheme interaction operator. If the system is pre- and post-selected, the conditional wave function of the device pointer variables is given by

$$\phi_m(p_1, p_2) \propto \int dq_1 dq_2 \langle\psi_f| \exp\left[i(q_1 A_1 + q_2 A_2)\right] |\psi_i\rangle \phi_i(q_1, q_2) e^{i\mathbf{q}\cdot\mathbf{p}}, \tag{17.48}$$

where for a strong measurement the initial device wave function $\phi_i(q_1, q_2)$ should be understood to be very wide (ideally a constant for an infinitely sharp measurement). The problem can now be cast as follows: Can we find non trivial examples of initial and final states and non-commuting A_1 and A_2, such that for all values of q_1 and q_2,

$$\langle\psi_f|\exp\left[i(q_1A_1+q_2A_2)\right]|\psi_i\rangle=\langle\psi_f|\psi_i\rangle\exp\left[i(q_1a_1+q_2a_2)\right], \quad (17.49)$$

where a_1 and a_2 are some c-numbers? If such is the case, then the conditional effect on the pointers will simply be the shift of the pointer wave function

$$\phi_i(p_1,p_2)\to\phi_m(p_1,p_2)=\phi_i(p_1-a_1,p_2-a_2), \quad (17.50)$$

and the conditional results of the measurement can be made arbitrarily sharp with a correspondingly sharp initial pointer wave function $\phi_i(p_1,p_2)$. Knowing the initial and final states, one should then be able to ascertain, precisely, the simultaneous outcomes of an intermediate joint measurement.

Surprisingly, there exist non-trivial solutions to the above problem, and as expected, the solution involves entangled initial and final states. For the operators A_1 and A_2, take the canonical variables X and P of a one-dimensional canonical degree of freedom. Consider also an ancillary canonical system with canonical variables X' and P'. Using these, we can define the collective variables X_+, P_+, X_- and P_- according to:

$$X_\pm=\frac{1}{\sqrt{2}}\left(X\pm X'\right),\qquad P_\pm=\frac{1}{\sqrt{2}}\left(P\pm P'\right). \quad (17.51)$$

Now we note that the linear combination q_1X+q_2P can be written in terms of the collective variables as

$$q_1X+q_2P=\frac{1}{\sqrt{2}}(q_1X_++q_2P_-)+\frac{1}{\sqrt{2}}(q_1X_-+q_2P_+). \quad (17.52)$$

As one easily verifies, the two operators

$$L\equiv\frac{1}{\sqrt{2}}(q_1X_++q_2P_-),\quad\text{and}\quad R\equiv\frac{1}{\sqrt{2}}(q_1X_-+q_2P_+) \quad (17.53)$$

commute. Thus one can perform the factorization

$$\exp\left[i(q_1X+q_2P)\right]=\exp(iL)\exp(iR). \quad (17.54)$$

Now we note that the state $|\tilde{X}_-,\tilde{P}_+\rangle$, a simultaneous eigenstate of X_- and P_+ with eigenvalues $\tilde{X}_-$,$\tilde{P}_+$ respectively, is also an eigenstate of R with eigenvalue $(q_1\tilde{X}_-+q_2\tilde{P}_+)/\sqrt{2}$. Similarly, the state $|\tilde{X}_+,\tilde{P}_-\rangle$, a simultaneous eigenstate of X_+ and P_- with eigenvalues $\tilde{X}_+$,$\tilde{P}_-$, is an eigenstate of L with eigenvalue $(q_1\tilde{X}_++q_2\tilde{P}_-)/\sqrt{2}$. Hence, we have that

$$\langle\tilde{X}_+,\tilde{P}_-|\exp\left[i(q_1X+q_2P)\right]|\tilde{X}_-,\tilde{P}_+\rangle=\langle\tilde{X}_+,\tilde{P}_-|\tilde{X}_-,\tilde{P}_+\rangle\exp\left[i(q_1\tilde{X}+q_2\tilde{P})\right], \quad (17.55)$$

where $\tilde{X}$ and $\tilde{P}$ are the c-numbers

$$\tilde{X}=\frac{1}{\sqrt{2}}\left(\tilde{X}_++\tilde{X}_-\right),\qquad\tilde{P}=\frac{1}{\sqrt{2}}\left(\tilde{P}_++\tilde{P}_-\right). \quad (17.56)$$

So indeed we find conditions under which the outcome of a joint simultaneous measurement of the two canonical variables can be inferred precisely if we are granted access to the system before and after the measurement.

There is also something remarkable about these initial and final conditions, when we consider instead intermediate weak measurements of arbitrary functions of X and P with the same boundary conditions. A correspondence between classical functions in phase-space and quantum operators can be established via the so-called Weyl Kernel [9]:

$$\hat{f} = \int dXdP\ \hat{\Delta}(X, P) f_{clas}(X, P), \tag{17.57}$$

where

$$\hat{\Delta}(X, P) = \frac{1}{(2\pi)^2} \int d^2q \exp\bigl[iq_1(\hat{X} - X) + iq_2(\hat{P} - P)\bigr], \tag{17.58}$$

and we use carets to denote quantum operators. With given initial and final states, the function

$$W_{if}(X, P) \equiv \frac{\langle\psi_f|\hat{\Delta}(X, P)|\psi_i\rangle}{\langle\psi_f|\psi_i\rangle} \tag{17.59}$$

is a "weak" generalization of the well-known Wigner function, and becomes the actual Wigner function when the initial and final states coincide. Evaluating $W_{if}(X, P)$ with the initial and final states $|\tilde{X}_-, \tilde{P}_+\rangle$ and $|\tilde{X}_+, \tilde{P}_-\rangle$ of our previous discussion, we readily find that

$$W_{if}(X, P) = \delta(X - \tilde{X})\delta(P - \tilde{P}), \tag{17.60}$$

where $\tilde{X}$ and $\tilde{p}$ are as defined in Eq. (17.56). Thus we can produce a "weak Wigner function" that is a delta-function in phase space; consequently, the weak value of any quantum operator that is a function of the canonical variables will be the corresponding classical function, evaluated at the phase-space point $(\tilde{X}, \tilde{P})$! In other words, quantum entanglement allows us to reproduce, in terms of weak values, the same observable results that would obtain for a classical system that is sharply localized in phase-space. This, indeed, is a remarkable quantum miracle!

17.4 Conclusion

We owe it to Yakir Aharonov to have shown us that buried under the noise of the standard measurement results of quantum mechanics, lies a wealth of surprising effects that only emerge when the ensembles are pre- and post-selected. In this contribution, we have tried to show just how rich the range of possible effects can be when the pre-and post-selections involve entangled states; namely, any assignment of weak values that is compatible with the linearity constraint is possible with suitable pre- and post-selections involving entangled states.

References

1. Y. Aharonov, L. Vaidman, Phys. Rev. A **41**, 11 (1990)
2. L. Vaidman, in *Compendium of Quantum Physics: Concepts, Experiments, History and Philosophy*, ed. by D. Greenberger, K. Hentschel, F. Weinert (Springer, Berlin, 2009), p. 802
3. W. Greiner, B. Müller, *Quantum Mechanics, Vol. II* (Springer, Berlin, 1989)
4. Y. Aharonov, D.Z. Albert, L. Vaidman, Phys. Rev. Lett. **60**, 1351 (1988)
5. O. Hosten, P.G. Kwiat, Science **319**, 787 (2008)
6. P. Dixon, D.J. Starling, A.N. Jordan, J.C. Howell, Phys. Rev. Lett. **102**, 173601 (2009)
7. L. Vaidman, Y. Aharonov, D.Z. Albert, Phys. Rev. Lett. **58**, 1385–1387 (1987)
8. A. Botero, Phys. Lett. A **374**, 823 (2010)
9. A.M. Ozorio de Almeida, Phys. Rep. **295**, 265 (1998)

Chapter 18
Weak Energy: Form and Function

Allen D. Parks

Abstract The equation of motion for a time-dependent weak value of a quantum mechanical observable contains a complex valued energy factor—the weak energy of evolution. This quantity is defined by the dynamics of the pre-selected and post-selected states which specify the observable's weak value. It is shown that this energy: (i) is manifested as dynamical and geometric phases that govern the evolution of the weak value during the measurement process; (ii) satisfies the Euler-Lagrange equations when expressed in terms of Pancharatnam (P) phase and Fubini-Study (FS) metric distance; (iii) provides for a PFS stationary action principle for quantum state evolution; (iv) time translates correlation amplitudes; (v) generalizes the temporal persistence of state normalization; and (vi) obeys a time-energy uncertainty relation. A similar complex valued quantity—the pointed weak energy of an evolving quantum state—is also defined and several of its properties in PFS coordinates are discussed. It is shown that the imaginary part of the pointed weak energy governs the state's survival probability and its real part is—to within a sign—the Mukunda-Simon geometric phase for arbitrary evolutions or the Aharonov-Anandan (AA) geometric phase for cyclic evolutions. Pointed weak energy gauge transformations and the PFS 1-form are defined and discussed and the relationship between the PFS 1-form and the AA connection 1-form is established. [*Editors note*: for a video of the talk given by Prof. Parks at the Aharonov-80 conference in 2012 at Chapman University, see quantum.chapman.edu/talk-25.]

18.1 Preamble

It is an honor and a pleasure to speak at Yakir's 80th birthday conference. Although I first met Yakir about ten years ago, I have been a disciple of his work for many more years than that. I commemorate Yakir's birthday today by reviewing certain aspects of my research which have been inspired by Yakir's insights into foundational quantum physics via weak measurements and weak value theory. In particu-

A.D. Parks (✉)
Electromagnetic and Sensor Systems Department, Naval Surface Warfare Center, Dahlgren, VA 22448, USA
e-mail: danparks649@gmail.com

D.C. Struppa, J.M. Tollaksen (eds.), *Quantum Theory: A Two-Time Success Story*,
DOI 10.1007/978-88-470-5217-8_18, © Springer-Verlag Italia 2014

lar, I will discuss the notion of weak energy—its relationships to both weak value measurements and quantum state evolution, as well as some of the formalism associated with it when it is expressed in terms of Pancharatnam phase and Fubini-Study metric distance.

18.2 Weak Measurements and Weak Values

The *weak value* A_w of a quantum mechanical observable A was introduced by Aharonov *et al.* a quarter century ago [1–3]. It is the statistical result of a standard measurement procedure performed upon a pre-selected and post-selected (PPS) ensemble of quantum systems when the interaction between the measurement apparatus and each system is sufficiently weak, i.e. when it is a *weak measurement*. Unlike a strong measurement of A which significantly disturbs the measured system (i.e., it "collapses" the wavefunction), a weak measurement of A does not appreciably disturb the quantum system and yields A_w as the observable's measured value. The peculiar nature of the virtually undisturbed quantum reality that exists between the boundaries defined by the PPS states is revealed by the eccentric characteristics of A_w, namely that A_w is complex valued and that the real $\mathrm{Re}A_w$ and imaginary $\mathrm{Im}A_w$ parts of A_w can be extremely large and lie far outside the eigenvalue spectral limits of $\hat{A}$. Although the interpretation of weak values remains somewhat controversial, experiments have verified several of the interesting unusual properties predicted by weak value theory [4–10].

Weak values arise in the von Neumann description of a quantum measurement at time t_0 of a time independent observable A that describes a quantum system in an initial pre-selected state $|\psi_i\rangle$ at t_0 and a final post-selected state $|\psi_f\rangle$ at t_0. If the initial normalized measurement pointer state is $|\phi\rangle$, then the pointer state immediately after the measurement is

$$|\Psi\rangle = \langle\psi_f|e^{-\frac{i}{\hbar}\gamma\hat{A}\hat{p}}|\psi_i\rangle|\phi\rangle,$$

where γ is the measurement interaction strength and $\hat{p}$ is the pointer momentum operator conjugate to the pointer position operator $\hat{q}$. When the measurement is weak, i.e., when γ is sufficiently small and the pointer uncertainty Δq is much larger than $\hat{A}$'s eigenvalue separation (qualifications required for weak measurements are discussed in [5, 11]), then the last equation becomes

$$|\Psi\rangle \approx \langle\psi_f|\psi_i\rangle\hat{S}(\gamma A_w)|\phi\rangle.$$

Here

$$A_w = \frac{\langle\psi_f|\hat{A}|\psi_i\rangle}{\langle\psi_f|\psi_i\rangle}, \quad \langle\psi_f|\psi_i\rangle \neq 0, \tag{18.1}$$

is the weak value of observable A and $\hat{S}(\gamma A_w) = e^{-\frac{i}{\hbar}\gamma A_w\hat{p}}$ is the translation operator for $|\phi\rangle$ defined by the action $\langle q|\hat{S}(\gamma A_w)|\phi\rangle = \phi(q - \gamma\mathrm{Re}A_w)$. Assuming that

$\phi(q)$ is real valued, then after the measurement the new mean pointer position and momentum are [12]

$$\langle\Psi|\hat{q}|\Psi\rangle = \langle\phi|\hat{q}|\phi\rangle + \gamma \mathrm{Re} A_w$$

and

$$\langle\Psi|\hat{p}|\Psi\rangle = \langle\phi|\hat{p}|\phi\rangle + 2\gamma\nu(p)\mathrm{Im} A_w,$$

where $\nu(p)$ is the variance of the initial pointer momentum.

18.3 Weak Energy of Evolution

A fundamental aspect of weak value theory is the tenet that although the measurement of $\hat{A}$ occurs at time t_0, the PPS states appearing in (18.1) are actually pre-selected and post-selected at times $t_i < t_0$ and $t_f > t_0$, respectively. PPS states selected at these times define past and future boundary conditions which influence A_w at measurement time t_0 via their unitary evolutions forward in time from t_i to t_0 and backward in time from t_f to t_0.

Let $T = [t_1, t_2]$ be a fixed closed time interval, Δt_i and Δt_f be fixed time intervals, and $\mathscr{A} = \{A_w(t) : t \in T\}$ be a time ordered set, where (i) $A_w(t)$ is the weak value of $\hat{A}$ at a measurement interaction time $t \in T$ defined by a state $|\psi_i(t_i)\rangle$ which has been pre-selected at time $t_i = t - \Delta t_i$ and by a state $|\psi_f(t_f)\rangle$ which has been post-selected at time $t_f = t + \Delta t_f$; and (ii) these PPS states continuously change from their initial states at t_i and t_f according to

$$\frac{d|\psi_i(t_i)\rangle}{dt_i} = -\frac{i}{\hbar}\hat{H}_i|\psi_i(t_i)\rangle, \quad t_i \in [t_1 - \Delta t_i, t_2 - \Delta t_i],$$

and

$$\frac{d|\psi_f(t_f)\rangle}{dt_f} = -\frac{i}{\hbar}\hat{H}_f|\psi_f(t_f)\rangle, \quad t_f \in [t_1 + \Delta t_f, t_2 + \Delta t_f].$$

When the Hamiltonians $\hat{H}_i$ and $\hat{H}_f$ are non-vanishing and explicitly time independent, then the evolutions of these PPS states to the time of measurement t are

$$|\psi_i(t)\rangle = e^{-\frac{i}{\hbar}\hat{H}_i\Delta t_i}|\psi_i(t_i)\rangle \equiv \hat{U}|\psi_i(t_i)\rangle$$

and

$$|\psi_f(t)\rangle = e^{\frac{i}{\hbar}\hat{H}_f\Delta t_f}|\psi_f(t_f)\rangle \equiv \hat{V}|\psi_f(t_f)\rangle$$

so that

$$A_w(t) = \frac{\langle\psi_f(t_f)|\hat{V}^\dagger\hat{A}\hat{U}|\psi_i(t_i)\rangle}{\langle\psi_f(t_f)|\hat{V}^\dagger\hat{U}|\psi_i(t_i)\rangle} = \frac{\langle\psi_f(t)|\hat{A}|\psi_i(t)\rangle}{\langle\psi_f(t)|\psi_i(t)\rangle}, \quad \langle\psi_f(t)|\psi_i(t)\rangle \neq 0. \tag{18.2}$$

Observe that since $[\hat{U}, \hat{H}_i] = 0 = [\hat{V}, \hat{H}_f]$, then the actions of $\hat{H}_i$ and $\hat{H}_f$ upon the PPS states at t_i and t_f are transformed into their actions upon the evolved PPS states at the measurement time t.

If $\dot{A}_w$ exists at each $t \in T$, then A_w is a continuous function over T and the equation of motion for $A_w(t)$ is obtained by taking the time derivative of (18.2):

$$\dot{A}_w = \frac{i}{\hbar}\{(H_f A - A H_i)_w - A_w (H_f - H_i)_w\}. \tag{18.3}$$

Here it is assumed that $\hat{A}$ is explicitly time independent. The peculiar complex valued factor $(H_f - H_i)_w$ appearing in the second term of (18.3) is the *weak energy of evolution* for the PPS system. This quantity is an artifact of the dynamics of the PPS states and is contemporaneous with the measurement time t.

Although $(H_f - H_i)_w$ is not directly measured during the measurement of $A_w(t)$, it defines phases and phase factors which are crucial to determining $A_w(t)$ at the measurement time. To see this, consider the general solution to (18.3) given by

$$A_w = e^{-\frac{i}{\hbar}\int_{t_1}^{t}(H_f - H_i)_w dt'}\left\{A_w(t_1) + \frac{i}{\hbar}\int_{t_1}^{t} e^{\frac{i}{\hbar}\int_{t_1}^{t'}(H_f - H_i)_w dt''}(H_f A - A H_i)_w dt'\right\}.$$

This solution clearly shows that the time integrated weak energy of evolution is an intrinsic attribute of $A_w(t)$ and that it determines and influences $A_w(t)$ through phase factors which have been introduced by the integrating factor $e^{\frac{i}{\hbar}\int_{t_1}^{t}(H_f - H_i)_w dt'}$.

The argument of this integrating factor is the phase sum [13]

$$\frac{1}{\hbar}\int_{t_1}^{t}(H_f - H_i)_w dt' = \delta_f(t) - \delta_i(t) + \beta_f(t) - \beta_i(t),$$

where

$$\delta_x(t) \equiv \frac{1}{\hbar}\int_{t_1}^{t}\langle\psi_x(t')|\hat{H}_x|\psi_x(t')\rangle dt', \quad x \in \{i, f\},$$

are real valued dynamical phases and

$$\beta_f(t) \equiv \frac{1}{\hbar}\int_{t_1}^{t}\Delta H_f \frac{\langle\psi_f^{\perp}(t')|\psi_i(t')\rangle}{\langle\psi_f(t')|\psi_i(t')\rangle}dt'$$

and

$$\beta_i(t) \equiv \frac{1}{\hbar}\int_{t_1}^{t}\Delta H_i \frac{\langle\psi_f(t')|\psi_i^{\perp}(t')\rangle}{\langle\psi_f(t')|\psi_i(t')\rangle}dt'$$

are complex valued geometric phases. Here, $|\psi_x^{\perp}(t)\rangle$ belongs to the Hilbert subspace which is orthogonal to the subspace containing $|\psi_x(t)\rangle, x \in \{i, f\}$, and satisfies $\langle\psi_x^{\perp}(t)|\psi_x(t)\rangle = 0$, as well as the energy uncertainty condition $\Delta H_x = \langle\psi_x^{\perp}(t)|\hat{H}_x|\psi_x(t)\rangle$.

That β_f is a geometric phase follows from the facts that β_f is invariant under (i) local U(1) gauge transformations $|\psi_x(t')\rangle \to e^{i\theta(t')}|\psi_x(t')\rangle, x \in \{i, f\}$, (which implies $|\psi_f^{\perp}(t')\rangle \to e^{i\theta(t')}|\psi_f^{\perp}(t')\rangle$) and (ii) the reparameterization $|\psi_f(t')\rangle =$

$|\psi'_f(\tau(t'))\rangle \equiv |\psi'_f(\tau)\rangle$ (which implies $|\psi_f^{\perp}(t')\rangle = |\psi_f^{\perp\prime}(\tau)\rangle$). Here $\tau(t')$ is monotone increasing over the interval $[\tau(t_1), \tau(t)]$ with endpoints $|\psi'_f(\tau(t_1))\rangle = |\psi_f(t_1)\rangle$ and $|\psi'_f(\tau(t))\rangle = |\psi_f(t)\rangle$ [14]. Similar arguments hold for β_i.

Thus, it may be concluded that *the time accumulation of the forward and backward time evolutions to the measurement time t of the actions of $\hat{H}_i$ and $\hat{H}_f$ upon the associated PPS states is physically manifested at t as the sum $\delta_f(t) - \delta_i(t) + \beta_f(t) - \beta_i(t)$ of dynamical and geometric phases which determines and influences A_w at t via the associated exponential phase factors.*

18.4 The Weak Energy of Evolution Stationary Action Principle

Let $\mathscr{H}$ be $A_w(t)$'s Hilbert space and $\mathscr{P}$ be the associated projective space consisting of all the rays of $\mathscr{H}$. Recall that a ray is an equivalence class $[\psi]$ of states $|\psi\rangle$ in $\mathscr{H}$ which differ only in phase. If $\Pi : \mathscr{H} \to \mathscr{P}$ is the projection map $\Pi(|\psi\rangle) = [\psi]$, then the evolutions of the PPS states which define $A_w(t)$ at any time $t \in T$ are represented by the two curves $|\psi_i(t)\rangle$ and $|\psi_f(t)\rangle$ in $\mathscr{H}$ such that their projections in $\mathscr{P}$ at any $t \in T$ are separated by the Fubini-Study (FS) metric distance defined by [15, 16]

$$s^2 \equiv s^2(t) = 4\big(1 - |\langle\psi_f(t)|\psi_i(t)\rangle|^2\big). \tag{18.4}$$

Also, if at any $t \in T$ the state $|\psi_f(t)\rangle$ is parallel transported along the unique path in $\mathscr{H}$ that is the pre-image under Π of the shortest geodesic joining $[\psi_f(t)]$ and $[\psi_i(t)]$ in $\mathscr{P}$, then the associated Pancharatnam(P) phase χ is given by [17]

$$\chi \equiv \chi(t) = \arg\frac{\langle\psi_f(t)|\psi_i(t)\rangle}{|\langle\psi_f(t)|\psi_i(t)\rangle|}. \tag{18.5}$$

Rearrangement of the time derivatives of (18.4) and (18.5) reveals that

$$\mathrm{Re}(H_f - H_i)_w = \hbar\dot{\chi}$$

and

$$\mathrm{Im}(H_f - H_i)_w = \hbar\left(\frac{s}{\sqrt{4 - s^2}}\right)\dot{s}.$$

This yields the following important identity which expresses the weak energy of evolution in terms of PFS coordinates [18]:

$$\mathscr{L}(s; \dot{s}, \dot{\chi}) \equiv (H_f - H_i)_w = \hbar\dot{\chi} + i\hbar\left(\frac{s}{\sqrt{4 - s^2}}\right)\dot{s}.$$

Since $\mathscr{L} \equiv \mathscr{L}(s; \dot{s}, \dot{\chi})$ satisfies the Euler-Lagrange equations

$$\frac{d}{dt}\left(\frac{\partial \mathscr{L}}{\partial \dot{x}}\right) = \frac{\partial \mathscr{L}}{\partial x}, \quad x \in \{\chi, s\}, \tag{18.6}$$

it is called the *PFS Lagrangian* and the first variation of the *weak energy of evolution action* $\int_{t_1}^{t_2} \mathcal{L} dt$ vanishes, i.e.

$$\delta \int_{t_1}^{t_2} \mathcal{L} dt = 0.$$

Thus, the following *weak energy of evolution stationary action principle* may be stated [19]:

The actual paths followed in $\mathcal{H}$ by $|\psi_f(t)\rangle$ and $|\psi_i(t)\rangle$ between their endpoint states at t_1 and t_2 during the closed time interval $[t_1, t_2]$ are such that the weak energy of evolution action is stationary for all variations of χ, s, and time that vanish at the interval endpoints t_1 and t_2.

18.5 Some Additional Properties of $\mathcal{L}$

It is easily determined that the equation of motion for the PPS correlation amplitude

$$\varphi_t \equiv \langle \psi_f(t) | \psi_i(t) \rangle, \quad t \in T = [t_1, t_2],$$

is given by

$$\dot{\varphi}_t = \frac{i}{\hbar} \mathcal{L} \varphi_t$$

and that it has as its solution

$$\varphi_t = e^{\frac{i}{\hbar} \int_{t_1}^{t} \mathcal{L} dt'} \varphi_{t_1}. \tag{18.7}$$

Thus, $\mathcal{L}$ defines a complex valued exponential multiplication factor which translates correlation amplitudes in time by capturing and transferring from one amplitude to another the essence of the state dynamics in $\mathcal{H}$ via the associated changes in $\mathcal{P}$ and the phase acquired from parallel transport in $\mathcal{H}$.

The transition probability at $t \in T$ associated with the PPS states is the square modulus of (18.7). This yields the identity

$$|\varphi_t|^2 = e^{-\frac{2}{\hbar} \int_{t_1}^{t} \mathrm{Im} \mathcal{L} dt'} |\varphi_{t_1}|^2$$

which shows that $\mathrm{Im}\mathcal{L}$ defines an exponential multiplication factor which time translates correlation probabilities and generalizes the temporal persistence of state normalization (because when $i = f$, then $s = 0$ and $|\varphi_t|^2 = |\varphi_{t_1}|^2$).

Consider the equation of motion for $\mathcal{L}$ by setting $\hat{A} = \hat{H}_f - \hat{H}_i$ in (18.3). Then

$$\dot{\mathcal{L}} = \frac{i}{\hbar} \{ (H_f^2 - 2 H_f H_i + H_i^2)_w - \mathcal{L}^2 \}.$$

When this derivative vanishes, then $\mathcal{L}$ is a constant of the motion and it serves as a good weak quantum number for the associated PPS system. If τ is the characteristic time needed for $\mathcal{L}$ to be changed by an amount equal to the width of its statistical distribution $\Delta_w \mathcal{L} \equiv |\Delta_w^2 \mathcal{L}|^{\frac{1}{2}}$, i.e. the weak energy uncertainty [3], then

$$\tau \approx \frac{\Delta_w \mathscr{L}}{|\dot{\mathscr{L}}|},$$

where $\Delta_w^2 \mathscr{L} = ((\hat{H}_f - \hat{H}_i)^2)_w - ((H_f - H_i)_w)^2$ is the weak variance of $\mathscr{L}$. This indicates that appreciable differences between correlation amplitudes should be expected only for times with differences significantly greater than τ (note that $\tau \to \infty$ when $\mathscr{L}$ is a good weak quantum number).

For the special case that $\hat{H}_i$ and $\hat{H}_f$ are mutual constants of the motion, i.e. $[\hat{H}_i, \hat{H}_f] = 0$ and $\frac{d\hat{H}_i}{dt} = 0 = \frac{d\hat{H}_f}{dt}$, then

$$\dot{\mathscr{L}} = \frac{i}{\hbar} \Delta_w^2 \mathscr{L}, \tag{18.8}$$

and changes in $\mathscr{L}$ are precisely due to its weak variance (clearly, in this case $\mathscr{L}$ is a constant of the motion when $\Delta_w^2 \mathscr{L} = 0$). Since

$$\Delta_w \mathscr{L} = \hbar^{\frac{1}{2}} |\dot{\mathscr{L}}|^{\frac{1}{2}}, \tag{18.9}$$

then the following associated time-energy uncertainty relation for $\mathscr{L}$ is readily obtained from the substitution of the square of (18.9) into the previous expression for τ:

$$\tau \Delta_w \mathscr{L} \approx \hbar$$

As a final point of interest, note from (18.6) that the generalized momentum p_χ that is conjugate to the coordinate χ is a constant of the motion for any PPS correlation amplitude. More specifically, $\dot{p}_\chi = 0$ since

$$p_\chi \equiv \frac{\partial \mathscr{L}}{\partial \dot{\chi}} = \hbar.$$

This suggests the following novel definition for Planck's constant:

$$h \equiv 2\pi \frac{\partial \mathscr{L}}{\partial \dot{\chi}},$$

where $\mathscr{L}$ is the PFS Lagrangian associated with any time dependent PPS correlation amplitude $\varphi_t, t \in T$.

18.6 Pointed Weak Energy, Pointed Probability Current, and Quantum Geometric Phase

Let $|\psi(t)\rangle$ be a state which is evolving in $\mathscr{H}$ under the action of the Hamiltonian $\hat{H}$. If $|\psi(0)\rangle$ and $|\psi(t)\rangle$ are used as the initial and fixed final states, respectively, and $\hat{H}_f - \hat{H}_i$ is replaced by $\hat{H}$, then the above development—with slight modifications—generally applies for this special case, yielding

$$\varphi_{t,0} \equiv \langle\psi(t)|\psi(0)\rangle = e^{\frac{i}{\hbar}\int_0^t \mathscr{L}_0 dt'} = \frac{1}{2}\sqrt{4 - s_0^2}\, e^{i\chi_0}, \quad \langle\psi(t)|\psi(0)\rangle \neq 0, \tag{18.10}$$

as the *pointed correlation amplitude* and

$$\mathscr{L}_0 \equiv \mathscr{L}_0(s_0; \dot{\chi}_0, \dot{s}_0) \equiv \frac{\langle\psi(t)|\hat{H}|\psi(0)\rangle}{\langle\psi(t)|\psi(0)\rangle} = \hbar\dot{\chi}_0 + i\hbar(\frac{s_0}{4-s_0^2})\dot{s}_0 = p_{\chi_0}\dot{\chi}_0 + p_{s_0}\dot{s}_0$$

as the *pointed weak energy* which is also a Lagrangian. Here use is made of the fact that $p_x = \frac{\partial \mathscr{L}_0}{\partial \dot{x}}, x \in \{\chi_0, s_0\}$.

From (18.10) it is seen that the *pointed correlation probability* Pr(s) is the survival probability for the initial state $|\psi(0)\rangle$ given by

$$Pr(s_0) \equiv \varphi_{t,0}^*\varphi_{t,0} = \frac{1}{4}(4 - s_0^2) \tag{18.11}$$

(so that $\varphi_{t,0} = \sqrt{Pr(s_0)}e^{i\chi_0}$) and that it defines the associated *pointed probability current* C_0 as

$$C_0 \equiv \frac{dPr(s_0)}{dt} = -\frac{1}{2}s_0\dot{s}_0.$$

It is clear from this that since C_0 satisfies the Euler-Lagrange equation

$$\frac{d}{dt}\left(\frac{\partial C_0}{\partial \dot{s}_0}\right) = \frac{\partial C_0}{\partial s_0}$$

the first variation δJ_0 of the *pointed probability current action* $J_0 \equiv \int_0^{t_2} C_0 dt$ vanishes. This is formalized as *the pointed probability current stationary action principle* which states that [20]: *The actual evolutionary path followed in $\mathscr{H}$ by the state $|\psi(t)\rangle$ between the end points $|\psi(0)\rangle$ and $|\psi(t_2)\rangle$ is such that J_0 is stationary for all variations in s_0 and time that vanish at the endpoints.*

The action of $\hat{H}$ upon $|\psi(t)\rangle$ can be uniquely written as [3]

$$\hat{H}|\psi(t)\rangle = \langle H\rangle|\psi(t)\rangle + \Delta H|\psi^{\perp}(t)\rangle, \tag{18.12}$$

where $\langle H\rangle = \langle\psi(t)|\hat{H}|\psi(t)\rangle$, $\Delta H = \sqrt{\langle H^2\rangle - \langle H\rangle^2}$, and $|\psi^{\perp}(t)\rangle$ satisfies the conditions $\langle\psi^{\perp}(t)|\psi(t)\rangle = 0$ and $\Delta H = \langle\psi^{\perp}(t)|\hat{H}|\psi(t)\rangle$. The following equivalent definition for the pointed weak energy is obtained when the dual form of (18.12) is first used to form the scalar product with the state $|\psi(0)\rangle$ and then this product is divided by $\langle\psi(t)|\psi(0)\rangle \neq 0$:

$$\mathscr{L}_0 \equiv \langle H\rangle + \Delta H\frac{\langle\psi^{\perp}(t)|\psi(0)\rangle}{\langle\psi(t)|\psi(0)\rangle}.$$

In this case

$$\varphi_{t,0} = e^{\frac{i}{\hbar}\int_0^t \mathscr{L}_0 dt'} = e^{i(\delta_0+\beta_0)},$$

where

$$\delta_0 \equiv \frac{1}{\hbar}\int_0^t \langle H\rangle dt'$$

is a real valued dynamical phase and

$$\beta_0 \equiv \frac{1}{\hbar}\int_0^t \Delta H \frac{\langle \psi^{\perp}(t')|\psi(0)\rangle}{\langle \psi(t')|\psi(0)\rangle} dt'$$

defines the complex valued *pointed phase* acquired by the system as a result of the evolution of $|\psi(t')\rangle$ over the interval $[0, t]$.

The pointed phase is invariant under the local U(1) gauge transformation $|\psi(t)\rangle \to e^{i\theta(t)}|\psi(t)\rangle$ (which implies that $|\psi^{\perp}(t)\rangle \to e^{i\theta(t)}|\psi^{\perp}(t)\rangle$), as well as under the reparameterization $|\psi(t)\rangle = |\psi'(t'(t))\rangle$ (which implies that $|\psi^{\perp}(t)\rangle = |\psi^{\perp\prime}(t'(t))\rangle$) over the interval $[t'(0), t'(t)]$. Here $t'(t)$ is monotone increasing with state endpoints $|\psi'(t'(0))\rangle = |\psi(0)\rangle$ and $|\psi'(t'(t))\rangle = |\psi(t)\rangle$ [14]. These two invariance properties imply that β_0 is a geometric phase [21]—the *pointed geometric phase*.

Since δ_0 is real valued, $\mathscr{L}_0$ and β_0 are complex valued, and

$$\frac{1}{\hbar}\int_0^t \mathscr{L}_0 dt' = \delta_0 + \beta_0,$$

then

$$\mathrm{Re}\beta_0 = \frac{1}{\hbar}\int_0^t \mathrm{Re}\mathscr{L}_0 dt' - \delta_0 = \chi_0(t) - \delta_0$$

and

$$\mathrm{Im}\beta_0 = \frac{1}{\hbar}\int_0^t \mathrm{Im}\mathscr{L}_0 dt' = \ln\frac{2}{\sqrt{4 - s_0^2(t)}}, \tag{18.13}$$

where use has been made of the fact that $\chi_0(0) = s_0(0) = 0$.

It can be inferred from (18.11) and (18.13) that since

$$e^{-\mathrm{Im}\beta_0} = \sqrt{Pr(s_0(t))},$$

then $\mathrm{Im}\beta_0$ governs the survival probability for $|\psi(0)\rangle$ and comparison of the expression for $\mathrm{Re}\beta_0$ with (4) in [22] identifies $-\mathrm{Re}\beta_0$ as the Mukunda-Simon phase. For the special case that the evolution of $|\psi(t)\rangle$ is cyclic, comparison of the expression for $\mathrm{Re}\beta_0$ with (3) in [23] reveals that $-\mathrm{Re}\beta_0$ is the Aharonov-Anandan phase.

18.7 Some Additional Properties of $\mathscr{L}_0$

Consider the transformation

$$\mathscr{L}_0^{\theta} = \mathscr{L}_0 - \hbar\dot{\theta}(\chi_0, s_0), \tag{18.14}$$

where $\theta \equiv \theta(\chi_0, s_0)$ is a dimensionless function with continuous derivatives. Since transformations of this form applied to Lagrangians are referred to as gauge transformations in the classical mechanics literature (e.g. [24]), then (18.14) is called a *pointed weak energy gauge transformation*. From (18.10) it is easy to see that

(18.14) corresponds to the local U(1) gauge transformation $|\psi(t)\rangle \to e^{i\theta_t}|\psi(t)\rangle$, i.e.

$$\varphi_{t,0}^{\theta} = e^{\frac{i}{\hbar}\int_0^t \mathscr{L}_0^{\theta} dt'} = e^{\frac{i}{\hbar}\int_0^t \mathscr{L}_0 dt'} e^{-i[\theta]_0^t} = \varphi_{t,0} e^{-i(\theta_t - \theta_0)} = \langle \psi(t)|e^{-i\theta_t} e^{i\theta_0}|\psi(0)\rangle,$$

where $\theta_t \equiv \theta(\chi_0(t), s_0(t)) = \theta$. It is interesting to note from (18.14) that the PFS 1-form $\omega_0 \equiv \frac{1}{\hbar}\mathscr{L}_0 dt$ transforms as a U(1) gauge potential, i.e.

$$\omega_0^{\theta} = \omega_0 - d\theta = \omega_0 + d(i\ln g),$$

where $g = e^{i\theta} \in \mathrm{U}(1)$.

Let $|\psi_0\rangle$ be a distinguished state in $\mathscr{H}$, $\mathscr{H}_{\sim\perp} \equiv \{|\psi\rangle \in \mathscr{H} : \langle\psi|\psi_0\rangle \neq 0\}$, $\mathscr{R}$ be the set of real numbers, and define the map $\Psi_0 : \mathscr{H}_{\sim\perp} \to \mathscr{R} \times \mathscr{R}$ by

$$\Psi_0(|\psi\rangle) \equiv \left(\arg\frac{\langle\psi|\psi_0\rangle}{|\langle\psi|\psi_0\rangle|}, 2\sqrt{1 - |\langle\psi|\psi_0\rangle|^2}\right) = (\chi_0, s_0).$$

Note that Ψ_0 provides an equivalence classification of the states in $\mathscr{H}_{\sim\perp}$. More specifically, $|\psi\rangle$ and $|\psi'\rangle$ are equivalent under Ψ_0, i.e. $|\psi\rangle \sim |\psi'\rangle$, when $\Psi_0(|\psi\rangle) = \Psi_0(|\psi'\rangle)$. *PFS configuration space* is the image set $\mathscr{B} \equiv im\Psi_0 = [0, 2\pi) \times [0, 2) \subset \mathscr{R} \times \mathscr{R}$.

If the map $\alpha : [0, \tau] \to \mathscr{H}_{\sim\perp}$ defines an evolutionary path with $|\psi(t)\rangle = \alpha(t)$, then the composition of maps $\rho \equiv \Psi_0 \circ \alpha$ is the curve in $\mathscr{B}$ which describes this evolution and has $\rho(0) = (\chi_0(0), s_0(0)) = (0, 0)$ as its *first point* and $\rho(\tau) = (\chi_0(\tau), s_0(\tau))$ as its *last point*. The curve ρ is *simple* if ρ is injective and is *closed* if $\rho(0) = \rho(\tau)$. The closed curve ρ is simple if the restriction of ρ to the domain $(0, \tau)$ is injective. An evolutionary path in $\mathscr{H}$ over $[0, \tau]$ for which $|\psi(t)\rangle \in \mathscr{H}_{\sim\perp}, t \in [0, \tau]$, and for which ρ is a smooth curve is said to be *proper* and ρ is the *proper evolution* in $\mathscr{B}$. Observe that ω_0 is exact since $\omega_0 = df$, $f = \chi_0 - \frac{i}{2}\ln(\frac{4-s_0^2}{4})$, so that for any proper evolution ρ in $\mathscr{B}$ which connects $(0, 0)$ to any other point (χ_0', s_0'),

$$\int_{\rho} \omega_0 = f(\chi_0', s_0'). \tag{18.15}$$

Thus, *the value of* $\int_{\rho}\omega_0$ *is independent of the path taken between (0,0) and* (χ_0', s_0') *in* $\mathscr{B}$.

Now suppose that $|\psi(t)\rangle$ and $|\psi'(t)\rangle$ are proper evolutionary paths over $[0, \tau]$ in $\mathscr{H}$ such that $|\psi(0)\rangle = |\psi'(0)\rangle$, $|\psi(\tau)\rangle \sim |\psi'(\tau)\rangle$, i.e. they are *last point equivalent paths*, and ρ and ρ' are their respective paths in $\mathscr{B}$. If the same transformation $e^{i\theta}$ is applied to both of these states, then

$$\int_{\rho} \omega_0^{\theta} - \int_{\rho'} \omega_0^{\theta} = \int_{\rho} \omega_0 - \int_{\rho'} \omega_0$$

or

$$\int_{\mathscr{C}} \omega_0^{\theta} = \int_{\mathscr{C}} \omega_0.$$

Here use has been made of the facts that negative signs preceding line integrals over curves reverse the orientation of the curves and that $\mathscr{C}$ is the simple smooth curve that is the union of ρ and $-\rho'$. It can therefore be concluded that *for any last point equivalent paths in $\mathscr{H}$ the value of ω_0 on $\mathscr{C}$ (i.e. $\int_{\mathscr{C}} \omega_0$) is invariant under pointed weak energy gauge transformations.* Furthermore, because of path independence and since ρ and ρ' have the same first and last points, then for any two last point equivalent paths in $\mathscr{H}$

$$\int_{\rho} \omega_0 = \int_{\rho'} \omega_0$$

so that

$$\int_{\mathscr{C}} \omega_0 = 0.$$

Let Ω be a simple closed curve in the projective space $\mathscr{P}$ of $\mathscr{H}$ over the time interval $[0, \tau]$ with $[\psi(0)] = [\psi(\tau)]$ and such that its image in $\mathscr{B}$ is smooth and simple. When the lift of Ω is closed, then $|\psi(0)\rangle = |\psi(\tau)\rangle$ and the image in $\mathscr{B}$ is a loop with $(0, 0)$ as its first and last point and which intersects the χ_0 and s_0 axes only at the origin. However, if the lift of Ω is not closed, then $|\psi(t)\rangle = e^{i\lambda}|\psi(0)\rangle$ and the image of the lift in $\mathscr{B}$ is a path σ which has $(\chi_0(\tau), 0) = (\lambda, 0)$ as its last point. It then follows from (18.15) that

$$\int_{\sigma} \omega_0 = f(\lambda, 0) = \lambda.$$

For the special case that the lift is a horizontal lift, then λ is the geometric phase and

$$\int_{\sigma} \omega_0 = \int_{\Omega} \mathsf{A},$$

where A is the associated Aharonov-Anandan connection 1-form [25]. It can therefore be concluded that *for each such horizontal lift of an Ω in $\mathscr{P}$ there exists a path σ in $\mathscr{B}$ for which the last equation is satisfied.*

Acknowledgements The preparation of this contribution to the Festschrift commemorating Yakir Aharonov's eightieth birthday was supported by a grant from the Naval Innovation in Science and Engineering program sponsored by the Naval Surface Warfare Center Dahlgren Division.

References

1. Y. Aharonov, D. Albert, D. Casher, L. Vaidman, Ann. N.Y. Acad. Sci. **480**, 417 (1986)
2. Y. Aharonov, D. Albert, L. Vaidman, Phys. Rev. Lett. **60**, 1351 (1988)
3. Y. Aharonov, L. Vaidman, Phys. Rev. A **41**, 11 (1990)
4. N. Ritchie, J. Storey, R. Hulet, Phys. Rev. Lett. **66**, 1107 (1991)
5. A. Parks, D. Cullin, D. Stoudt, Proc. R. Soc. A **454**, 2997 (1998)
6. K. Resch, J. Lundeen, A. Steinberg, Phys. Lett. A **324**, 125 (2004)
7. Q. Wang, F. Sun, Y. Zhang, J. Li, Y. Huang, G. Guo, Phys. Rev. A **73**, 023814 (2006)
8. O. Hosten, P. Kwiat, Science **319**, 787 (2008)
9. Y. Yokota, T. Yamamoto, M. Koashi, N. Imoto, New J. Phys. **11**, 033011 (2009)
10. P. Dixon, D. Starling, A. Jordan, J. Howell, Phys. Rev. Lett. **102**, 173601 (2009)

11. I. Duck, P. Stevenson, E. Sudarshan, Phys. Rev. D **40**, 2112 (1989)
12. R. Jozsa, Phys. Rev. A **76**, 044103 (2007)
13. A. Parks, J. Phys. A, Math. Theor. **41**, 335305 (2008)
14. N. Mukunda, R. Simon, Ann. Phys. **228**, 205 (1993)
15. J. Anandan, Found. Phys. **21**, 1265 (1991)
16. A. Pati, J. Phys. A, Math. Gen. **25**, L1001 (1992)
17. J. Anandan, Y. Aharonov, Phys. Rev. D **38**, 1863 (1988)
18. A. Parks, J. Phys. A, Math. Gen. **33**, 2555 (2000)
19. A. Parks, J. Phys. A, Math. Gen. **36**, 7185 (2003)
20. A. Parks, J. Phys. A, Math. Gen. **39**, 601 (2006)
21. A. Parks, J. Phys. A, Math. Theor. **40**, 2137 (2007)
22. A. Pati, Phys. Rev. A **52**, 2576 (1995)
23. Y. Aharonov, J. Anandan, Phys. Rev. Lett. **58**, 1593 (1987)
24. H. Rund, *The Hamilton-Jacobi Theory in the Calculus of Variations: Its Role in Mathematics and Physics* (Van Nostrand, London, 1966), p. 163
25. A. Bohm, A. Mostafazadeh, H. Koizumi, Q. Niu, J. Zwanziger, *The Geometric Phase in Quantum Systems: Foundations, Mathematical Concepts, and Applications in Molecular and Condensed Matter Physics* (Springer, Berlin, 2003), pp. 53–60

Chapter 19
Weak Values Beyond Post-selection

Lars M. Johansen

Abstract Weak values are usually encountered in weak measurements on pre- and post-selected systems. Here we consider experiments where a final post-selection is replaced with an initial projector measurement. We show how weak values may be observed both for strong and weak measurements of the projector. We also show that the real part of the weak value emerges uniquely as a generalized conditional expectation from a symmetry principle applied to strong measurements. [*Editors note*: for a video of the talk given by Prof. Johansen at the Aharonov-80 conference in 2012 at Chapman University, see quantum.chapman.edu/talk-26.]

19.1 Introduction

Weak values appear as a result of weak measurements with post-selection [1]. The post-selection is essential in order to observe weak values. If one ignores the post-selection, the same experiment simply gives the expectation value. In this note, we will give a brief review of some other methods of observing the real part of the weak value. We also derive the real part of the weak value from a symmetry principle.

We will first give a brief review of the standard procedure for observing weak values. One first prepares the system in a state $|\psi\rangle$ and a probe is prepared in a state $|\chi\rangle$ (Fig. 19.1). The state of the probe $|\chi\rangle$ is usually taken to be a real gaussian.

The interaction between the system and the probe is usually taken to be the impulsive von Neumann interaction [2]

$$H = \delta(t) A P, \tag{19.1}$$

where A is the system observable we want to measure and P is the momentum of the probe. This interaction produces a displacement in the probe position Q conjugate to the momentum P, and this displacement gives information about the system observable A. We assume that this interaction dominates during the short interaction period, so that other parts of the Hamiltonian may be ignored. Assuming that

L.M. Johansen (✉)
Faculty of Technology, Buskerud University College, 3603 Kongsberg, Norway
e-mail: lars.m.johansen@hibu.no

D.C. Struppa, J.M. Tollaksen (eds.), *Quantum Theory: A Two-Time Success Story*,
DOI 10.1007/978-88-470-5217-8_19, © Springer-Verlag Italia 2014

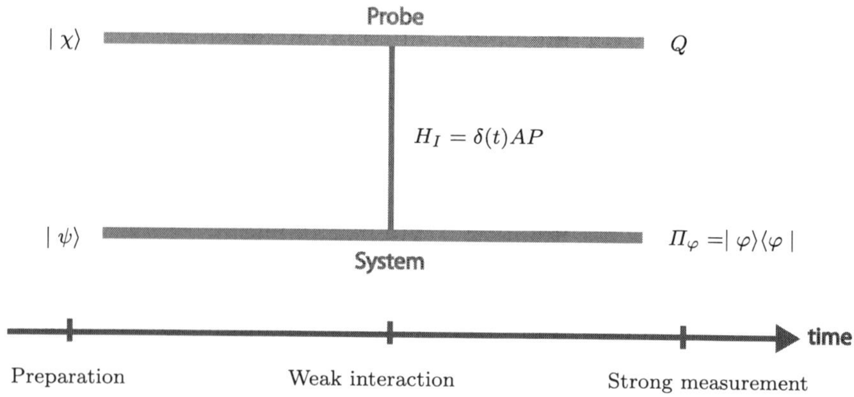

Fig. 19.1 Weak measurement with post-selection

Fig. 19.2 Marginal distribution of probe position for the interaction (19.1). The average displacement gives the expectation of A

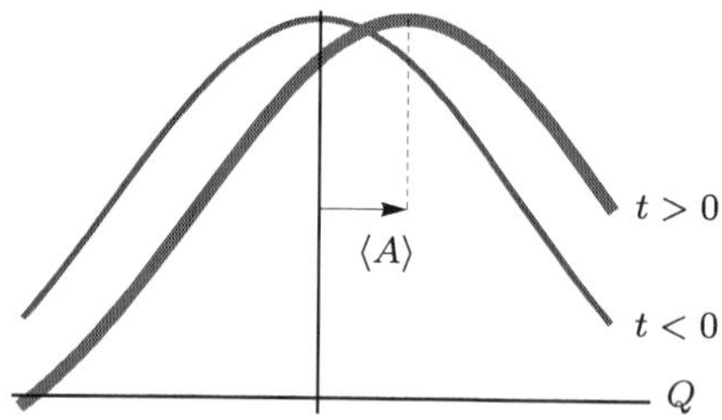

Fig. 19.3 Marginal distribution of projector eigenvalues. In this example, the probability of an eigenvalue 1 is small. This corresponds to a post-selection being a rare event

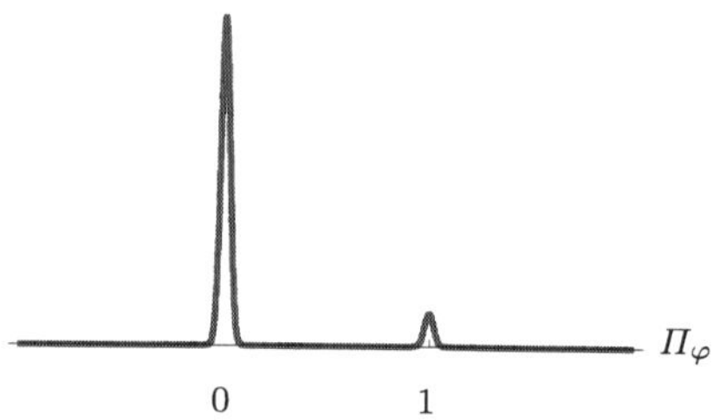

the position distribution of the probe is sufficiently broad, the measurement will be weak. As a consequence, one cannot observe individual eigenvalues. Nevertheless, if no post-selection is performed, the average displacement of the probe position Q gives the expectation value of the observable A (see Fig. 19.2)

After the interaction, a post-selection on a state $|\phi\rangle$ is performed. This is equivalent to a strong measurement of a projector $\Pi_\varphi = |\phi\rangle\langle\phi|$, where the sub-ensemble corresponding to the eigenvalue 1 is selected (Fig. 19.3).

It then can be shown [1] that the expectation of the probe position Q after the interaction, and conditional on a successful post-selection, is the real part of the

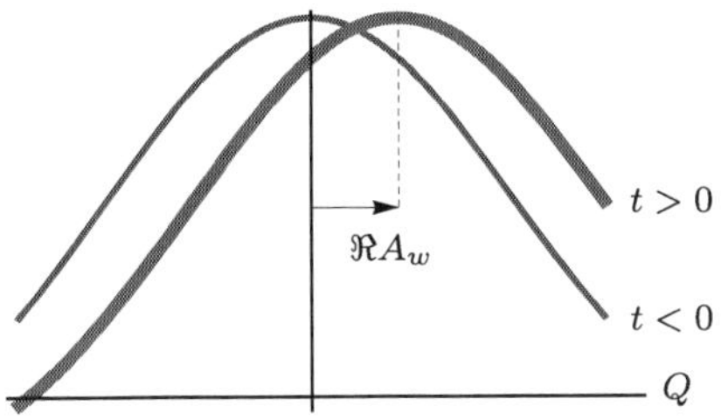

Fig. 19.4 Distribution of probe position for the interaction (19.1) conditional on a successful post-selection

weak value (see Fig. 19.4)

$$A_w = \frac{\langle \varphi | A | \psi \rangle}{\langle \varphi | \psi \rangle}. \tag{19.2}$$

19.2 Conditional Expectations

To begin with, we give a brief derivation of an expression for conditional expectations in quantum mechanics. Consider two commuting observables X and Y with spectral resolution

$$X = \sum_x x \Pi_x, \tag{19.3}$$

$$Y = \sum_y y \Pi_y, \tag{19.4}$$

where

$$\Pi_x \Pi_{x'} = \delta_{xx'} \Pi_x, \quad \sum_x \Pi_x = 1, \tag{19.5}$$

$$\Pi_y \Pi_{y'} = \delta_{yy'} \Pi_y, \quad \sum_y \Pi_y = 1. \tag{19.6}$$

By definition, the conditional expectation of X given $Y = y$ is

$$E(X \mid Y = y) = \sum_x x P(X = x \mid Y = y) = \sum_x x \frac{P(X = x, Y = y)}{P(Y = y)}. \tag{19.7}$$

Since

$$P(Y = y) = \langle \Pi_y \rangle, \tag{19.8}$$

$$P(X = x, Y = y) = \langle \Pi_x \Pi_y \rangle, \tag{19.9}$$

we have

$$E(X \mid Y = y) = \frac{\sum_x x \langle \Pi_x \Pi_y \rangle}{\langle \Pi_y \rangle} = \frac{\langle X \Pi_y \rangle}{\langle \Pi_y \rangle}. \tag{19.10}$$

Alternatively, we could write

$$E(X \mid \Pi_y = 1) = \frac{\langle X\Pi_y \rangle}{\langle \Pi_y \rangle}. \tag{19.11}$$

Measuring a conditional expectation of an observable X is equivalent to measuring a correlation between X and a projector representing the condition.

The observable A that we have been considering for weak measurements does not in general commute with the projector Π_φ. But in a weak measurement, this observable is measured indirectly through interaction with a probe. In the Schrödinger picture, where the time evolution is reflected in the state, all probe and system observables commute at all times.

In a weak measurement of an observable A with postselection on a state $|\varphi\rangle$, the expectation of the probe position Q conditional on $\Pi_\varphi = 1$ gives the real part of the weak value, i.e.

$$\frac{\langle Q\Pi_\varphi \rangle}{\langle \Pi_\varphi \rangle} = \Re A_w, \tag{19.12}$$

where brackets indicate averaging over the combined system + probe after the measurement interaction.

It is interesting to note that the weak value (19.2) may be written as a complex conditional expectation of A

$$A_w = \frac{\langle \psi | \Pi_\varphi A | \psi \rangle}{\langle \psi | \Pi_\varphi | \psi \rangle}. \tag{19.13}$$

19.3 Weak Measurements Without Post-selection

In a weak measurement with post-selection, we use a weak interaction with a probe to get information about the system observable A. This disturbs the system very little, and we may afterwards get information about the projector Π_φ through a strong measurement. But could we do it the other way around? What would be the result if we use a weak interaction with a probe which is designed to get information about the projector Π_φ and then measure the observable A using a strong measurement?

The measurement interaction would then take the form

$$H = \delta(t)\Pi_\varphi P. \tag{19.14}$$

This experiment is depicted in Fig. 19.5. In this case, the probe position Q conjugate to P will give information about the projector.

This time, if we assume that the position distribution of the probe is sufficiently broad, we cannot observe the eigenvalues of the *projector*. Nevertheless, the average displacement of the probe position Q gives the expectation value of the projector (Fig. 19.6). However, this time we may perform a final and strong measurement of the observable A, hence we may observe also the eigenvalues of A (Fig. 19.7).

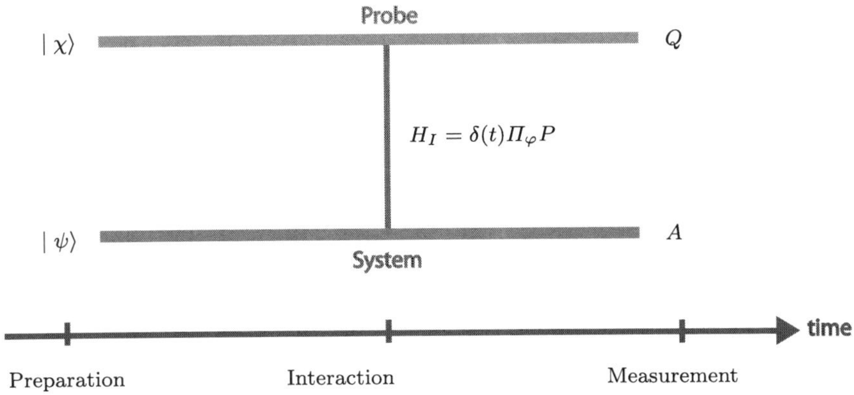

Fig. 19.5 Measurement where projector is measured first

Fig. 19.6 Marginal distribution of probe position for the interaction (19.14). The average displacement gives the expectation of Π_φ

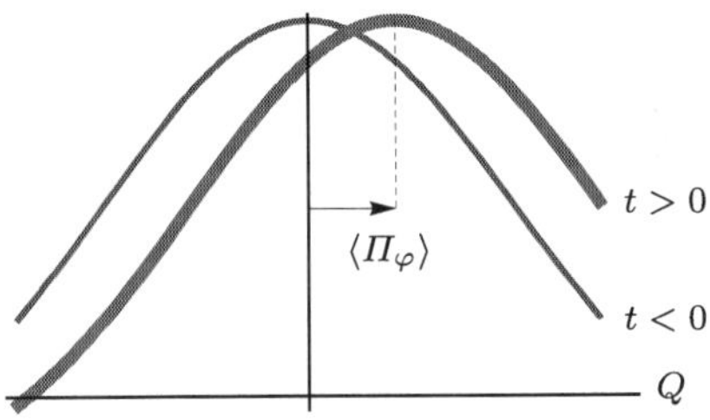

Fig. 19.7 Distribution of eigenvalues of A

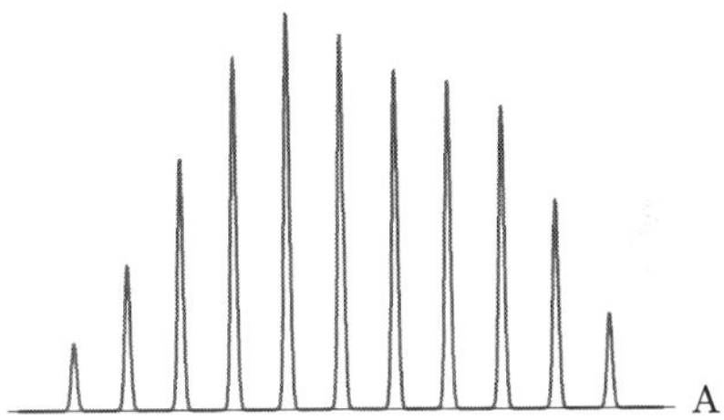

But can we obtain the weak value in this experiment? Yes, in this case the weak value emerges as a correlation between the probe and the system observable A [3],

$$\frac{\langle QA \rangle}{\langle Q \rangle} = \Re A_w. \tag{19.15}$$

So there is a symmetry here. The order of operations does not matter as long as the first measurement is weak. We can either do a weak measurement of A or a weak measurement of the projector. In both cases, weak values emerge from a correlation between probe and system.

19.4 Strong Measurements Without Post-selection

Next, we consider an experiment where the projector Π_φ is measured first, but this time with a strong interaction [4]. This measurement disturbs the subsequent measurement of A. Since we are only considering strong measurements, we don't need to include a description of the probe in modelling the experiment.

In this section, we need to introduce density operators in order to represent the state after non-selective measurements. Our argument is based on the identity

$$\Re A_w = \mathrm{Tr}\,\rho_s A + \frac{1}{2}\left\{\frac{\mathrm{Tr}\,\rho A - \mathrm{Tr}\,\rho_n A}{\mathrm{Tr}\,\rho \Pi_\varphi}\right\}, \tag{19.16}$$

which is easily verified [4]. Here we have used the notation

$$\rho = |\psi\rangle\langle\psi|, \tag{19.17}$$

$$\rho_s = \frac{\Pi_\varphi \rho \Pi_\varphi}{\mathrm{Tr}\,\rho \Pi_\varphi}, \tag{19.18}$$

$$\rho_n = \Pi_\varphi \rho \Pi_\varphi + (1 - \Pi_\varphi)\rho(1 - \Pi_\varphi). \tag{19.19}$$

ρ represents the initial state, whereas ρ_s and ρ_n represent the state after a selective and a non-selective strong measurement, respectively, of Π_φ. These measurements should be of the Lüders type [5].

Equation (19.16) suggests a way to measure the real part of the weak value using only strong measurements of the Lüders type [5]. A similar relation exists for the imaginary part [4].

The first term on the r.h.s. of Eq. (19.16) is the expectation of A on a system that has been *pre-selected* in the state $|\phi\rangle$. The second term is proportional to the *change* imposed on the expectation of A by a non-selective measurement of Π_φ. This change will not take place in ideal classical measurements.

19.5 Derivation from a Symmetry Principle

Let us now consider the experiment in Sect. 19.4 once more, i.e., let us consider an initial strong measurement of the Lüders type of a projector Π_φ followed by a strong measurement of the observable $A = \sum_i a_i \Pi_i$ where $\Pi_i \Pi_j = \delta_{ij}$ and $\sum_i \Pi_i = 1$. But this time we shall not assume that weak values have anything to do with this experiment. We seek to *derive* the conditional expectation from this experiment.

We start by analyzing a simpler experiment where the final measurement is made only of one of the projectors Π_i in the spectral resolution of A. The joint probability of first observing $\Pi_\varphi = 1$ and thereafter $\Pi_i = 1$ is

$$\mathrm{Tr}(\rho \Pi_\varphi)\,\mathrm{Tr}(\rho_s \Pi_i), \tag{19.20}$$

where ρ_s is defined in Eq. (19.18). This joint probability contains information not only about the initial state ρ but also the reduced state ρ_s. As is well known, there does not exist any joint probability for two non-commuting projectors in quantum mechanics. However, we may seek a "generalized probability" F which may exceed the bounds of classical probabilities. There are many such generalized probabilities in quantum mechanics, including the Wigner distribution [6]. We require that this generalized probability should still satisfy a usual condition on joint probabilities, namely that it should give the correct marginal probabilities;

$$F(\Pi_\varphi, \Pi_i \mid \rho) + F(\Pi_\varphi, 1 - \Pi_i \mid \rho) = \mathrm{Tr}\, \rho \Pi_\varphi, \tag{19.21}$$

$$F(\Pi_\varphi, \Pi_i \mid \rho) + F(1 - \Pi_\varphi, \Pi_i \mid \rho) = \mathrm{Tr}\, \rho \Pi_i. \tag{19.22}$$

F will change if the state ρ changes. In particular, F may change after a non-selective measurement of Π_φ. Thus, we have

$$F(\Pi_\varphi, \Pi_i \mid \rho) = F(\Pi_\varphi, \Pi_i \mid \rho_n) + \Delta F(\Pi_\varphi, \Pi_i \mid \rho). \tag{19.23}$$

This equation defines ΔF. A simple argument [7] shows that the first term on the r.h.s. is determined by Eq. (19.20). Next, we shall determine ΔF by a symmetry requirement, namely that ΔF should be invariant under orthogonal projector complementation

$$\Delta F(\Pi_\varphi, \Pi_i \mid \rho) = \Delta F(1 - \Pi_\varphi, \Pi_i \mid \rho). \tag{19.24}$$

This means that F is insensitive to whether we perform a non-selective measurement of Π_φ or of the orthogonal complement $1 - \Pi_\varphi$. Such a requirement is not unnatural, since we obtain the same information whether we measure Π_φ or $1 - \Pi_\varphi$. It may be noted that the same invariance under non-selective measurements applies to the state itself (including the change of the state), as reflected in Eq. (19.19).

From this requirement it follows [7] that

$$F(\Pi_\varphi, \Pi_i \mid \rho) = \Re\left\{\mathrm{Tr}\, \rho \Pi_\varphi \Pi_i\right\}. \tag{19.25}$$

This generalized joint probability was first studied in this general form by Dirac [8]. We may now construct the generalized conditional probability for Π_i given that $\Pi_\varphi = 1$,

$$F(\Pi_i \mid \Pi_\varphi, \rho) = \frac{F(\Pi_\varphi, \Pi_i \mid \rho)}{\mathrm{Tr}\, \rho \Pi_\varphi}, \tag{19.26}$$

and we find that the generalized conditional expectation of A given $\Pi_\varphi = 1$ is

$$\sum a_i F(\Pi_i \mid \Pi_\varphi, \rho) = \Re A_w. \tag{19.27}$$

References

1. Y. Aharonov, D.Z. Albert, L. Vaidman, Phys. Rev. Lett. **60**(14), 1351 (1988)

2. J. von Neumann, *Mathematical Foundations of Quantum Mechanics* (Princeton University Press, Princeton, 1955)
3. L.M. Johansen, P.A. Mello, eprint (2009). arXiv:0907.5437
4. L.M. Johansen, Phys. Lett. A **366**(4–5), 374 (2007)
5. G. Lüders, Ann. Phys. **8**(6), 322 (1951)
6. E. Wigner, Phys. Rev. **40**, 749 (1932)
7. L.M. Johansen, eprint (2008). arXiv:0804.4379
8. P.A.M. Dirac, Rev. Mod. Phys. **17**(2/3), 195 (1945)

Part VII
Mathematics of Weak Measurements

Chapter 20
On Superoscillations Longevity: A Windowed Fourier Transform Approach

Y. Aharonov, F. Colombo, I. Sabadini, D.C. Struppa, and J. Tollaksen

Abstract In this paper we prove that the evolution of a superoscillating sequence of functions $Y_N(x)$, which we take as the initial value of the Schrödinger equation, remains superoscillating for all values of the time $t < \sqrt{N}$. We prove this by using a windowed Fourier transform approach. [*Editor's note*: for a video of the talks given by Prof. Aharonov at the Aharonov-80 conference in 2012 at Chapman University, see quantum.chapman.edu/talk-3 and quantum.chapman.edu/talk-30. For the subject of super-oscillations, the editors also recommend the video of the talk given by Prof. Berry at the Aharonov-80 conference in 2012 at Chapman University, see quantum.chapman.edu/talk-6 along with the talk by Prof. Casher, see quantum.chapman.edu/talk-21.]

20.1 Introduction

A universally accepted truth in spectral analysis is that signals, be they space dependent, as in optical imaging, or time dependent, cannot have details on a scale shorter than the shortest wavelength or shortest time period of their Fourier components. This applies to all wave phenomenon: optical, sonic, electrical, etc. Yet, Aharonov and his collaborators, followed shortly thereafter by Berry [1], have described a phenomenon known as superoscillations which, at least apparently, seemed to violate this principle.

This phenomenon is indeed very general. In particular, a superoscillatory function is "band-limited," that is, its Fourier representation has a maximum wavenum-

Y. Aharonov · J. Tollaksen
Institute for Quantum Studies and Schmid College of Science and Technology, Chapman University, Orange 92866, CA, USA

F. Colombo · I. Sabadini
Dipartimento di Matematica, Politecnico di Milano, Via E. Bonardi, 9 20133 Milano, Italy

D.C. Struppa (✉)
Institute for Quantum Studies and Schmid College of Science, Chapman University, Orange 92866, CA, USA
e-mail: struppa@chapman.edu

D.C. Struppa, J.M. Tollaksen (eds.), *Quantum Theory: A Two-Time Success Story*,
DOI 10.1007/978-88-470-5217-8_20,

ber (or spatial frequency), say $|k_{\max}| \leq 1$, but it can nevertheless oscillate, over a substantial region, at a much faster rate, say by a factor a, arbitrarily larger than 1 and therefore well outside the "allowed" spectrum. Specifically, one can show that it is possible to approximate e^{iax} by a sequence $\{\sum_{j=1}^{N} c_j(N)e^{ik_j(N,a)x}\}_{N=1}^{+\infty}$ or in brief,

$$\sum_j c_j e^{ik_j x} \to e^{iax}. \tag{20.1}$$

At first glance, one would think that waves of the form given by the left-hand side of Eq. (20.1) cannot distinguish features smaller than its smallest wavelength which is about $\pi/k_{\max} = \pi$, and yet a much better resolution of $\Delta x = \frac{\pi}{a}$ can be achieved within the superoscillating region. The careful reader will immediately recognize that this apparently paradoxical phenomenon is made possible by the circumstance that the coefficients c_j and the frequencies k_j all depend on N, so that the approximating sequence changes its terms with the change of N. In this, one clearly sees that this is not the usual Fourier expansion, as we discuss in more detail in [2, 3].

The superoscillation literature has been growing rapidly [4–14]. Of great interest is the recent rapid application of superoscillations as a tool to obtain resolutions better than allowed by the diffraction limit in a purely classical optical setting [15]. Normally, the focusing of light by a lens has limited resolution determined by about one-half the wavelength of light used. This is not due to any imperfection in the lens, but due to the laws of physics so that, for example, an optical microscope can only resolve structures several hundred nanometers in size. To observe smaller structures such as DNA, proteins, or viruses, one must use smaller wavelengths, such as X-rays. But such probes have a variety of experimental limitations and can damage living organisms.

In the past, attempts to achieve super-resolution of very fine features was achieved with a very different type of phenomena, namely with evanescent waves which only occur in the near-field. But superoscillations have many advantages over evanescent waves: they do not require a media-substrate and can therefore penetrate much deeper into the media than evanescent waves. The probe does not have to be in contact with the specimen, the measurement process works very quickly, and it is much less invasive.

These practical applications were of course preceded by many theoretical, mathematical developments. E.g., many of the mathematical foundations for superoscillations have been clarified in a series of papers [2, 3, 16–18]. It is also known that regions of superoscillations are typical in random fields [19], i.e. in superpositions of plane waves with a random selection of amplitude and direction.

Superoscillations are not just an intriguing mathematical phenomenon: they have turned out to be very useful. A recent *Nature* article [20] concluded that "superoscillation-based imaging has unbeatable advantages over other technologies. It is non-invasive, allowing the object to be at a substantial distance from the lens, and can operate at any wavelength from X-rays to microwaves." New devices will allow for in-vivo nanometer-size imaging and manipulation of structures inside living cells. In this Festschrift, Chap. 21, Michael Berry has used super-oscillations to create arbitrarily narrow radiation patterns produced by antennas.

In general, the regions of superoscillations are created at the expense of having the function grow exponentially in other regions [1, 9, 10]. This apparently "universal" behavior thus points to one of the key disadvantages for various applications, namely the comparatively low intensity of the super-oscillating region. With this exponential behavior, it would therefore seem natural to expect that the superoscillations would be quickly "over-taken" by tails coming from the exponential regions and would thus be short-lived. However, it has been shown that superoscillations are remarkably robust [15] and can last for a surprisingly long time. We have added to this line of inquiry in a number of ways [2, 3, 16–18] and continue in this paper by using a windowed Fourier transform approach. We have proven that the phenomenon actually arises in a larger context that has never before been foreseen [17]. We have also developed a rigorous treatment of the superoscillatory phenomenon in terms of the Taylor and the Fourier coefficients of a superoscillating sequence [2, 3, 16–18]. We have used this treatment to deduce important properties of these functions and have also shown how superoscillatory functions can be applied to sequences of operators [18].

In a broader context, superoscillations are examples of unusual weak values [21] which can be obtained for pre- and post-selected quantum systems. The impact of weak values [21–23] and derivative works (e.g. superoscillations [24], quantum random walk [25]), has been significant, with broad participation from many physics and mathematics disciplines. More than 20 laboratories around the world have performed experiments to verify the many novel predictions. These phenomenon have proven to be very useful tools for analyzing various physical phenomena, for constructing efficient devices for high precision measurements, etc. Phenomenon which were thought to be unmeasureable, have now been seen using this new approach. As a new paradigm for the design of sensors, it is being broadly applied to, precision Doppler frequency measurements, gravitational detectors, etc.

AAV and superoscillations also led to the notion of the quantum random walk [25]. If the coefficients for a step to the left or right were probabilities, as would be the case in a classical random walk, then N steps of step size 1 could generate an average displacement of $\sqrt{N}$, but never a distance larger than N. However when the steps are superposed with probability *amplitudes*, as with (20.9), then the random walk can produce **any** displacement. In other words, instead of saying that a "quantum step" is made up of probabilities, we say that a quantum step is a superposition of the amplitude for a step "to the left" and the amplitude for a "step to the right," then one can superpose small Fourier components and obtain a large shift. This has proven to be a most useful tool in quantum information. E.g., it was recently shown that implementation of the quantum random walk would lead to a universal quantum computer as well as a quantum simulator (to study, e.g., phase transitions). Recent experimental realizations of the quantum random walk have been successful (trapped atom with optical lattice and ion trap; photons in linear optics).

20.2 Preliminaries

20.2.1 Overview of the Mathematics of the Aharonov-Albert-Vaidman Effect

The original insights which eventually led to superoscillations started with the observation by Aharonov, Bergmann and Lebowitz [26] (ABL) that, as a result of the uncertainty principle, the initial boundary condition of a quantum mechanical system can be selected independently of the final boundary conditions. Subsequently it was demonstrated by Aharonov, Albert and Vaidman (AAV) [22, 23, 27–35] that if non-disturbing measurements are performed on such pre- and post-selected systems, then strange outcomes will be obtained during the intermediate time. Such outcomes depend on both the pre- and the post-selection, and can lie outside the usually allowed eigenvalue spectrum. This was subsequently developed as the notion of superoscillation [24] and by Berry as the concept of SuperFourier [1].

Traditionally, it was believed that if a measurement interaction is weakened so that there is no disturbance on the system, then no information will be obtained. However, it has been shown that information can be obtained even though not a single particle (in an ensemble) was disturbed [36].

To begin with, we will recall a general theorem for Hilbert spaces. We repeat the proof for sake of completeness.

Theorem 20.1 *Let $|\psi\rangle$ be a vector in a Hilbert space H, and let $\hat{A}$ be an operator on H. Then*

$$\hat{A}|\psi\rangle = \langle\hat{A}\rangle|\psi\rangle + \Delta A|\psi_\perp\rangle$$

where $\langle\hat{A}\rangle = \langle\psi|\hat{A}|\psi\rangle$, $\Delta A^2 = \langle\psi|(\hat{A} - \langle\hat{A}\rangle)^2|\psi\rangle$, and $|\psi_\perp\rangle$ is a vector such that $\langle\psi|\psi_\perp\rangle = 0$.

Proof One can always write the result of any operator $\hat{A}$ on any vector $|\psi\rangle$ as a combination of $|\psi\rangle$ and a perpendicular vector $|\psi_\perp\rangle$, i.e.

$$\hat{A}|\psi\rangle = a|\psi\rangle + b|\psi_\perp\rangle.$$

In order to evaluate a we left multiply by $\langle\psi|$ and obtain that $a = \langle\hat{A}\rangle$; as to b we left multiply by $\langle\psi|A^+$ and by noticing that $|(A - \langle A\rangle)|\psi\rangle|^2 = \Delta A^2$ we conclude the proof. □

Now, the average of any operator $\langle\hat{A}\rangle \equiv \langle\Psi|\hat{A}|\Psi\rangle$ can be measured as the "eigenvalue" of a single "collective operator," $\hat{A}^{(N)} \equiv \frac{1}{N}\sum_{i=1}^{N}\hat{A}_i$ without causing a disturbance (with $\hat{A}_i$ the same operator $\hat{A}$ acting on the i-th particle). To see this, we apply Theorem 20.1 to the N particle product state $|\Psi^{(N)}\rangle = |\psi\rangle_1|\psi\rangle_2\ldots|\psi\rangle_N$ with all particles in the same state $|\psi\rangle$. We see that:

$$\hat{A}^{(N)}|\Psi^{(N)}\rangle = \frac{1}{N}\left[N\langle\hat{A}\rangle|\Psi^{(N)}\rangle + \Delta A\sum_i|\Psi_\perp^{(N)}(i)\rangle\right] \tag{20.2}$$

where $\langle\hat{A}\rangle$ is the average for any one particle and the N states

$$|\Psi_{\perp}^{(N)}(i)\rangle = |\psi\rangle_1|\psi\rangle_2\dots|\psi_{\perp}\rangle_i\dots|\psi\rangle_N$$

are mutually orthogonal. With a normalized state, $|\Psi_{\perp}^{(N)}\rangle = \sum_i \frac{1}{\sqrt{N}}|\Psi_{\perp}^{(N)}(i)\rangle$, the last term of Eq. (20.2) is $\frac{\Delta A}{\sqrt{N}}|\Psi_{\perp}^{(N)}\rangle$ and

$$\left|\frac{\Delta A}{\sqrt{N}}|\Psi_{\perp}^{(N)}\rangle\right|^2 \propto \frac{1}{N}.$$

The probability that measuring $\hat{A}_i/N$ changes the state of the i-th system is proportional to $1/N^2$ and therefore the probability that it changes the state of any system is proportional to $1/N$. Thus, as $N \to \infty$, $|\Psi^{(N)}\rangle$ becomes an eigenstate of $\hat{A}^{(N)}$ with value $\langle\hat{A}\rangle$ and not even a single particle has been disturbed.

To actually make a measurement of an observable such as $\hat{A}^{(N)}$, we switch on an interaction $H_{\text{int}} = \lambda g(t)\hat{Q}_{\text{md}}\hat{A}^{(N)}$, where $\hat{Q}_{\text{md}}$ is an observable of the measuring device (i.e. position), λ is a coupling constant which determines the strength of the measurement, and $g(t)$ is a normalized time profile $\int g(t)dt = 1$ which determines the duration of the measurement (setting $\hbar = 1$). We fix $\Delta P_{\text{md}} = 1$ for the distribution in the momentum $\hat{P}_{\text{md}}$ (i.e. the pointer) which is conjugate to $\hat{Q}_{\text{md}}$. We can then take $\lambda \gg 1$, in order to distinguish the shift, $\lambda\langle\hat{A}\rangle$ from the width. In addition, fixing $\lambda \leq \sqrt{N}$ along with $|\hat{A}_i| < 1$ ensures that the measurement does not shift any particle into an orthogonal state. While the coupling to any individual member of the ensemble is reduced by $\frac{1}{N}$ and therefore the probability that a measurement will disturb any member of the ensemble approaches zero as $\frac{1}{N}$, nevertheless, information about the average is obtained.

By adding a post-selection to these ordinary—yet weakened—von Neumann measurements, the measuring device will register a weak value [22, 23]:

$$\hat{A}_w = \frac{\langle\Psi_{\text{fin}} \mid \hat{A} \mid \Psi_{\text{in}}\rangle}{\langle\Psi_{\text{fin}} \mid \Psi_{\text{in}}\rangle} \tag{20.3}$$

with $|\Psi_{\text{in}}\rangle$ and $|\Psi_{\text{fin}}\rangle$ the initial and final (post-selected) states. The weak-value, $\hat{A}_{\text{w}}$, is an unusual quantity and is not in general an eigenvalue of $\hat{A}$. Equation (20.3) can also be motivated by inserting a complete set of states $\{|\Psi_{\text{fin}}\rangle_j\}$ into $\langle\hat{A}\rangle$

$$\langle\hat{A}\rangle = \langle\Psi_{\text{in}}|\hat{A}|\Psi_{\text{in}}\rangle = \sum_j |\langle\Psi_{\text{fin}}|_j\, \Psi_{\text{in}}\rangle|^2 \underbrace{\frac{\langle\Psi_{\text{fin}}|_j\, \hat{A} \mid \Psi_{\text{in}}\rangle}{\langle\Psi_{\text{fin}}|_j\, \Psi_{\text{in}}\rangle}}_{A_{\text{w}}^j \equiv weak\ value} \tag{20.4}$$

with $|\Psi_{\text{fin}}\rangle_j$ the states corresponding to the outcome of a final ideal measurement on the system (i.e. the post-selection). The average $\langle\hat{A}\rangle$ over all post-selections j is thus constructed out of pre- and post-selected sub-ensembles in which the weak value (A_{w}^j) is multiplied by a probability to obtain the particular post-selection $|\Psi_{\text{fin}}\rangle_j$.

Having weak values outside the spectrum of the operators involved has been discussed at length in the past and has most comprehensively been investigated for spin-1/2 systems [1, 15, 37–42]. In addition, Berry [38] looked at superweak statistics for much more general situations, and the authors proved that if the Hilbert space is sufficiently high in dimensions, and if the pre- or post-selection are, in a sense, 'generic', then the existence of superweak values becomes common or typical.

We have used such limited disturbance measurements to explore many paradoxes (see, e.g. [43–47]).

While AAV called such measurements "weak measurements" (after their non-disturbing nature) [21], these measurements can be quite precise. Consider a large number N of spin 1/2 particles with pre- and post-selection $|\hat{S}_{\mathrm{z}}=\frac{N}{2}\rangle=\prod_{\mathrm{j}=1}^{N}|\uparrow_z\rangle_j$ and $|\hat{S}_{\mathrm{x}}=\frac{N}{2}\rangle=\prod_{\mathrm{j}=1}^{N}|\uparrow_x\rangle_j$, respectively. One may measure the magnetic field with an error of $\sqrt{N}$, while not disturbing more than $\sqrt{N}$ of the spins. By way of example, one may consider measuring the spin in a direction $\xi=45^\circ$ relative to the $x-z$ plane during $t\in[t_{\mathrm{in}},t_{\mathrm{fin}}]$ using such weak measurements. This can be modeled by a collective observable $\hat{S}_{\xi}^{(N)}\equiv\frac{1}{N}\sum_{\mathrm{i}=1}^{N}\{\frac{\hat{S}_x^i+\hat{S}_z^i}{\sqrt{2}}\}$. In this regime, $\hat{S}_{\mathrm{Z}}^{(N)}$ and $\hat{S}_{\mathrm{X}}^{(N)}$ can both be measured without "disturbing" the pre- and post-selection (since they effectively commute). AAV therefore predicted that the weak measurement of $\hat{S}_{45}^{(N)}$ will yield the weak value:

$$\hat{S}_{45}{}^{(N)}{}_w=\frac{\prod_{k=1}^{N}\langle\uparrow_z|_k\{\hat{S}_{\mathrm{Z}}^{(N)}+\hat{S}_{\mathrm{X}}^{(N)}\}\prod_{\mathrm{j}=1}^{N}|\uparrow_x\rangle_j}{\sqrt{2}(\langle\uparrow_z|\uparrow_x\rangle)^N}$$
$$=\frac{\frac{N}{2}+\frac{N}{2}}{\sqrt{2}}=\frac{\sqrt{2}}{2}N\pm O(\sqrt{N}) \tag{20.5}$$

i.e. a value completely outside the spectrum of the spin operator. The possible values for $\hat{S}_{45}^{(N)}$ extend only from $-\frac{N}{2}$ to $\frac{N}{2}$, while the weak measurement registers a result bigger than the maximum allowed value.

We can see the phenomenon of superoscillation in this example if we focus on the measuring device rather than the system. How can a superposition of shifts in the pointer of the measuring device by amounts within the eigenvalue spectrum $[-\frac{N}{2},\frac{N}{2}]$ result in a shift of the pointer that is arbitrarily far outside this spectrum (e.g. $\sqrt{2}N/2$)? The answer is that the pointer states of the measuring device interfere constructively around the "impossible" value, and destructively for all other values. This is a superoscillation in the *Fourier transform* of the pointer basis of the measuring device. To be more precise, the final state of the measuring device is:

$$|\Phi_{\mathrm{fin}}^{\mathrm{MD}}\rangle=\prod_{j=1}^{N}\langle\uparrow_z|_j\exp\left\{\frac{\lambda}{N}\hat{Q}_{\mathrm{md}}\sum_{\mathrm{k}=1}^{N}\hat{S}_{\xi}^{k}\right\}\prod_{i=1}^{N}|\uparrow_x\rangle_i|\Phi_{\mathrm{in}}^{\mathrm{MD}}\rangle \tag{20.6}$$

$$=[\langle\uparrow_z|\uparrow_x\rangle]^N\left\{\cos\frac{\lambda\hat{Q}_{\mathrm{md}}}{N}-i\alpha_w\sin\frac{\lambda\hat{Q}_{\mathrm{md}}}{N}\right\}^N|\Phi_{\mathrm{in}}^{\mathrm{MD}}\rangle \tag{20.7}$$

$$=\underbrace{\left\{1-\frac{\lambda^2(\hat{Q}_{\mathrm{md}})^2}{N^2}-\frac{i\lambda\alpha_w\hat{Q}_{\mathrm{md}}}{N}\right\}^N}_{\equiv\psi(Q_{md})}|\Phi_{\mathrm{in}}^{\mathrm{MD}}\rangle\approx e^{i\lambda\alpha_w\hat{Q}_{\mathrm{md}}}|\Phi_{\mathrm{in}}^{\mathrm{MD}}\rangle \tag{20.8}$$

where we have substituted the weak value $\alpha_w \equiv (\hat{S}_\xi)_w = \frac{\langle\uparrow_z|\hat{S}_\xi|\uparrow_x\rangle}{\langle\uparrow_z|\uparrow_x\rangle}$. When projected onto the pointer P_{md}, we see it shifted by the weak value $\hat{S}_{45}^{(N)} = \frac{\sqrt{2}}{2}N \pm O(\sqrt{N})$.

Alternatively, we can also view the expression in the brackets of Eq. (20.8) (i.e. $\psi(Q_{md})$) in a very different way, by performing a binomial expansion:

$$\psi(Q_{md}) = \sum_{n=0}^{N} c_n \exp\left\{\frac{i\lambda \hat{Q}_{\text{md}}(2n-N)}{N}\right\} \tag{20.9}$$

Since the exponentials in $\psi(Q_{md})$ act as translation operators on the wavefunction of the measuring device, we see that this wavefunction is a superposition of waves with small wavenumbers $|k| \leq 1$ ($k = \frac{(2n-N)}{N}$). For a small region (which can include several wavelengths $2\pi/\alpha_w$, depending on how large one chooses N), $\psi(Q_{md})$ appears to have a very large momentum, since α_w (from Eq. (20.6)) can be arbitrarily large. This is an example of superoscillatory phenomenon.

20.2.2 Overview of the Mathematics of Superoscillations

The work reviewed in the previous sections led to the notion of a super-oscillating sequence [2]. Mathematically, superoscillatory sequences demonstrated that a superposition of small Fourier components with a bounded Fourier spectrum, each with modulus less or equal to 1, can nevertheless result in an oscillation by an arbitrarily large a, well outside the spectrum. They can be thought of as an approximation of e^{iax} in terms of a sequence of the form

$$\left\{\sum_{j=0}^{N} C_j(N,a)e^{ik_j(N)x}\right\}_{N=0}^{+\infty},$$

where $a > 1$, $|k_j(N)| \leq 1$. The example which is usually considered prototypical derives from the sequence of functions:

$$F_N(x,a) = \left[\cos\left(\frac{x}{N}\right) + ia\sin\left(\frac{x}{N}\right)\right]^N = \left(\frac{1+a}{2}e^{i\frac{x}{N}} + \frac{1-a}{2}e^{-i\frac{x}{N}}\right)^N \tag{20.10}$$

where $a \in \mathbb{R}$, $a > 1$. This sequence can be written as

$$\sum_{j=0}^{N} C_j(N,a)e^{i(1-2j/N)x}$$

for suitable coefficients $C_j(N,a)$. If we perform a binomial expansion of $F_N(x,a)$, we see that the smallest wavelength in the expansion is 1. However, around $|x| < \sqrt{N}$, $F_N(x,a)$ can be approximated as $F_N(x,a) \approx e^{iax}$, that is, with a wavelength much shorter than one. This phenomenon is very general and holds for a wide range of functions and coefficients.

Formally, we give the following definition:

Definition 20.1 We call a sequence a *superoscillating sequence* if it is of the form:

$$Y_N(x,a) := \sum_{j=0}^{N} C_j(N,a) e^{ik_j(N)x} \tag{20.11}$$

where $a \in \mathbb{R}$, $C_j(N,a)$ and $k_j(N)$ are real valued functions of the variables N, a and N, respectively, and such that

- $a > 1$, $|k_j(N)| \leq 1$;
- there exists a compact subset of $\mathbb{R}$, which will be called *a superoscillation set*, on which $Y_N(x,a)$ converges uniformly to $e^{ig(a)x}$ where g is a continuous function such that $|g(a)| \geq a$.

In [3], we demonstrated that the evolution of a superoscillating sequence of functions which are taken as the initial value of the Schrödinger equation

$$i\frac{\partial\psi(x,t)}{\partial t} = H\psi(x,t), \quad \psi(x,0) = Y_N(x),$$

where

$$H\psi(x,t) := -\frac{1}{2}\frac{\partial^2\psi(x,t)}{\partial x^2},$$

and where we have set $\hbar = m = 1$, remains superoscillating for all values of the time $t < \sqrt{N}$. This extends our recent results [16] where we considered the special case in which the initial datum is $\psi(x,0) = F_N(x,a)$ as in (20.10). The arguments we used in [16, 48] and in [3] take advantage of some refined functional analysis techniques, which highlight an interesting connection between superoscillating functions and some convolutors, which generalize infinite order differential operators.

In this paper, we prove the same result, but following a different approach which may be considered more direct and which allows a more intuitive physical interpretation. Let us briefly discuss the strategy that we will follow to prove the result. The first step is to represent the superoscillating sequence $F_N(x)$ (we will omit the dependence on a unless necessary) as a suitable double integral of a kernel multiplied by a modified gaussian. This will be accomplished by means of the windowed Fourier transform. The second step consists of studying an auxiliary problem, namely the time evolution of the modified gaussians when taken as initial value for the Schrödinger equation. The third step consists of evolving $F_N(x)$ by taking its integral representation and evolving the modified gaussians inside it. As a final step, we will recognize that the evolution of the modified gaussians for any fixed t and large enough N, results again in a gaussian with asympotically the same width and with its center translated in such a way that its tails do not interfere with the original gaussian. In summary, the reason for the longevity of the superoscillations is the following. The size of the tails grows at a rate proportional to a^N. The exponential tails oscillate at a rate proportional to $\frac{1}{a}$. The exponential tail to the left of the superoscillating region is both too slow and too far away to overcome and destroy the superoscillating region. The exponential tails to the right of the superoscillating region are more of a threat to the longevity of the superoscillating region because the faster moving superoscillating region could "catch-up" with the slower-moving exponential tail. In this paper, we prove that the superoscillating region can survive as long as the time $t < \sqrt{N}$.

Let us therefore begin by writing the functions in the superoscillating sequence as the inverses of their windowed Fourier transform, as introduced by Gabor in [49]. According to the uncertainty principle, the energy spread of a function and of its Fourier transform cannot be arbitrarily small simultaneously and it is well known that the minimum of the product of these two spreads is achieved by gaussians. Motivated by this principle in quantum mechanics, in 1946 the physicist Gabor defined what is now known as the Gabor chirp, i.e. the modified gaussian function

$$g_{u,\xi}(t) := e^{i\xi t} g(t-u)$$

where $g(t) = \frac{1}{\sqrt{\pi}} e^{-t^2}$. Using the Gabor chirp, one can define the so-called windowed Fourier transform as follows:

Definition 20.2 Let f be in $L^2(\mathbb{R})$, then its windowed Fourier transform (also known as the short time Fourier transform) is defined as

$$S(f)(u,\xi) = \int_{-\infty}^{+\infty} f(t) g(t-u) e^{-i\xi t} dt.$$

The following theorem is well known (see [50])

Theorem 20.2 *If f is a function in $L^2(\mathbb{R})$ then $Sf \in L^2(\mathbb{R})$, $\|Sf\| = \|f\|$ and*

$$f(t) = \frac{1}{2\pi} \int_{-\infty}^{+\infty} \int_{-\infty}^{+\infty} Sf(u,\xi) g(t-u) \exp(i\xi t) d\xi du.$$

In particular, we have:

Corollary 20.1 *Let K be compact in $\mathbb{R}$ and let χ_K be its characteristic function, then there are functions $g_N(x_0, k_0)$ such that*

$$Y_N(x,a) = \int_K \int_{-\infty}^{+\infty} g_N(x_0,k_0,a) \exp\left(\frac{-(x-x_0)^2}{2\Delta(0)^2}\right) \exp(ik_0 x) dx_0\, dk_0.$$

Proof After a trivial change of variable, this is an immediate consequence of Theorem 20.2. □

20.3 Evolution of Superoscillations

In this section we show how to solve the auxiliary Cauchy problem for the Schrödinger equation when the initial datum is the Gabor chirp centered at x_0

and with initially a single wave number k_0, and we use this result to conclude the longevity of superoscillations.

Specifically we consider the following problem:

$$i\frac{\partial\phi(x,t)}{\partial t}=-\frac{\partial^2\phi(x,t)}{\partial x^2},\qquad \phi(x,0)=\exp\left(-\frac{(x-x_0)^2}{2\Delta_0^2}+ik_0x\right),\tag{20.12}$$

where Δ_0 is the initial spread. We will solve this problem with a standard use of the Fourier and anti-Fourier transforms defined by

$$\mathcal{F}\big[\phi(x,t)\big]:=\int_{\mathbb{R}}\phi(x,t)e^{-ipx}dx,\qquad \mathcal{F}^{-1}\big[g(p,t)\big]:=\frac{1}{2\pi}\int_{\mathbb{R}}g(p,t)e^{ipx}dp.$$

For simplicity we will write $\mathcal{F}[\phi(x,t)]=\hat{\phi}(p,t)$. Taking the Fourier transform of the Schrödinger equation we get

$$i\frac{d\hat{\phi}(p,t)}{dt}=p^2\hat{\phi}(p,t)$$

and integrating we obtain

$$\hat{\phi}(p,t)=C(p)e^{-ip^2t}$$

where the arbitrary function $C(p)$ will be determined by the initial condition. Recall the well known Fourier transform

$$\int_{\mathbb{R}}e^{-\frac{x^2}{2\Delta_0^2}}e^{-ipx}\,dx=\sqrt{2\pi\,\Delta_0^2}\,e^{-\Delta_0^2p^2}$$

and the fact that $\mathcal{F}[g(x-x_0)]=\mathcal{F}[g(x)]e^{-ix_0p}$ so that we obtain

$$C(p)=\hat{\phi}(p,0)=\int_{\mathbb{R}}e^{-\frac{(x-x_0)^2}{2\Delta_0^2}+ik_0x}e^{-ipx}dx=\sqrt{2\pi\,\Delta_0^2}\,e^{-ipx_0-\Delta_0^2(p-k_0)^2}.$$

Taking now the anti-Fourier transform $\mathcal{F}^{-1}$ we have

$$\hat{\phi}(p,t)=\frac{1}{2\pi}\sqrt{2\pi\,\Delta_0^2}\int_{\mathbb{R}}e^{-ipx_0-\Delta_0^2(p-k_0)^2}e^{ipx}dp.$$

Now we finally obtain (see also formula (5.4) in [51] in which we have set $\hbar=m=1$):

$$\phi(x,t,x_0,k_0)=\frac{1}{(1+2it)^{1/2}}\exp\left(-i\frac{k_0^2}{2}t\right)\exp(ik_0x)\exp\left(-\frac{(x-x_0-2k_0t)^2}{2(\Delta_0^2+2it)}\right)$$

which can be written as

$$\phi(x,t,x_0,k_0)=\frac{1}{(1+2it)^{1/2}}\exp\left[i\left(\frac{k_0^2}{2}t+k_0x+t\frac{(x-x_0-2k_0t)^2}{\Delta_0^4+4t^2}\right)\right]$$
$$\times\exp\left(-\frac{(x-x_0-2k_0t)^2}{2(\Delta_0^2+4\frac{t^2}{\Delta_0^2})}\right).$$

Since the evolution according to the Schrödinger equation of the functions in the superoscillatory sequence is given by

$$Y_N(x,t) = \int\int g_N(x_0,k_0)\phi(x,t,x_0,k_0)dx_0\,dk_0,$$

we are now ready to prove that the functions $Y_N(x,t)$ preserve the superoscillatory behavior of $Y_N(x)$, and therefore that superoscillations persist for large values of t, when evolved according to the Schrödinger equation.

Theorem 20.3 *Let $Y_N(x)$ be a superoscillatory sequence. Then, for any fixed time t, its evolution $Y_N(x,t)$ obtained by solving the Cauchy problem for the Schrödinger equation with initial datum $Y_N(x)$ is still a superoscillatory sequence on any arbitrary large set in $\mathbb{R}$.*

Proof For any time t we can choose an N such that $t \approx N^{\frac{1}{2}-\varepsilon}$. Now we know that the time evolution of $\phi(x)$ is, up to the factor $(1+2it)^{-1}$, the product of an oscillatory function (with an amplitude of 1)

$$\exp\left[i\left(\frac{k_0^2}{2}t + k_0 x + t\frac{(x-x_0-2k_0t)^2}{\Delta_0^4+4t^2}\right)\right]$$

and of a translated gaussian

$$\exp\left(-\frac{(x-x_0-2k_0t)^2}{2(\Delta_0^2+4\frac{t^2}{\Delta_0^2})}\right).$$

We note that the spreads of these new gaussians are given by $\Delta^2(t) = \Delta_0^2 + t^2/\Delta_0^2$ which, by the assumption on t and if we choose $\Delta_0 \approx N^{\frac{1}{2}}$, are approximatively the same at any given moment. Now observe that this wave packet has width approximately equal to $\sqrt{N}$ and therefore if we consider it centered in the point $x_0 + 2k_0 t$ and so in the interval $[\lambda\sqrt{N}, (\lambda+1)\sqrt{N}]$ for some λ, we see that its contribution outside its spread does not interfere with the original superoscillatory region. □

References

1. M.V. Berry, Faster than Fourier, in *Quantum Coherence and Reality*, ed. by J.S. Anandan, J.L. Safko (World Scientific, Singapore, 1994), pp. 55–65 (in celebration of the 60th Birthday of Yakir Aharonov)
2. Y. Aharonov, F. Colombo, I. Sabadini, D.C. Struppa, J. Tollaksen, Some mathematical properties of superoscillations. J. Phys. A **44**, 365304 (2011)
3. Y. Aharonov, F. Colombo, I. Sabadini, D.C. Struppa, J. Tollaksen, The mathematics of superoscillations, in preparation
4. M.V. Berry, J. Phys. A **27**, L391–L398 (1994)
5. M.V. Berry, J. Phys. A **43**, 415302 (2010)
6. M.V. Berry, N. Brunner, S. Popescu, P. Shukla, J. Phys. A **44**, 492001 (2011)
7. M.V. Berry, J. Phys. A **45**, 185308 (2012)

8. Y. Aharonov, J. Anandan, S. Popescu, L. Vaidman, Phys. Rev. Lett. **65**, 2965 (1990)
9. P.J.S.G. Ferreira, A. Kempf, J. Phys. A **37**, 12067–12076 (2004)
10. P.J.S.G. Ferreira, A. Kempf, IEEE Trans. Signal Process. **54**, 3732–3740 (2006)
11. J. Lindberg, Mathematical concepts of optical superresolution. J. Opt. **14**, 083001 (2012)
12. W. Qiao, J. Phys. A **29**, 2257–2258 (1996)
13. M.V. Berry, J. Opt. **15**(4), 044006 (2013)
14. M.V. Berry, J. Phys. A: Math. Theor. **46**(20), 205203 (2013)
15. M.V. Berry, S. Popescu, J. Phys. A **39**, 6965–6977 (2006)
16. Y. Aharonov, F. Colombo, I. Sabadini, D.C. Struppa, J. Tollaksen, On the Cauchy problem for the Schrödinger equation with superoscillatory initial data. J. Math. Pures Appl. **99**, 165–173 (2013)
17. Y. Aharonov, F. Colombo, S. Nussinov, I. Sabadini, D.C. Struppa, J. Tollaksen, Superoscillations phenomenona in SO(3). Proc. R. Soc. A, Math. Phys. Eng. Sci. **468**, 3587–3600 (2012)
18. Y. Aharonov, F. Colombo, I. Sabadini, D.C. Struppa, J. Tollaksen, On some operators associated to superoscillations. Complex Anal. Oper. Theory **7**, 1299–1310 (2013)
19. M.R. Dennis, C. Hamilton, Opt. Lett. **33**, 2976–2978 (2008)
20. E.T.F. Rogers, J. Lindberg, T. Roy, S. Savo, J.E. Chad, M.R. Dennis, N.I. Zheludev, Nat. Mater. **11**, 432–435 (2012)
21. Y. Aharonov, S. Popescu, J. Tollaksen, A time-symmetric formulation of quantum mechanics. Phys. Today **63**(11), 27–32 (2010)
22. Y. Aharonov, D.Z. Albert, L. Vaidman, Phys. Rev. Lett. **60**, 1351–1354 (1988)
23. M. Büttiker, Phys. Rev. B **27**, 6178 (1983)
24. Y. Aharonov, S. Popescu, D. Rohrlich. Preprint, TAUP 184790, Tel Aviv University (1990)
25. Y. Aharonov, L. Davidovich, N. Zagury, Phys. Rev. A **48**, 1687 (1993)
26. Y. Aharonov, P.G. Bergmann, J.L. Lebowitz, Phys. Rev. B **134**, 1410 (1964)
27. Y. Aharonov, L. Vaidman, Phys. Rev. A **41**, 11–20 (1990)
28. Y. Aharonov, L. Vaidman, Time in Quantum Mechanics, ed. by J. Muga, R. Sala Mayato, I. Egusquiza. quant-ph/0105101
29. N. Ritchie, J. Story, R. Hulet, Phys. Rev. Lett. **66**, 1107 (1991)
30. A. Parks, D. Cullen, D. Stoudt, Proc. R. Soc. Lond. A **454**, 2997 (1998)
31. K. Resch, J. Lundeen, A. Steinberg, Phys. Lett. A **324**, 125 (2004)
32. Q. Wang, F. Sun, Y. Zhang, J. Li, Y. Huang, G. Guo, Phys. Rev. A **73**, 023814 (2006)
33. O. Hosten, P. Kwiat, Science **319**, 787 (2008)
34. K. Yokota, T. Yamamoto, M. Koashi, N. Imoto, New J. Phys. **11**, 033011 (2009)
35. P. Dixon, D. Starling, A. Jordan, J. Howell, Phys. Rev. Lett. **102**, 173601 (2009)
36. J. Tollaksen, in *Quantum Information and Computation V*, ed. by E. Donkor, A. Pirich, H. Brandt. Proc. of SPIE, vol. 6573 (2007) (CID 6573-33)
37. M.V. Berry, M.R. Dennis, J. Phys. A **42**, 022003 (2009)
38. M.V. Berry, P. Shukla, Pragya, J. Phys. A **45**, 015301 (2012)
39. M.V. Berry, M.R. Dennis, B. McRoberts, P. Shukla, J. Phys. A **44**, 205301 (2011)
40. Y. Aharonov, A. Botero, Quantum averages of weak values. Phys. Rev. A **72**(5), 052111 (2005)
41. A. Botero, Sampling weak values: a non-linear Bayesian model for non-ideal quantum measurements. arXiv:quant-ph/0306082
42. M.V. Berry, P. Shukla, J. Phys. A **43**, 354024 (2010)
43. Y. Aharonov, S. Massar, S. Popescu, J. Tollaksen, L. Vaidman, Phys. Rev. Lett. **77**, 983 (1996)
44. Y. Aharonov, A. Botero, S. Popescu, B. Reznik, J. Tollaksen, Phys. Lett. A **301**, 130–138 (2002)
45. S.N. Nussinov, J. Tollaksen, Phys. Rev. D **78**, 036007 (2008)
46. J. Tollaksen, Robust weak measurements on finite samples. J. Phys. Conf. Ser. **70**, 012015 (2007)
47. J. Tollaksen, J. Phys. A **40**, 9033–9066 (2007)
48. Y. Aharonov, F. Colombo, I. Sabadini, D.C. Struppa, J. Tollaksen, Superoscillating sequences as solutions of generalized Schrödinger equations. Preprint (2013)

49. D. Gabor, Theory of communication. J. Inst. Electr. Eng., Lond. **93**, 429–457 (1946)
50. S. Mallat, *A Wavelet Tour of Signal Processing* (Academic Press, San Diego, 1998)
51. Y. Aharonov, D. Rohrlich, *Quantum Paradoxes: Quantum Theory for the Perplexed* (Wiley-VCH, Weinheim, 2005)

Chapter 21
Superoscillations, Endfire and Supergain

M.V. Berry

Abstract Superoscillatory functions vary faster than their fastest Fourier component. Here they are employed to give an alternative description and explicit recipe for creating endfire arrays with supergain, that is antennas with radiation patterns concentrated in an arbitrarily narrow angular range and of arbitrary form. Two examples are radiation patterns described by sinc and Gaussian functions. [*Editor's note*: for a video of the talk given by Prof. Berry (titled 'Weak Value Probabilities') at the Aharonov-80 conference in 2012 at Chapman University, see quantum.chapman.edu/talk-6.]

Dedicated to Yakir Aharonov on his 80th birthday: still quick, still deep, still subtle.

21.1 Introduction

The three elements of the title are connected. My aim here is to show how, in the hope of giving further insight into them.

A central ingredient of Aharonov's weak measurement scheme [1–3] is the concept of *superoscillations*. This is a property of band-limited functions: they can oscillate arbitrarily faster than their fastest Fourier component, over arbitrarily long intervals [4–6]. The price paid for this apparently paradoxical behaviour is that the functions are exponentially larger outside the superoscillatory region than in it.

An *endfire array* is an antenna in the form of a set of radiating sources arranged on a line (Fig. 21.1), the object of interest being the radiation pattern, that is, the angular distribution of wave intensity in the far field. It has been known at least since the 1940s [7–10] that if the array contains sufficiently many sources their strengths and phases can be so arranged that the radiation pattern is confined to an arbitrarily narrow region near the forward direction, even if the line containing all the sources

M.V. Berry (✉)
H H Wills Physics Laboratory, Tyndall Avenue, Bristol BS8 1TL, UK
e-mail: asymptotico@bristol.ac.uk

D.C. Struppa, J.M. Tollaksen (eds.), *Quantum Theory: A Two-Time Success Story*,
DOI 10.1007/978-88-470-5217-8_21, © Springer-Verlag Italia 2014

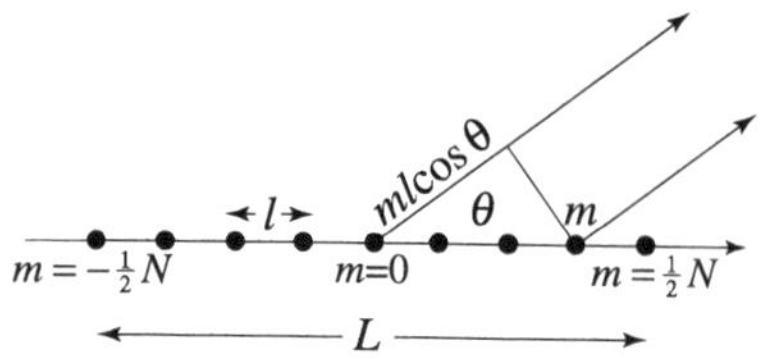

Fig. 21.1 Geometry and notation for endfire array with $N+1$ radiating sources

is arbitrarily smaller than the wavelength of the radiation. This phenomenon, apparently contradicting diffraction-theory folklore about resolving power, is *supergain* (also called superdirectivity). The concept, originating in radar physics, has been applied in optics [11–17].

Section 21.2 gives a general theory of supergain in terms of superoscillations, leading to a simple recipe for creating radiation patterns of any form, and connects this with the traditional explanation involving complex-variable theory. Section 21.3 gives two explicit examples: radiation patterns in the form of sinc functions and gaussians.

21.2 General Theory

Consider $N+1$ sources (N even) arranged uniformly on a line of length L (Fig. 21.1), so their spacing is $l = L/N$. The sources are point radiators of monochromatic scalar waves with wave length λ and wavenumber $k = 2\pi/\lambda$. With excitation amplitudes A_m, the far-field angular amplitude is

$$\begin{aligned}\psi(\theta) &= \sum_{m=-\frac{1}{2}N}^{\frac{1}{2}N} A_m \exp(-imkl\cos\theta + imkl) \\ &= \sum_{m=-\frac{1}{2}N}^{\frac{1}{2}N} A_m \exp\left(2imkl\sin^2\frac{1}{2}\theta\right). \end{aligned} \tag{21.1}$$

The phases $imkl$ have been included so that if all the A_m have the same phase the sources add coherently in the forward direction $\theta = 0$.

In this simplest case—all the A_m equal—the angular distribution is

$$A_m = \frac{1}{N+1}, \qquad \psi(\theta) = \frac{\sin(\frac{(N+1)}{N}kL\sin^2\frac{1}{2}\theta)}{(N+1)\sin(\frac{1}{N}kL\sin^2\frac{1}{2}\theta)} \approx \frac{\sin(kL\sin^2\frac{1}{2}\theta)}{kL\sin^2\frac{1}{2}\theta}. \tag{21.2}$$

The zero-amplitude directions ('cones of silence') are

$$\theta_n = 2\sin^{-1}\sqrt{\frac{\pi N n}{(N+1)kL}}, \qquad 1 \le |n| \le \frac{kL}{\pi} = \frac{2L}{\lambda}. \tag{21.3}$$

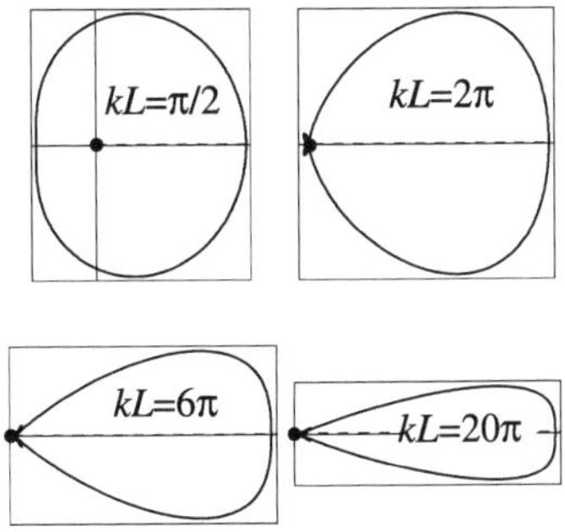

Fig. 21.2 Radiation patterns without supergain: polar plots of $\psi^2(\theta)$ given by (21.2) for the indicated values of kL. The *dots* indicate the origin $\psi^2(\theta)$, and the *dashed lines* indicate the forward direction $\theta = 0$

In particular, for large N, representing a line source, and $kL \gg 1$, the first zero is

$$\theta_1 \approx 2\sqrt{\frac{\pi}{kL}}, \tag{21.4}$$

as expected on the basis of the Rayleigh resolution criterion. As is well known, and as Fig. 21.2 illustrates, a narrow radiation pattern requires $kL \gg 1$.

However, by suitable choice of the strengths and phases of the amplitudes A_m it is possible to create radiation patterns as narrow as desired and indeed of arbitrary shape, even for $kL \ll 1$, i.e. $L \ll \lambda$—that is, to get supergain. To achieve supergain using superoscillatory functions, we first note that θ is not the natural variable, because $\psi(\theta)$ in (21.1), is not band-limited. This follows from the Bessel-Fourier relation

$$\exp(it\cos\theta) = \sum_{n=-\infty}^{\infty} i^n J_n(t)\exp(in\theta), \tag{21.5}$$

indicating that each term in (21.1) contains infinitely many angular Fourier components (i.e. angular momenta). However, defining the new variable x by

$$x \equiv kL\sin^2\frac{1}{2}\theta \tag{21.6}$$

gives the function

$$f(x) = \psi(\theta) = \sum_{m=-\frac{1}{2}N}^{\frac{1}{2}N} A_m \exp\left(\frac{2imx}{N}\right), \tag{21.7}$$

which is band-limited, with largest Fourier components proportional to $\exp(\pm ix)$.

It is possible to create such functions with arbitrarily fast oscillations for small $|x|$, e.g.

$$f_{\text{super}}(x,a) \approx \exp(iax) \quad \big(a > 1, |x| \ll 1, N \gg 1\big). \tag{21.8}$$

The most familiar and much-studied [13] superoscillatory function, to be used extensively in the following, is

$$f_{\text{super}}(x) = \left(\cos\frac{x}{N} + \mathrm{i}a\sin\frac{x}{N}\right)^N. \tag{21.9}$$

This is periodic in x with period N. The small-x approximation $\exp(\mathrm{i}ax)$ oscillates rapidly but does not lead to a concentrated radiation pattern.

However, the simple procedure of integrating over the scale variable a, with a suitable weight function, enables any radiation pattern to be created, including the narrow ones corresponding to supergain. (This is a simplified variant, suggested to me by Professor Sandu Popescu, of a procedure previously used in an acoustic example in Sect. 4 of [4].) Let $g(a)$ be the Fourier transform of the desired angular amplitude, and construct the function

$$f(x) = \int_{-\infty}^{\infty} \mathrm{d}a f_{\text{super}}(x, a)g(a). \tag{21.10}$$

Then, for $|x| \ll 1$, $f(x)$ has, from (21.8), the desired form

$$f(x) \approx \int_{-\infty}^{\infty} \mathrm{d}a \exp(\mathrm{i}ax)g(a). \tag{21.11}$$

If $g(a)$ is even, as it will be in the examples in Sect. 21.3, $f(x)$ is a real function. The radiation pattern thus constructed is

$$\psi(\theta) = f\left(kL\sin^2\frac{1}{2}\theta\right). \tag{21.12}$$

As θ increases, this samples $f(x)$ from $x = 0$, corresponding to $\theta = 0$, to $x_{\text{max}} = kL$, corresponding to the backward direction $\theta = \pi$. It is necessary to take N large enough to ensure that x_{max} lies in the superoscillatory interval where (21.8) applies, and this will be illustrated in Sect. 21.3. For $x > x_{\text{max}}$ ('beyond backwards') the values of θ given by (21.6) would be complex.

The formula (21.12) represents the far field. We will see that the coefficients A_m, whose near-cancellation is responsible for supergain, take enormously large values (see Fig. 21.5 later). This implies that the wave intensity in the near field is correspondingly large, a phenomenon responsible for the well-known inefficiency of supergain antennas [8–10].

The conventional way of understanding supergain [7, 11] is via the complex variable

$$z = \exp\left(\frac{2\mathrm{i}x}{N}\right) = \exp\left(2\mathrm{i}kl\sin^2\frac{1}{2}\theta\right), \tag{21.13}$$

and the corresponding function

$$F(z) = \psi(\theta) = f(x) = \sum_{m=-\frac{1}{2}N}^{\frac{1}{2}N} A_m z^m. \tag{21.14}$$

Physical directions $-\pi \le \theta \le \pi$ (i.e. real θ, so $\sin^2(\theta/2) \le 1$) correspond to z on the unit circle and $|\arg z| \le 2kl$; for fixed L, this region shrinks towards the forward direction $z = 1$ as N increases. The zeros of $F(z)$ correspond to the cones of silence. In the simplest case discussed earlier, in which all A_m are in phase, we have

$$A_m = \frac{1}{N+1}, \qquad F(z) = \frac{(z^{N+1} - 1)}{(z-1)}, \tag{21.15}$$

whose zeros, uniformly distributed round the circle, are the $(N+1)$th roots of unity with $z = 1$ excluded. Supergain [11] corresponds to choosing the A_m so that the zeros, whose density increases with N, get concentrated into the physical region near $\arg z \ll 1$, that is, near the forward direction.

21.3 Examples: Sinc and Gaussian Radiation Patterns

21.3.1 Sinc Radiation Pattern

The simplest implementation of (21.10) is to make $g(a)$ constant on a finite interval and zero outside, that is, the Fourier transform of the sinc function. Then if the width of the interval is $2/w$, the function $f(x)$ given by (21.10) is

$$\begin{aligned} f(x, w) &= 2w \int_{-1/w}^{1/w} da f_{\text{super}}(x, a) \\ &= \frac{w}{(N+1)\sin\frac{x}{N}} \operatorname{Im}\left[\left(\cos\frac{x}{N} + \frac{\mathrm{i}}{w}\sin\frac{x}{N}\right)^{N+1}\right], \end{aligned} \tag{21.16}$$

and the radiation pattern is

$$\begin{aligned} \psi(\theta, w) = &\frac{w}{(N+1)\sin(kl\sin^2\frac{1}{2}\theta)} \\ &\times \operatorname{Im}\left[\left(\cos\left(kl\sin^2\frac{1}{2}\theta\right) + \frac{\mathrm{i}}{w}\sin\left(kl\sin^2\frac{1}{2}\theta\right)\right)^{N+1}\right]. \end{aligned} \tag{21.17}$$

When $w = 1$, this reproduces the radiation pattern of a line of coherently radiating sources (Eq. (21.2)). Otherwise, the analogous relation to (21.8) for small x and

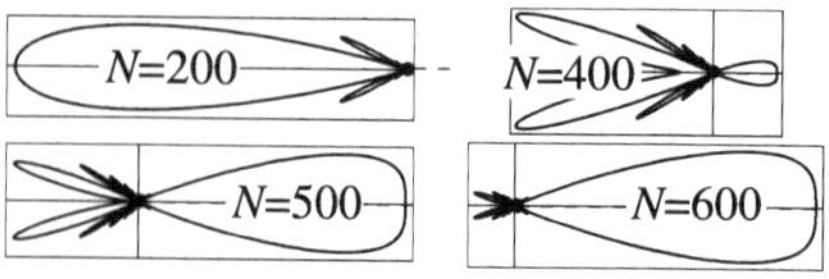

Fig. 21.3 Radiation patterns with supergain: polar plots of the sinc-based pattern $\psi^2(\theta)$ given by (21.17) for $kL = \pi/2$ (that is, an endfire array of length $L = \lambda/4$), and $w = 1/40$, for the indicated values of N. The *dots* indicate the origin $\psi = 0$, and the *dashed lines* indicate the forward direction $\theta = 0$

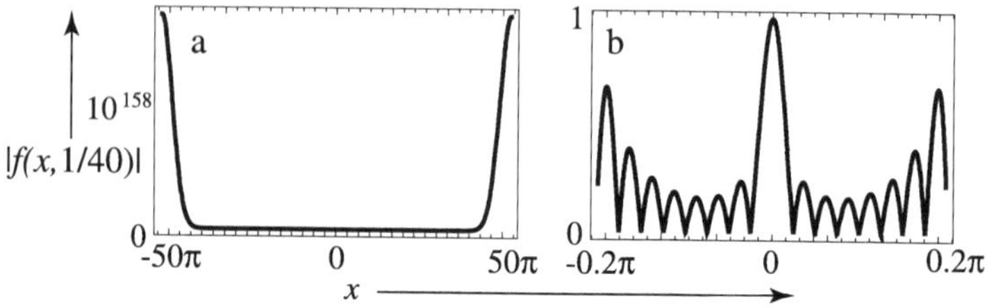

Fig. 21.4 Modulus of sinc-based function $|f(x, 1/40)|$ (Eq. (21.16)) for $N = 100$. (**a**) includes a complete period $|x| \leq 50\pi$, and the sinc function near the origin is invisible; (**b**) includes the much smaller range $|x| \leq 0.2\pi$, revealing the sinc

large N, namely

$$f(x, w) \approx \frac{\sin(x/w)}{(x/w)} \quad (x \ll 1, N \gg 1), \tag{21.18}$$

gives the approximate pattern

$$\psi(\theta, w) \approx \frac{w \sin(\frac{kL}{w} \sin^2 \frac{1}{2}\theta)}{kL \sin^2 \frac{1}{2}\theta}, \tag{21.19}$$

and hence supergain if $w < 1$. Figure 21.3 shows how supergain emerges as N increases, in the form of a narrow sinc radiation pattern for a short endfire array with length $L = \lambda/4$, and the small value $w = 1/40$.

When N is too small, the radiation is concentrated near the backward direction (e.g. for $N = 200$ in Fig. 21.3), because the limiting value $x_{\text{max}} = kL$ sampled in $f(x, w)$ (Eq. (21.16)) is outside the region near $x = 0$ where the desired sinc function occurs, and instead lies in the exponentially large region beyond. Figure 21.4 illustrates the enormous disparity of values between these regions. For a quantitative understanding, we approximate $\psi(\theta)$ in (21.17) near $\theta = \pi$ and for $w < kL$:

$$\begin{aligned}\psi(\pi - \mu) \approx \left(\frac{w}{kL}\right) \sin\left(\frac{kL}{w}\left(1 - \frac{1}{4}\mu^2\right)\right) \\ \times \exp\left(\frac{1}{2N}\left(\frac{kL}{w}\right)^2 (1 - w^2)\left(1 - \frac{1}{2}\mu^2\right)\right) \quad (\mu \ll 1). \end{aligned} \tag{21.20}$$

This is exponentially large unless N exceeds a critical value N_c, which can be defined by requiring ψ^2, averaged over the oscillations near $\theta = \pi$, to equal the value $\psi(\theta = 0) = 1$. For $N > N_c$, the value near $\theta = \pi$ is exponentially small, corresponding to supergain. From (21.20),

$$N_c = \left(\frac{kL}{w}\right)^2 \frac{(1-w^2)}{2\log\frac{\sqrt{2}kL}{w}}. \tag{21.21}$$

For the parameters of Fig. 21.3, $N_c = 440$, in fair agreement with the crossover to supergain.

The zeros of the radiation pattern are the directions

$$\theta_n = 2\sin^{-1}\sqrt{\frac{N}{kL}\tan^{-1}\left(w\tan\frac{n\pi}{N+1}\right)} \quad \left(1 \le |n| \le \frac{1}{2}N\right). \tag{21.22}$$

When $w = 1$ these cones of silence coincide with those of the equally phased array in (21.3). For small w they cluster closer to $\theta = 0$:

$$\theta_n \approx 2\sqrt{\frac{wn\pi}{kL}} \quad (w \ll 1, N \gg 1, n \ll N). \tag{21.23}$$

In terms of the variable z of conventional supergain theory, the zeros cluster closer to $z = 1$ on the unit circle.

The remarkable superdirectivity accomplished by superoscillation is a consequence of near-destructive interference between successive sources. For the pattern (21.17), the excitation amplitudes in (21.16) are

$$\begin{aligned} A_m(w) &= \frac{wN!}{2^{N+1}(\frac{1}{2}N+m)!(\frac{1}{2}N-m)!}\int_{-1/w}^{1/w} \mathrm{d}a\,(1-a^2)^{\frac{1}{2}N}\left(\frac{1+a}{1-a}\right)^m \\ &= \frac{wN!}{2^{\frac{1}{2}N-m+1}(\frac{1}{2}N-m)!} \\ &\quad\times \sum_{k=0}^{\frac{1}{2}N+m} \frac{(-1)^{\frac{1}{2}N-m}(\frac{1}{w}-1)^{\frac{1}{2}N-m+k+1} + (-1)^k(\frac{1}{w}+1)^{\frac{1}{2}N-m+k+1}}{2^k k!(\frac{1}{2}N+m-k)!(\frac{1}{2}N-m+k+1)}. \end{aligned} \tag{21.24}$$

(The sum can be expressed in terms of incomplete beta functions.) It is clear that $A_{-m}(w) = A_m(w)$, so we need study only $m \ge 0$. The simplest case is $w = 1$, for which $A_m(1) = 1/(N+1)$, corresponding to all sources in phase. For general w, the coefficients for the first few values of N are shown in Table 21.1.

The case of interest for supergain is $N \gg 1$, $w < 1$. Then Stirling's formula, and expansion of the integrands about their maxima, gives the asymptotics in terms of

Table 21.1 Coefficients $A_m(w)$ for $N = 2, 4, 6$

	$N=2$	$N=4$	$N=6$
$m=0$	$\frac{1}{6w^2}(-1+3w^2)$	$\frac{1}{40w^4}(3-10w^2+15w^4)$	$\frac{1}{112w^6}(-5+7w^2(3-5w^2+5w^4))$
$m=1$	$\frac{1}{12w^2}(1+3w^2)$	$\frac{1}{20w^4}(-1+5w^4)$	$\frac{1}{448w^6}(15+7w^2(-3-5w^2+15w^4))$
$m=2$		$\frac{1}{80w^4}(1+10w^2+5w^4)$	$\frac{1}{224w^6}(-3+7w^2(-3+5w^2+3w^4))$
$m=3$			$\frac{1}{448w^6}(1+7w^2(3+5w^2+w^4))$

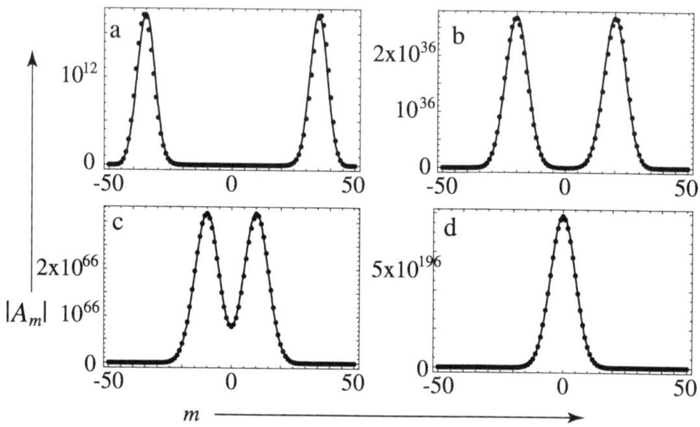

Fig. 21.5 Points: moduli of excitation coefficients A_m (Eq. (21.24)) for endfire array with $N+1 = 101$ radiating sources, for (**a**) $w = 0.7$, (**b**) $w = 0.4$, (**c**) $w = 0.2$, (**d**) $w = 0.01$. The *curves* show the approximation (21.25)

two Gaussians:

$$A_m(w) \approx \frac{(-1)^{\frac{1}{2}N+m}}{N^{\frac{3}{2}} w^N \sqrt{2\pi(1-w^2)}} \left[\exp\left(-\frac{2(m-\frac{1}{2}Nw)^2}{N(1-w^2)} \right) + \exp\left(-\frac{2(m+\frac{1}{2}Nw)^2}{N(1-w^2)} \right) \right]. \tag{21.25}$$

Note the alternating signs, indicating the near-cancellation responsible for supergain. Figure 21.5 shows the moduli of the coefficients. The agreement with the asymptotic formula appears excellent. However, using (21.25) instead of the exact coefficients in (21.1) leads to numerical instability and no hint of supergain: the interference is far too delicate to be captured by this lowest level of asymptotics. (For $w > 1$, the asymptotics is different: all amplitudes are in phase and there is no supergain.)

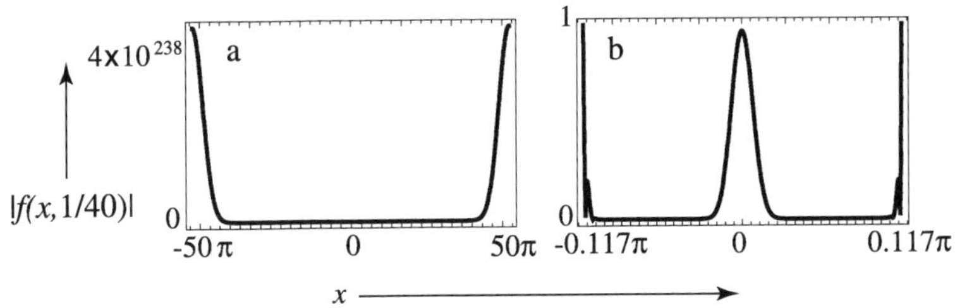

Fig. 21.6 Modulus of Gaussian-based function $|f(x,1/40)|$ (Eq. (21.26)) for $N=100$. (**a**) includes a complete period $|x|\leq 50\pi$, and the Gaussian near the origin is invisible; (**b**) includes the much smaller range $|x|\leq 0.117\pi$, revealing the Gaussian

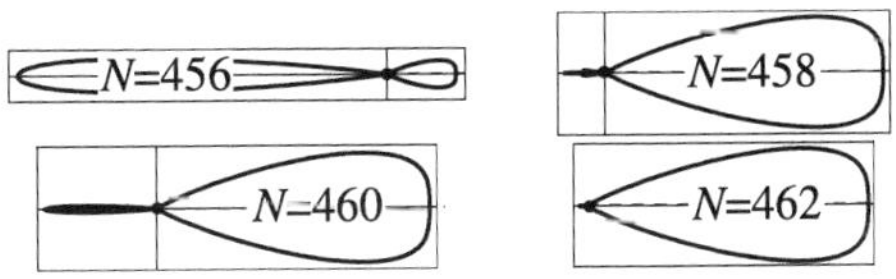

Fig. 21.7 As Fig. 21.3, for the Gaussian-based function generated from (21.26) using (21.12), with $w=1/20$

21.3.2 Gaussian Radiation Pattern

The simplest example of a desired radiation pattern without zeros is the Gaussian, for which (21.10) is

$$\begin{aligned} f(x,w) &= \frac{w}{\sqrt{2\pi}}\int_{-\infty}^{\infty} da\, f_{\mathrm{super}}(x,a)\exp\left(-\frac{1}{2}a^2w^2\right) \\ &= \frac{\sin^N(x/N)}{(N+1)w^N 2^{N/2}}\left(\frac{w}{\sqrt{2}}\cot\left(\frac{x}{N}\right)H_{N+1}\left(\frac{w}{\sqrt{2}}\cot\left(\frac{x}{N}\right)\right)\right. \\ &\quad \left.-\frac{1}{2}H_{N+2}\left(\frac{w}{\sqrt{2}}\cot\left(\frac{x}{N}\right)\right)\right), \end{aligned} \tag{21.26}$$

where H denotes the Hermite polynomials. For small x, this gives the desired Gaussian with width w, and supergain for $w<1$:

$$f(x,w)\approx \exp\left(-\frac{x^2}{2w^2}\right) \quad (x\ll 1). \tag{21.27}$$

(This limiting form, obvious from (21.11), can also be obtained from (21.26) by a large N resummation of the power series for the Hermite polynomials.) As in the case of the sinc pattern, enormous magnification is required to see the Gaussian, as Fig. 21.6 illustrates.

Some of the corresponding radiation patterns, obtained by the substitution (21.12), are shown in Fig. 21.7. In this case, the crossover to supergain as N increases is much more sensitive than in the sinc case (cf. Fig. 21.3).

Acknowledgements I thank Professor Stephen Lipson for introducing me to supergain, Professor Sandu Popescu for a helpful suggestion, Chapman University for generous hospitality while this work was begun, and the Leverhulme Trust for research support—and of course Yakir Aharonov for continuing inspiration.

References

1. Y. Aharonov, D.Z. Albert, L. Vaidman, How the result of a measurement of a component of the spin of a spin 1/2 particle can turn out to be 100. Phys. Rev. Lett. **60**, 1351–1354 (1988)
2. Y. Aharonov, D. Rohrlich, *Quantum Paradoxes: Quantum Theory for the Perplexed* (Wiley-VCH, Weinheim, 2005)
3. Y. Aharonov, S. Popescu, J. Tollaksen, A time-symmetric formulation of quantum mechanics. Phys. Today **63**(11), 27–33 (2010)
4. M.V. Berry, in *Faster than Fourier in Quantum Coherence and Reality; in Celebration of the 60th Birthday of Yakir Aharonov*, ed. by J.S. Anandan, J.L. Safko (World Scientific, Singapore, 1994), pp. 55–65
5. A. Kempf, P.J.S.G. Ferreira, Unusual properties of superoscillating particles. J. Phys. A **37**, 12067–12076 (2004)
6. Y. Aharonov, F. Colombo, I. Sabadini, D.C. Struppa, J. Tollaksen, Some mathematical properties of superoscillations. J. Phys. A **44**, 365304 (2011). (16 pp.)
7. S.A. Schelkunoff, A mathematical theory of linear arrays. Bell Syst. Tech. J. **22**, 80–107 (1943)
8. R.C. Hansen, Linear arrays, ed. by A.W. Rudge, K. Milne, A.D. Olver, P. Knight, *The Handbook of Antenna Design*, vol. 2 (Peter Peregrinus, London, 1983), pp. 1–134
9. H. Cox, R.M. Zeskind, T. Kooij, Practical supergain. IEEE Trans. Acoust. Speech Signal Process. **ASSP-34**, 393–398 (1986)
10. R.P. Haviland, Supergain antennas: possibilities and problems. IEEE Antennas Propag. Mag. **37**, 13–26 (1995)
11. G. Toraldo di Francia, SuperGain antennas and optical resolving power. Suppl. Nuovo Cim. **9**, 426–438 (1952)
12. I. Leiserson, S.G. Lipson, V. Sarafis, Superresolution in far-field imaging. Opt. Lett. **25**, 209–211 (2000)
13. M.V. Berry, S. Popescu, Evolution of quantum superoscillations, and optical superresolution without evanescent waves. J. Phys. A **39**, 6965–6977 (2006)
14. J. Lindberg, Mathematical concepts of optical superresolution. J. Opt. **14**, 083001 (2012). (23 pp.)
15. M.R. Dennis, A.C. Hamilton, J. Courtial, Superoscillation in speckle patterns. Opt. Lett. **33**, 2976–2978 (2008)
16. E.T.F. Rogers, J. Lindberg, T. Roy, S. Savo, J.E. Chad, M.R. Dennis, N.I. Zheludev, A superoscillatory lens optical microscope for subwavelength imaging. Nat. Mater. **11**, 432–435 (2012)
17. M.V. Berry, M.R. Dennis, Natural superoscillations in monochromatic waves in D dimensions. J. Phys. A **42**, 022003 (2009)

Chapter 22
Relating Local Time Evolutions with Bipartite States: An Exact Map Manifested by Weak Measurements

S. Marcovitch and B. Reznik

Abstract We suggest a natural mapping between bipartite states and quantum evolutions of local states, which is a Jamiolkowski map. It is shown that spatial correlations of weak measurements in bi-partite systems precisely coincide with temporal correlations of local systems. This mapping has several practical and conceptual implications on the correspondence between Bell and Leggett-Garg inequalities, the statistical properties of evolutions in large systems, temporal decoherence and computational gain, in evaluation of spatial correlations of large systems. [*Editor's note*: for a video of the talk given by Prof. Reznik at the Aharonov-80 conference in 2012 at Chapman University, see quantum.chapman.edu/talk-23.]

22.1 Introduction

The quantum mechanical nature of bipartite states is well known. Among bipartite states the maximally entangled ones manifest maximal nonlocality, a unique property of quantum theory. These states may be regarded as possessing the highest degree of quantumness. One may expect that in the time domain unitary evolutions would manifest the highest degree of quantumness too, as they describe evolutions without environment interference. But how do we test the quantum mechanical nature of evolutions? Using tomography of states at different times the evolution and environment traces can be tracked given a known pure state to begin with. However, this method is indirect. In particular, with tomography the evolution of the maximally mixed state cannot be tested.

The temporal correlations of observables before and after the tested dynamics, however, provide a direct examination of the evolution. In this letter we construct a Jamiolkowski map [1] between the space of bipartite systems $\rho_{AB} \in H_A \otimes H_B$ and the space of time evolutions transforming systems from Hilbert spaces H_A to H_B. In this map spatial correlations of two separated operators in bipartite systems precisely coincide with the temporal correlations of the mapped operators in local

S. Marcovitch · B. Reznik (✉)
School of Physics and Astronomy, Raymond and Beverly Sackler Faculty of Exact Sciences, Tel-Aviv University, Tel-Aviv 69978, Israel
e-mail: benni.reznik@gmail.com

D.C. Struppa, J.M. Tollaksen (eds.), *Quantum Theory: A Two-Time Success Story*,
DOI 10.1007/978-88-470-5217-8_22,

Table 22.1 one-to-one map between time evolutions and bipartite states

Temporal (assuming the system is maximally mixed $\rho_S = I/d_A$)	Spatial
Pure evolution	Pure bipartite system
$M = \sqrt{d_A}\sum_{ij}\alpha^*_{ji}\|i\rangle\langle j\|$ where $\rho_S \to M\rho_S M^\dagger$	$\|\psi_{AB}\rangle = \sum_{ij}\alpha_{ij}\|i\rangle \otimes \|j\rangle$
$\mathrm{Tr}[M^\dagger M\rho_S] = 1$	$\langle\psi_{AB}\|\psi_{AB}\rangle = 1$
Unitary evolution $M = U,\ UU^\dagger = I$	Maximally entangled $\|\psi_{AB}\rangle$
Selective projector $M = \sqrt{d_A}\|i\rangle\langle j\|$	Product state $\|\psi_{AB}\rangle = \|i\rangle \otimes \|j\rangle$
Mixed evolution ($\sum p_\mu = 1$)	Mixed bipartite system ($\sum p_\mu = 1$)
$\{M_\mu,\ p_\mu\}:\ \rho_S \to \sum_\mu p_\mu M_\mu \rho_S M^\dagger_\mu$	$\{\|\psi^\mu_{AB}\rangle,\ p_\mu\}:\ \rho_{AB} = \sum_\mu p_\mu \|\psi^\mu_{AB}\rangle\langle\psi^\mu_{AB}\|$
Non-selective environment $\sum_\mu p_\mu M^\dagger_\mu M_\mu = I$	$\rho_A = I/d_A,\ \rho_B = I/d_B$
Representation transformation with unitary U	Representation transformation with unitary U
$\tilde{N}_\nu = U_{\nu\mu}\sqrt{p_\mu}M_\mu$ (un-normalized)	$\|\tilde{\phi}^\nu_{AB}\rangle = U_{\nu\mu}\sqrt{p_\mu}\|\psi^\mu_{AB}\rangle$ (un-normalized)
$N_\nu = \tilde{N}_\nu/\sqrt{\mathrm{Tr}(\tilde{N}^\dagger_\nu \tilde{N}_\nu \rho_S)},\ q_\nu = \mathrm{Tr}(\tilde{N}^\dagger_\nu \tilde{N}_\nu \rho_S)$	$\|\phi_\nu\rangle = \|\tilde{\phi}_\nu\rangle/\sqrt{\langle\tilde{\phi}_\nu\|\tilde{\phi}_\nu\rangle},\ q_\nu = \langle\tilde{\phi}_\nu\|\tilde{\phi}_\nu\rangle$
$\rho_S \to \sum_\nu q_\nu N_\nu \rho_S N^\dagger_\nu$	$\rho_{AB} = \sum_\nu q_\nu \|\phi^\nu_{AB}\rangle\langle\phi^\nu_{AB}\|$

systems. Thus, the entanglement between A and B is mapped to a correlation between the past and the future, which characterize the evolutions of systems and their quantum mechanical nature.

The suggested one-to-one map provide interesting physical consequences, as shown schematically in Table 22.1. In particular, the maximally entangled states are mapped to unitary evolutions—making explicit the expectation that unitary evolutions in dynamical processes are as maximal entangled states in bipartite systems. Non-maximally entangled states correspond to evolutions under the influence of selective measurements wherein the environment is observed and one particular outcome is selected. Specifically, pure product states correspond to selective projector measurements. Mixed bipartite systems are mapped to mixtures of the corresponding evolutions which are described by Kraus representation [2]. Closed systems with non-selective environment correspond to bipartite states in which the reduced density matrices are the maximally mixed ones.

The map we suggest has no natural generalization to the multipartite case. Nevertheless, we find that this limit strengthens the physical nature of the map, as it is sensitive to the uniqueness of the one dimensional time with respect to space. In particular, a tripartite state has no inherent order, whereas three events in time are not symmetric under all permutations. Another feature of multipartite states which is not satisfied in the temporal setting is the monogamy of entanglement [3]. If two qubits A and B are maximally entangled, they cannot be correlated at all with a third qubit C. In the temporal case, however, we can arrange simple scenarios in which any pair of instances among t_1, t_2, t_3 *etc.* is maximally correlated.

It may be argued that in general two sequentially measured operators do not commute and effect each other due to the uncertainty principle. Therefore, the temporal

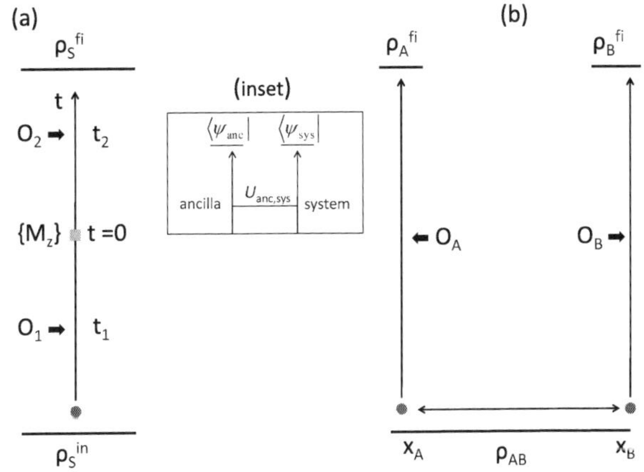

Fig. 22.1 Mapping of bipartite states to time evolutions. (**a**) A local state with initial and final boundary states ρ_S^{in} and ρ_S^{fi} respectively is weakly measured at $t_1 < 0$ and $t_2 > 0$ by O_1 and O_2 respectively, where the system undergoes an instantaneous evolution given by Kraus operators $\{M_\mu\}$ at $t = 0$. (**b**) A and B share a bipartite state ρ_{AB} and weakly measure it by O_A and O_B respectively. The system has local final boundary states ρ_A^{fi} and ρ_B^{fi}. ρ_{AB}, $O_A(x_A)$, $O_B(x_B)$, ρ_A^{fi} and ρ_B^{fi} are mapped to $\{M_\mu\}$, $O_1(t_1)$, $O_2(t_2)$, ρ_S^{in} and ρ_S^{fi}, respectively. *Inset.* Realization of post-selection to a mixed state by interaction with ancilla and post selecting both the system and the ancilla to pure states

correlations are limited to the formalistic analysis, and cannot be measured. However, as is well known, there is a trade-off between the accuracy of the measurement and the disturbance caused to the system [4]. The limit in which individual measurements provide vanishing information gain was first analyzed by Aharonov *et al.* [5] and was termed *weak measurements*. Since weak measurements only slightly disturb the systems, they provide a non-destructive and operational method for measuring temporal correlations by which the effect of evolutions can be measured directly.

One may have noticed that the system in the temporal setting has no counterpart in the spatial setting—it is the evolution that maps to a bipartite system. In Table 22.1 we indeed assume a trivial initial system in the temporal setting $\rho_S^{in} = I/d_A$. To overcome this we extend the mapping to include any initial system in the temporal setting. Surprisingly, this state is mapped to a local final boundary condition in the spatial setting, that is a boundary conditions of only one of the parties. Pleasantly, a final boundary condition in the temporal setting then corresponds to a final boundary condition of the second party in the spatial setting. This construction is illustrated by the correspondence of Fig. 22.1 (a) and (b).

The paper is organized as follows. In the following section we set the ground for the mapping by introducing generalized time evolutions. In Sect. 22.3 we define the map between time evolutions and bipartite states. In this section we provide the main result regarding the equality of temporal and spatial correlations and show that a non-selective environment corresponds to a bipartite state in which any reduced density matrix equals the maximally mixed state. In addition we show in Sect. 22.3 that the temporal (and spatial) correlations can be measured by utilizing weak measurements. In Sect. 22.4 we generalize the map to include boundary conditions. In Sect. 22.5 we provide proofs of the results in Sects. 22.3–22.6. In Sect. 22.6 we suggest several implications of the suggested map: manifestation of state independent decoherence, the correspondence between Bell [6] and Leggett-Garg [7] inequalities, the statistical properties of evolutions in large systems, and computational gain, in evaluation of spatial correlations in large systems. We conclude in Sect. 22.7.

22.2 Generalized Time Evolutions

To set the ground for the mapping let us first discuss generalized time evolutions. The evolution of a system ρ_S, subject to interaction with a larger system, is most generally described as a completely positive map given by Kraus operators $\{\mathcal{M}_\mu\}$:

$$\rho_S \to \sum_{\mu=1}^{K} \mathcal{M}_\mu \rho_S \mathcal{M}_\mu^\dagger, \tag{22.1}$$

where

$$\sum_\mu \mathcal{M}_\mu^\dagger \mathcal{M}_\mu = I. \tag{22.2}$$

Beyond the trivial unitary evolutions $\rho_S \to U\rho_S U^\dagger$, Kraus operators describe evolutions due to the interaction with an environment.

We will adopt a normalization in which the Kraus operators M_μ are normalized such that

$$\mathrm{Tr}\big(M_\mu^\dagger M_\mu\big) = d_A \tag{22.3}$$

transforming Eqs. (22.1, 22.2) to

$$\rho_S \to \sum_{\mu=1}^{K} p_\mu M_\mu \rho_S M_\mu^\dagger, \tag{22.4}$$

$$\sum_\mu p_\mu M_\mu^\dagger M_\mu = I. \tag{22.5}$$

where $M_\mu \equiv \mathcal{M}_\mu / \sqrt{p_\mu}$ and $p_\mu \equiv \mathrm{Tr}(\mathcal{M}_\mu^\dagger \mathcal{M}_\mu)/d_A$.

Equations (22.4, 22.5) (or 22.1, 22.2) provide the most general description of the evolution of a system which is part of a larger *closed* system. However, these Eqs. do not describe the scenario in which an external observer measures the environment and selects a single outcome. In order to describe the effect of a selective measurement of Kraus operator M_μ, we remove the constraint imposed in Eq. (22.5), set

$p_\mu = 1$ and describe the evolution by $\rho_S \to M'_\mu \rho_S M'^\dagger_\mu$, where M'_μ is normalized to preserve the trace of the density matrix $M'_\mu = M_\mu / \sqrt{\mathrm{Tr}(M^\dagger_\mu M_\mu \rho_S)}$.

Moreover, the selective setting can be generalized to more than a single selection since now the p_μ are not fixed by the constraint imposed in Eq. (22.5). One may choose to select M_μ with probability p_μ, and to select one or many other M_ν's with probabilities p_ν. In this case the normalization of the M_μ changes according to

$$M'_\mu = \frac{M_\mu}{\sqrt{\mathrm{Tr}[\sum_\nu p_\nu M^\dagger_\nu M_\nu \rho_S]}}, \tag{22.6}$$

in order to preserve the trace of $\rho_S \to \sum_\mu p_\mu M'_\mu \rho_S M'^\dagger_\mu$. Note that a set of normalized Kraus operators and probabilities $\{M_\mu, p_\mu\}$ is unique, where in case Eq. (22.5) is satisfied, the set corresponds to a non-selective environment. Otherwise one should renormalize the M_μ to M'_μ according to Eq. (22.6).

Clearly, a set of Kraus operators and probabilities $\{M'_\mu,\ p_\mu\}$ is not unique. It can be transformed to a set $\{N'_\nu,\ q_\nu\}$ with a unitary transformation $U_{K\times K}$, producing the un-normalized operator $\tilde{N}'_\nu = U_{\nu\mu}\sqrt{p_\mu} M'_\mu$. Then $q_\nu = \mathrm{Tr}(\tilde{N}'^\dagger_\nu \tilde{N}'_\nu \rho_S)$ and $N'_\nu = \tilde{N}'_\nu / \sqrt{q_\nu}$. Again one can represent the set $\{N'_\nu, q_\nu\}$ by the canonical set $\{N_\nu, q_\nu\}$ in which the N_ν are normalized according to Eq. (22.3).

22.3 The Map

In the temporal setting we assume an initially prepared system ρ_S^{in} with dimension d_A and internal Hamiltonian H_0 is subject to an evolution described by Kraus operators (as normalized in Eq. 22.6), which without loss of generality we take as instantaneous at time $t = 0$. In addition, for clarity we assume the internal Hamiltonian in the spatial setting vanishes.

Following a formal definition of the map between time evolutions and bipartite states. Any pure bipartite state is mapped to a single (normalized) Kraus operator by

$$|\psi_\mu\rangle = \sum_{ij} \alpha^\mu_{ij} |i\rangle \otimes |j\rangle \quad \Leftrightarrow \quad M_\mu = \sqrt{d_A} \sum_{ij} \alpha^{z*}_{ji} |i\rangle\langle j|. \tag{22.7}$$

where d_A is the dimension of H_A, $1 \le i \le d_A$, and d_B of H_B, $1 \le j \le d_B$. The map extends to mixed states/evolutions by convex combinations:

$$\rho_{AB} = \sum_\mu p_\mu |\psi_\mu\rangle\langle\psi_\mu| \quad \Leftrightarrow \quad \rho_S \to \sum_\mu p_\mu M'^\dagger_\mu \rho_S M'_\mu, \tag{22.8}$$

where M'_μ are given by Eq. (22.6). Note that the map does not define a spatial correspondence to the initial state in the temporal setting ρ_S^{in}. We shall assume (for now) that this state is maximally mixed $\rho_S = I/d_A$. Nontrivial initial (and final) boundary conditions are discussed in the following.

In Table 22.1 the correspondence between evolutions and states is given for several cases. In particular, it is evident from Eqs. (22.7), (22.8) that maximally entangled states are mapped to unitary evolutions.

Lemma 22.1 *Non-selective environment in the temporal setting maps to a state ρ_{AB} in which each of the reduced states is maximally mixed.*

Lemma 22.1 strengthens the physical essence of the map. A non-selective environment is one in which the future is still not known. This is mapped in the spatial setting to the scenario in which locally each party is maximally ignorant regarding her state.

Before stating our main result we would like to discuss temporal correlations.[1] Let there be two Hermitian operators $O_1(t_1)$, $O_2(t_2)$ given in the Heisenberg representation

$$O_i(t_i) = e^{iH_0t_i} O_i e^{-iH_0t_i} \quad \text{(taking } \hbar = 1\text{)}. \tag{22.9}$$

Clearly, the effect of Kraus operators $\mathcal{M}_\mu$ on the evolution of observable O is exactly as that of the density operator: $O \to \sum_\mu \mathcal{M}_\mu O \mathcal{M}_\mu^\dagger$. Therefore, the most straightforward definition of the temporal correlation of two operators O_1 and O_2 at instances t_1 and t_2 respectively is given by

$$E\big[O_2(t_2)O_1(t_1)\big] = \frac{1}{d_A}\mathrm{Tr}\Big[O_2(t_2)\sum_\mu p_\mu M'_\mu O_1(t_1) M'^\dagger_\mu\Big], \tag{22.10}$$

where d_A^{-1} is a normalization factor.

Theorem 22.1 *Let there be two Hermitian operators in the temporal setting $O_1(t_1)$, $O_2(t_2)$, $t_1 < 0 < t_2$, and in the spatial setting $O_A(x_A)$, $O_B(x_B)$, such that $O_1(t_1) = O_A(x_A)$, $O_2(t_2) = O_B(x_B)^*$. Given the mapping defined in Eqs. (22.7, 22.8), the temporal and spatial correlations equal:*

$$\frac{1}{d_A} Tr\Big[O_2 \sum_\mu p_\mu M'_\mu O_1 M'^\dagger_\mu\Big] = Tr[O_A \otimes O_B \rho_{AB}], \tag{22.11}$$

where we omit the t and x parameters from now on. This mapping is illustrated in Fig. 22.1, where for now we assume that in the temporal setting the state is $\rho_S = I/d_A$ and no final boundary conditions are assumed.

---

[1]Throughout the paper the term *correlation* corresponds to the expectation of the product of operators, sometimes regarded as correlator, without subtracting the first moments as required in the statistical definition of the term.

Corollary 22.1 *The expectation values of the single operators equal as well*:

$$E(O_1) = E(O_A), \qquad E(O_2) = E(O_B). \tag{22.12}$$

Equation (22.11) is symmetric to the exchange of indices A and B, given that we take $M_\mu'^{t}$ instead of M_μ' and $t \to -t$. That is, given a point in spacetime the mapping is symmetric under time and space reflections. The map above has no natural generalization to the multipartite case, as can be seen from the structure of Eq. (22.7). Nevertheless, we find that this limit strengthens the physical nature of the map, as it is sensitive to the uniqueness of the one dimensional time with respect to space. In particular, a tripartite state has no inherent order, whereas three events in time are not symmetric under all permutations. Another feature of multipartite states which is not satisfied in the temporal setting is the monogamy of entanglement [3]. If two qubits A and B are maximally entangled, they cannot be correlated at all with a third qubit C. In the temporal case, however, we can choose $M = I$. Then any pair of instances among t_1, t_2, t_3 *etc.* is maximally correlated.

22.3.1 Measuring the Temporal Correlation with Weak Measurements

Let us now utilize *weak measurements* [5] to show that the temporal correlation in the LHS of Eq. (22.11) can be measured. We assume that the system is measured weakly (and instantaneously) at $t_1 < 0$ and $t_2 > 0$ by operators $O_1(t_1)$ and $O_2(t_2)$ with two pointer readings q_1 and q_2 respectively, as illustrated in Fig. 22.1(a).

Lemma 22.2 *The correlation of the instruments' pointers is given by*:

$$E(q_1 q_2) = \frac{1}{2} Tr\left[O_2 \sum_\mu p_\mu M_\mu' \{O_1, \rho_S^{\text{in}}\} M_\mu'^{\dagger} \right], \tag{22.13}$$

which includes both selective and non-selective measurements. Eq. (22.13) *reduces to the temporal correlation in the LHS of Eq.* (22.11) *given that* $\rho_S^{\text{in}} = I/d_A$. *See related results in the context of unitary evolutions for correlations of two-level system with continuous weak measurements* [8], *in the context of post-selection* [9–11] *and of two sequential measurements* [12, 13].

In the spatial setting we assume that the initially prepared bipartite system ρ_{AB} is measured by parties A and B with operators O_A and O_B with pointers q_A and q_B respectively, as illustrated in Fig. 22.1(b). Note that the spatial correlation in the RHS of Eq. (22.11) can be measured with regular strong measurements. But an immediate consequence of Eq. (22.13) is that given weak measurements

$$E\left(q_A^{\text{weak}} q_B^{\text{weak}}\right) = \text{Tr}[O_A \otimes O_B \rho_{AB}]. \tag{22.14}$$

Weak measurements in the spatial setting will become essential in the following generalization of the map to include boundary conditions.

Notice that if ρ_S^{in} is the initial state in the temporal setting and $M = I$, three weak measurements O_1, O_2 and O_3 at times $t_1 < t_2 < t_3$ yield

$$E(q_1 q_2 q_3) = \frac{1}{4}\text{Tr}\big[O_3, \big\{O_2, \big\{O_1, \rho_S^{\text{in}}\big\}\big\}\big]. \tag{22.15}$$

In contrast to weak measurements at two times, Eq. (22.15) implies that the correlation of three depends on their order, in contradiction with the multipartite spatial scenario.

22.4 Generalizing the Map to Include Boundary Conditions

Remarkably, the mapping above can be generalized to the case that the initial state in the temporal setting is not maximally mixed. In this case the state corresponds to a local final boundary condition of one of the parties in the spatial setting. In addition, a final boundary condition in the temporal setting corresponds to a local final boundary condition of the second party in the spatial setting. Explicitly, in the case of an initial state ρ_S and a final boundary state ρ_S^{fi} with dimension d_B in the temporal setting, Eq. (22.13) generalizes to

$$E(q_1 q_2) = \frac{1}{4}\text{Tr}\left[\rho_S^{\text{fi}}\left\{O_2, \sum_\mu p_\mu M'_\mu \{O_1, \rho_S^{\text{in}}\} M'^{\dagger}_\mu\right\}\right], \tag{22.16}$$

where

$$M'_\mu = \frac{M_\mu}{\sqrt{\text{Tr}[\rho_S^{\text{fi}} \sum_\nu p_\nu M_\nu \rho_S^{\text{in}} M_\nu^\dagger]}}. \tag{22.17}$$

Physically, one can create a final mixed boundary state by post-selecting the state and an ancilla as described in the inset of Fig. 22.1. Note that no final boundary state is equivalent to having a maximally distributed final state $\rho_S^{\text{fi}} = I/d_B$.

In the spatial setting, having boundary states ρ_A^{fi} in A and ρ_B^{fi} in B generalizes Eq. (22.14) using Eq. (22.16) to

$$E\big(q_A^{\text{weak}} q_B^{\text{weak}}\big) = \frac{\text{Tr}[\rho_A^{\text{fi}} \otimes \rho_B^{\text{fi}} \{I_A \otimes O_B, \{O_A \otimes I_B, \rho_{AB}\}\}]}{4\text{Tr}[(\rho_A^{\text{fi}} \otimes \rho_B^{\text{fi}})\rho_{AB}]}. \tag{22.18}$$

Note that given final boundary states in the spatial setting weak and strong measurements provide different results, where the mapping applies only to the weak measurement regime.

Theorem 22.2 *Given that* $\rho_S^{\text{in}} = \rho_A^{\text{fi}}$, *and* $\rho_S^{\text{fi}} = \rho_B^{fi*}$, *Theorem* 22.1 *generalizes to* $E(q_1 q_2) = E(q_A^{\text{weak}} q_B^{\text{weak}})$ *or explicitly,*

$$\frac{1}{4}Tr\left[\rho_S^{\rm fi}\left\{O_2, \sum_\mu M'_\mu\{O_1, \rho_S^{\rm in}\}M'^{\dagger}_\mu\right\}\right]$$
$$= \frac{1}{4}\frac{Tr[\rho_A^{\rm fi}\otimes\rho_B^{\rm fi}\{I_A\otimes O_B, \{O_A\otimes I_B, \rho_{AB}\}\}]}{Tr[(\rho_A^{\rm fi}\otimes\rho_B^{\rm fi})\rho_{AB}]}. \tag{22.19}$$

The map is fully illustrated in Fig. 22.1.

22.5 Proofs of the Results

We proceed by proving the results to the cases including the boundary conditions, where reduction to the case of no(/totally mixed) boundary states is straightforward. O_i in the temporal setting is given by Eq. (22.9).

Proof of Lemma 22.2 Observables O_1, O_2 are measured sequentially on system $\rho_S^{\rm in}$ at times t_1, t_2 where $t_1 < 0 < t_2$. In addition, $\rho_S^{\rm in}$ evolves at $t = 0$ with Kraus operators $\{M_\mu\}$. The von-Neumann interaction measurement corresponding to O_1 and O_2 is

$$H_{int} = \delta(t - t_1)p_1 O_1 + \delta(t - t_2)p_2 O_2, \tag{22.20}$$

where $[q_i, p_i] = i$ $(\hbar = 1)$. We assume identical initial Gaussian wavepackets $\phi(q_1)$ and $\phi(q_2)$ for the pointers:

$$\rho_i = \phi(q_i)\phi(q_i') = \int dq_i dq_i' \sqrt{\frac{\epsilon}{2\pi}} e^{-\epsilon(q_i^2+q_i'^2)/4} \quad (i = 1, 2). \tag{22.21}$$

The initial state of the system and the apparatuses $\rho_S^{\rm in}\otimes\rho_1\otimes\rho_2$, evolves to

$$U_2\left[\sum_\mu p_\mu M'_\mu(U_1\rho_S^{\rm in}\otimes\rho_1\otimes\rho_2 U_1^\dagger)M'^{\dagger}_\mu\right]U_2^\dagger, \tag{22.22}$$

where $U_i = e^{-ip_i O_i}$. Each operation of p yields an order of ϵ, where in the limit of weak measurements $\epsilon \to 0$. By expanding U_i to second order $(i = 1, 2)$ $U_i = 1 - ip_i O_i - \frac{1}{2}p_i^2 O_i^2 + o(\epsilon^3)$, one can compute the composite state of the system and pointers:

$$\rho_S^{\rm in}\otimes\rho_1\otimes\rho_2 \to \sum_\mu p_\mu M'_\mu \rho_S^{\rm in}\rho_1\rho_2 M'^{\dagger}_\mu$$
$$-\sum_\mu p_\mu M'_\mu\{O_1, \rho_S^{\rm in}\}M'^{\dagger}_\mu \phi'(q_1)\phi(q_1')\rho_2$$
$$+\frac{1}{2}\sum_\mu p_\mu M'_\mu\{O_1^2, \rho_S^{\rm in}\}M'^{\dagger}_\mu \phi''(q_1)\phi(q_1')\rho_2$$

$$
\begin{aligned}
&-\Big\{O_2,\sum_\mu p_\mu M'_\mu \rho_S^{\text{in}} M'^{\dagger}_\mu\Big\}\phi'(q_2)\phi(q'_2)\rho_1\\
&+\frac{1}{2}\Big\{O_2^2,\sum_\mu p_\mu M'_\mu \rho_S^{\text{in}} M'^{\dagger}_\mu\Big\}\phi''(q_2)\phi(q'_2)\rho_1\\
&+\Big\{O_2,\sum_\mu p_\mu M'_\mu \{O_1,\rho_S^{\text{in}}\} M'^{\dagger}_\mu\Big\}\phi'(q_1)\phi(q'_1)\phi'(q_2)\phi(q'_2).
\end{aligned}
\tag{22.23}
$$

First note that since $\int q\phi^2(q)dq = 0$ and $\int \phi(q)\phi'(q)dq = 0$, all terms in Eq. (22.23) except the last do not contribute. By tracing out the system ρ_S^{in} and using $\int q\phi(q)\phi'(q)dq = -1/2$, we obtain

$$
E(q_1q_2) = \frac{1}{4}\text{Tr}\Big[\Big\{O_2,\sum_\mu p_\mu M'_\mu \{O_1,\rho_S^{\text{in}}\} M'^{\dagger}_\mu\Big\}\Big],
\tag{22.24}
$$

which coincide with Eq. (22.13). □

Proof of Eq. (22.16) Preparation of a mixed state is realized by projecting a system to a pure state $|\psi_S^{\text{in}}\rangle$ which then interacts with an ancilla in a known state. Correspondingly, post selection to a mixed state ρ_S^{fi} is realized in the reversed order:

$$
\rho_S^{\text{fi}} = U_{int}^\dagger |0_{anc}\rangle\langle 0_{anc}| \otimes |\psi_S^{\text{fi}}\rangle\langle\psi_S^{\text{fi}}| U_{int}
\tag{22.25}
$$

(as illustrated in the inset of Fig. 22.1). Then the proof follows the same steps as that of Lemma 22.1 where instead of tracing out the system, one projects the system to the final state and renormalizes the remaining state. In case $M = I$ the normalization yields a factor of

$$
\frac{1}{\text{Tr}[\langle 0_{anc}|\langle\psi_S^{\text{fi}}|U_{int}\rho_S^{\text{in}} \otimes I_{anc} U_{int}^\dagger|\psi_S^{\text{fi}}\rangle|0_{anc}\rangle]} = \frac{1}{\text{Tr}[\rho_S^{\text{in}}\rho_S^{\text{fi}}]}.
\tag{22.26}
$$

The generalization to arbitrary evolution is straightforward. Note that Wizeman [14] analyzed a similar case for a single weak measurement. □

Proof of Theorem 22.2 Let us first show the correspondence for a pure bipartite state $|\psi\rangle$, which is mapped to a single Kraus operator M' (with $p = 1$). We show the equality of the temporal and spatial denominators D_T, D_S and nominators N_T and N_S of Eq. (22.16) and (22.18) respectively. From Eqs. (22.17, 22.7) up to a factor of 4:

$$
\begin{aligned}
D_T &= \mathrm{Tr}\big[\rho_S^{\mathrm{fi}} M \rho_S^{\mathrm{in}} M^\dagger\big] = \sum_{i,j,k,l} \alpha_{ij}\alpha_{kl}^* \rho_{S_{ki}}^{\mathrm{in}} \rho_{S_{ij}}^{\mathrm{fi}}, \\
D_S &= \mathrm{Tr}\big[\big(\rho_A^{\mathrm{fi}} \otimes \rho_B^{\mathrm{fi}}\big)|\psi\rangle\langle\psi|\big] = \sum_{i,j,k,l} \alpha_{ij}\alpha_{kl}^* \rho_{A_{ki}}^{\mathrm{fi}} \rho_{B_{lj}}^{\mathrm{fi}}, \\
N_T &= \mathrm{Tr}\big[\rho_S^{\mathrm{fi}}\big\{O_2, M\big\{O_1, \rho_S^{\mathrm{in}}\big\}M^\dagger\big\}\big] \\
&= \sum_{i,j,k,l} \alpha_{ij}\alpha_{kl}^* \big\{\rho_S^{\mathrm{in}}, O_1\big\}_{ki} \big\{\rho_S^{\mathrm{fi}}, O_2\big\}_{jl}, \\
N_S &= \mathrm{Tr}\big[\big(\rho_A^{\mathrm{fi}} \otimes \rho_B^{\mathrm{fi}}\big)\big\{I_A \otimes O_B, \{O_A \otimes I_B, |\psi\rangle\langle\psi|\}\big\}\big] \\
&= \sum_{i,j,k,l} \alpha_{ij}\alpha_{kl}^* \big\{\rho_A^{\mathrm{fi}}, O_A\big\}_{ki} \big\{\rho_B^{\mathrm{fi}}, O_B\big\}_{jl},
\end{aligned}
\tag{22.27}
$$

where we use the notation $A_{ki} = \langle k|A|i\rangle$ for matrix elements. In correspondence with the mapping defined in Theorems 22.1, 22.2, $D_T = D_S$ and $N_T = N_S$. By proving $D_S = D_T$ we have explicitly confirmed that the mapping corresponds to Jamiolkowski isomorphism [1]. To extend to a set of Kraus operators $\{M'_\mu, p_\mu\}$, note that D_T, D_S, N_T, N_S become now a convex combinations of p_μ, which respects their equality. This concludes the proof of Theorems 22.1 and 22.2. □

22.6 Implications

The suggested mapping has several important implications which we discuss in some detail.

22.6.1 State Independent Decoherence

A common framework of decoherence [15] deals with the transition of a state to a one with higher level of mixing. By the suggested mapping one can distinguish decoherence of states from *decohering dynamics*. Decohering dynamics can be observed by detecting the temporal decay of correlations in case of non exact unitary evolution, even on the maximally distributed mixed state.

Let us discuss a particular manifestation. We can write the evolution of system $\rho_S(t)$ according to the Lindblad master equation:

$$
\dot{\rho}_S = \frac{1}{i\hbar}\big[\hat{H}, \rho_S(0)\big] + \sum_k \left[\hat{L}_k \rho_S(0) \hat{L}_k^\dagger - \frac{1}{2}\big\{\rho_S(0), \hat{L}_k^\dagger \hat{L}_k\big\}\right], \tag{22.28}
$$

where $\hat{L}_k$ are Lindblad operators satisfying $\hat{K} = -\frac{1}{2}\sum_k \hat{L}_k^\dagger \hat{L}_k$ such that $\hat{K}$ is Hermitian. For simplicity, consider a single Linbdlad operator $L = \sigma_z$ which starts operating at $t = 0$ where the free Hamiltonian vanishes. This simplifies Eq. (22.28) at $t > 0$ to

$$\dot{\rho}_S = \sigma_z \rho_S \sigma_z - \rho_S. \tag{22.29}$$

In addition we assume $\rho_S = I/2$, thus it does not change in time.

We manifest the decoherence model by studying a Leggett-Garg inequality [7], the Bell inequality in time. In this test a single observer measures four observables $O_1,\ O_2, O_3,\ O_4$ at four instances t_1, t_2, t_3, t_4 $(t_1 < t_2 < t_3 < t_4)$ and look on a combination of their correlations, which structurally corresponds to the spatial CHSH inequality [16]:

$$B_{LG} = \frac{1}{2}\mathrm{Re}[O_1O_3 + O_1O_4 + O_2O_3 - O_2O_4]. \tag{22.30}$$

We choose to measure $\sigma_z(t = t_1)$, $\sigma_x(t = t_2)$, $\sigma_{\pi/4}(t = t_3)$ and $\sigma_{3\pi/4}(t = t_4)$, where

$$\sigma_{\frac{\pi}{4},\frac{3\pi}{4}} = \frac{1}{\sqrt{2}}(\sigma_x \pm \sigma_z).$$

Given the trivial evolution $M = I$ $B_{LG} = 2\sqrt{2}$, $B_{LG} = 2\sqrt{2}$ for any initial state. But now let us apply the evolution imposed in Eq. (22.29) at $t = 0$, where we assume $t_1, t_2 < 0$ and $t_3, t_4 > 0$ so the effect of decoherence takes place in the last two measurements. Transforming to the Heizenberg representation each Pauli operator σ transforms as $\dot{\sigma} = \sigma_z \sigma \sigma_z - \sigma$. Therefore off-diagonal terms decays in time as $\sigma_{12}(t) = \sigma_{12} e^{-\sqrt{2}\sigma_{12} t}$. Using Eq. (22.13) we obtain

$$B_{LG} = \frac{\sqrt{2}}{2}\left(2 + e^{-\sqrt{2}t_3} + e^{-\sqrt{2}t_4}\right). \tag{22.31}$$

For $t_3 = t_4$ B_{LG} decays to 2 as $t_3 \sim 0.623$. By this simple application decoherence is manifested on the observables due to the non-unitary evolution, whereas the state ρ_S is constantly the maximally mixed state.

22.6.2 *The Correspondence Between Bell and Leggett-Garg Inequalities*

The temporal inequalities suggested by Leggett and Garg [7] have the same bounds as the corresponding spatial Bell inequalities [6]. For example, CHSH inequality [16] and the corresponding temporal inequality (Eq. 2b in [7]) are bounded by $2\sqrt{2}$ [17]. In a previous paper [13] we have shown that Bell's inequalities can be maximally violated using weak measurements even if all observables are measured for each member of the ensemble. A similar result for Leggett-Garg inequalities was given in [18, 19]. By our mapping the correspondence between these two type of inequalities becomes clear. Leggett-Garg inequalities are distinguished from the Bell inequalities as their maximal violation depends only on the measured observables and not on the state of the system. By the mapping above we see that this is a consequence of unitary evolutions which correspond to maximally entangled bipartite states. By having non unitary evolutions Leggett-Garg inequalities are less violated.

In particular, the CHSH inequality is maximally violated by the maximally entangled (Bell) state, for example $|\psi\rangle = \frac{1}{\sqrt{2}}(|00\rangle + |11\rangle)$. This state is mapped in the temporal setting to $M = I$ (the trivial evolution). The case in which no post-selection is performed in the spatial case corresponds in the temporal test to having the maximally mixed initial state $\rho_S = I/2$. The $2\sqrt{2}$ bound on Leggett-Garg inequality is then obtained by half the trace of the matrix

$$B_{LG} = \mathrm{Re}[\sigma_x\sigma_{\pi/4} + \sigma_x\sigma_{3\pi/4} + \sigma_z\sigma_{\pi/4} - \sigma_z\sigma_{3\pi/4}] = 2\sqrt{2}I, \qquad (22.32)$$

where $\sigma_{\pi/4,3\pi/4} = (\sigma_x \pm \sigma_z)/\sqrt{2}$. Since post-selection of a single party does not change the bound on CHSH inequality, the same bound is obtained for any initial state in the temporal setting.

Partial violation of CHSH inequality is obtained by non-maximally entangled pure or mixed states. The case of pure bipartite states corresponds to having a single non-unitary M in the temporal setting, where the initial state is $I/2$. Having a different initial state corresponds to post-selection of one of the parties. The non-entangled pure product state corresponds to a selective projector imposing "disentanglement" between the past and the future.

The case of mixed states is richer. Let us discuss the example of Werner states [20]. Since Werner states are convex combinations of Bell states, they cannot produce higher violation in case one of the parties post-selects. Therefore, the temporal violation does not depend on the initial state. By explicit computation and as a corollary of the proof of Theorem 22.1, the Werner states that violate CHSH inequality are exactly mapped to the same mixtures of unitary evolutions which violate Leggett-Garg inequality.

Since our mapping is exact all the results concerning bipartite Bell inequalities are valid in the corresponding temporal inequalities. An example is the anomaly of nonlocality in bipartite systems with dimension greater than two [21], in which there are Bell inequalities that are not maximally violated by the maximally entangled state. In particular, let us explore Collins-Gissin-Linden-Massar-Popescu (CGLMP) inequality [22] (see also [23]). This generalized Bell inequality corresponds to a setting in which every party measures two observables, each having d outcomes (instead of the dichotomic observables in the CHSH test). In a local hidden variable model CGLMP inequality is bounded by 2 for any d. It is maximally violated by the non-maximally entangled state for $d > 2$. One can explicitly show that the same anomaly appears in the temporal setting, where maximal violation is obtained with the corresponding non-unitary evolution.

Let us show this result explicitly (even though it is a corollary of the proof of Theorem 22.1). As shown in [21] the Bell operator corresponding to CGLMP inequality in case $d = 3$ is

$$B_{CGLMP} = \begin{pmatrix} 0 & 0 & 0 & 0 & \frac{2}{\sqrt{3}} & 0 & 0 & 0 & 2 \\ 0 & 0 & 0 & 0 & 0 & \frac{2}{\sqrt{3}} & 0 & 0 & 0 \\ 0 & 0 & 0 & 0 & 0 & 0 & 0 & 0 & 0 \\ 0 & 0 & 0 & 0 & 0 & 0 & 0 & \frac{2}{\sqrt{3}} & 0 \\ \frac{2}{\sqrt{3}} & 0 & 0 & 0 & 0 & 0 & 0 & 0 & \frac{2}{\sqrt{3}} \\ 0 & \frac{2}{\sqrt{3}} & 0 & 0 & 0 & 0 & 0 & 0 & 0 \\ 0 & 0 & 0 & 0 & 0 & 0 & 0 & 0 & 0 \\ 0 & 0 & 0 & \frac{2}{\sqrt{3}} & 0 & 0 & 0 & 0 & 0 \\ 2 & 0 & 0 & 0 & \frac{2}{\sqrt{3}} & 0 & 0 & 0 & 0 \end{pmatrix}. \tag{22.33}$$

For the maximally entangled state $|\psi\rangle = \frac{1}{\sqrt{3}}(|00\rangle + |11\rangle + |22\rangle)$ the violation is $\langle\psi|B_{CGLMP}|\psi\rangle = \frac{4}{9}(3 + 2\sqrt{3}) \sim 2.87293$. However, the maximal eigenvalue of B_{CGLMP} is $\frac{1}{3}(3 + \sqrt{33}) \sim 2.91485$ with eigenstate $|\psi_m\rangle = \frac{1}{\sqrt{n}}(|00\rangle + \gamma|11\rangle + |22\rangle)$, where $\gamma = (\sqrt{11} - \sqrt{3})/2$ and $n = 2 + \gamma^2$.

Correspondingly, in the temporal setting we study the case of maximally mixed initial state $I/3$, as no post-selection is assumed in the spatial setting. In case the evolution corresponds to the maximally entangled state $M = 1$, we obtain

$$B_{LG}^3 = \begin{pmatrix} 2 + \frac{2}{\sqrt{3}} & 0 & 0 \\ 0 & \frac{4}{\sqrt{3}} & 0 \\ 0 & 0 & 2 + \frac{2}{\sqrt{3}} \end{pmatrix}, \tag{22.34}$$

where $\mathrm{Tr}[B_{LG}]/3 = \frac{4}{9}(3 + 2\sqrt{3})$. Now let us set the non-unitary evolution corresponding to $|\psi_m\rangle$

$$M^m = \eta \begin{pmatrix} 1 & 0 & 0 \\ 0 & \gamma & 0 \\ 0 & 0 & 1 \end{pmatrix}, \tag{22.35}$$

where $\eta = \sqrt{3/(11/2 - \sqrt{33}/2)}$ such that $\mathrm{Tr}[M^\dagger M]/3 = 1$. Then

$$B_{LG}^{3m} = \begin{pmatrix} \frac{1}{2} + \frac{7}{2\sqrt{33}} & 0 & 0 \\ 0 & \frac{4}{\sqrt{33}} & 0 \\ 0 & 0 & \frac{1}{2} + \frac{7}{2\sqrt{33}} \end{pmatrix}, \tag{22.36}$$

such that $\mathrm{Tr}[B_{LG}^m]/3 = \frac{1}{3}(3 + \sqrt{33}) \sim 2.91485$ with the same bound as in the spatial setting.

Table 22.2 Anomaly of nonlocality in the temporal and spatial settings

Temporal		Spatial
$\rho_S = I/d$. LG inequality maximally violated with non-unitary evolution	$\Leftarrow$	Bell inequality maximally violated with non-maximally entangled state
$\Downarrow$		$\Uparrow$
ρ_S general. LG inequality maximally violated with unitary evolution	$\Rightarrow$	Local post-selection. Bell inequality maximally violated with maximally entangled state

Interestingly, given a maximally entangled state in the spatial setting, post-selection of a single party produces higher violation of CGLMP inequality for $d > 2$ (in contrast to the $d = 2$ case). This can be easily seen by checking the temporal setting and having a state $|\psi\rangle$ different from the maximally mixed state $I/3$. Since the evolution is unitary $M = 1$ one need not renormalize $B^3_L G$. Maximal violation corresponds to the maximal eigenvalue of B^3_{LG}: $2 + \frac{2}{\sqrt{3}} \sim 3.1457$, where the corresponding eigenstates are $|0\rangle$ or $|2\rangle$. This result shows that in the usual context of Leggett-Garg inequalities in which only unitary evolutions are considered, the quantum-mechanical bound is in general distinguished from the one in the corresponding Bell inequality. The ordinary bound of Leggett-Garg inequalities with a unitary evolution and an arbitrary initial state corresponds to the bound on the corresponding Bell inequality on the maximally entangled state given post-selection of a single party.

Another example is the I_{3322} inequality suggested by Collins and Gisin [24]. In [24] a (non-optimal) quantum mechanical bound of $\frac{1}{4}$ is found using maximally entangled Bell states. However, in [25, 26] it is shown that the violation is higher for non-maximally entangled states, thus providing another example of the anomaly of nonlocality.

Now let us discuss the corresponding temporal inequality. In case the state is the maximally mixed state and the evolution is unitary with $M = I$ it can be shown that $I_{3322} \leq \frac{1}{4}$. The bound is satisfied by taking the operators $A_1, A_2, A_3, B_1, B_2, B_3$ along the xy plane with $\phi_1^A = 0, \phi_2^A = \pi/3, \phi_3^A = -\pi/3, \phi_1^B = \pi/3, \phi_2^B = 0, \phi_3^B = 2\pi/3$. In the spatial case the bound of $\frac{1}{4}$ does not depend on the dimension for maximally entangled states. Therefore, the same characteristic is mapped to the temporal test.

For arbitrary initial state the optimal violation is $B_{3322}^{LG} \leq \frac{3\sqrt{5}}{8} - \frac{1}{2} \sim 0.3385$, which is satisfied already at $d = 2$ with $\phi_1^A = 0, \phi_2^A = 2\pi/5, \phi_3^A = -2\pi/5, \phi_1^B = \pi/5, \phi_2^B = -\pi/5, \phi_3^B = 3\pi/5$. Through the map I_{3322} has the same bound 0.3385 in case the state is a maximally entangled one and one of the parties post-selects to the initial state of the temporal setting. This can be explicitly shown with the maximally entangled state $\frac{1}{\sqrt{2}}(|00\rangle + |11\rangle)$, where $\phi_1^A = 0, \phi_2^A = 2\pi/5, \phi_3^A = -2\pi/5, \phi_1^B = -\pi/5, \phi_2^B = \pi/5, \phi_3^B = -3\pi/5$.

Beyond the curiosities above being elucidated, we believe our mapping provides a new perspective on the anomaly of nonlocality, as illustrated in Table 22.2.

22.6.3 Statistical Characteristics of Large Systems

In the work by Hayden *et al.* [27] correlation properties of random high-dimensional bipartite pure systems were examined in the context of the Haar measure. They showed that there exist large subspaces in which almost all pure states are close to maximally entangled. Through the mapping the space of bipartite states maps to "pure evolutions" on local systems where a pure evolution corresponds to a single Kraus operator. By mapping the Haar measure to the temporal setting, it follows that there exist large subspaces of evolutions in which all pure evolutions are close to unitary ones.

22.6.4 Computational Gain

In numerical computations of two point correlation function of bipartite states, needed for instance in evaluating Bell inequality bounds, one can utilize the corresponding Leggett-Garg inequalities. For example, given $d_A = d_B = N$, instead of manipulating $N^2 \times N^2$ matrices, one can use only $N \times N$ matrices.

22.7 Discussion and Conclusions

We would like to remark that a notion of *entanglement in time* was introduced in a different context by Brukner *et al.* [28], who analyze correlations of successive ± 1 strong measurements. These temporal correlations violate Leggett-Garg inequalities [7], the Bell inequalities [6] in time. Brukner *et al.* also show that there are no genuine multi-time correlations and that the monogamy of spatial correlations is violated in the temporal setting. However, there are crucial differences between temporal correlations of strong and weak measurements as correlations of successive strong measurements do not depend on the state and are a particular feature of ± 1 observables. The suggested mapping does not apply for strong measurements.

Leifer has shown an extension of Jamiolkowski isomorphism to include two POVMs before and after the evolution in the temporal setting and in parallel in the spatial setting [29], to manifests the correspondence of no cloning/no broadcasting theorems and monogamy of entanglement (see also [30]). Jamiolkowski mapping has also been used to analyze channel capacity [31–34]. Manifestation of "superposition of unitary operations" is given in [35].

To conclude, space and time are distinguished in the formalism of quantum theory. A system that is separated in two parts of space is described by a positive semidefinite operator that lies in a tensor product of two Hilbert spaces. The time evolution of a local system is described by a trace preserving complete positive map from one Hilbert space to another Hilbert space. Mathematically, one can define a Jamiolkowski map between the space of bipartite states and the space of time evolutions, which is defined by the Hilbert-Schmidt scalar product. We extended Jamiolkowski map in a physical setting in which the entanglement of bipartite states finds exact correspondence with temporal correlations between the past and future. The fact that the mapping holds only for the bipartite case strengthens the uniqueness of the one dimensional time which can only be bisected once to unordered parts, whereas space may be sectioned into many parts with no internal order. We use the tool of weak measurements to show that these correlations can be observed. One can show that the map can be tested with a single ion and two ions in an ion-trap respectively, where instead of the displacement operator the pointer observables correspond to the phonon number operator. By having an exact mapping between spatial and temporal correlations, non-relativistic quantum-mechanics manifests a structural unification of time and space.

Acknowledgements We are deeply grateful to Y. Aharonov whose insights initiated this work. We also thank A. Botero and P. Skrzypczyk. This work has been supported by the Israel Science Foundation grant number 920/09, the German-Israeli foundation, and the European Commission (PICC).

References

1. A. Jamiolkowski, Rep. Math. Phys. **3**, 275 (1972)
2. K. Kraus, *States, Effects and Operations: Fundamental Notions of Quantum Theory* (Springer, Berlin, 1983)
3. V. Coffman, J. Kundu, W.K. Wootters, Phys. Rev. A **61**, 052306 (2000)
4. J. von Neumann, *Mathematical Foundations of Quantum Mechanics* (Princeton University Press, Princeton, 1955)
5. Y. Aharonov, D.Z. Albert, A. Casher, L. Vaidman, Phys. Lett. A **124**, 199–203 (1987)
6. J.S. Bell, Physics **1**, 195–200 (1964)
7. A.J. Leggett, A. Garg, Phys. Rev. Lett. **54**, 857–860 (1985)
8. A.N. Korotkov, D.V. Averin, Phys. Rev. B **64**, 165310 (2001)
9. K.J. Resch, A.M. Steinberg, Phys. Rev. Lett. **92**, 130402 (2004)
10. G. Mitchison, R. Jozsa, S. Popescu, Phys. Rev. A **76**, 062105 (2007)
11. J.S. Lundeen, A.M. Steinberg, Phys. Rev. Lett. **102**, 020404 (2009)
12. L.M. Johansen, P.A. Mello, Phys. Lett. A **372**, 5760–5764 (2008)
13. S. Marcovitch, B. Reznik. arXiv:1005.3236
14. H.M. Wiseman, Phys. Rev. A **65**, 032111 (2002)
15. G. Lindblad, Commun. Math. Phys. **48**, 199 (1976)
16. J.F. Clauser, M.A. Horne, A. Shimony, R.A. Holt, Phys. Rev. Lett. **23**, 880–884 (1969)
17. B.S. Cirel'son, Lett. Math. Phys. **4**, 93–100 (1980)
18. R. Ruskov, A.N. Korotkov, A. Mizel, Phys. Rev. Lett. **96**, 200404 (2006)
19. A.N. Jordan, A.N. Korotkov, M. Büttiker, Phys. Rev. Lett. **97**, 026805 (2006)
20. R.F. Werner, Phys. Rev. A **40**, 4277 (1989)
21. A. Acin, T. Durt, N. Gisin, J.I. Latorre, Phys. Rev. A **65**, 052325 (2002)
22. D. Collins, N. Gisin, N. Linden, S. Massar, S. Popescu, Phys. Rev. Lett. **88**, 040404 (2002)
23. D. Kaszlikowski, P. Gnacinski, M. Zukowski, W. Miklaszewski, A. Zeilinger, Phys. Rev. Lett. **85**, 4418 (2000)
24. D. Collins, N. Gisin, J. Phys. A, Math. Gen. **37**, 1775 (2004)
25. T.H. Yang, M. Navascués, L. Sheridan, V. Scarani, Phys. Rev. A **83**, 022105 (2011)
26. T. Vidick, S. Wehner. arXiv:1011.5206

27. P. Hayden, D.W. Leung, A. Winter, Commun. Math. Phys. **265**, 95 (2006)
28. C. Brukner, S. Taylor, S. Cheung, V. Vedral. arXiv:quant-ph/0402127
29. M.S. Leifer, Phys. Rev. A **74**, 042310 (2006)
30. R.B. Griffiths, Phys. Rev. A **71**, 042337 (2005)
31. F. Verstraete, H. Verschelde. arXiv:quant-ph/0202124v2
32. P. Arrighi, C. Patricot, Ann. Phys. **311**, 26–52 (2004)
33. W. Dür, J.I. Cirac, Phys. Rev. A **64**, 012317 (2001)
34. M.L. Nowakowski1, P. Horodecki. arXiv:quant-ph/0503070
35. J. Oppenheim, B. Reznik, Phys. Rev. A **70**, 022312 (2004)

Chapter 23
Anatomy of Quantum Tunneling

Neil Turok

Abstract We use complex classical solutions to provide a simple and transparent treatment of quantum tunneling. We show that the imaginary part of the tunneling coordinate may be weakly measured, and that it yields a quantum-anomalous contribution to the classically-conserved momentum of a classical von Neumann "pointer." Further developments and applications are mentioned. [*Editor's note*: for a video of the talk given by Prof. Turok at the Aharonov-80 conference in 2012 at Chapman University, see quantum.chapman.edu/talk-19.]

23.1 Introduction and Summary

The Feynman path integral for the Schrödinger kernel,

$$\begin{aligned} \Psi(x_f, t_f) &= \int dx_i K(x_f, t_f; x_i, t_i)\Psi(x_i, t_i), \\ K(x_f, t_f; x_i, t_i) &= \int_{x_i, t_i}^{x_f, t_f} \mathscr{D}x e^{i\mathscr{S}/\hbar} \end{aligned} \tag{23.1}$$

where $\mathscr{S}$ is the classical action, provides the closest and most useful connection between classical and quantum dynamics. In the limit as $\hbar \to 0$ (or, more properly, in the limit that a corresponding dimensionless quantity tends to zero), one expects the path integral typically to be dominated by a saddle point solution, representing a classical path. Generically, such paths are complex: in this talk I shall describe how these complex classical paths may be used to describe quantum tunneling.

Traditionally, tunneling processes have been described using classical instanton solutions which are real functions of imaginary time [1, 2]. The main advantage is that such instanton solutions take a simple analytical form. However, there are significant limitations to their use. First, in quantum theory a metastable state is inherently ill-defined. In the instanton approach, the initial state is defined rather

N. Turok (✉)
Perimeter Institute for Theoretical Physics, 35 Caroline St N, Waterloo, ON N2L2Y5, Canada
e-mail: nturok@perimeterinstitute.ca

D.C. Struppa, J.M. Tollaksen (eds.), *Quantum Theory: A Two-Time Success Story*,
DOI 10.1007/978-88-470-5217-8_23, © Springer-Verlag Italia 2014

implicitly, at imaginary time equal to minus infinity. Second, imaginary time instantons do not easily provide a real-time description of the tunneling process, nor can they describe tunneling in processes which are inherently time-dependent, such as when the potential is time-dependent. These drawbacks underly some persistent controversies, for example about the "tunneling time" [3].

My main claim in this talk is that when one is careful to describe tunneling using the language of pre- and postselection, at finite real times, then in the semiclassical (small $\hbar$) limit there is a real-time description within which the tunneling coordinate is complex. At first, it may seem strange that a real, Hermitian quantum operator be described by a complex classical solution, but there is no contradiction. The fact that the semiclassical approximation to the coordinate is complex may even be experimentally verified by coupling the tunneling system to a measuring device and performing a weak measurement. The imaginary part of the tunneling coordinate leads to a quantum anomaly in a classically conserved quantity for the combined system, which can be measured (weakly) with arbitrary precision. My main conclusion is that within the framework of pre- and post-selection, weak measurement and classicality, the quantum world is typically described using complex numbers rather than real ones.

23.2 Example

In this talk I shall focus on a very simple one dimensional tunneling problem, for the coordinate x described by the classical action

$$\mathscr{S} = \int_{t_i}^{t_f} \left(\frac{1}{2}\dot{x}^2 - V(x)\right), \qquad V(x) = \frac{1}{2}\kappa x^2 - \frac{1}{2}\lambda x^4. \tag{23.2}$$

It is convenient to redefine t, x to be dimensionless, $t \to t\sqrt{m/\kappa}$ and $x \to x\sqrt{\kappa/\lambda}$, so that the action becomes

$$\mathscr{S} = \frac{\kappa^{\frac{3}{2}} m^{\frac{1}{2}}}{\lambda} \int_{t_i}^{t_f} \frac{1}{2}\left(\dot{x}^2 - x^2 + x^4\right) \equiv C\hat{\mathscr{S}}, \tag{23.3}$$

where the prefactor $C = \frac{\kappa^{\frac{3}{2}} m^{\frac{1}{2}}}{\lambda}$ governs the overall value of the action but is immaterial to the classical equations.

The traditional approach to tunneling is based upon the following "Euclidean bounce" instanton solution:

$$\tau = i\left(t + \frac{\pi}{2}\right); \qquad x_B(\tau) = (\cosh\tau)^{-1} = -(\sin t)^{-1}, \tag{23.4}$$

which describes the particle emerging from the potential at $x = 1$, with zero velocity, at time $t = -\frac{\pi}{2}$, and running off to infinite x at $t = 0$. Before emerging, the solution is described in imaginary time τ, with x tending to the classical

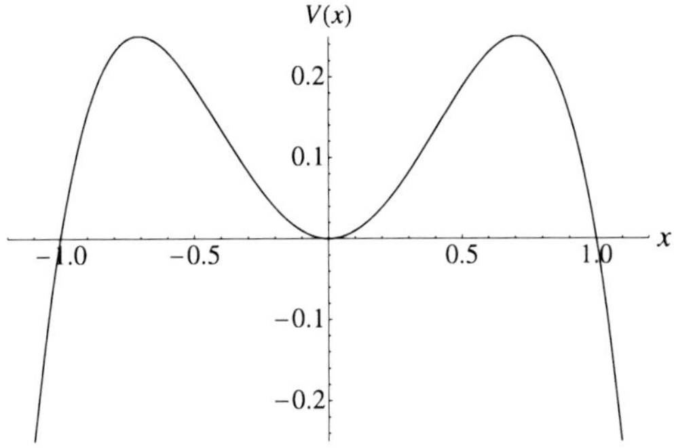

Fig. 23.1 The dimensionless potential $V(x) = \frac{1}{2}(x^2 - x^4)$ with false vacuum at $x = 0$ and two classically allowed regions at $|x| \geq 1$

false vacuum $x = 0$ at $\tau = -\infty$. The tunneling rate is exponentially suppressed by the factor $e^{-C\hat{\mathscr{S}}_B/\hbar}$, with $\hbar$ Planck's constant and the Euclidean bounce action $\hat{\mathscr{S}}_B \equiv -i\hat{\mathscr{S}} = \int_{-\infty}^{\infty} d\tau \frac{1}{2}(x_{B,\tau}^2 - x_B^2 + x_B^4) = \frac{2}{3}$. (See Fig. 23.1.)

While this picture of tunneling is elegant, as I have already mentioned, any dependence on the initial metastable state is very implicit and it is hard to answer real-time questions such as how long the tunneling process took, or what happened during the tunneling. It is natural to ask whether we can do better. When the action is large in units of $\hbar$, we can expect the path integral to be well-approximated by a classical saddle point solution which, as with analogous ordinary integrals, may well be complex.

23.3 Complex Classical Solutions in Real Time

Let us begin, then, by describing the general (complex) classical solution to the equations of motion following from the action $\hat{\mathscr{S}}$. The general solution is specified by two complex numbers: the energy E and a time delay t_0. We are used to considering both to be real and in particular t_0 to be a trivial shift in the time. However, for tunneling solutions, as we shall see, both are generally complex and the imaginary part of t_0 is significant. First, consider the case of real E but complex t_0. In this case, one may show that the solutions are periodic in t, but for each E they vary as one changes the imaginary part of t_0 (see Fig. 23.2).

The general solution, for arbitrary complex E and t_0, is easily obtained:

$$x(t) = -\frac{1}{\sqrt{1+m}\,\mathrm{sn}(\frac{t-t_0}{\sqrt{1+m}}|m)}; \qquad E = \frac{m}{2(1+m)^2} \tag{23.5}$$

where sn is a Jacobi elliptic function. For an initial "false vacuum" state, *i.e.* the ground state wavefunction for a harmonic oscillator with potential $\frac{1}{2}x^2$, we shall be interested in small imaginary energy E. The Jacobi elliptic function is doubly periodic in the complex plane of the argument $u = (t - t_0)/\sqrt{1+m}$, with quarter periods $K(m)$ and $iK'(m)$, where $K'(m) = K(1-m)$.

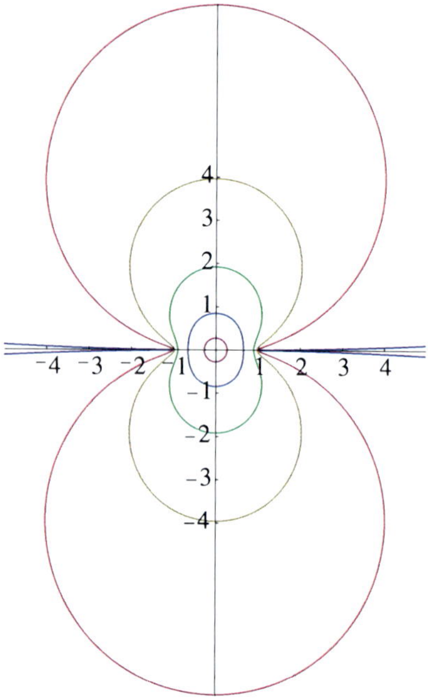

Fig. 23.2 The general complex solution for real energy $E = 0$ is $x = -1/\sin(t - t_0)$. In the figure, this is plotted in the complex x-plane for various values of t_0. The real part of t_0 gives a trivial shift in the time t. The imaginary part is significant, however: from the outer curve inward, the plots show the zero-energy solutions for $t_0 = i/200, i/8, i/4, i/2, i, 2i$

The solution (23.5) has a useful expansion in terms of ordinary sine functions:

$$x(t) = -\frac{\pi}{2K\sqrt{1+m}}\left(\frac{1}{\sin u} + 4\sum_0^\infty \frac{q^{2n+1}}{1-q^{2n+1}}\sin(2n+1)u\right); \qquad u = \frac{t-t_0}{\sqrt{1+m}}, \tag{23.6}$$

where the "nome" $q = e^{-\pi K'/K}$. For small m, K, K' and q have the expansions

$$K(m) \approx \frac{\pi}{2}\left(1 + \frac{m}{4} + \frac{9m^2}{64}\dots\right); \qquad K'(m) \approx -\frac{1}{2}\ln\frac{m}{16} + o(m\ln m); \qquad q \approx \frac{m}{16} + \frac{m^2}{32}\dots. \tag{23.7}$$

Let us turn now to the boundary conditions on the classical solution for tunneling from the initial false vacuum. At the initial time t_i, by assumption the wavefunction is a Gaussian centered on $x = 0$, so $\Psi_i \propto e^{-x^2/4\Delta x^2}$. This obeys

$$\left(\frac{x}{\Delta x} + i\frac{p}{\Delta p}\right)\Psi_i = 0, \tag{23.8}$$

where $p = -i\hbar\frac{\partial}{\partial x}$, and $\Delta p = \hbar/(2\Delta x)$ saturating Heisenberg's uncertainty principle bound. For the false vacuum state in our problem, we have $\Delta x = 1/\sqrt{2}$ which, in (23.8), provides the initial condition on the classical solution. The final condition is simply that the position of the particle equals the argument x_f of the final wavefunction Ψ_f at time t_f. The classical solution is thus determined by the two boundary conditions:

$$x + i\dot{x} = 0 \quad \text{at } t = t_i; \qquad x = x_f \quad \text{at } t = t_f \equiv 0. \tag{23.9}$$

Time translation invariance allows us to take the final time t_f to be zero, so that all times t of interest are negative.

We want to describe tunneling in a situation where it is rare. So, let us choose the initial time to be large and negative, $t_i = -T$ with $T \gg 1$. The particle's motion will become classical when it has tunneled far from the false vacuum, so we also set $x_f \gg 1$. Noticing that the Jacobi sn function becomes infinite when t tends to t_0, it is clear that we should choose the time delay t_0 to be small. In this case, we can express x_f as a series in t_0, or vice versa:

$$t_0 \approx x_f^{-1} + \frac{1}{6}x_f^{-3} + \frac{3 + 2m + 3m^2}{40(1+m)^2}x_f^{-5} \cdots \ll 1. \tag{23.10}$$

We determine the energy E from the initial condition in (23.9) as follows. Setting $m = i\epsilon$, and defining $z = e^{iu}$, from (23.6) we obtain

$$x \approx \frac{2i}{z - z^{-1}} + \frac{m}{8i}\left(z - z^{-1}\right)\ldots, \qquad t \ll 0, \tag{23.11}$$

where $u \approx (1 - 3i\frac{\epsilon}{4})(t - t_0) \approx t - \frac{3}{4}i\epsilon t$ for large negative t. With positive ϵ, z becomes large at large negative t, so that

$$x \approx -\frac{1}{8}imz + 2iz^{-1} + 2iz^{-3}\ldots, \qquad t \ll 0, \tag{23.12}$$

and the initial condition in (23.9) gives

$$x + i\dot{x} \approx -\frac{1}{4}imz - 4iz^{-3}\ldots, \qquad t \ll 0. \tag{23.13}$$

Substituting $m = i\epsilon$, we obtain the transcendental relation

$$3\epsilon T e^{3\epsilon T} \approx -48iTe^{-4iT}, \qquad \text{with } T \gg 1. \tag{23.14}$$

The solution of this equation is given by Lambert's function; qualitatively one has $\epsilon \approx \ln(T)/(3T)$, so that ϵ is indeed small and positive for large T. It is not hard to check that the solution for ϵ is actually unique.

The fact that the energy E has a small imaginary part allows us to understand qualitatively the behavior of the solution, as compared to the solutions for zero energy shown in Fig. 23.2. The small imaginary part mediates a slow transition

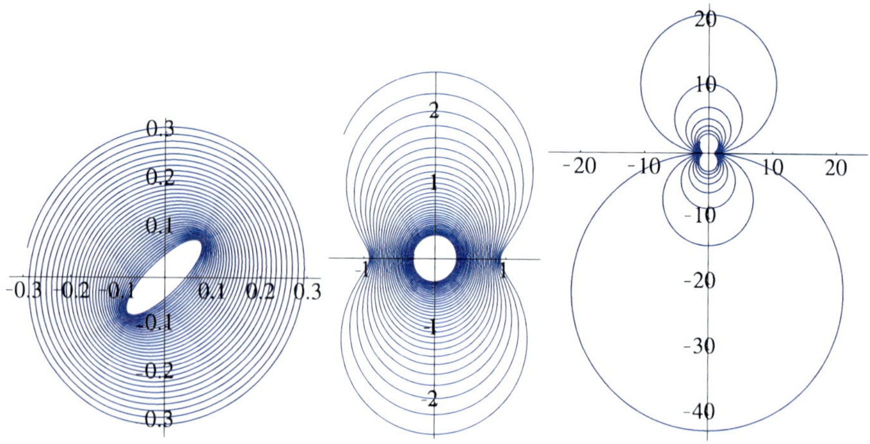

Fig. 23.3 The complex solution describing finite-time tunneling from a false-vacuum Gaussian centered on $x = 0$ to a position $x = x_f$ over a time $T \gg 1$. The solution for $m = i/100$, and $x_f \gg 1$ is plotted in the complex x-plane. The *left plot* shows the solution for times $-450 \le t \le -250$, the *center plot* $-250 \le t \le -50$ and the *right plot* $-50 \le t \le 0$. At earlier times, the solution spirals outwards into the complex x-plane again, due to the complex periodicity of the Jacobi sn function

from the solutions for large imaginary t_0 (and hence small absolute magnitude of x) to small imaginary t_0, tending to the real classical solution describing the particle rolling down the hill to large x_f at late times.

Now that we have found the two integration constants t_0 and E (or $m = i\epsilon$), when T and x_f are both large, we can now plot the relevant classical solutions in the complex x-plane. The pictures below show the solution following a tight spiral around the origin at very early times, growing into a double-lobed spiral at intermediate times and finally shooting out nearly parallel to the real x-axis to large positive x_f.

It is also interesting to understand the analytic structure of the solutions in the complex t-plane (or u-plane). As we have already mentioned, the Jacobi sn function is doubly periodic in its argument u, considered as a complex variable. This is illustrated in Fig. 23.4. Zeros and poles of the sn function correspond to poles and zeros of the solution, forming a lattice in the complex u-plane, whose fundamental domain is $(0, K, iK', K + iK')$ (see Fig. 23.4). Considered as a function of real time, the classical tunneling solution we have given starts near a zero and follows a straight line path (with slope $1/\sqrt{1+m}$) towards a pole, for example the pole at the origin.

We can now understand the relationship between the classical solution we have found, and the instanton solution traditionally used to describe tunneling. First, notice from (23.7) that as we take the time T to be large, since m tends to zero, the quarterperiod iK' becomes large and imaginary. The solution is then very nearly zero along a line of zeros related to its initial zero, and so one can deform the time contour from the black arrow on the left of Fig. 23.4 to the series of blue arrows shown on the right. For large T, the solution is almost zero all along the first

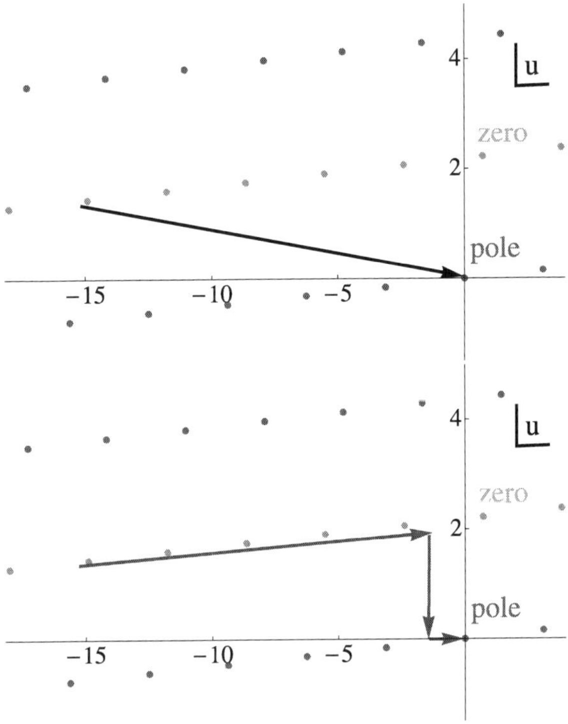

Fig. 23.4 The solution describing finite-time tunneling shown in the complex u-plane, where u is the argument of the Jacobi sn function. The latter is doubly-periodic, with K and iK' being the two quarter-periods. The plots show the solution and quarter-periods for $m = i/5$. The *left hand plot* shows the real time solution, and the *right hand plot* shows the deformation of the contour used to obtain the imaginary-time instanton description

arrow. Along the second arrow, the time is imaginary and runs from large imaginary value to the point $t \approx -\frac{\pi}{2}$ at which $x = 1$. The solution along the second arrow is therefore just the imaginary time bounce solution $x_B(\tau) = (\cosh\tau)^{-1}$, with $t = -\frac{\pi}{2} - i\tau$ and $-\infty < \tau \le 0$. The solution is nearly real along the third arrow, for which u is nearly real (see Fig. 23.4). In the leading semiclassical approximation, the wavefunction $\Psi(x_f, t_f) \sim e^{i\mathscr{S}/\hbar}$. Evaluating the action for the classical solution as a contour integral along the path shown by the arrows in the right hand diagram, the first part of the path gives $\mathscr{S} \approx 0$. The second part gives the Euclidean bounce action, $i\mathscr{S}/\hbar \approx -\mathscr{S}_B/(2\hbar)$. (The factor of two arises because the full bounce solution runs from $-\infty < \tau < +\infty$.) Along the third part of the path, the solution is nearly real so that the action contributes an unimportant overall phase to the wavefunction. In this way, we see that for large T and x_f, we recover the usual expression for the suppression of the tunneling rate, $\Psi^2 \propto e^{-\mathscr{S}_B/\hbar}$.

23.4 Making Weak Measurements During Tunneling

Just as Aharonov's concept of postselection is useful in identifying the unique classical solution responsible for finite-time tunneling, his concept of weak measurement is useful in understanding what this solution means. The basic idea is that we would like to understand where the tunneling particle is, but without performing a strong measurement which would significantly disturb the wavefunction. As Aharonov pointed out, we can do that by performing a weak measurement, which in our case means allowing the particle to interact weakly with a "pointer," and then performing the measurement many times on identically-prepared systems to obtain an accurate average result.

In order to measure the position of the particle x at some measurement time t_m, we introduce a simple von Neumann pointer Hamiltonian of the form

$$\mathcal{H}_P = \frac{P^2}{2M} + gPx\delta(t), \tag{23.15}$$

where P and M are the pointer momentum and mass and g is a small dimensionless coupling. Classically, the pointer momentum P is conserved as a consequence of translation invariance, by Noether's theorem, and the sole effect of this Hamiltonian is to shift the pointer position by the quantity $gPx(t_m)$. Thus, from the movement of the pointer one can infer the position of the particle at the measurement time. Quantum mechanically, the effect of the interaction is easily seen in the P-representation of the wavefunction, in which it simply multiplies the total wavefunction by a factor,

$$\Psi\left(x, P, t_m^+\right) = e^{-igPx/\hbar}\Psi\left(x, P, t_m^-\right). \tag{23.16}$$

If the pointer wavefunction just before the measurement is a Gaussian with width ΔX, then to first order in g, in the semiclassical approximation the effect on the average pointer position and momentum is just

$$\langle X\rangle \to \langle X\rangle + g\mathrm{Re}\big(x_{Cl}(t_m)\big); \qquad \langle P\rangle \to \langle P\rangle + \frac{g\hbar}{2\Delta X^2}\mathrm{Im}\big(x_{Cl}(t_m)\big), \tag{23.17}$$

where $x_{Cl}(t)$ is the classical tunneling solution. For small $g\mathrm{Im}(x)/\Delta X$, the shift in pointer momentum is a small fraction of the uncertainty in the pointer position. Nevertheless, the shift is measurable if one repeats the experiment many times.

It would be very interesting to perform an experiment like this on a tunneling particle, and to observe the resulting quantum shift in the pointer momentum. As we have already mentioned, the pointer momentum is classically conserved—the shift in the momentum proportional to the imaginary part of the measured particle's position is proportional to $\hbar$ and should thus be thought of as a quantum anomaly. One could in principle imagine a very precise experiment which would observe the fine structure of the complex classical tunneling solution, shown in Fig. 23.3. As an example of universal phenomena, the classical solution at early times, $t \ll 1$, as the particle is preparing for tunneling, exhibits exponential growth in its distance from the origin of the complex x-plane, $r \propto e^{t \ln T/(4T)}$, which is presumably independent of the details of the potential at large x.

23.5 Further Developments and Applications

The technique described here is extremely simple, conceptually. Many possible applications and developments and applications can be anticipated.

i. I have chosen a polynomial potential $x^2 - x^4$ for which the only moveable singularities in the classical solutions are simple poles. The absence of branch cuts in the complex t-plane means that the classical solutions are single valued, considerably simplifying the analysis. The same property holds for a cubic polynomial potential $x^2 - x^3$, for which the only singularities in the classical solutions are double poles.
ii. The classical solutions give the saddle point to the Feynman path integral with the boundary conditions (23.9). The classical action gives the leading exponential behavior of the wavefunction, $\Psi \sim e^{-A/\hbar}$, in the semiclassical ($\hbar \to 0$) limit. It is of interest to calculate the prefactor as an expansion in $\hbar$. The first step is to compute the path integral in the leading (Gaussian) approximation, giving a functional determinant. As I shall explain elsewhere, the functional determinant turns out to be given straightforwardly in terms of the complex classical solution and its derivatives.
iii. Within the method given here, it is straightforward to vary the parameters specifying the initial quantum state such as the centre, width or momentum of the Gaussian wavepacket. One could also vary the shape of the initial wavefunction or include time-dependent potentials $V(x,t)$.
iv. The method given may be extended to quantum field theory, where it describes bubble nucleation [4] in Minkowski spacetime for a theory with a metastable false vacuum state. One of the features of the Euclidean, imaginary-time formulation is that it cannot describe the nucleation of more than one bubble since there is no two-instanton solution. The real-time formulation given here resolves this problem.
v. The method may similarly be extended to bubble nucleation in quantum field theories coupled to gravity [4]. Here, it is notable that the general $O(3,1)$-invariant classical solutions for certain scalar field potentials (including some with false vacua) coupled to gravity, and describing the nucleation of bubbles, have been recently obtained, by Bars, Chen, Steinhardt and me [5]. They can be used to describe the nucleation of Coleman-DeLuccia bubbles in the corresponding theories. The present technique may enable the resolution of certain paradoxes about such bubbles in the context of "multiverse" scenarios for the global structure of the universe.

Acknowledgements I would like to thank Yakir Aharonov, Carl Bender, Michael Berry and Curtis Callan for encouragement, and other participants of this exciting meeting for many useful discussions.

References

1. S.R. Coleman, Phys. Rev. D **15**, 2929 (1977). (Erratum) Ibid., **16**, 1248 (1977)

2. C.G. Callan Jr., S.R. Coleman, Phys. Rev. D **16**, 1762 (1977)
3. P.C.W. Davies, Am. J. Phys. **73**, 23 (2005)
4. S.R. Coleman, F. De Luccia, Phys. Rev. D **21**, 3305 (1980)
5. I. Bars, S.-H. Chen, P.J. Steinhardt, N. Turok, Phys. Rev. D **86**, 083542 (2012). arXiv:1207.1940 [hep-th]

Part VIII
Weak Measurement Experiments

Chapter 24
Experimental Implementations of Quantum Paradoxes

G.A.D. Briggs

Abstract Remarkable progress is being made in experiments that highlight the distinctive predictions of quantum mechanics. The Leggett-Garg inequality was devised to test for macrorealism (Leggett and Garg in Phys. Rev. Lett. 54:857–860, 1985). Various experiments have been performed, including one with non-invasive measurements in the kind of way that was originally envisaged, using spins in phosphorous impurities in silicon (Knee et al. in Nat. Commun. 3:606, 2012). This has led to fresh understanding of what kind of realism is excluded by the result. The quantum three-box paradox (Aharonov and Vaidman in J. Phys. A, Math. Gen. 24:2315–2328, 1991) provides a further test, which can be re-expressed in terms of the Leggett-Garg inequality. This has been experimentally implemented with projective measurements using an NV^- centre in diamond, yielding results 7.8 standard deviations beyond a classical bound (George et al. in Proc. Natl. Acad. Sci. USA 110:3777–3781, 2013). [*Editor's note*: for a video of the talk given by Prof. Briggs at the Aharonov-80 conference in 2012 at Chapman University, see quantum.chapman.edu/talk-18.]

More than a quarter of a century ago, Yakir Aharonov published, with David Albert and Susan D'Amato, a suggestion for an experiment in which a particle may be located within any one of three small impenetrable boxes [1]. The experiment was devised with a pre-selection of the state $(|1\rangle + |2\rangle)/\sqrt{2}$, and post-selection of $(|2\rangle + |3\rangle)/\sqrt{2}$. This produced a paradox. Local measurement would lead to finding the particle in Box 2 with certainty, whereas a nonlocal variable with eigenstates $(|1\rangle + |3\rangle)/\sqrt{2}$, $|2\rangle$, $(|1\rangle - |3\rangle)/\sqrt{2}$ gives a probabilistic outcome. Subsequently he developed with Lev Vaidman a new N-box experiment [2], now often presented as the quantum three-box paradox. As a gift to Yakir Aharonov in celebration of his eightieth birthday, I wish to present an implementation of that experiment and unwrap the results [3].

The experiment provides a brilliant illustration of how a quantum state can be described by two state vectors, one evolving forward from the past and one evolv-

G.A.D. Briggs (✉)
Department of Materials, University of Oxford, 16 Parks Road, Oxford, OX1 3PH, UK
e-mail: andrew.briggs@materials.ox.ac.uk

D.C. Struppa, J.M. Tollaksen (eds.), *Quantum Theory: A Two-Time Success Story*,
DOI 10.1007/978-88-470-5217-8_24, © Springer-Verlag Italia 2014

ing backward from the future [2]. This does not produce new physics which could not be derived from conventional quantum theory, but it can lead to new insights. It suggests a family of pre- and post-selection (PPS) experiments, in which there is symmetry between the pre- and post-selection events, in a way that can sometimes allow remarkable conclusions to be drawn about the intermediate state. A well known example of this arises from a weak measurement of the intermediate state, in which the effect of the PPS can be to select preferentially measurements that lie outside the normal range of eigenstates [4]. By using projective measurements, the three-box paradox can provide a test of macrorealism, raising fresh questions about how measurements disturb a quantum system.

24.1 The Three-Box Paradox and Its Implementation

Imagine a gambling game between two players, Alice and Bob, in which in the long run if quantum theory is correct Alice will win, and if classical realism is correct then Bob will win. The three box paradox can be expressed in those terms [5, 6]. Alice, who in this implementation is the younger and cleverer of the two, offers Bob a wager with odds better than 50 % in Bob's favour, that she will know when he saw the ball in the box that he opened; Bob will win when she accepts a game in which he did not see the ball. Alice puts a ball in one of three boxes and then shuffles them to make it random which box the ball is in. This is the pre-selection. She will then leave the room. While she is not watching, Bob may open either box 1 or box 2, and look to see whether the ball is in the one that he opens. This is the intermediate measurement. If he does not see a ball, then he wins, but only if Alice subsequently accepts the wager during this round of the game. Bob then closes the box. There is an impartial judge to ensure that Bob does not cheat. Alice returns. Bob does not tell Alice which box he opened or what he saw. Alice rearranges the boxes and looks in box 3. If she finds the ball in box 3, then she accepts the game. This is the post-selection. Bob reckons according to classical realism that at best Alice can look in one of the two boxes which he might have opened, and therefore at best she can know the result of his observation half the time. Alice has asserted that she will identify the rounds in which Bob found his box empty with $>$ 50 % probability, so he accepts. In a perfect quantum implementation Alice would win every time. Even in our imperfect experiment we find that Alice's success cannot be explained in terms of a realist interpretation.

The pre-selection is achieved by initializing the system in the state $(|1\rangle + |2\rangle + |3\rangle)/\sqrt{3}$. The observation then projects the system into, supposing Box 1 is opened, either $|1\rangle$ if the ball is in that box, or $(|2\rangle + |3\rangle)/\sqrt{2}$ if not. The post-selection is performed by testing for the state $(|1\rangle + |2\rangle - |3\rangle)/\sqrt{3}$, which is achieved by transforming that state into $|3\rangle$ and then looking for the ball in Box 3. The state $(|2\rangle + |3\rangle)/\sqrt{2}$, which Bob prepares if he finds Box 1 empty, has zero projection onto $(|1\rangle + |2\rangle - |3\rangle)/\sqrt{3}$, and therefore if Alice finds the ball in Box 3, Bob must have seen it when he looked in Box 1. Since there is symmetry between Boxes 1 and 2,

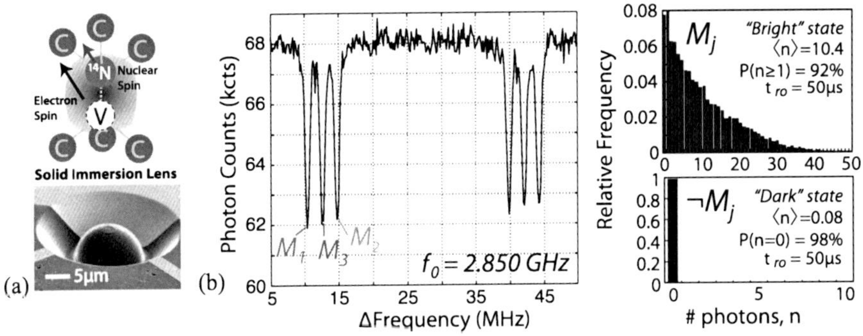

Fig. 24.1 (**a**) NV^- centre and solid immersion lens. (**b**) Spectrum of microwave transitions, and statistics of photon counting to detect presence (*top*) or absence of fluorescence [3]

the same argument applies if Bob looked in Box 2. If Alice sees the ball in Box 3, she can conclude that Bob saw it whichever Box he opened, and she can accept the game confident of winning every time. Bob is flummoxed. One interpretation is that between the pre- and post-selection, there was unit probability that the ball would have been found in Box 1, and unit probability that the ball would have been found in Box 2. If the total probability is required to sum to 1, then there must have been a probability of -1 that the ball would have been found in Box 3. This could alternatively be expressed as unit probability that there was -1 ball in Box 3, or more simply the weak value of the projection on $|3\rangle$ was -1.

The practical implementation with spins is illustrated in Fig. 24.1. Previous implementations with photons have used a three-path Mach-Zender interferometer or a three-slit interference mask [7, 8]. The spin implementation used a nitrogen-vacancy defect in a diamond crystal. The defect is charged with a single electron, to make an NV^- centre. The ^{14}N nitrogen nucleus of the NV^- has three angular momentum eigenstates that play the role of the three boxes: A classical rotating body would have to occupy one of these three allowed states in any given trial of the three box protocol, whilst a quantum mechanical object may exist in superpositions between them. The electronic structure of the NV^- allows for probing the nuclear angular momentum state via optical fluorescence. A solid immersion lens is used to achieve a high efficiency in collecting photons during fluorescence. The centre and lens are illustrated in Fig. 24.1(a). The spectrum of microwave transitions is given in Fig. 24.1(b), together with examples of photon statistics from 10,000 trials in each case, showing the difference between detecting and not detecting the corresponding optical fluorescence. The coloured lettering in Fig. 24.1(b) relates to the transitions between energy levels shown in Fig. 24.2, which also shows how the energy levels map onto the quantum and classical three box states of the game.

The experiment proceeds as follows. First, the system is prepared in the $|3\rangle$ state by an optical pulse followed by verification as in Fig. 24.1(b). When the state $|3\rangle$ is found, it is transformed into $(|1\rangle + |2\rangle + |3\rangle)/\sqrt{3}$ by two radiofrequency pulses. Alice then hands the system over to Bob, who performs a measurement M_1 or M_2 by an appropriate microwave pulse followed by a fluorescence measurement, again as

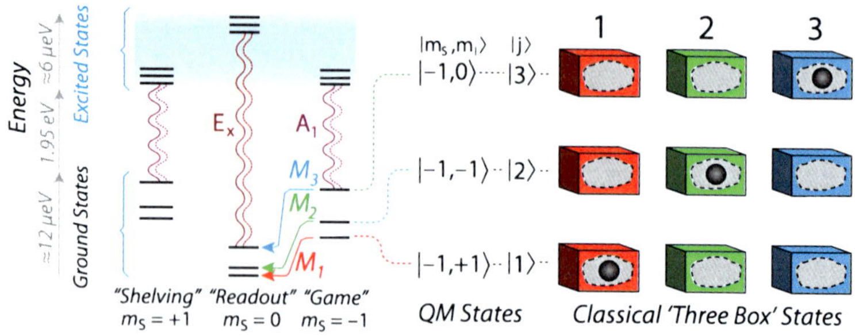

Fig. 24.2 Energy levels of the NV^- centre, showing the microwave transitions between different spin states in the spectrum of Fig. 24.1(a), and the fluorescence transition used in measurements. The corresponding quantum (QM) and classical three box states of the game are indicated. The experiment proceeds as follows. First, the system is prepared in the $|3\rangle$ state by an optical pulse followed by verification as in Fig. 24.1(b). This is transformed into $(|1\rangle + |2\rangle + |3\rangle)/\sqrt{3}$ by two radiofrequency pulses. Alice then hands the system over to Bob, who performs a measurement M_1 or M_2 by an appropriate microwave pulse followed by a fluorescence measurement, again as in Fig. 24.1(b). Bob then closes the box by an identical microwave pulse at the same frequency, and he checks that the box is closed by a further optical measurement in which he should see no fluorescence. Alice then uses a sequence of radiofrequency pulses which would transform the state $(|1\rangle + |2\rangle - |3\rangle)/\sqrt{3}$ into $|3\rangle$, and she performs a measurement M_3. She post-selects depending on the value which she obtains [3]

in Fig. 24.1(b). Bob then closes the box by an identical microwave pulse at the same frequency, and he checks that the box is closed by a further optical measurement in which he should see no fluorescence. Alice then uses a sequence of radiofrequency pulses which would transform the state $(|1\rangle + |2\rangle - |3\rangle)/\sqrt{3}$ into $|3\rangle$, and she performs a measurement M_3. She post-selects depending on the value which she obtains.

In order to check that Alice is not cheating, on his own classical terms, Bob can satisfy himself that there is one and only one ball, that it is placed at random in the three boxes, that having seen or not seen it in one box he will get the same result if he looks again, and that apart from the post-selection Alice has no other way of knowing which measurement Bob performed, or even whether he measured at all.

Bob's checks are illustrated in Fig. 24.3. Figure 24.3(a) shows that Bob finds after Alice's initialization, the probability of the ball being in each box are 1/3 to within experimental error. If, having found not the ball in Box 2 Bob then looks for it in Box 1 or 3, he finds it with probability a little less than the ideal 50 %, and fails to see it on repeat examination of Box 2, with probability close to the ideal of zero. If having found the ball in Box 3 Bob measures again, he gets the same result with approximately 75 % probability. There is about 10 % probability that the measurement will change to one of the other boxes, and about 15 % probability that the outcome will be undetermined. The undetermined outcomes correspond to branching to the $m_S = +1$ levels which are not measured. Finally, in Fig. 24.3(b), Bob sees what Alice finds in her final M_3. Her overall probability, about 15 %, is higher than the value of 1/9 which she would find in a perfect experiment, probably due to nuclear spin dephasing following heating by the radiofrequency NMR

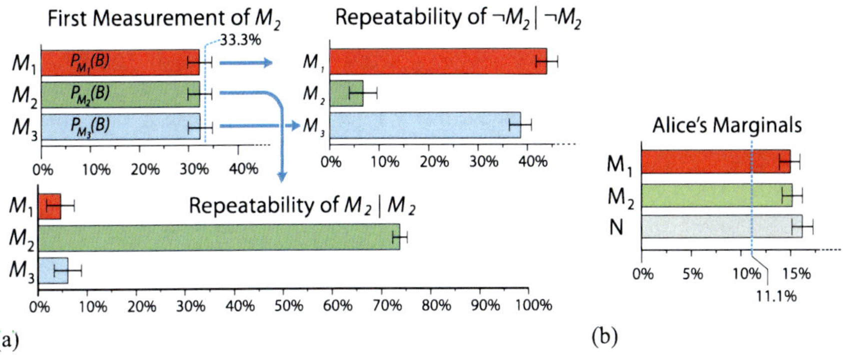

Fig. 24.3 (**a**) Bob verifies that there is one and only one ball with equal probability in each box, and that repeated measurements give consistent results, using measurements within the $m_S = -1$ manifold only. Bob's measurement results when observing the state prepared by Alice in the $|j\rangle$ basis are independent of the j-value selected to within experimental error. The repeatability of a second M_j measurement conditioned on the result M_2 or on the result not-M_2 is less than perfect in each case, but is adequate for our purpose. (**b**) Alice's measurement M_3 cannot tell whether Bob chose to perform measurement M_1, M_2, or neither (N) [3]

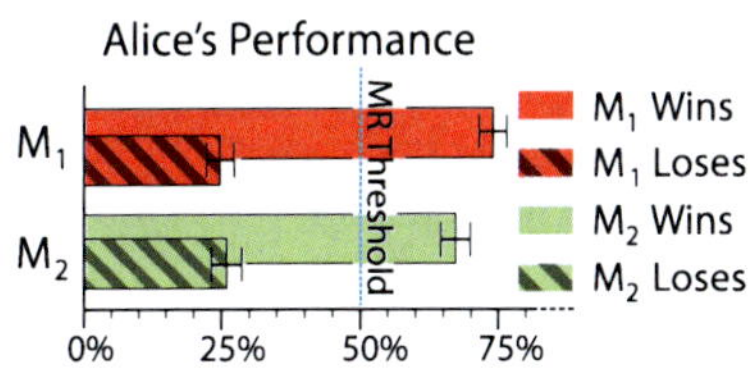

Fig. 24.4 In the games which Alice post-selects to play, the probability that Bob has ('Wins') or has not ('Loses') seen state M_j, given that Alice has seen M_3. Alice's probability of winning exceeds 50 % by much more than the experimental error regardless of whether Bob chooses M_1 or M_2 [3]

pulses, but the differences between the three cases are less that the experimental uncertainty. Alice still does not know which of Boxes 1 and 2 Bob chose, and whether he opened it. Her probability of finding the ball in box 3, and subsequently accepting the round, are unaltered when Bob choses to inspect box 1, box 2 or neither box.

It is therefore inconceivable for Bob, on the basis of classical phenomena, that Alice should be able to guess that he saw the ball with greater that 50 % probability. But that is what she does, as presented in Fig. 24.4. In the cases where she accepts to play the game, on the basis of her final measurement M_3, in nearly 75 % of the cases Bob had indeed seen the ball in his measurement. This is less than the 100 % success which would be achieved in a perfect experiment, but greater by well over 7 standard deviations than the best that she could achieve by classical gamesmanship.

24.2 The Leggett-Garg Inequality and Its Significance

In the same year that Albert, Aharonov, and D'Amato published the idea that became the Three Box Problem [1], Anthony Leggett and Anupam Garg published a paper with a criterion to distinguish experimentally between quantum mechanics and macroscopic realism [9]. Their paper was provocatively subtitled, "Is the flux there when nobody looks?", referring to a conversation in which Albert Einstein asked Abraham Pais whether he really believed that the moon exists only when he looks at it [10]. Distinctions can be made between interpretations of quantum mechanics which do and do not describe some external reality, and the further possibility that at some level between the quantum world of the small and the apparently classical world of our us-sized lives, other non-quantum principles may intervene [11]. The Leggett-Garg inequality tests for the conjunction of two postulates:

1. Realism (R): the system is always in one of its available states;
2. Non-invasive measurability (NIM): it is possible in principle to determine the state of the system without altering its subsequent evolution.

To these can be added the inductive postulate that the state cannot be affected by subsequent intervention, which is equivalent to making the second postulate time-symmetric. It is apparent that quantum mechanics is incompatible with both of these postulates. The Leggett-Garg inequality can be expressed either in a CHSH-like form [12] as a function of correlations between observations Q_i of a single system at four successive times t_i, or as the sum of correlations of observations at three times, t_1, t_2, t_3. With the definition

$$\langle K \rangle = \langle Q_1 Q_2 \rangle + \langle Q_2 Q_3 \rangle + \langle Q_1 Q_3 \rangle$$

and the understanding that each realization of Q_j has two possible outcomes, -1 or $+1$, then for realism to be tenable

$$-1 \leq \langle K \rangle \leq +3$$

For nearly a quarter of a century no one was able to devise a practical experiment to implement this, but in the past few years tests have been performed in several systems: superconducting qubits, polarized photons, and electron and nuclear spins [13–16]. Some of these required either weak measurements or an assumption of stationarity. The experiment of Knee *et al.* [16] required neither of these. It used an ensemble in silicon of phosphorous nuclei as the system. Each nucleus had an associated electron spin which was used as an ancilla to measure that nuclear spin through either a CNOT (controlled NOT) or an anti-CNOT gate to perform a non-invasive measurement. It was assumed that if the ancilla spin did not undergo a spin flip, then the measurement was non-invasive, and it would no doubt have been possible to conduct an explicit ideal-negative-result (INR) measurement as originally proposed by Leggett [17] and subsequently refined to close the so-called clumsiness loophole [18]. Since the experiments were performed at finite temperature and magnetic field, the ancilla electrons were imperfectly polarized. It proved possible to

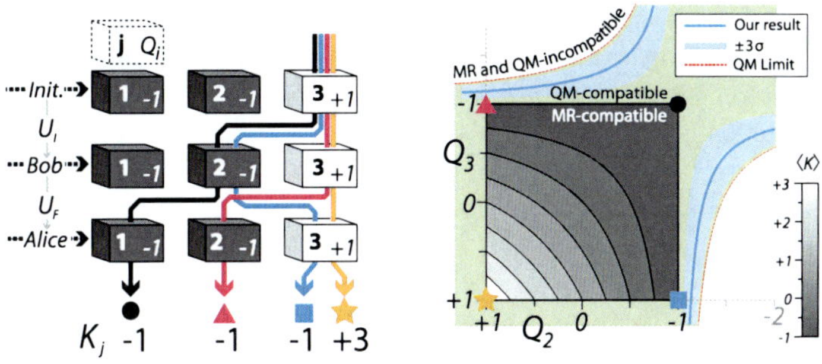

(a) (b)

Fig. 24.5 (**a**) NV^- centre and solid immersion lens. (**b**) Spectrum of microwave transitions, and statistics of photon counting to detect presence (*top*) or absence of fluorescence [3]

violate the inequality in a way that would convince both a moderate realist who accepts that the thermal distribution leads to random correlations, and also an aggressive realist (dubbed 'neurotic realist' by Serge Haroche) who asserts that by some unidentified process all the imperfectly initialised spins conspire to give worst-case correlations. These experiments provide a methodology to violate the inequality in increasingly macroscopic systems, including ensembles, but they do not determine which of the two postulates above is experimentally ruled out.

The quantum three-box experiment can be cast as a test of the Leggett-Garg inequality, in a way that allows a stronger conclusion to be drawn. We assign a value $Q = -1$ if the ball is in Box 1 or 2, and $Q = +1$ if the ball is in Box 3. This is presented schematically in Fig. 24.5(a), where it can be seen that a classical sequence entirely through the $j = 3$ (white) boxes yields $K = +3$, whereas if the ball is one of the other boxes either when Bob looks (whether or not he sees it) or when Alice looks inside Box 3 then $K = -1$. Thus in all these cases the value of K lies within the Leggett-Garg bound. This is represented as a square in Fig. 24.5(b), in which contours of K are plotted for values of Q_2 and Q_3, given that the experiments are intitialised with $Q_1 = +1$. Macrorealist (MR-compatible) values of K lie within the shaded square. The range of K permitted by quantum mechanics (QM-compatible) is shown as the green shaded area, within the dotted red limit. This limit is

$$\langle K \rangle \geq -\frac{13}{9} = -1.4\dot{4}$$

The experimental results in Fig. 24.4 yielded $\langle K \rangle = -1.265 \pm 0.023$, which is shown as the blue curves in Fig. 24.5(b). For a moderate realist assumption of fair sampling this gives a violation of macrorealism by 11.3 sigma, and even for the aggressive realist assumption, corresponding to Alice cheating, this still exceeds the Leggett-Garg bound by 7.8 standard deviations.

There is a crucial difference in the non-invasive measurability (NIM) between experiments in 2-D and 3-D Hilbert space. In the 2-D case it is not possible to violate the Leggett-Garg inequality using three projective measurements in a single run and still retain NIM. Defining the disturbance D as the difference in $\langle K \rangle$ if all three measurements are made in a single run compared with measuring the three pairs of measurements in three runs (or six allowing for both CNOT and anti-CNOT NIM), it can be shown that if $D = 0$, then inevitably $-1 \leq \langle K \rangle \leq +3$ [19]. Thus any measurement of all three values in a single experiment which violates Leggett-Garg (L-G) also fails to satisfy NIM. The situation is different in 3-D, where all three measurements are always made in each experiment, without even needing anti-CNOT gates.

That Alice cannot detect the effect of Bob's measurements statistics is evident in Fig. 24.3(b). This may be contrasted with the 2-D case, where the act of measuring the ancilla would affect the statistics of the subsequent measurement of Q in the same experiment. In private correspondence, Tony Leggett has described Bob's measurement of M_1 or M_2 in the three box experiment as a partial measurement of M_3, since a positive result implies that $Q = -1$, whereas a negative result allows $Q = -1$ or $Q = +1$ thus corresponding to 'no measurement'. This might suggest an interpretation in which $\langle K \rangle$ is a weighted average of something constrained by the inequality and something characteristic of 2-D. However, this would not affect the game, since the post-selected runs all correspond to Bob having obtained a positive result. It remains to be investigated whether there might be adversarial games in which the measurements might be even more robustly non-invasive, for example by using different manipulations and larger Hilbert space. This certainly promises to be fruitful in the further development of no-go theorems of epsistemic interpretations of reality [20].

The consequence is profound of demonstrating that the measurements in the quantum three box are indeed non-invasive. If the NIM postulate is tenable, and if the L-G inequality is not satisfied, then it follows that realism, as defined here, is not tenable. In these particular experiments that is perhaps not surprising, since most people in the field would accept that the behaviour of an NV^- centre in diamond is accurately described by quantum theory. Yakir Aharonov and his colleagues have been at pains to emphasise that while their formulation of quantum states at a given time may provide new insights and new ways of deriving results which might not otherwise have been foreseen, it "neither contradicts ordinary quantum theory nor extends it in the sense of new physical laws" [2]. Nevertheless, like the earlier 2-D L-G experiments, the quantum three-box experiment provides a methodology for more macroscopic systems where the quantum implications for reality have yet to be established. What is meant by more macroscopic is an open question. Does macroscopicness lie in greater numbers of atoms or photons, or in greater mass or spatial size, or in greater complexity (if so how should that be quantified?), or in a greater number of dimensions in Hilbert space? Each of these could be explored in pursuit of extending the tests of macroscopic realism towards the realm of human experience and mental processes.

I personally find it remarkable that philosophical questions about the nature of reality are open to experimental investigation. Conventionally reality is that which is the case, independent of the observer. Critical realism allows for a dialogue between the observer and the observed, in a way that admits of refinement of knowledge. There is still no consensus on how this is constrained by quantum theory. A recent poll of physicists, philosophers, and mathematicians revealed that the foundations of quantum mechanics remain hotly debated in the scientific community, with a divergence of views on some fundamental questions [21]. But the implications go further, all the way to how concepts such as truth, morality, responsibility, and belief, have their basis in reality. Yakir Aharonov has provided new ideas for experiments to push the limits of 'quantumness' conceptually and experimentally, with benefits both for technologies and for foundational questions. I offer these results to him with congratulations on his 80th birthday and best wishes for many happy returns.

Acknowledgements I gladly acknowledge my co-authors of [3] and also [16] for the experiments and results and their interpretation, and for many stimulating discussions about the further implications. Those papers should be taken as definitive in the event of any inadvertent discrepancy, though I take responsibility for additional views which go beyond the papers. I thank Richard George for helpful comments on the manuscript, and the John Templeton Foundation, together with the other agencies acknowledged in the papers, for funding the research.

References

1. D.Z. Albert, Y. Aharonov, S. D'Amato, Curious new statistical prediction of quantum-mechanics. Phys. Rev. Lett. **54**, 5–7 (1985)
2. Y. Aharonov, L. Vaidman, Complete description of a quantum system at a given time. J. Phys. A, Math. Gen. **24**, 2315–2328 (1991)
3. R.E. George et al., Opening up three quantum boxes causes classically undetectable wavefunction collapse. Proc. Natl. Acad. Sci. USA **110**, 3777–3781 (2013)
4. Y. Aharonov, D.Z. Albert, L. Vaidman, How the result of a measurement of a component of the spin of a spin-1/2 particle can turn out to be 100. Phys. Rev. Lett. **60**, 1351–1354 (1988)
5. N. Aharon, L. Vaidman, Quantum advantages in classically defined tasks. Phys. Rev. A **77**, 052310 (2008)
6. Y. Aharonov, D. Rohrlich, *Quantum Paradoxes: Quantum Theory for the Perplexed* (Wiley-VCH, New York, 2008)
7. K.J. Resch, J.S. Lundeen, A.M. Steinberg, Experimental realization of the quantum box problem. Phys. Lett. A **324**, 125–131 (2004)
8. P. Kolenderski et al., Aharon-Vaidman quantum game with a Young-type photonic qutrit. Phys. Rev. A **86**, 012321 (2012)
9. A.J. Leggett, A. Garg, Quantum-mechanics versus macroscopic realism: is the flux there when nobody looks? Phys. Rev. Lett. **54**, 857–860 (1985)
10. A. Pais, Einstein and the quantum theory. Rev. Mod. Phys. **51**, 863–914 (1979)
11. A.J. Leggett, The quantum measurement problem. Science **307**, 871–872 (2005)
12. J.F. Clauser, M.A. Horne, A. Shimony, R.A. Holt, Proposed experiment to test local hidden-variable theories. Phys. Rev. Lett. **23**, 880–884 (1969)
13. A. Palacios-Laloy et al., Experimental violation of a Bell's inequality in time with weak measurement. Nat. Phys. **6**, 442–447 (2010)
14. M.E. Goggin et al., Violation of the Leggett-Garg inequality with weak measurements of photons. Proc. Natl. Acad. Sci. USA **108**, 1256–1261 (2011)
15. G. Waldherr, P. Neumann, S.F. Huelga, F. Jelezko, J. Wrachtrup, Violation of a temporal Bell inequality for single spins in a diamond defect center. Phys. Rev. Lett. **107**, 090401 (2011)
16. G.C. Knee et al., Violation of a Leggett-Garg inequality with ideal non-invasive measurements. Nat. Commun. **3**, 606 (2012)

17. A.J. Leggett, Experimental approaches to the quantum measurement paradox. Found. Phys. **18**, 939–952 (1988)
18. M.M. Wilde, A. Mizel, Addressing the clumsiness loophole in a Leggett-Garg test of macrorealism. Found. Phys. **42**, 256–265 (2012)
19. R.E. George et al., Opening up three quantum boxes causes classically undetectable wavefunction collapse (Supporting Information). Proc. Natl. Acad. Sci. USA **110**, 3777–3781 (2013)
20. M.F. Pusey, J. Barrett, T. Rudolph, On the reality of the quantum state. Nat. Phys. **8**, 474–477 (2012)
21. M. Schlosshauer, J. Kofler, A. Zeilinger, The interpretation of quantum mechanics: from disagreement to consensus? Ann. Phys. **525**(4), A51–A54 (2013)

Chapter 25
Standard and Null Weak Values

Oded Zilberberg, Alessandro Romito, and Yuval Gefen

Abstract Weak value (WV) is a quantum mechanical measurement protocol, proposed by Aharonov, Albert, and Vaidman. It consists of a weak measurement, which is weighed in, conditional on the outcome of a later, strong measurement. Here we define another two-step measurement protocol, null weak value (NVW), and point out its advantages as compared to WV. We present two alternative derivations of NWVs and compare them to the corresponding derivations of WVs.

25.1 Introduction

This contribution is dedicated to Yakir Aharonov, on his 80th birthday. His seminal work in quantum mechanics, and the stimulating discussions we have had with him, have influenced our own work in physics in a deep way. Y.G. is indebted to Yakir Aharonov for the close interaction, and his continuous support over the years.

The von Neumann formulation of measurement in quantum mechanics invokes a generic Hamiltonian of the form [1],

$$H = H_{\mathrm{S}} + H_{\mathrm{D}} + H_{\mathrm{int}}, \tag{25.1}$$

where H_{S} is the Hamiltonian of the system to be measured, H_{D} is the detector's Hamiltonian, and H_{int} represents the coupling between the two:

$$H_{\mathrm{int}} = \lambda g(t)\hat{p}\hat{A}. \tag{25.2}$$

O. Zilberberg (✉) · Y. Gefen
Department of Condensed Matter Physics, Weizamnn Institute of Science, Rehovot 76100, Israel
e-mail: oded.zilberberg@weizmann.ac.il

Y. Gefen
e-mail: yuval.gefen@weizmann.ac.il

A. Romito
Dahlem Center for Complex Quantum Systems and Fachbereich Physik, Freie Universität Berlin, 14195 Berlin, Germany
e-mail: alessandro.romito@fu-berlin.de

D.C. Struppa, J.M. Tollaksen (eds.), *Quantum Theory: A Two-Time Success Story*,
DOI 10.1007/978-88-470-5217-8_25,

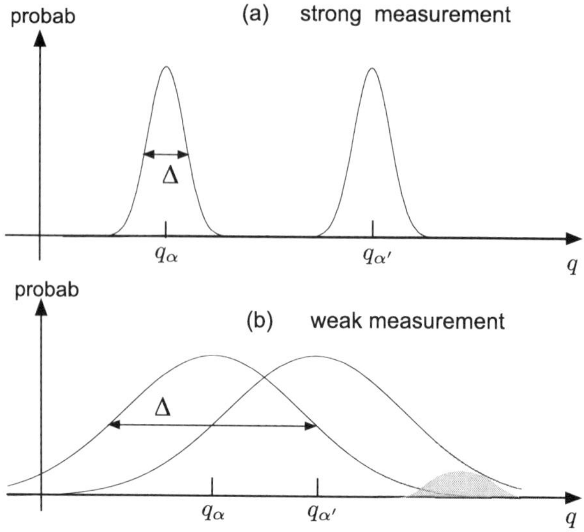

Fig. 25.1 Strong and weak measurements. When the system is in a given eigenstate, $|\alpha\rangle$, of the measured observable, $\hat{A}$, the coordinate shift in the detector, q, has a certain probability distribution, $P_\alpha(q)$. (**a**) When such distributions are well peaked around different values of q_α, i.e. $\Delta \ll |q_{\alpha'} - q_\alpha|$, one can uniquely associate a value of the coordinate, q_α, to each eigenstate of $\hat{A}$, $|\alpha\rangle$, which corresponds to a strong measurement. (**b**) In the opposite case of a weak measurement, $\Delta \gg |q_{\alpha'} - q_\alpha|$, the small shift in the detector's coordinate (due to the weak system-detector coupling) is blurred by the distribution uncertainty. The *shaded (green) area on the right* represents the part of the distribution selected by the WV protocol

Here $\hat{p}$ is the momentum canonically conjugate to the position of the detector's pointer, $\hat{q}$, $g(t)$ represents the time window during which the measurement (system-detector coupling) takes place, and λ is the dimensionless strength of the measurement. The measured observable is $\hat{A} = \sum_i a_i |a_i\rangle\langle a_i|$.

Strong measurement is associated with the collapse of the wave function dogma [1]. In an ideal strong measurement there is a one-to-one correspondence between the observed value of the detector's coordinate, q_α, and the eigenstates $|\alpha\rangle$ of the system's measured operator, $\hat{A}$ [cf. Fig. 25.1(a)]. In a weak measurement ($\lambda \ll 1$), instead, the ranges of values of q that correspond to two distinct eigenstates of $\hat{A}$, $|\alpha\rangle$ and $|\alpha'\rangle$, are described by two strongly overlapping probability distribution functions, $P_\alpha(q)$ and $P_{\alpha'}(q)$, respectively. Hence the measurement of q provides only partial information on the state of the system [cf. Fig. 25.1(b)] and changes its state only slightly. Nonetheless the mean value of a weak measurement (averaged over many repetition of the measurement) coincides with that of a strong measurement.

One may extend the notion of weak measurement by referring to a sequence of correlated (especially conditional) measurements. Conditional quantum measurements can lead to results that cannot be interpreted in terms of classical probabilities, due to the quantum correlations between measurements. An intriguing example for correlated quantum measurements outcome is the so called *weak value* (WV).

It is the outcome of a measurement scheme originally developed by Aharonov, Albert and Vaidman [2]. The WV measurement protocol consists of (i) initializing the system in a certain pure state $|i\rangle$ (preselection; generalization to a mixed state is possible [3] but will not be discussed here), (ii) weakly measuring observable $\hat{A}$ of the system, (iii) retaining the detector output only if the system is eventually measured to be in a chosen final state $|\mathrm{f}\rangle$ (postselection). The average signal monitored by the detector will then be proportional to the real (or imaginary) part of the complex WV,

$$
{}_f\langle \hat{A} \rangle_i = \frac{\langle \mathrm{f}|\hat{A}|i\rangle}{\langle \mathrm{f}|i\rangle}, \tag{25.3}
$$

rather than to the standard average value, $\langle i|\hat{A}|i\rangle$ [cf. Fig. 25.1(b)]. Further discussion of the context in which WVs should be understood has been provided [4–6].

Going beyond the peculiarities of WV protocols, recent series of works explored the potential of WVs in quantum optics [7–13] and solid-state physics [14–17], ranging from experimental observation to their application to hypersensitive measurements. In the latter, a measurement, performed by a detector *entangled* with a system whose states can be preselected and postselected, leads to an amplified signal in the detector that enables sensing of small quantities [9–13, 17]. Quite generally, within a WV-amplification protocol, only a subset of the detector's readings, associated with the tail of the signal's distribution, is accounted for. Notwithstanding the scarcity of data points, the large value of ${}_f\langle \hat{A} \rangle_i$, leads to an amplification [11, 17] of the signal-to-noise ratio (SNR) for systems where the noise is dominated by an external (technical) component.

The amplification originating from WV protocols is non-universal. The specifics of such amplification are diverse and system-dependent. In fact, for statistical (inherent) noise, SNR amplification resulting from large WVs is generally suppressed due to the reduction in the statistics of the collected data: postselection restricts us to a small subset of the readings at the detector. The upside of the WV procedure has several facets: if we try to enhance the statistics by increasing the intensity of input signal through the system (e.g., intensity of the impinging photon beam), possibly entering a non-linear response regime, postselection will effectively reduce this intensity back to a level accessible to the detector sensitivity [10]. Alternatively, amplification may originate from the imaginary component of the WV [9], or from the different effect of the noise and the measured variable on the detector's signal [17]. However, as long as quantum fluctuations (leading to inherent statistical noise) dominate, the large WV is outweighed by the scarcity of data points, failing to amplify the signal-to-statistical-noise ratio [17, 18].

We have recently presented an alternative measurement protocol dubbed *null weak value* (NWV), that overcomes the SNR problem of WV-protocols [19]. Like the WV-protocol, NWV consists of a two-step conditional measurement. The recipe goes as follows: (i) We prepare the system in a given pure state $|i\rangle$ (a generalization to a mixed state will not be discussed here). (ii) We perform a strong measurement of the observable $\hat{A}$; we arrange the setup such that the probability, p, that our detector "clicks" (hence collapse is taking place), is small ($p \ll 1$). If the detector clicks, the system's state has collapsed, and we start our measurement all over again, with a new replica of the system. If no click has occurred, $|i\rangle$ has been (by way of back-action) modified, $|i\rangle \to |i_p\rangle$. (iii) We now let the system evolve in time, possibly manipulating $|i_p\rangle$ in a controlled way. (iv) We perform a strong measurement of a certain observable $\hat{B}$. Formulating the results of the first and second measurements in an anti-causal manner, the outcome of the first measurement (the detector clicking) is conditional on the outcome of the second measurement (of $\hat{B}$). The NWV of $\hat{A}$ is then

$$\langle \hat{A} \rangle_{\mathrm{NWV}} = \frac{\langle i|\hat{A}|i\rangle}{|\langle f|i\rangle|^2}, \tag{25.4}$$

to be compared with the WV of $\hat{A}$, Eq. (25.3).

In principle, the derivation of standard WVs, as well as that of NWVs, can be done by following the dynamics of the measurement process in the extended system-detector Hilbert space. Here we discuss the derivation of NWVs and compare it to the derivation of WVs, taking two different paths: (i) analyzing the the effect of the detector on states and amplitudes in the system's subspace. We refer to this as "derivation in terms of quantum states". (ii) We derive expressions for WVs/NWVs analyzing the probabilities and conditional probabilities involved in the various steps of the protocols.

The outline of this paper is as follows. In Sect. 25.2 we present a derivation of a standard WV in terms of quantum states. This follows by a derivation in Sect. 25.3, which employs conditional probabilities. Section 25.4 addresses NWV from the viewpoint of conditional probabilities, and Sect. 25.5 presents a derivation of NWV in terms of quantum states. In Sect. 25.6 we outline a few conclusions.

25.2 Standard WV Derivation

Weak values describe the outcome read in a detector when the measured system is subsequently found to be in a specific state. The expression for WVs can be derived most simply through an argument due to Yakir Aharonov based on a one-line identity for the average value of an observable $\hat{A}$:

$$\langle \hat{A} \rangle = \langle i|\hat{A}|i\rangle = \sum_n \langle i|f_n\rangle \langle f_n|\hat{A}|i\rangle = \sum_n |\langle f_n|i\rangle|^2 \frac{\langle f_n|\hat{A}|i\rangle}{\langle f_n|i\rangle} \equiv \sum_n P_{i\to n}\, {}_{f_n}\langle \hat{A}\rangle_i . \tag{25.5}$$

The identity is obtained by inserting a complete set of states $1 = \sum_n |f_n\rangle\langle f_n|$. Also in the last equality we introduced the notation ${}_{f_n}\langle \hat{A}\rangle_i \equiv \langle f_n|\hat{A}|i\rangle/\langle f_n|i\rangle$, $P_{i\to n} \equiv |\langle f_n|i\rangle|^2$. The reasoning goes as follows: The states $\{|f_n\rangle\}$ above can be interpreted as the possible states obtained by measuring an observable $\hat{B}$ after $\hat{A}$. If one can assume that the measurement $\hat{A}$ *leaves the initial state unchanged*, $P_{i\to n}$ can be interpreted as the probability that the system is (finally) to be found in $|f_n\rangle$.

Equation (25.3) gives then a natural interpretation of the WV, ${}_{f_n}\langle\hat{A}\rangle_i$, as the average of $\hat{A}$ conditional to a postselection on $|f_n\rangle$. The so obtained expression for the WV is universal in the sense that it does not depend on the the specific detector or its coupling to the system, as long as the state of the system is unaffected by the detection process. Not modifying the state of the system has to do with the measurement back-action and its precise meaning is what defines the weakness of the measurement, hence the name weak value. The weakness of the measurement can be addressed in a specific model of the system-detector coupling. One may, of course, reproduce the correct expression for the WV in a treatment involving the system-detector Hilbert space.

The weak coupling between a system and a detector is performed by an ideal von Neumann measurement [1], described by the Hamiltonian in Eqs. (25.1) and (25.2) with $\lambda \to 0$. We assume for simplicity that the free Hamiltonians of the system and the detector vanish and that $g(t) = \delta(t - t_0)$.

The system is initially prepared in the state $|i\rangle$, and the detector in the state $|\phi_0\rangle$. The latter is assumed to be a Gaussian wave-packet centered at $q = q_0$, $|\phi_0\rangle = Ce^{-(q-q_0)^2/4\Delta^2}$. After the interaction with the detector the entangled state of the two is

$$|\psi\rangle = e^{-i\lambda\hat{p}\hat{A}}|i\rangle|\phi_0\rangle. \tag{25.6}$$

In a regular measurement the signal in the detector, i.e. the pointer's position $\langle q\rangle = q_0 + \lambda\langle\hat{A}\rangle$, is read. From the classical signal, $\langle q\rangle$, one can infer the average value of the observable $\hat{A}$.

In a WV protocol the signal in the detector is kept provided that the system is successfully postselected to be in a state $|f\rangle$. Hence, the detector ends up in the state

$$|\psi\rangle = |\phi_0\rangle - i\lambda\big[\langle f|\hat{A}|i\rangle/\langle f|i\rangle\big]\hat{p}|\phi_0\rangle \approx e^{-i\lambda\hat{p}\,{}_f\langle\hat{A}\rangle_i}|\phi_0\rangle, \tag{25.7}$$

that corresponds to a shift in the position of the pointer proportional to $Re[{}_f\langle A\rangle_i]$. Hence the expectation value of the coordinate of the pointer is given by

$${}_f\langle\hat{q}\rangle_i = q_0 + \lambda\, Re\big[{}_f\langle\hat{A}\rangle_i\big], \tag{25.8}$$

where the conditional average value of $\hat{A}$ is inferred from the detector's reading.

We note that the approximation in Eq. (25.7) is valid when $\Delta \gg \lambda \max_{i,j}|a_i - a_j|$. This means that the initial detector's wave function and the shifted one due to the interaction with the system are strongly overlapping. In turn, this means that for *any* outcome of the detector the state of the system is weakly changed. This corresponds to a weak measurement. As long as the measurement of the observable, $\hat{A}$, is weak, the WV is system independent and does not depend on the details of the coupling or the specific choice of the detector.

25.3 Standard Weak Values in Terms of Conditional Probabilities

In this section we provide a correspondence between a conditional probability notation and the standard state-vector notation used in most WV works.

Let us consider the case of a qubit system weakly coupled to another two-level detector. Their respective states are initially

$$|i\rangle = \alpha|0\rangle + \beta|1\rangle \quad \text{Hilbert space of system,} \tag{25.9}$$

$$|d\rangle = \gamma|0\rangle + \delta|1\rangle \quad \text{Hilbert space of detector.} \tag{25.10}$$

Once the system and detector are coupled their resulting entangled state is

$$|i, d\rangle = \mathcal{N}\big(\alpha\gamma|0,0\rangle + \alpha\delta|0,1\rangle + \beta\tilde{\gamma}|1,0\rangle + \beta\tilde{\delta}|1,1\rangle\big), \tag{25.11}$$

where we assumed that if the system is in state $|0\rangle$ the detector remains unaffected, $\tilde{\gamma}, \tilde{\delta}$ represent the detector amplitudes when the system is in state $|1\rangle$, and $\mathcal{N}$ is a normalization factor.

Measuring the detector state and applying calibration yields an observable of the system:

$$\langle i|\hat{A}|i\rangle = \frac{P(\hat{A}') - |\delta|^2}{|\tilde{\delta}|^2 - |\delta|^2} = |\beta|^2, \tag{25.12}$$

where $\hat{A} = |1\rangle\langle 1|$ operates in the system's Hilbert space and $\hat{A}' = |0,1\rangle\langle 0,1| + |1,1\rangle\langle 1,1|$ operates in the joint Hilbert space of system and detector.

Taking the probability of this weak measurement outcome conditional on the outcome of a subsequent postselection, yields the detector response to a standard WV protocol

$$\frac{P(\hat{A}'|\hat{B}) - |\delta|^2}{|\tilde{\delta}|^2 - |\delta|^2} \sim \big(|\tilde{\delta}|^2 - |\delta|^2\big)\mathrm{Re}\{{}_f\langle\hat{A}\rangle_i\}, \tag{25.13}$$

where $\hat{B} = |f\rangle\langle f|$ is a strong postselection in the system space, and we used the weakness of the first measurement $|\tilde{\delta}|^2 - |\delta|^2 \ll 1$. The counter-causal conditional probability $P(\hat{A}'|\hat{B})$ is calculated using Bayes theorem and using the causal conditional probabilities appearing in the tree diagram in Fig. 25.2(a).

25.4 Null Weak Values in Terms of Conditional Probabilities

We now turn to describe our new measurement protocol (null-WV) [cf. Fig. 25.2(b)]. The qubit state is measured twice. The first measurement $\hat{A} = p_0|0\rangle\langle 0| + p_1|1\rangle\langle 1|$ is a strong (projective) measurement which is performed on the system with small

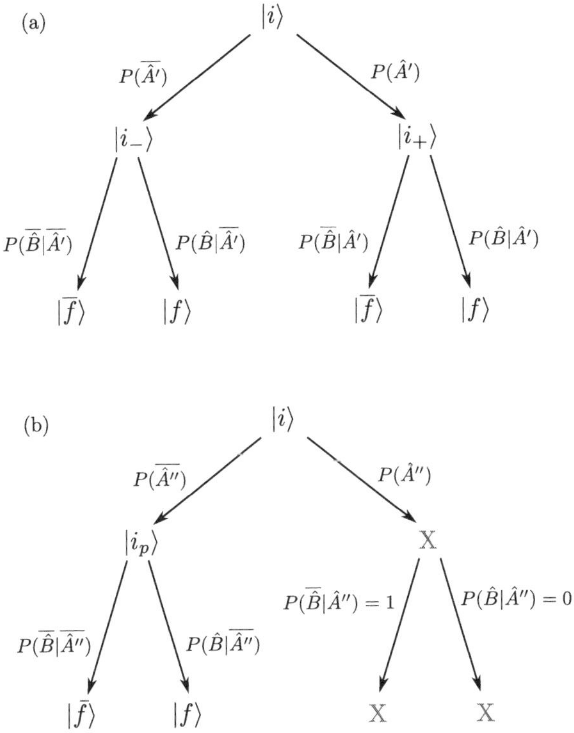

Fig. 25.2 Tree diagrams of the state evolution during the (**a**) weak value protocol and (**b**) null weak value protocol. The respective probabilities of events are indicated. (**a**) Upon the first weak measurement the detector "clicks" [no "click"] with probability $P(\hat{A}')$ $[P(\overline{\hat{A}'})]$. Due to the measurement back-action, the system evolves into the state $|i_+\rangle$ $[|i_-\rangle]$. Subsequently the system if strongly measured to be in the $|f\rangle$ [or not]. (**b**) The first partial-collapse measurement "clicks" [no "click"] with probability $P(\hat{A})$ $[P(\overline{\hat{A}})]$. If a "click" occurred the system is destroyed [marked by X (*red*)]. If not, due to the measurement back-action, the system evolves into the state $|i_p\rangle$. Subsequently the system if strongly measured to be in the $|f\rangle$ [or not]

probability. Here the states $\{|0\rangle, |1\rangle\}$ are measured with probabilities $\{p_0, p_1\}$, respectively. For simplicity, hereafter, we assume that only the state $|1\rangle$ is measured with probability $p_1 = p$ and $p_0 = 0$. If the detector "clicks" (the measurement outcome is positive), the qubit state is destroyed. Very importantly, having a "null outcome" (no click) still results in a weak back-action on the system. We refer to this stage of the measurement process as "weak partial-collapse". Subsequently the qubit state is (strongly) measured a second time (postselected), $\hat{B}$, to be in the state $|f\rangle$ (click) or $|\bar{f}\rangle$ (no click).

Similarly to the previous case, readout of the number of "clicks" in this first detector yields an observable on the system

$$\langle i|\hat{A}|i\rangle = \frac{P(\hat{A})}{p} = |\beta|^2. \tag{25.14}$$

Studying the correlation between this first partial-collapse measurement and a "no click" postselection yields the null value

$$\frac{P(\hat{A}|\overline{\hat{B}})}{p} = \frac{\langle i|\hat{A}|i\rangle}{P(\overline{\hat{B}})} \sim \frac{\langle i|\hat{A}|i\rangle}{|\langle f|i\rangle|^2}, \tag{25.15}$$

where the last approximation is for small p.

Our protocol takes advantage of the correlation between the two measurements. To shed some light on its outcome we calculate explicitly the conditional probabilities following the measurement procedure sketched in Fig. 25.2(b). For example, if the first measurement results in a "click" the system's state is destroyed and any subsequent measurement on the system results in a null-result, implying $P(\hat{B}|\hat{A}) = 0$, and $P(\overline{\hat{B}}|\hat{A}) = 1$. This represents a classical correlation between the two measurements. By contrast, $P(\overline{\hat{B}}|\overline{\hat{A}})$ embeds non-trivial quantum correlations. The first partial-collapse measurement of a given preselected state $|i\rangle$ results in the detector clicking with probability $P(\hat{A}) = p|\beta|^2$. If no click occurs [with probability $P(\overline{\hat{A}}) = 1 - P(\hat{A})$], the qubit's state is modified by the measurement back-action into $|i_p\rangle = [\alpha|0\rangle + \sqrt{1-p}\beta|1\rangle]/\sqrt{P(\overline{\hat{A}})}$. The second strong measurement, $\hat{B}$, yields a click [no click] with probability $P(\hat{B}|\overline{\hat{A}}) = |\langle f|i_p\rangle|^2$ $[P(\overline{\hat{B}}|\overline{\hat{A}}) = |\langle \bar{f}|i_p\rangle|^2]$. Finally, using Bayes theorem, we can write $P(\hat{A}|\overline{\hat{B}}) = P(\hat{A})/[P(\hat{A}) + P(\overline{\hat{A}})P(\overline{\hat{B}}|\overline{\hat{A}})]$. This correlated outcome is useful in obtaining amplified SNR in a quantum state discrimination problem [19].

25.5 Description of Standard and Null Weak Values in Terms of Quantum States

The result of the NWV protocol, Eq. (25.4), though emanating from a weak measurement, is different from the standard WV, Eq. (25.3). On the face of it, the derivation that leads to Eq. (25.5) appears to be universally adapted to any two-step (the first is weak) measurement procedure. The fact that the expression for NWV is different from that of WV may then seem paradoxical. It is therefore instructive to understand how the NWV relates to the standard WV in the framework of the derivation of Eq. (25.5).

The idea behind Eq. (25.5) is quite general: one writes an identity for the standard quantum mechanical expectation value in terms of a sum of probabilities to reach the possible postselected states. These probabilities are weighted-in with the appropriate coefficients, namely

$$\langle \hat{A}\rangle = \sum_n P_{i\to n}\ {}_{f_n}\langle \hat{A}\rangle_i, \tag{25.16}$$

where $P_{i\to n}$ is the probability to obtain the postselected state $|f_n\rangle$ given the initial (prepared) state of the system is unchanged. The coefficients ${}_{f_n}\langle\hat{A}\rangle_i$ obtained from this identity are then naturally interpreted as the conditional averages one is after. In the (standard WV) derivation of Eq. (25.5), the first weak measurement does not affect the state significantly, and this interpretation comes natural with $P_{i\to n} = |\langle f_n|i\rangle|^2$. For NWVs, the factor $|\langle f_n|i\rangle|^2$ can no longer be interpreted as the probability to find the system in $|f_n\rangle$, following the second strong measurement (postselection). In fact, (without extending our Hilbert space) the partial-collapse measurement cannot be directly described by a von Neumann-like measurement on which Eq. (25.5) is implicitly based. Therefore, despite the fact that Eq. (25.5) is an identity, which holds true also in the NWV case, it does not allow for an interpretation of the conditional outcome as the NWV.

Nevertheless, it is possible to write other identities in the spirit of Eq. (25.16) which can be useful in the present case. For example, following the steps (i) using $\hat{A} = \sum_j a_j \hat{\Pi}_j$, where $\{a_j\}$ are the eigenvalues of $\hat{A}$ and $\{\hat{\Pi}_j\}$ the corresponding projection operators onto the states $\{|j\rangle\}$, (ii) inserting the projector identity $\hat{\Pi}_j \equiv \hat{\Pi}_j^2$, and (iii) inserting the identity in terms of postselected states, $\mathbb{1} = \sum_n |f_n\rangle\langle f_n|$, one obtains

$$\langle\hat{A}\rangle = \sum_n \left(\sum_j |\langle f_n|\Pi_j|i\rangle|^2\right) \frac{\sum_j a_j |\langle f_n|\Pi_j|i\rangle|^2}{\sum_j |\langle f_n|\Pi_j|i\rangle|^2}. \tag{25.17}$$

If the measurement of $\hat{A}$ is strong, the term in parenthesis in Eq. (25.17) corresponds to the probability of postselection. Therefore the conditional average is the remaining expression outside of the parenthesis. Equation (25.17) is evidently an identity, but it does not lend itself to any physically meaningful interpretation if the protocol involves a weak measurement.

Since partial-collapse measurements are in fact strong measurements that occur with a small probability, Eq. (25.17) is particularly useful for NWVs. Indeed, one may effectively describe a partial-collapse measurement as a von Neumann measurement in an extended Hilbert space. To do so, we formally extend the system's Hilbert space to include an extra ancilla state, $|R\rangle$. The idea is to describe the partial-collapse measurement as a combination of a weak transition to the ancilla state followed by a strong projective measurement of this newly added state. Let us describe this more precisely.

The initial state one is interested in is $|i\rangle = \alpha|0\rangle + \beta|1\rangle$. Allowing the state $|1\rangle$ to be transferred to $|R\rangle$ with transition rate Γ for a time window t, evolves the initial state into $\hat{U}|i\rangle = \alpha|0\rangle + \sqrt{1-p}\beta|1\rangle + \sqrt{p}\beta|R\rangle$, where $p = 1 - \exp(-\Gamma t)$ is the probability to undergo this transition over time t, and $\hat{U}|i\rangle$ is in the extended Hilbert space spanned by $|0\rangle, |1\rangle, |R\rangle$. It is now apparent that the partial-collapse measurement can be written as $\hat{A} = \hat{U}^\dagger|R\rangle\langle R|\hat{U}$. Subsequently we postselect on $|f\rangle$ (a state within the system's Hilbert space, $|0\rangle, |1\rangle$). Hence, in the extended Hilbert space the measurement can be formulated according to the standard measurement theory.

Let us adjust Eq. (25.17) to the NWV case,

$$
\begin{aligned}
\langle \hat{A} \rangle &= \frac{\langle \hat{A} \rangle}{p} \\
&= \frac{1}{p} \sum_{f_n = f, \bar{f}} \left(\sum_{j=\bar{R},R} |\langle f_n | \Pi_j \hat{U} | i \rangle|^2 \right) \frac{|\langle f_n | \Pi_R \hat{U} | i \rangle|^2}{\sum_{j=\bar{R},R} |\langle f_n | \Pi_j \hat{U} | i \rangle|^2} \quad \text{(inappropriate!)},
\end{aligned}
\tag{25.18}
$$

where $\Pi_{\bar{R}} = |0\rangle\langle 0| + |1\rangle\langle 1|$ is the projector onto the Hilbert space of the original system, i.e. the subspace orthogonal to $|R\rangle$. However, owing to the fact that the postselection states $|f\rangle$ and $|\bar{f}\rangle$ are within the reduced Hilbert space, the numerator on the right hand side is identically zero.

In order to harness this approach to describe the procedure at hand, we introduce a crucial modification in our scheme. One may think of the postselection as performing an additional "partial-collapse" measurement with the state $|f\rangle$ having probability $p \equiv 1$ to become transmitted into state $|R\rangle$, i.e. the postselection is with respect to the operator (observable) $\hat{\hat{B}} = \hat{\hat{U}}^\dagger |R\rangle\langle R| \hat{\hat{U}}$, with $\hat{\hat{U}}[a|\bar{f}\rangle + b|f\rangle] = [a|\bar{f}\rangle + b|R\rangle]$. Due to the specific nature of the partial-collapse measurement, the state $|R\rangle$ is not affected by this transformation. In particular, it does *not* couple back to the original Hilbert space during the second tunneling event. Note that this renders the evolution *non-unitary* in the extended Hilbert space. This prescription yields,

$$
\begin{aligned}
\langle \hat{A} \rangle &= \frac{1}{p} \sum_{f_n = R, \bar{f}} \left(\sum_{j=\bar{R},R} |\langle f_n | \hat{\hat{U}} \Pi_j \hat{U} | i \rangle|^2 \right) \frac{|\langle f_n | \hat{\hat{U}} \Pi_R \hat{U} | i \rangle|^2}{\sum_{j=\bar{R},R} |\langle f_n | \hat{\hat{U}} \Pi_j \hat{U} | i \rangle|^2} \\
&\equiv P_{i \to f} \frac{|\langle R | \hat{U} | i \rangle|^2}{p \, P_{i \to f}} \quad \text{(appropriate for NWV)},
\end{aligned}
\tag{25.19}
$$

with $P_{i \to f} \equiv |\langle R | \hat{\hat{U}} \Pi_{\bar{R}} \hat{U} | i \rangle|^2 + |\langle R | \hat{U} | i \rangle|^2$. Indeed, the expression on the left hand side corresponds to the probability to end up in state "$|f\rangle$"$\equiv \hat{\hat{U}}^\dagger |R\rangle$. Note that for $p \ll 1$, the probability $P_{i \to f}$ reduces to the form $|\langle f | i \rangle|^2$, the same as in the standard WV case [cf. Eq. (25.3)]. Last but not least, the multiplicative term in the middle equality of Eq. (25.19) (multiplying the parenthesis), is identical to the expression for the NWV [r.h.s. of Eq. (25.15)]. Hence, we have cast the NWV in the form of Eq. (25.16).

25.6 Conclusions

We have presented here a novel measurement protocol. Similarly to standard weak values, the outcome of this protocol—null weak value—is the result of a first (weaker) measurement correlated with a strong postselection. Ostensibly, as long as a single measurement is concerned, the first measurement in both protocols yields the same outcome. However, the substantial difference between the standard—and null—WVs comes to show that back-action on the system is profoundly different. Hence, involving a postselection leads to qualitatively different correlated results.

Acknowledgements This work was supported by GIF, the Israel science foundation, Minerva Foundation of the DFG, Israel-Korea MOST grant, and EU GEOMDISS.

References

1. J. von Neumann, *Mathematische Grundlagen der Quantenmechanik* (Springer, Berlin, 1932)
2. Y. Aharonov, D. Albert, L. Vaidman, Phys. Rev. Lett. **60**, 1351 (1988)
3. A. Romito, Y. Gefen, Physica E **42**, 343 (2010)
4. H. Wiseman, Phys. Rev. A **65**, 032111 (2002)
5. R. Jozsa, Phys. Rev. A **76**, 044103 (2007)
6. J. Dressel, S. Agarwal, A.N. Jordan, Phys. Rev. Lett. **104**, 240401 (2010)
7. N.W.M. Ritchie, J.G. Story, R.G. Hulet, Phys. Rev. Lett. **66**, 1107 (1991)
8. G. Pryde, J. O'Brien, A. White, T. Ralph, H. Wiseman, Phys. Rev. Lett. **94**, 220405 (2005)
9. O. Hosten, P. Kwiat, Science **319**, 787 (2008)
10. P.B. Dixon, D.J. Starling, A.N. Jordan, J.C. Howell, Phys. Rev. Lett. **102**, 173601 (2009)
11. D. Starling, P. Dixon, A. Jordan, J. Howell, Phys. Rev. A **80**, 041803 (2009)
12. N. Brunner, C. Simon, Phys. Rev. Lett. **105**, 010405 (2010)
13. D.J. Starling, P.B. Dixon, N.S. Williams, A.N. Jordan, J.C. Howell, Phys. Rev. A **82**, 011802 (2010)
14. N.S. Williams, A.N. Jordan, Phys. Rev. Lett. **100**, 4 (2008)
15. A. Romito, Y. Gefen, Y.M. Blanter, Phys. Rev. Lett. **100**, 056801 (2008)
16. V. Shpitalnik, Y. Gefen, A. Romito, Phys. Rev. Lett. **101**, 226802 (2008)
17. O. Zilberberg, A. Romito, Y. Gefen, Phys. Rev. Lett. **106**, 080405 (2011)
18. X. Zhu, Y. Zhang, S. Pang, C. Qiao, Q. Liu, S. Wu, Phys. Rev. A **84**, 052111 (2011)
19. O. Zilberberg, A. Romito, D.J. Starling, G.A. Howland, C.J. Broadbent, J.C. Howell, Y. Gefen, Phys. Rev. Lett. **110**, 170405 (2013)

Chapter 26
Increase of Signal-to-Noise Ratio in Weak Value Measurements

C. Byard, T. Graham, A. Danan, L. Vaidman, A.N. Jordan, and P. Kwiat

Abstract We present a method for improving the signal-to-noise ratio of optical weak measurement techniques by recycling the non-post selected light. We designed a system that allows input photons to cycle through a Sagnac interferometer twice, effectively doubling the detected intensity. A deflection as small as 20 picoradians is discernible. The expected improvement for a shot-noise-limited system is a factor of $\sqrt{2}$; however, we observed a larger signal-to-noise ratio enhancement (1.73 ± 0.46) which can occur for systems limited by technical noise. [*Editor's note*: for a video of the talk given by Prof. Kwiat at the Aharonov-80 conference in 2012 at Chapman University, see quantum.chapman.edu/talk-5.]

26.1 Introduction

Quantitative measurement is the most basic tool for comparing expectation and reality. We know from fundamental quantum mechanics that repeated measurements of some property on a quantum system must give an average value between the largest and smallest of its eigenvalues. With the use of weak value measurements, however, one can manipulate the state of a system to effectively achieve an amplification of that property. Aharonov, Albert, and Vaidman [1] first showed in 1988 the unexpected result that through such weak measurement techniques, one can measure a

C. Byard · T. Graham · P. Kwiat (✉)
Department of Physics, University of Illinois at Urbana-Champaign, Urbana, IL 61801, USA
e-mail: kwiat@illinois.edu

A. Danan · L. Vaidman
Raymond and Beverly Sackler School of Physics and Astronomy, Tel-Aviv University, Tel-Aviv 69978, Israel

L. Vaidman · A.N. Jordan
Institute for Quantum Studies, Chapman University, Orange 92866, CA, USA

A.N. Jordan
Department of Physics and Astronomy, University of Rochester, Rochester, NY 14627, USA

D.C. Struppa, J.M. Tollaksen (eds.), *Quantum Theory: A Two-Time Success Story*,
DOI 10.1007/978-88-470-5217-8_26,

value that is far outside the allowed range of the system's initial possible eigenvalues.

Weak measurements are possible through a pre-selection and post-selection method [1]. The measuring device is first prepared in an initial state, $|\psi_i\rangle$, which then interacts impulsively with the system operator **A** we wish to measure. This interaction couples **A** with an operator (such as momentum) of the measuring device, known as the meter. The state of the measuring device is essentially unchanged in this process. Finally, a state $|\psi_f\rangle$ nearly orthogonal to the initial state is post-selected and the perturbation on the measuring device is monitored. The outcome of the meter is then given by $A_{weak} = \langle\psi_f|\mathrm{A}|\psi_i\rangle/\langle\psi_f|\psi_i\rangle$ [1]. We see that the low probability of post-selection ($\langle\psi_f|\psi_i\rangle \ll 1$) causes a much larger change in the meter than would be possible with a strong measurement.

Since the first experimental demonstration [2], weak value measurements have proven to be a powerful tool in precision metrology, as the amplifying effect makes it possible to detect smaller changes than conventional measurements could. The technique has been utilized to demonstrate the spin Hall effect of light [3], for example, to an accuracy of about 1 angstrom, i.e. the system can reliably detect about a 1-angstrom displacement of a 1-mm diameter laser beam. It has also been shown that weak measurements can have practical applications in the telecom industry [4].

In a relevant experiment, Dixon et al. [5] used weak-value amplification in a Sagnac interferometer to detect the tiny tilt of a mirror. Light was split with equal probability into the clockwise (CW) or counterclockwise (CCW) directions around the interferometer, creating the pre-selected state. A birefringent element was placed inside the interferometer so that a controllable phase shift (φ) could be introduced between the two paths, thus allowing a small amount of light to leak out of the dark port. One mirror was equipped with a piezo-electric crystal so that it could be tilted slightly. The small momentum shift caused by the piezo crystal acts as the weak perturbation, shifting the relative positions of the CW and CCW beams (though by much less than the intrinsic beam divergence). The beam profile from the dark port is "warped" to one side by an amount determined by φ, and can be hundreds to thousands of times larger than the shift caused in each individual beam by the mirror tilt alone. Dixon et al. were able to detect an angular shift of the mirror of about 400 femtoradians, corresponding to a change in the piezo length of only 15 fm. We have also chosen to work with a tilted mirror in a Sagnac interferometer in our current study.

26.2 Recycling

While weak measurements do offer a significant improvement over unamplified deflections [6] (particularly in situations dominated by technical noise), there is still the potential for improvement. Most of the light is discarded in the process of post-selection, so that only a small fraction of the available light is utilized in the experiment. We propose *recycling* the escaped light, so that some or all of it is eventually detected.

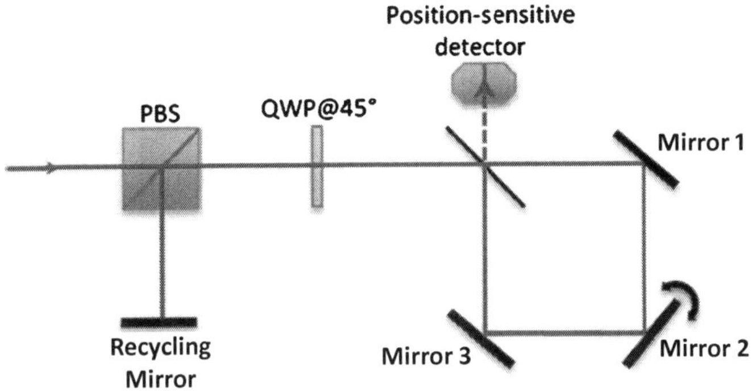

Fig. 26.1 Double-pass scheme. A PBS passes H-polarized light incident from a fiber-coupled laser. Light is then switched to L polarization by a QWP. After passing through the Sagnac, the QWP switches the polarization to V. Light is now reflected at the PBS to another mirror, which sends the beam back through the system for a second pass

As shown in Fig. 26.1, the addition of a quarter-wave plate (QWP), polarizing beamsplitter (PBS), and recycling mirror allows the unused portion of the light that exits the bright port of the Sagnac interferometer to be redirected back inside. Because only a small fraction of the light (ε) escapes the dark port in the first pass, essentially the same amount exits on the second ($(1-\varepsilon)*\varepsilon \sim \varepsilon$). The total number of photons incident on the detector is then nearly doubled with this configuration. We expect the signal, which is proportional to the number of incident photons, to double as well. In the shot-noise-limit the noise scales as the square root of the number of photons, and hence we expect the noise to only increase by a factor of 1.4 ($=\sqrt{2}$) over the single-pass configuration. Thus, the SNR should also be increased by a factor of 1.4. We will use the measurement operator formalism [7] to describe in more detail how the double pass experiment changes the physics.

26.3 Theory

Such a situation can be theoretically modeled by considering an input state, $|\Psi\rangle = |w\rangle|\chi\rangle|p\rangle$, consisting of a direct product of three states; a which-path interferometer state, $|w\rangle$, a transverse profile state, $|\chi\rangle$, and a polarization state, $|p\rangle$. According to our geometry in Fig. 26.1, the polarization starts in horizontal ($|H\rangle$) polarization, passes through the PBS, and is rotated to left circular polarization, ($|L\rangle$) by the QWP. It then enters the interferometer creating a which-path state, $|+\rangle = (1/\sqrt{2})(|\circlearrowright\rangle + i|\circlearrowleft\rangle)$; a linear combination of clockwise and counter-clockwise propagating states in the Sagnac interferometer. State $|+\rangle$ describes the output (post-selected) state of the bright port of the interferometer to be discussed shortly; the dark-port of the interferometer is described with the state $|-\rangle = (1/\sqrt{2})(|\circlearrowright\rangle - i|\circlearrowleft\rangle)$. We define the which-path operator to be $\mathbf{W} = |\circlearrowright\rangle\langle\circlearrowright| - |\circlearrowleft\rangle\langle\circlearrowleft|$, noting that $\mathbf{W}$ squares to the identity. The action of the tilted mirror, providing transverse momentum kick k to the transverse state, can be expressed as unitary operator, $\mathbf{U}_P = \exp[-ik\mathbf{W}\mathbf{x}]$, where $\mathbf{x}$ is the transverse position operator. $\mathbf{U}_P$ acts on both which-path and transverse states. Similarly, the birefringent element will give a relative phase shift φ to the clockwise and counter-clockwise paths, $\mathbf{U}_{BE} = \exp[i\varphi\mathbf{W}/2]$. Taken together, acting on which-path and transverse states, these operators give the state after one traversal, $|\Psi_1\rangle = \mathbf{U}_P\mathbf{U}_{BE}|\Psi\rangle$. The light then exits the 50/50 beamsplitter, projecting the state onto the post-selection state $|+\rangle$ or $|-\rangle$. The effect of this on the transverse beam profile can be represented as one of two measurement operators, $|\chi\rangle \to \mathbf{M}_\pm|\chi\rangle$, where $\mathbf{M}_\pm = \langle\pm|\mathbf{U}_P\mathbf{U}_{BE}|+\rangle$, describing post-selection on either the bright $(+)$ or dark port $(-)$. For our case, the measurement operators take on the simple form, $\mathbf{M}_+ = \cos(\frac{\varphi}{2} - k\mathbf{x})$, and, $\mathbf{M}_- = \mathrm{i}\sin(\frac{\varphi}{2} - k\mathbf{x})$.

In the first pass, the photon can either exit through the dark port, which has the state $|\Psi_{d,1}\rangle = \mathbf{M}_-|\chi\rangle|L\rangle$, or it can exit the bright port creating the state $|\Psi_{b,1}\rangle = \mathbf{M}_+|\chi\rangle|L\rangle$. This bright-port photon state will subsequently pass through the QWP (switching the polarization to vertical, $(|V\rangle)$), reflect off the PBS, mirror, PBS, pass again through the QWP (switching the polarization to right circular $(|R\rangle)$), and once again enter the interferometer with state $|+\rangle(\mathbf{M}_+|\chi\rangle)|R\rangle$. This process is now repeated again, but with the initial state profile changed to $\mathbf{M}_+|\chi\rangle$ and a new polarization state. On the second round, the photon will exit to the dark port with the state $|\Psi_{d,2}\rangle = (\mathbf{M}_-\mathbf{M}_+|\chi\rangle)|R\rangle$, and to the bright port with the state $|\Psi_{b,2}\rangle = (\mathbf{M}_+\mathbf{M}_+|\chi\rangle)|R\rangle$, the polarization of which will be rotated back to $|H\rangle$, which will pass through the PBS and travel back to the laser, ending this analysis.

We are now interested in the properties of the light incident on the position-sensitive detector from the dark port. This will be a combination of the photons that have made one and two passes. We note that because the polarization is different for the two passes (either left or right circular polarization), these will not interfere with each other, so we can simply sum the two probabilities. If the initial transverse profile is $n_0(x) = |\langle x \mid \chi\rangle|^2$, then the final profile on the detector is $n_{\mathrm{tot}}(x) = n_1(x) + n_2(x)$, where $n_1 = |\langle x \mid \mathbf{M}_- \mid \chi\rangle|^2$, and $n_2 = |\langle x \mid \mathbf{M}_-\mathbf{M}_+ \mid \chi\rangle|^2$. Since the measurement operators are diagonal in the position basis, we have the following expressions for n_1, n_2

$$n_1(x) = n_0(x)\sin^2\left[\frac{\varphi}{2} - kx\right] \tag{26.1}$$

$$n_2(x) = n_0(x)\sin^2\left[\frac{\varphi}{2} - kx\right]\cos^2\left[\frac{\varphi}{2} - kx\right] \tag{26.2}$$

In the weak value regime, the initial distribution n_0 is usually taken to be a Gaussian with width σ, so that the parameters are ordered as $k\sigma \ll \varphi \ll 1$. In this range, the fraction of photons in the transverse plane that arrives at the detector is $N_d/N_0 = \int dx n_{\mathrm{tot}}/\int dx n_0$, and we have for the double pass,

$$N_{d2}/N_{02} = (\varphi^2/2)(1 - \varphi^2/16) + O(ka)^2 \tag{26.3}$$

as opposed to

$$N_{d1}/N_{01} = (\varphi^2/4)(1 - \varphi^2/12) + O(ka)^2 \tag{26.4}$$

for the one pass alone. Thus, so long as the phase shift remains small, we have an effective doubling of collected photons for a double pass. It is straightforward to check that the split-detector signal and variance are left unchanged when normalized to the total photon number (up to corrections of order φ^2 as shown in Eqs. (26.3) and (26.4) above),

$$\langle x \rangle = 2k\sigma^2 \operatorname{Im} W_{weak} = -4k\sigma/\varphi \tag{26.5}$$

$$\operatorname{Var}[x] = \langle x^2 \rangle - \langle x \rangle^2 = \sigma^2 \tag{26.6}$$

where the averages are taken with respect to the post-selected distribution [5]. Here, W_{weak} is the weak value of the which-path operator $\mathbf{W}$, where the pre- and post-selected states are $|+\rangle$ and $\mathbf{U}_{BE}|-\rangle$. This leaves the doubling of photon number $N \to 2N$ as the only important effect in this limit.

The signal-to-noise ratio (SNR) (for a split-detector signal measuring photons that are uncorrelated in position [6]) is defined as the ratio of the average signal on the detector, $S = N_d\langle x \rangle$, to the square-root of the variance, $N^2 = N_d \operatorname{Var}[x]$. Thus,

$$\text{SNR} = S/N = \sqrt{N_d}\langle x \rangle/\sqrt{\operatorname{Var}[x]} \tag{26.7}$$

As shown in Eqs. (26.5) and (26.6), both $\langle x \rangle$ and $\langle x^2 \rangle$ are unchanged up to corrections of order φ^2 between the single and double pass, and thus in the comparison of SNR, we see that both signal and variance increase by a factor of 2; consequently, for a shot-noise limited system, the SNR will increase by a factor of $\sqrt{2} \approx 1.4$ over the single pass. If, however, the system has technical noise as the dominant noise source, the variance will remain unchanged from single pass to double pass and the SNR will then increase by a factor of 2.

26.4 Experiment

A 633-nm laser was coupled to a single-mode fiber and the output then collimated to a $1/e^2$ radius of 1 mm. A quarter-wave plate and half-wave plate were used to adjust the intensity. As shown in Fig. 26.1, the beam was then passed through a PBS and QWP, switching the polarization to L. A 50–50 beamsplitter split the light into CW and CCW arms of the Sagnac interferometer. Mirror 1 was tipped out of the plane of the beam path, causing a slight path length difference (since the CW path experiences the tilt first) which was used to control the relative phase. Mirror 2 was tilted in the plane of the interferometer with a piezoelectric crystal (Thorlabs model KC1-PZ).

Mirror 2 was driven with a 100-Hz sinusoidal voltage at a range of amplitudes. The resulting shift was measured with a quadrant cell detector (New Focus model

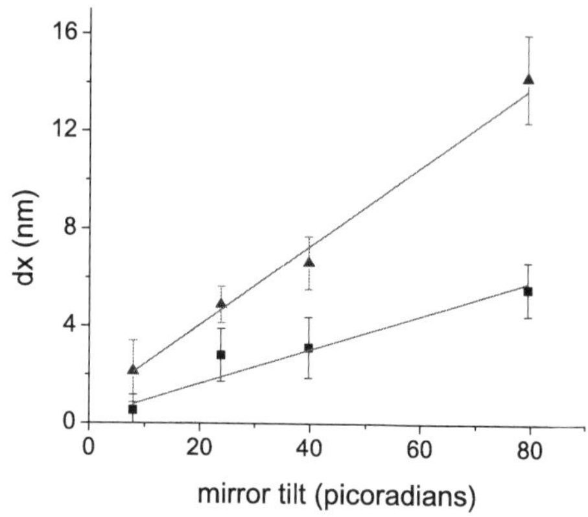

Fig. 26.2 Measured signal for double- (*triangles*) and single-pass (*squares*) systems. Double-pass measurements (when normalized to the single-pass power) show twice the signal of the single pass data

2921) and the output signal filtered with a lock-in amplifier (SR830). Position shifts were determined by normalizing the voltage difference between the left and right photodiodes by the total voltage. Results are shown in Fig. 26.2. Light that exited the bright port, now R polarized, again passed through the QWP and was switched to V. At the PBS, the light was then reflected to a mirror to be sent through the system a second time.

Measurements were taken with the interferometer in two slightly different configurations. The interferometer was square with sides about 5 cm, with a path length of 22.5 cm (26 cm) from mirror 2 to the detector. The maximum power achieved at the detector output port of the interferometer was 1.9 mW; the minimum output power was 6.3 μW (4.8 μW) corresponding to a visibility of 99.3 % (99.5 %). When the relative phase was shifted to the chosen post-selection probability, 15 μW (19.7 μW) exited to the quadrant cell detector for the single-pass data (corresponding to a relative phase difference of 8 (10) degrees), while 29 μW (39 μW) exited for the double-pass data.

26.5 Discussion

As expected, our measurements show that the signal of the double-pass system from the position-sensitive detector, when normalized to the same power as a single pass, is approximately doubled (see Fig. 26.2). Our preliminary results show that the SNR is consistently larger for the double cycle case over the single cycle; however, we observed an average SNR increase by a factor of 1.73 ± 0.46. (To calculate the SNR ratio, we normalized the double-pass signal by the total double-pass power. Because the small deflection and uncertainty in the measurements makes the error bars on the SNR ratio quite large, we weighted 13 measurements by the reciprocal of the standard deviation.) This effect is consistent with measurements that are not shot-noise limited, for which we would expect an increase of only 1.4. For a system limited by technical noise that is independent of the number of cycles (e.g., electrical noise from the detector itself), the doubling of the signal is not accompanied by an increase in noise, and results in a doubling of the SNR. Hence, in some circumstances recycling can give an even larger benefit than would be observed in the shot-noise limit.

The current resolution of our system is about 20 picoradians (for comparison, without weak measurement amplification, the smallest distinguishable deflection for this same configuration would be about 3 nanoradians), presently limited by electrical noise from the detector and mechanical jitter in the laser cavity. With a shot-noise-limited system, we expect to be able to reach a resolution of about 0.5 picoradians with the current configuration, though the possibility exists to go even lower by enlarging the beam, since the amplification increases as the square of the radius. If we increase the number of cycles so that all of the light is eventually detected, we anticipate increasing the SNR by an order of magnitude, thus pushing the resolution that much lower.

26.6 Conclusion

We have demonstrated a technique for recycling light through a weak-measurement experiment in order to increase the signal-to-noise ratio of the amplified signal. Our results show an improvement of the SNR, and highlight the fact that the advantage can range from a factor of 1.4 (for a shot-noise limited system) to 2 (for a technical noise limited system). If all photons incident on the interferometer could be collected, quantum states of light, such as squeezed states or entangled photons, could be utilized along with the weak value measurements to expand the limit of detection beyond what is possible with quantum states or weak measurements alone.

Acknowledgements This work was supported by Army Research Office "Advanced Quantum Sensing" W911NF-12-1-0237 and the Binational Science Foundation Grant No. 32/08. The authors would like to thank B. Dixon for helpful discussions.

References

1. Y. Aharonov, D. Albert, L. Vaidman, How the result of a measurement of a component of the spin of a spin-1/2 particle can turn out to be 100. Phys. Rev. Lett. **60**, 1351–1354 (1988)
2. N.M.W. Ritchie, J.G. Story, R.G. Hulet, Realization of a measurement of a "weak value". Phys. Rev. Lett. **66**, 1107–1110 (1991)
3. O. Hosten, P. Kwiat, Observation of the spin hall effect of light via weak measurements. Science **319**, 787–790 (2008)
4. N. Brunner, A. Acin, D. Collins, N. Gisin, V. Scarani, Optical telecom networks as weak quantum measurements with postselection. Phys. Rev. Lett. **91**, 180402 (2003)
5. B.P. Dixon, D.J. Starling, A.N. Jordan, J.C. Howell, Ultrasensitive beam deflection measurement via interferometric weak value amplification. Phys. Rev. Lett. **102**, 173601 (2009)
6. D.J. Starling, B.P. Dixon, A.N. Jordan, J.C. Howell, Optimizing the signal-to-noise ratio of a beam-deflection measurement with interferometric weak values. Phys. Rev. A **80**, 041803 (2009). arXiv:0910.2410
7. J. Dressel et al., Unpublished

Part IX
Yakir Aharonov: Thinking Quantum

Chapter 27
Yakir Aharonov: From A to B

A. Pines

Abstract Twenty years ago, I was accorded the privilege and pleasure of presenting the after banquet speech—reprinted below—at a conference celebrating the sixtieth year of my dear friend Yakir Aharonov. Does what I said then still hold today? You bet it does . . .the now adult Yakir has maintained the same burning curiosity, fiery enthusiasm and explosive genius for physics that characterized the life of the adolescent sixty year old. There may be no free lunch, but Yakir is living proof that there is free will, and he has exercised his free will to make ever more creative contributions to quantum physics, future and present. My admiration of Yakir as a scientist and mensch has only grown over the years. Sorry I can't be with you all at Chapman to participate in this marvelous eightieth year celebration. Dear Yakir, Ditsa and I wish you, Nilli and your family many more joyous years of health, science and friendship. . .' til 120.

Following the dictates of David Mermin, I have prepared some spontaneous remarks:

Ladies and Gentiles,

You see before you a most reluctant after-dinner speaker. Someone once said that if you took all the after-dinner speakers and laid them head-to-toe at the equator, . . . that would be a very good thing. In fact, some years ago, my friend Anatole Abragam warned me—Alex, when they start asking you to give after-dinner speeches, it might be an indication that you are no longer on the way up. So when

Reprinted with permission from "After-Banquet Talk in Honor of Aharonov's 60th Birthday: Yakir Aharonov: From A to B," Alex Pines, *Fundamental Problems in Quantum Theory*, editors J.A. Anandan and J. Safko, Copyright 1994, World Scientific, Columbia, South Carolina, December 10–12, 1992.

A. Pines (✉)
California Institute for Quantitative Biosciences, Materials Sciences Division, Lawrence Berkeley National Laboratory, Berkeley, CA, USA
e-mail: a_pines@lbl.gov

A. Pines
Department of Chemistry, University of California, Berkeley, Berkeley, CA, USA

D.C. Struppa, J.M. Tollaksen (eds.), *Quantum Theory: A Two-Time Success Story*, DOI 10.1007/978-88-470-5217-8_27, © Springer-Verlag Italia 2014

I was asked to talk about Aharonov tonight, the first two words that carne to my mind were—oy vey.

But, ladies and gentlemen, this is no ordinary occasion—Yakir Aharonov is not only a truly great scientist and one of the most brilliant and stimulating people I have ever known, he is an extraordinary colleague and dear friend, and it is a privilege and a pleasure for me to say a few words about him. You might well ask, why me, a chemist, talking about a physicist.

Well, Aharonov himself once paid me what he considers the greatest compliment you guys can give a chemist—Come on, Alex, you're not really a chemist, you're too smart, . . . you're a physicist. Yakir, it's your birthday, let me return the compliment—you don't look seventy.

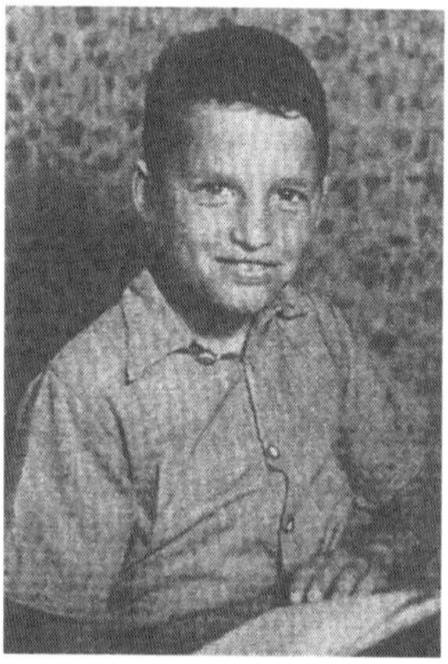

Yakir Aharonov was born in 1932, in Haifa Israel, to Russian parents. He grew up, so to speak, in Kiryat Haim, where, already at age five, it was abundantly clear that he was a mathematical prodigy. The residents of Kiryat Haim soon became accustomed to the apparition of the boy Aharonov accosting and threatening them in the streets, challenging them to give him a problem—a novel concept of mathematical mugging, your problem or your life.

Because his parents were unwilling to teach him chess (a waste of time), Aharonov traded some strawberries from his yard to a neighbor, an older child, who taught him the game. When not playing with his friend, Aharonov would play by himself, one hand against the other, one playing white and the other black. It is not known which hand was stronger, his left hand or his other left hand. As many of you know, Aharonov had a natural aptitude for the game and became a very strong player, today an Israeli candidate master. During his period as Miller Professor at Berkeley, Aharonov made an unforgettable impression not only on the scientists, but also on the nationally renowned Berkeley chess community. As a young man, Aharonov had a gift not only for math and chess; he was good at all sorts of games and puzzles. He discovered, to his joy, that his prowess at backgammon made him almost irresistible to middle-eastern women.

The last time I played blitz chess against Aharonov, he again asked if I wanted a handicap. I related to him the (perhaps apocryphal) story told to me recently by John Rowlinson about Max Euwe, the former world chess champion. Euwe was on a train analyzing a game on his pocket chess set. A fellow traveler in the compartment asked him if he played chess, to which Euwe replied that yes, he did. Would you like to play a game, asked the other fellow; sure, said Euwe, who proceeded to set up the pieces and then removed one of his rooks. What are you doing, asked his partner. I'm giving you a rook, replied Euwe. You're giving me a rook? You've never played against me, you don't know who I am, how can you give me a rook? If I couldn't give you a rook, said Euwe, I'd know who you are.

Well, Aharonov doesn't give me a rook, but he does give me a differential time handicap in order to imbue the game with some semblance of balance. In other words, he beats the hell out of me. It is because of Aharonov that I have now resorted to playing for money against small children. But Aharonov too is fallible—about twenty five years ago, in New York, he played, and lost, three games against Bobby Fischer. Aharonov maintains that this is pretty good; he lost only three games, so he did better than the famous Russian, Taimanov, and the great Dane, Larsen, who each lost six games against Fischer.

At age eleven, Aharonov took up the violin, an instrument that he cherishes to this very day. He soon discovered that the best acoustics for his instrument were in the kitchen and bathroom. It was later, after he read how Einstein had independently made the same discovery, that Aharonov decided he would become a physicist.

After graduation from high school, Aharonov was inducted into the army, into the artillery division. Yes, the artillery division. He soon lost interest in experimental artillery after he proved that quantum corrections to the ballistic trajectories were insignificant and, much to the relief of the commanding authorities, he volunteered for an army research unit. The only legacy of Aharonov's army experience was his occasional, misguided tendency to force himself upon his friends as a bodyguard.

After his discharge from the army, Aharonov studied at the Technion, the Israel Institute of Technology, where he met the late David Bohm. Here Aharonov is shown at the Technion with a co-student whom he identifies as Tsachi Gozani. Gozani allegedly spent much of his time begging Aharonov to stay away from the apparatus. After discussions with faculty members who feared for their lives, Aharonov seriously contemplated becoming a theoretician. He moved with Bohm to Bristol to do his Ph.D. and it was there that the famous Aharonov-Bohm effect *was* conceived, elucidated and published.

One of the external examiners for Aharonov's Ph.D. was Rudolph Peierls, who claimed he did not believe some argument that Aharonov had formulated about energy-time uncertainty, but Peierls could not find an error. He invited Aharonov to Birmingham, where they sat and argued for days, after which Peierls was convinced and said that he now believed. But Yakir tells me that just two years ago, Peierls was in Israel for the Landau Symposium—he ran into Aharonov and said hey, aren't you Aharonov? Yes, I am. Well, said Peierls, now I don't believe you again.

It was during his time in England that Aharonov became concerned about his Israeli accent, because he felt that it was hindering his chances with women. He arranged for intensive tutoring sessions in elocution, seeking to acquire not just any old accent, but an Oxford accent, and devoting considerable time and effort to

the enterprise. On the day of the first experiment with his new accent, an excited Aharonov ventured into the streets of Bristol and asked for directions to go somewhere; I imagine that we can all sympathize with his frustration when the answer came back in Hebrew.

Following his Ph.D., Aharonov spent several years at Brandeis and Yeshiva Universities in the United States. In 1962, he created a sensation when he talked about the Aharonov-Bohm effect at the Cincinnati Conference on quantum theory (the other participants included Dirac, Furry, Podolsky, Rosen and Wigner). The conference made headlines despite the many other exciting events in Cincinnati at the time.

In 1966, Aharonov joined the faculty at South Carolina and, in 1967, he became Full Professor at Tel-Aviv University. He was subsequently honored with chairs in physics both at Tel-Aviv and here in South Carolina, where, I understand, he is again contemplating changing his accent. His colleagues here know that, for Aharonov, physics is not just a job—it is a passion, like chess. That Tel Aviv University and

the University of South Carolina pay him to indulge in his passion remains for him unfathomable. Yakir, may it become yet more unfathomable.

Over the years, Aharonov has further cultivated, carefully and successfully, his image as a shlemiel, thereby shielding him from annoying appeals to help around the lab, the department or the house, and leaving him time to do what he loves and does best—to think. And, as many of us know, Aharonov thinks best in an atmosphere composed of ten percent oxygen, forty percent nitrogen and fifty percent cigar smoke. What kind of cigar smoke? Well, let's just say that many years ago, I gave him one of my prized Montecristos from Havana, and he was able to exchange it for a year's supply of his beloved White Owls. Aharonov continues with his tradition of visiting Berkeley whenever he runs out of cigars, much to the delight of my children, by whom he is much admired.

Yakir Aharonov is a giant of modern physics. Prom his Ph.D. with Bohm to his work on geometric phases, he has made monumental contributions to quantum theory, and he has profoundly advanced our understanding of electromagnetism and other gauge theories of fundamental interactions. On two occasions, John Maddox, the editor of Nature (the science magazine), suggested, justifiably, many of us believed, that Aharonov, Bohm and Berry should get the Nobel Prize for physics. In his first editorial on the subject, in 1989, Maddox writes about Abrahamov and the Abrahamov-Bohm effect; in his second editorial on the subject, this year, he makes a slightly better approximation, writing about Aharanov and the Aharanov-Bohm effect. And listen to the perverse, yet quaint 1989 description of the effect—Abrahamov and Bohm, independently of M. J. Berry, have shown that the supposedly insignificant complex phase of Maxwell's electromagnetic potential is measurable.

Well, Yakir Aharonov is no stranger to honor and to ceremony. He is a member of the Israel and U.S. National Academies of Sciences, and amongst his many awards are the prestigious Israel Prize in exact sciences and the Elliot Cresson Medal of the Franklin Institute in Philadelphia. But Aharonov is particularly proud of the knighthood bestowed upon him by his friends on the occasion of his fiftieth birthday which, he calculates, was ten years ago. I guess the citation reads—why is this knight different from all other knights?

Ladies and gentlemen, I was asked to make my remarks either witty or brief so I must come to a close.

פיזיקה של עולם קפריזי
הכל בראש
לפני שלושים שנה נרשמה על־שמו של פרופסור אהרונוב, חתן פרס ישראל השנה, אחת התגליות החשובות בתורת הקוואנטים. בינתיים תרמה למספר חידושים טכנולוגיים כמו: טרנזיסטור, מיקרוסקופ ומרכיבי מחשב מסוגים חדשים. יש לו רעיונות מקוריים גם בתחום החברתי: הספקת סמים לקשישים, פיתוח קריירה שניה ושינוי חלוקת ההכנסות במסלול חיי האדם. החיים במשחק שח

Yakir Aharonov is a man with a legendary hunger for science and for life. But beyond his genius and his accomplishments, Aharonov has that rarest of human qualities—he is a mensch. Dear Yakir, I am sure that I speak on behalf of everyone here when I say that you have earned our respect. On the occasion of your sixtieth birthday, permit me to offer a toast to you and your family—the Aharonovs, the Abrahamovs and the Aharanovs—Yakir and Nilli, to another sixty years.

THE

PHYSICAL REVIEW

A journal of experimental and theoretical physics established by E. L. Nichols in 1893

SECOND SERIES, VOL. 115, No. 3 AUGUST 1, 1959

Significance of Electromagnetic Potentials in the Quantum Theory

Y. AHARONOV AND D. BOHM
H. H. Wills Physics Laboratory, University of Bristol, Bristol, England
(Received May 28, 1959; revised manuscript received June 16, 1959)

In this paper, we discuss some interesting properties of the electromagnetic potentials in the quantum domain. We shall show that, contrary to the conclusions of classical mechanics, there exist effects of potentials on charged particles, even in the region where all the fields (and therefore the forces on the particles) vanish. We shall then discuss possible experiments to test these conclusions; and, finally, we shall suggest further possible developments in the interpretation of the potentials.

1. INTRODUCTION

IN classical electrodynamics, the vector and scalar potentials were first introduced as a convenient mathematical aid for calculating the fields. It is true that in order to obtain a classical canonical formalism, the potentials are needed. Nevertheless, the fundamental equations of motion can always be expressed directly in terms of the fields alone.

In the quantum mechanics, however, the canonical formalism is necessary, and as a result, the potentials cannot be eliminated from the basic equations. Nevertheless, these equations, as well as the physical quantities, are all gauge invariant; so that it may seem that even in quantum mechanics, the potentials themselves have no independent significance.

In this paper, we shall show that the above conclusions are not correct and that a further interpretation of the potentials is needed in the quantum mechanics.

2. POSSIBLE EXPERIMENTS DEMONSTRATING THE ROLE OF POTENTIALS IN THE QUANTUM THEORY

In this section, we shall discuss several possible experiments which demonstrate the significance of potentials in the quantum theory. We shall begin with a simple example.

Suppose we have a charged particle inside a "Faraday cage" connected to an external generator which causes the potential on the cage to alternate in time. This will add to the Hamiltonian of the particle a term $V(x,t)$ which is, for the region inside the cage, a function of time only. In the nonrelativistic limit (and we shall assume this almost everywhere in the following discussions) we have, for the region inside the cage, $H=H_0+V(t)$ where H_0 is the Hamiltonian when the generator is not functioning, and $V(t)=e\phi(t)$. If $\psi_0(x,t)$ is a solution of the Hamiltonian H_0, then the solution for H will be

$$\psi=\psi_0 e^{-iS/\hbar},\quad S=\int V(t)dt,$$

which follows from

$$i\hbar\frac{\partial\psi}{\partial t}=\left(i\hbar\frac{\partial\psi_0}{\partial t}+\psi_0\frac{\partial S}{\partial t}\right)e^{-iS/\hbar}=[H_0+V(t)]\psi=H\psi.$$

The new solution differs from the old one just by a phase factor and this corresponds, of course, to no change in any physical result.

Now consider a more complex experiment in which a single coherent electron beam is split into two parts and each part is then allowed to enter a long cylindrical metal tube, as shown in Fig. 1.

After the beams pass through the tubes, they are combined to interfere coherently at F. By means of time-determining electrical "shutters" the beam is chopped into wave packets that are long compared with the wavelength λ, but short compared with the length of the tubes. The potential in each tube is determined by a time delay mechanism in such a way that the potential is zero in region I (until each packet is well inside its tube). The potential then grows as a function of time, but differently in each tube. Finally, it falls back to zero, before the electron comes near the

Printed by Publishers' Graphics LLC
KSO140508.23.35.27